Plant Metabolism and Biotechnology

Plant Metabolism and Biotechnology

Edited by

HIROSHI ASHIHARA

Department of Biological Sciences, Graduate School of Humanities and Sciences, Ochanomizu University, Tokyo, Japan

ALAN CROZIER

Plant Products and Human Nutrition Group, School of Medicine, College of Medical, Veterinary and Life Sciences, University of Glasgow, Glasgow, UK

ATSUSHI KOMAMINE

The Research Institute of Evolutionary Biology, Tokyo, Japan

WILEY

A John Wiley and Sons, Ltd., Publication

This edition first published 2011

Registered office
John Wiley & Sons Ltd, The Atrium, Southern Gate, Chichester, West Sussex, PO19 8SQ, United Kingdom

For details of our global editorial offices, for customer services and for information about how to apply for permission to reuse the copyright material in this book, please see our website at www.wiley.com.

Library of Congress Cataloging-in-Publication Data

Plant metabolism and biotechnology / edited by Hiroshi Ashihara, Alan Crozier, Atsushi Komamine.
p. cm.
Includes bibliographical references and index.
ISBN 978-0-470-74703-2 (hardback)
1. Plants–Metabolism. 2. Plant biotechnology. I. Ashihara, Hiroshi. II. Crozier, Alan.
III. Komamine, Atsushi, 1929–
QK881.P534 2011
660.6–dc22 2010049393

A catalogue record for this book is available from the British Library.

Print ISBN: 9780470747032
ePDF ISBN: 9781119991328
oBook ISBN: 9781119991311
ePub ISBN: 9781119993223
eMobi ISBN: 9781119993230

Typeset in 10/12pt Times by Aptara Inc., New Delhi, India.

MIX
Paper from responsible sources
FSC® C013604

Contents

List of Contributors

Yutaka Abe, Division of Food Additives, The National Institute of Health Sciences, Setagaya-ku, Tokyo, 158-8501 Japan.

Hiroshi Ashihara, Department of Biological Sciences, Graduate School of Humanities and Sciences, Ochanomizu University, 2-1-1, Otsuka, Bunkyo-ku, Tokyo, 112-8610 Japan.

Frederik Börnke, Department Biologie, Lehrstuhl für Biochemie, Friedrich-Alexander-Universität Erlangen-Nürnberg, Staudtstraße 5, 91058 Erlangen, Germany.

Alan Crozier, Plant Products and Human Nutrition Group, Graham Kerr Building, School of Medicine, College of Medical, Veterinary and Life Sciences, University of Glasgow, Glasgow G12 8QQ, UK.

Vincenzo De Luca, Department of Biological Sciences, Brock University, 500 Glenridge Avenue, St. Catharines, Ontario, Canada L2S 3A1.

Isabel Desgagné-Penix, Department of Biological Sciences, University of Calgary, 2500 University Drive, Calgary, Alberta, Canada, T2N 1N4.

Peter J. Facchini, Department of Biological Sciences, University of Calgary, 2500 University Drive, Calgary, Alberta, Canada, T2N 1N4.

Takashi Hashimoto, Graduate School of Biological Sciences, Nara Institute of Science and Technology, 8916-5 Takayama-cho, Ikoma, Nara, 630-0101 Japan.

David Hildebrand, 403 Plant Science Building, 1405 Veterans Drive, University of Kentucky, Lexington, KY 40546-0312, USA.

Indu B. Jaganath, Biotechnology Research Centre, Malaysian Agricultural Research Institute, 43400 Serdang Selangor, Malaysia.

Atsushi Komamine, The Research Institute of Evolutionary Biology, Setagaya-ku, Tokyo, 158-0098 Japan

Hiroshi Kouchi, Department of Plant Sciences, National Institute of Agrobiological Sciences, Tsukuba, 305-8602 Japan.

Miyako Kusano, RIKEN Plant Science Center, Tsurumi-ku, Yokohama, 230-0045 Japan.

Yuki Matsuba, Department of Biotechnology and Life Science, Faculty of Engineering, Tokyo University of Agriculture and Technology, Koganei, Tokyo, 184-8588 Japan.

Fumio Matsuda, RIKEN Plant Science Center, Tsurumi-ku, Yokohama, 230-0045 Japan.

Shinjiro Ogita, Biotechnology Research Center, Toyama Prefectural University, Toyama, 939-0398 Japan.

Akira Oikawa, RIKEN Plant Science Center, Tsurumi-ku, Yokohama, 230-0045 Japan.

Yozo Okazaki, RIKEN Plant Science Center, Tsurumi-ku, Yokohama, 230-0045 Japan.

Akemi Ohmiya, National Institute of Floricultural Science, Fujimoto 2-1, Tsukuba, Ibaraki, 305-8519 Japan.

Yoshihiro Ozeki, Department of Biotechnology and Life Science, Faculty of Engineering, Tokyo University of Agriculture and Technology, Koganei, Tokyo, 184-8588 Japan.

Dae-Kyun Ro, Department of Biological Sciences, University of Calgary, 2500 University Drive, Calgary, Alberta, Canada T2N IN4.

Kazuki Saito, Graduate School of Pharmaceutical Sciences, Chiba University, Inage-ku, Chiba 263-8522, Japan, and RIKEN Plant Science Center, Tsurumi-ku, Yokohama, 230-0045 Japan.

Masaaki Sakuta, Department of Biological Sciences, Graduate School of Humanities and Sciences, Ochanomizu University, 2-1-1, Otsuka, Bunkyo-ku, Tokyo, 112-8610 Japan.

Nobuhiro Sasaki, Department of Biotechnology and Life Science, Faculty of Engineering, Tokyo University of Agriculture and Technology, Koganei, Tokyo, 184-8588 Japan.

Tsubasa Shoji, Graduate School of Biological Sciences, Nara Institute of Science and Technology, 8916-5 Takayama-cho, Ikoma, Nara, 630-0101 Japan.

Sophia Sonnewald, Department Biologie, Lehrstuhl für Biochemie, Friedrich-Alexander-Universität Erlangen-Nürnberg, Staudtstraße 5, 91058 Erlangen, Germany.

Hideki Takahashi, Department of Biochemistry and Molecular Biology, Michigan State University, 209 Biochemistry Building, East Lansing, MI 48824-1319, USA.

Naoyuki Umemoto, Central Laboratories for Frontier Technology, Kirin Holding CO., Ltd., Sakura, Tochigi, 329-1414 Japan.

Rita Zrenner, Leibniz-Institute of Vegetable and Ornamental Crops, 14979 Großbeeren, Germany.

Preface

There have been significant advances in the basic and applied studies related to plant metabolism over the last two decades. Most of the metabolic pathways are now elucidated at the molecular level using genomic information. This book describes current knowledge on the plant metabolism, primary metabolism and the biosynthetic pathways of various secondary plant metabolites. The function of individual pathways *in planta* and applications for biotechnology are included, where appropriate examples are available. Although not all chapters contain a description of biotechnology, the results from basic studies that are discussed will be useful for future application. Emphasis is placed on explanation of metabolism of various metabolites in plants which are important *in planta*, as well as for potential exploitation for human use.

We selected several metabolic pathways of important compounds in plant metabolism. Chapters 1 to 5 cover basic metabolism including carbohydrate metabolism, lipid biosynthesis, nitrogen fixation, sulphur metabolism and nucleotide metabolism. Some examples of the use of genetic engineering to alter the content and quality of starch and to make industrial-use oilseeds are included. The following two chapters (Chapters 6 and 7) describe the biosynthesis of caffeine and nicotine, which are popular secondary metabolites. The production of transgenic decaffeinated coffee beans is discussed, as well as caffeine-producing tobacco plants in which the purine alkaloid acts as a natural pesticide. Possible mechanisms to reduce nicotine production are also proposed. Reviews on the biosynthesis of terpenoid (Chapter 8), benzylisoquinoline alkaloids (Chapter 9) and monoterpenoid indole alkaloids (Chapter 10) include actual and potential applications in the production of medicines. The biosynthesis of flavonoids and anthocyanins, as well as betacyanins and carotenoids, are reviewed in Chapters 11, 12 and 13, respectively. Finally, the use of metabolomics in plant biotechnology is described in Chapter 14.

There are several books on plant biotechnology but these mainly describe techniques, and there is a lack of basic information on plant metabolism. This book provides concise descriptions of various aspects of plant metabolism and biotechnology by experts in the field. The book will, therefore, be more attractive to students and lecturers than the traditional textbooks on plant metabolism. We anticipate that it will also be of value to researchers in applied fields such as agriculture, biotechnology, and medical and pharmacological sciences.

1

Biosynthesis and Metabolism of Starch and Sugars

Frederik Börnke and Sophia Sonnewald

1.1 Introduction

Regulation of photosynthetic carbon metabolism is central for plant growth and development. In plants, carbohydrates are produced in photosynthetically active tissues, primarily in the chloroplast-containing cells of source leaves. The conversion of photoassimilates into sucrose allows the transport via the phloem from these source tissues to support growth of sink tissues such as young leaves, roots, fruits or tubers which themselves are unable to produce assimilates. During development, sink to source ratios change, which implies that assimilate production must be adjusted to the changing needs of distant tissues. Research on this subject has recently undergone a renaissance, driven by the significance of photosynthetic carbon metabolism in determining crop yield. Rising demand for food and bioenergy makes it imperative to devise novel strategies based on biotechnology and conventional breeding, respectively, for increased crop yield. In order to achieve these goals, a thorough understanding of the regulatory properties of individual enzymes as well as of the regulatory networks linking entire pathways of primary photosynthetic metabolism is required. The widespread adoption of transgenic plants, the availability of plant genome information and the rise of plant functional genomics research has greatly advanced our understanding of the factors controlling the synthesis and degradation of carbohydrates and their partitioning within and between organs.

In this chapter, we summarise the current understanding of the central pathways of carbohydrate metabolism in plants, before describing approaches to exploit this knowledge for

Plant Metabolism and Biotechnology, First Edition. Edited by Hiroshi Ashihara, Alan Crozier, and Atsushi Komamine.

plant metabolic engineering. Finally, we briefly introduce the rising field of systems biology as an approach toward a holistic understanding of central carbon metabolism in plants.

1.2 Carbon Partitioning in Mesophyll Cells

During CO_2 assimilation the reductive pentose phosphate cycle (Calvin-Benson cycle) generates triose phosphate (triose-P) at the expense of energy (ATP) and reducing equivalents (NADPH) that were generated by photosynthetic light reactions. Triose-P can either be retained with the plastid in order to either regenerate the primary CO_2 acceptor ribulose-1,5-bisphosphate (RuBP) or to be fed into the synthesis of starch, respectively, or it can be exported into the cytosol to make soluble sugars (Figure 1.1). The rate of consumption

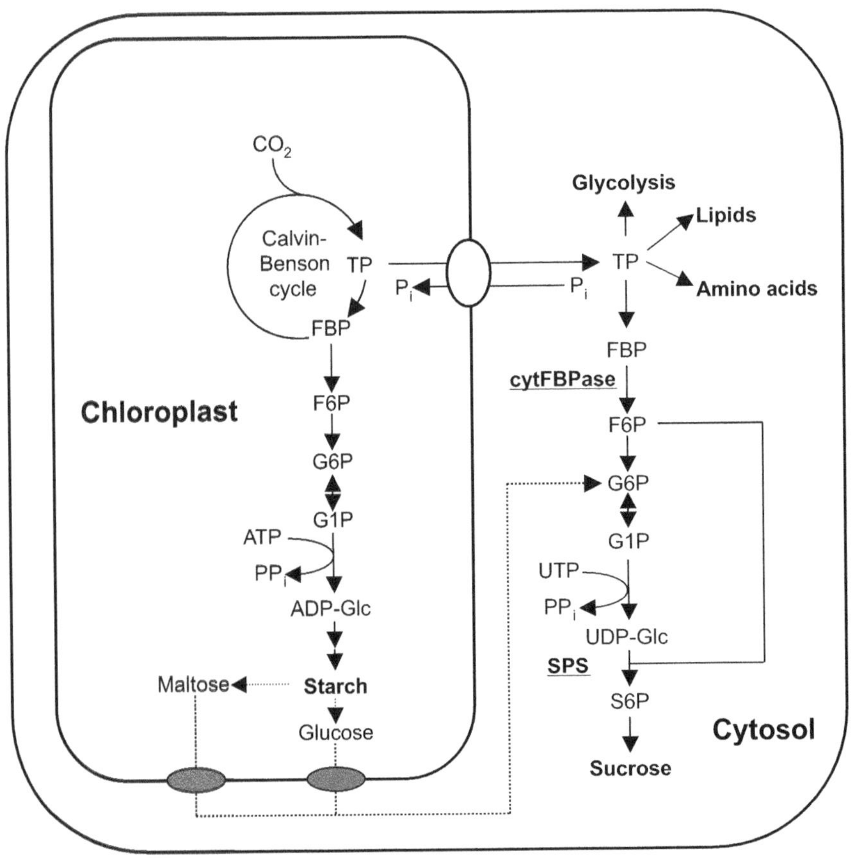

Figure 1.1 Photosynthetic carbon metabolism in mesophyll cells of source leaves. Dotted lines indicate the night path of carbon export from the chloroplast when starch mobilisation occurs. Abbreviations: FBP, fructose-1,6-bisphosphate; F6P, fructose-6-phosphate; G6P, glucose-6-phosphate; G1P, glucose-1-phosphate; ADP-Glc, ADP-glucose; UDP-Glc, UDP-glucose; S6P, sucrose-6-phosphate; cytFBPase, cytosolic fructose-1,6-bisphosphatase; SPS, sucrose-phosphate synthase; TP, triose phosphate

of triose-P by either of these processes has to be balanced with the momentary rate of photosynthesis to ensure that the correct portion is recycled back into the regeneration of RuBP. Therefore, communication between the stromal compartment of the chloroplast and the cytosol is necessary to adjust the rate of photosynthesis to the demands of various parts of the plant for photoassimilates. A specific transport system, the triose-P translocator (TPT), located in the inner membrane of the chloroplast envelope, mediates the export of triose-P into the cytosol in strict counter-exchange for P_i (Flügge, 1998). The rate of triose-P export catalysed by the TPT is thought to be primarily regulated by the availability of P_i liberated from phosphorylated intermediates by cytosolic processes, particularly sucrose synthesis. When the export of triose-P from the chloroplast is limited by the availability of P_i, for example, due to decreased sucrose synthesis, fixed carbon can be deposited in the chloroplast in the form of transitory starch, which is mobilised during the following dark period and used to sustain carbon export from source tissue.

Transgenic plants with reduced activity of the TPT or mutants defective in TPT function do not show a substantial growth phenotype but profound changes in the allocation of carbohydrates and the distribution of metabolites between the chloroplast and the cytosol (Riesmeier *et al.*, 1993; Häusler *et al.*, 1998; Schneider *et al.*, 2002). The patterns of carbohydrate synthesis in such plants suggest that they compensate metabolically for the reduced levels of TPT by diverting assimilates into starch and thereby releasing P_i required for continuous photosynthesis. This additional starch is largely absent by the end of the night, indicating that there is an increased rate of starch breakdown during the night that provides the substrates needed to sustain sucrose export to the rest of the plant. Because starch degradation preferentially uses an amylolytic pathway the leads to the formation of glucose and maltose (see discussion below), sugars not requiring TPT activity for export from the chloroplast to the cytosol, the reduced availability of carbohydrates during the day is compensated for by an increase export rate during the night. Only plants in which triose-P export and starch synthesis are inhibited simultaneously show a severe growth phenotype (Schneider *et al.*, 2002).

1.3 Sucrose Biosynthesis in Source Leaves

As outlined in the previous section, triose-P leaving the Calvin-Benson cycle can be exported from the plastid into the cytosol via the TPT and is then distributed between glycolysis and amino acid, lipid and sucrose synthesis (Figure 1.1). Sucrose is the major transport form of photoassimilates in higher plants, and as such forms the interface between photosynthetically active source tissue and heterotrophic sink tissue, where it serves as an energy source for growth and provides building blocks for storage metabolism. During the light period, sucrose synthesis proceeds from triose-P and comprises seven enzymatic steps. The entry molecule is fructose-1,6-bisphosphate (FBP), which is formed by the condensation of two molecules of triose-P catalysed by the enzyme aldolase. In a subsequent reaction, a phosphate group is removed from the C1 atom of FBP by the cytosolic isoform of fructose-1,6-bisphosphatase (cytFBPase) to yield fructose-6-phosphate (F6P). This reaction is essentially irreversibly and is supposed to represent one of the key regulatory steps of sucrose biosynthesis (see below). Glucose-6-phosphate (G6P) and glucose-1-phosphate (G1P) are maintained in equilibrium with F6P by the action of phosphoglucoseisomerase

(PGI) and phosphoglucomutase (PGM), respectively. Uridine diphosphate glucose (UDP-Glc) and pyrophosphate are then formed from UTP and G1P by the action of UDP-glucose pyrophosphorylase (UGPase). UDP-Glc is then combined with F6P to form sucrose-6^F-phosphate (Suc6P), catalysed by sucrose-6-phosphate synthase (SPS). Subsequently, Suc6P is hydrolysed to form sucrose and P_i by a specific sucrose phosphatase (SPP). The reaction catalysed by SPP is irreversible *in vivo* and displaces the reversible SPS reaction from equilibrium into the direction of net sucrose synthesis (Stitt *et al.*, 1987). SPS has been identified as the second major control point of the whole pathway, and the enzyme is subject to complex regulation on multiple levels.

The 'night path' of sucrose synthesis slightly differs from that operating during the light period. It is now generally accepted that the primary products of starch mobilisation appearing in the cytosol are maltose and, to a lesser extent, Glc, which are initially metabolised to G6P and then further to F6P and G1P, respectively (Smith *et al.*, 2005). Thus, the night path of sucrose synthesis involves neither triose-P as intermediate nor the enzymatic step catalysed by cytFBPase (Figure 1.1; dotted lines).

1.3.1 Regulatory Enzymes of the Pathway

In source leaves, sucrose synthesis has to be balanced with the momentary rate of photosynthesis. If sucrose synthesis operates too quickly, photosynthesis is inhibited because intermediates are withdrawn from the Calvin-Benson cycle too rapidly and RuBP regeneration is inhibited. On the other hand, if sucrose synthesis is too slow, P_i is sequestered from the chloroplast into phosphorylated intermediates, and ATP synthesis, 3PGA production and photosynthesis are inhibited (Stitt *et al.*, 1987).

Three enzymes of the pathway of sucrose synthesis are known to catalyse reactions removed from thermodynamic equilibrium *in vivo*. At these sites, the flux depends on the current activity of the enzyme, and such enzymes often possess regulatory properties, including allosteric regulation and/or post-translational modification. Hence, they are viewed as potential control sites. The current view of how sucrose synthesis is regulated by coordination of these enzymatic activities is discussed below.

1.3.1.1 Cytosolic Fructose-1,6-Bisphosphatase

CytFBPase irreversibly converts FBP into F6P and thus catalyses the first committed step of sucrose synthesis. The enzyme is also active in non-photosynthetic tissues where it controls the rate of F6P production in gluconeogenesis. The crucial role of cytFBPase for photosynthetic carbon partitioning has been established in mutant plants or transformants with a reduction in enzyme activity (Serrato *et al.*, 2009), although differences exist in the metabolic response between species. The inhibition of cytFBPase below 20% of the wild type activity led to an accumulation of triose-P, 3PGA and F1,6BP in source leaves of potato antisense transformants, but caused no major phenotypic effects (Zrenner *et al.*, 1996). This led to a massive shift of carbon partitioning towards starch, yielding starch levels which were three times higher at the end of the light period in transgenic plants than in the wild type. The surplus of starch is efficiently remobilised during the next night when sucrose synthesis is not dependent on cytFBPase activity, thereby circumventing the limitation of sucrose synthesis caused by a reduced cytFBPase activity. The situation is somewhat different in rice mutants carrying a T-DNA insertion in the cytFBPase gene (Lee *et al.*, 2008). These plants showed a marked decrease in photosynthesis, reduced growth rates and eventually died when grown in soil. Biochemical analyses revealed that the impaired synthesis of sucrose in the cytosol of the cytFBPase rice mutants cannot be compensated for by an increase in carbon partitioning towards starch (Lee *et al.*, 2008). Obviously, the capacity to synthesise and store starch is limited in rice, and it appears that sucrose represents the major transitory carbon storage compound in rice leaves.

The regulatory properties of cytFBPase are highly complex and distinct from those described for the plastidic isoform. While the latter is mainly regulated through thioredoxin mediated thiol modification, regulation of cytFBPase involves the action of several metabolites (Daie, 1993). In the absence of effector molecules, the enzyme has a high affinity for its substrate FBP (K_m = 4–6 μM). It is strictly Mg^{2+} dependent, weakly inhibited by F6P, P_i and AMP, and strongly inhibited by the regulatory metabolite fructose-2,6-bisphosphate (F2,6BP). F2,6BP is effective already at nano- to micromolar concentrations and substantially lowers the affinity of cytFBPase for its substrate FBP by shifting the sigmoidal substrate saturation curve of the tetrameric enzyme. The concentration of F2,6BP is determined by the relative *in vivo* activities of fructose-6-phosphate,2-kinase (F6P,2-K) and fructose-2,6-bisphosphate phosphatase (F2,6BPase). Both of these activities reside on a single bifunctional polypeptide (F2KP), and the ratio between the two activities is allosterically regulated through intermediates of primary metabolism (Nielsen *et al.*, 2004). F6P,2-K activity is enhanced by F6P and P_i and, in turn, is inhibited by 3PGA and triose-P, whereas F2,6BPase activity is inhibited by F6P and P_i. Upon the onset of photosynthesis, rising 3PGA and triose-P and decreasing P_i lead to a rapid decline in F2,6BP levels which, together with rising F1,6BP, stimulate cytFBPase activity.

The proposed role of F2,6BP in the regulation of photosynthetic carbon partitioning has gained additional support from the analysis of transgenic plants with altered expression of the bifunctional F2KP. Potato and *Arabidopsis* plants harbouring F2KP antisense constructs have a decreased F2,6BP content and show a two-fold increase in the sucrose to starch ratio in $^{14}CO_2$-labelling experiments (Draborg *et al.*, 2001; Rung *et al.*, 2004). In *Arabidopsis*, this resulted in a 20–30% higher level of sucrose and a delay in diurnal starch accumulation (Draborg *et al.*, 2001).

1.3.1.2 SPS and SPP

The last two steps in sucrose synthesis are unique to the pathway; they are catalysed by SPS and SPP. Owing to the rapid removal of Suc6P by SPP, the reaction catalysed by SPS is considerably displaced from equilibrium *in vivo* (Stitt *et al.*, 1987), and thus it is thought that SPS contributes to control of flux into sucrose. The enzyme is regulated by hierarchy mechanisms that operate at different levels and in different time frames (Huber and Huber, 1996; Winter and Huber, 2000). Allosteric regulation, involving activation by G6P and inhibition by P_i, allows sucrose synthesis to be immediately increased in response to increased availability of precursors. The regulation of SPS from spinach (*Spinacia oleracea*) has been particularly well characterised. The enzyme contains three phosphorylation sites, Ser-158, Ser-229 and Ser-424, which are involved in light/dark regulation (Huber and

Huber, 1996), 14-3-3 protein binding (Toroser *et al.*, 1998), and osmotic stress activation (Toroser and Huber, 1997), respectively. In darkness, phosphorylation of Ser-158 inactivates SPS by decreasing the enzyme's affinity for its substrates and by making it less prone to activation by G6P. After the onset of photosynthesis, rising G6P and falling P_i levels activate SPS allosterically, and rising G6P inhibits SPS-kinase, leading to dephosphorylation and post-translational activation of SPS (Huber and Huber, 1996).

There is growing evidence that higher plants contain more than one gene encoding SPS. An inspection of the *Arabidopsis* genome revealed the presence of four genes putatively encoding SPS enzymes, all of which appear to be both transcribed (Lunn and MacRae, 2003) and enzymatically active (F. Börnke, unpublished data). Phylogenetically, the four SPS sequences from *Arabidopsis* and all those known from other dicot species fall into three families: A, B and C (Lunn and MacRae, 2003). Recently, it was shown that monocot species contain an additional D family that probably arose after monocots and dicots diverged (Castleden *et al.*, 2004). The number of SPS genes in plants raises the question about functional specialisation of particular isoforms. A-family members have been the subject of most expression studies and most of the expressed sequence tags (ESTs) examined belong to the A-family, implying that A-family genes are more abundantly expressed than those belonging to other families. Antisense repression of SPS A in *Arabidopsis* led to a reduction of SPS activity of about 60–70%. While this led to an inhibition of sucrose synthesis in leaves, photosynthetic carbon partitioning was surprisingly not altered towards starch (Strand *et al.*, 2000). This approach targeted only one of the SPS gene families, and the extent to which other SPS genes were targeted or may have been up-regulated to compensate has not been investigated. Expression analysis of SPS family members in tobacco (*Nicotiana tabacum*) revealed that the A and C family members constitute the major SPS isoforms expressed in source leaves (Chen *et al.*, 2005a). Specific down-regulation of tobacco SPSA or SPSC by RNA-interference in transgenic plants affected overall SPS activity only slightly, indicating that during the day A and C can be mutually substituted by each other or, alternatively, by SPS B. However, transgenic tobacco plants with reduced SPS C displayed dramatically increased starch content in their leaves. Further analysis revealed that starch accumulation in NtSPSC silenced plants was not due to an increased partitioning of carbon into starch, but rather showed that starch mobilisation was impaired. The transgenic plants were unable to mobilise their transitory leaf starch efficiently during a prolonged period of darkness, and accumulated maltose as a major intermediate of starch breakdown. The NtSPSC mRNA level increased appreciable during the dark period, while transcript levels of the other isoforms showed no diurnal changes (Chen *et al.*, 2005a). Together, these results suggest that NtSPSC is specifically involved in the synthesis of sucrose during starch mobilisation in the dark, and future studies will have to investigate differences in the kinetic and regulatory properties of the different SPS isoforms to understand the biochemical basis for the functional specification of this protein family.

The role of SPP in the regulation of photosynthetic carbon partitioning has been relatively neglected. A recent study on transgenic tobacco plants with reduced SPP expression revealed that SPP is abundant in source leaves and does not exert significant control over sucrose synthesis under standard growth conditions (Chen *et al.*, 2005b). The SPP activity in those transformants could be reduced to 10% of the wild-type value before any effect on sucrose synthesis and photosynthetic carbon partitioning was observed.

1.4 Starch Metabolism in Source Leaves

1.4.1 Starch Synthesis within the Chloroplast

Starch is the major higher plant storage carbohydrate and is made up of glucose molecules that are linked in two different forms. Amylose is an essentially linear polymer in which the glucose moieties are linked end-to-end by $\alpha(1\rightarrow4)$ linkages. Amylopectin is a much larger branched molecule, in which about 5% of the glucose units are joined by $\alpha(1\rightarrow6)$ linkages. Despite its simple composition, starch forms complex semi-crystalline structures, starch granules, in plastids.

In leaves, as much as half of the triose-P produced during photosynthesis is partitioned at a linear rate into starch during the day (Zeeman and ap Rees, 1999). Leaf starch represents a transient store for assimilates, which is mobilised during the following night also at a linear rate to support leaf metabolism, and continued synthesis and export of sucrose. Initially, starch was viewed as an overflow product that is synthesised when the rate of photosynthesis exceeds the rate of other end products such as sucrose. However, the amount of starch synthesised appears to be precisely regulated to match the length of the dark period and is affected by developmental and sink/source effects, as well as by environmental factors including light intensity and photoperiod. If plants are grown in a short photoperiod enabling less photosynthesis per day, a larger proportion of photosynthate is temporarily stored as starch (reviewed by Smith and Stitt, 2007). This prevents carbon limitation at the end of a prolonged dark period, which can have serious consequences for plant growth.

Starch synthesis requires three consecutive enzymatic reactions catalysed by ADP-glucose pyrophosphorylase (AGPase), starch synthase (SS) and starch branching enzyme (SBE). The first committed step in the pathway of starch synthesis is the formation of ADP-glucose (ADP-Glc) from G1P and ATP in a reaction that is catalysed by AGPase and liberates pyrophosphate. In chloroplasts, ATP is derived from photosynthesis, and G1P can be supplied by the Calvin-Benson cycle via the plastidial isoforms of PGI and PGM. The reaction catalysed by AGPase is reversible under physiological conditions, but the high activity of plastidial alkaline pyrophosphatase (hydrolysing pyrophosphate to P_i) is assumed to drive the reaction towards ADP-Glc production.

AGPase in higher plants is a heterotetramer that consists of two 'regulatory' subunits (AGPS; ~51 kDa) and two slightly smaller 'catalytic' subunits (AGPB; ~50 kDa) (Okita *et al.*, 1990). AGPase activity is subject to allosteric regulation, with 3PGA acting as an activator and P_i as an inhibitor (Preiss, 1988). The ratio of these two metabolites changes according to supply of photoassimilates and the demand for them. For example, feedback regulation of sucrose synthesis in the cytosol will lead to the accumulation of phosphorylated intermediates, depletion of P_i and activation of AGPase by the rising 3PGA:P_i ratio in the chloroplast, eventually resulting in a compensatory stimulation of starch synthesis.

It is now apparent that AGPase activity is also redox-regulated (Fu *et al.*, 1998) in both non-photosynthetic (Tiessen *et al.*, 2002) and photosynthetic (Hendriks *et al.*, 2003) tissues. The mechanism involves a redox-regulated formation of an intermolecular Cys bridge between the two small AGPB subunits of the AGPase heterotetramer. Reduction of the enzyme breaks the bridge and results in activation. In chloroplasts, redox-activation is probably mediated by the ferredoxin/thioredoxin system, allowing the Calvin-Benson cycle and starch synthesis to be coordinately regulated (Hendriks *et al.*, 2003). In addition to light, redox-activation of AGPase is sensitive to metabolic changes. Activation of AGPase and hence increased starch synthesis can be triggered by high levels of glucose, sucrose or trehalose (Tiessen *et al.*, 2002; Hendriks *et al.*, 2003; Kolbe *et al.*, 2005). It appears that sugars signal redox-activation of AGPase through different, apparently independently operating pathways. Kolbe *et al.* (2005) provided evidence that sucrose-dependent activation might be mediated by the newly identified signalling metabolite trehalose-6-phosphate (Tre6P) and additionally involves SnRK1 (sucrose-non-fermenting-1-related protein kinase). The treatment of isolated chloroplast with Tre6P led to a complete reductive activation of AGPase (Kolbe *et al.*, 2005). The factors controlling Tre6P levels in plants are currently unclear. In *Arabidopsis*, it has been demonstrated that Tre6P amounts increase at the onset of the light period in parallel to a redox-activation of AGPase (Lunn *et al.*, 2006). The increase in Tre6P content was even greater when the dark period was extended by 6h. An extended night leads to carbon starvation towards the end of the dark period which considerably affects growth (Gibon *et al.*, 2004). The elevation of Tre6P level at the start of the following day is preceded by a rapid rise in sucrose level which together is assumed to lead to redox-activation of AGPase (Lunn *et al.*, 2006). This results in an increased partitioning of photoassimilate into starch as compared with that in the previous light period. Thus a single night during which starch supply was not adequate to meet demand could trigger a mechanism ensuring that a greater reserve of starch is built up during the subsequent light period, sufficient to allow for an extension of the subsequent night (Lunn *et al.*, 2006).

Recently, Michalska *et al.* (2009) demonstrated that NADP-thioredoxin reductase C (NTRC) mediates NADPH-dependent reduction of AGPase *in vitro*. NTRC is an unusual plastid-localised bifunctional protein that contains a NADP-thioredoxin reductase (NTR) and a thioredoxin domain in a single enzyme (Serrato *et al.*, 2004). Leaves of an *Arabidopsis* NTRC knock-out mutant showed a decrease of both in the extent of AGPase redox-activation and in the enhancement of starch synthesis either in the light or after treatment with external sucrose (Michalska *et al.*, 2009). The authors propose that NTRC serves as an alternative system to transferring reducing equivalents to AGPase in leaves, thereby enhancing starch synthesis. In the light, NTRC is mainly linked to photoreduced ferredoxin via ferredoxin-NADP reductase. In the dark or under conditions in which light reactions are impaired, NTRC is primarily linked to sugar oxidation via the initial reactions of the oxidative pentose phosphate pathway and regulates the redox-status of AGPase in this way, independently of the ferredoxin/thioredoxin system (Michalska *et al.*, 2009)

The ADP-Glc produced by AGPase is used as a substrate by SSs which catalyse the formation of new glycosidic linkages by transferring the glucose residue of ADP-Glc to the non-reducing end of an existing α-1,4-linked glucan chain, thereby elongating it. Higher plant SSs are encoded by five gene classes, designated GBSS (for granule bound starch synthase), SSI, SSII, SSIII and SSIV. GBSS is found exclusively bound to, and buried within, the starch granule. It is responsible for the synthesis of the linear amylose fraction of starch and is thought to elongate either soluble malto-oligosaccharides or the side-chains of amylopectin. The SS families SSI, SSII and SSIII are localised in the plastid stroma and appear to all be involved in the elongation of the amylopectin chains. The function of the SSIV class has not been established, although this isoform might be involved in starch-granule initiation (Roldán *et al.*, 2007).

The branching of amylopectin proceeds concurrently with chain elongation. SBE forms the $\alpha(1\rightarrow6)$ linkages found in amylopectin via a glucosyltransferase reaction in which an existing α(1→4) linked chain is cut and a glucan segment of six or more glucose residues is transferred to the same, or an adjacent chain. Higher plant SBEs fall into two classes, designated class I and II. Class I enzymes preferentially transfer longer chains than class II enzymes (for a recent review, see Zeeman *et al.*, 2007).

In addition to SSs and SBEs, other glucan-modifying enzymes participate in the starch biosynthetic process. Debranching enzymes (DBEs; a-1,6-glucanohydrolase), which cleave branch points, are important determinants of amylopectin morphology and structure. Mutations affecting particular classes of DBEs result in the complete replacement of semi-crystalline starch granules by soluble phytoglycogen. Thus DBE activity has been proposed to be necessary for crystallisation (Zeeman *et al.*, 2007).

Although the enzymatic steps within the starch biosynthetic pathway succeeding the AGPase reaction were regarded as unregulated, there is increasing evidence for the occurrence of multi-enzyme complexes comprising SSs, SBEs and other enzymes (Kötting *et al.*, 2010). The functional significance of complex formation and whether it occurs in all starch-synthesising organs is not clear yet, but it could influence both enzyme properties and the product of their combined action.

1.4.2 Starch Breakdown in Leaves and Metabolism of its Degradation Products in the Cytosol

For a long time our understanding of how transitory starch is remobilised at night and how these processes are regulated considerably lagged behind that concerning the complex metabolic network regulating photoassimilate partitioning between the different pathways during the photoperiod. During the last decade, the identification and analysis of *Arabidopsis* mutants defective in key enzymes of starch degradation have introduced a radically new picture of the pathway resulting in nocturnal degradation of leaf starch. These so-called starch-excess (*sex*) mutants accumulate starch in their leaves over repeated diurnal cycles and thus can be identified in large populations with relative ease by simple iodine staining. This approach constitutes a prime example of how *Arabidopsis* genetics and genomics have revolutionised pathway discovery also in a relatively mature research field such as photosynthetic primary metabolism (Stitt *et al.*, 2010).

One of the surprising findings during the last years was that starch degradation requires the prior incorporation of phosphoryl groups into the polyglucan (Figure 1.2) (reviewed by Fettke *et al.*, 2009). Relevant to starch phosphorylation is a glucan-water dikinase (GWD, formerly known as R1; Lorberth *et al.*, 1998; Ritte *et al.*, 2002) that transfers the β-phosphate of ATP to the C6 position of a glycosyl residue of amylopectin (Ritte *et al.*, 2002). Antisense suppression of GWD activity in potato substantially lowers the phosphorylation level of amylopectin and leads to a starch excess phenotype in leaves, but also prevents cold-induced starch degradation in tubers (Lorberth *et al.*, 1998). Similarly, loss of GWD in *Arabidopsis* (the *sex 1* mutant) leads to a very severe *sex* phenotype (Yu *et al.*, 2001). The activity of GWD was recently found to be dependent on the reduction by thioredoxin in a reaction that also alters its starch-binding capacity (Mikkelsen *et al.*, 2005).

A second glucan-water dikinase was identified in *Arabidopsis* (PWD, phosphoglucan water dikanase) that phosphorylates the C3 position of glucosyl residues and is also required for starch degradation, as judged from the weak *sex* phenotype of *Arabidopsis pwd*

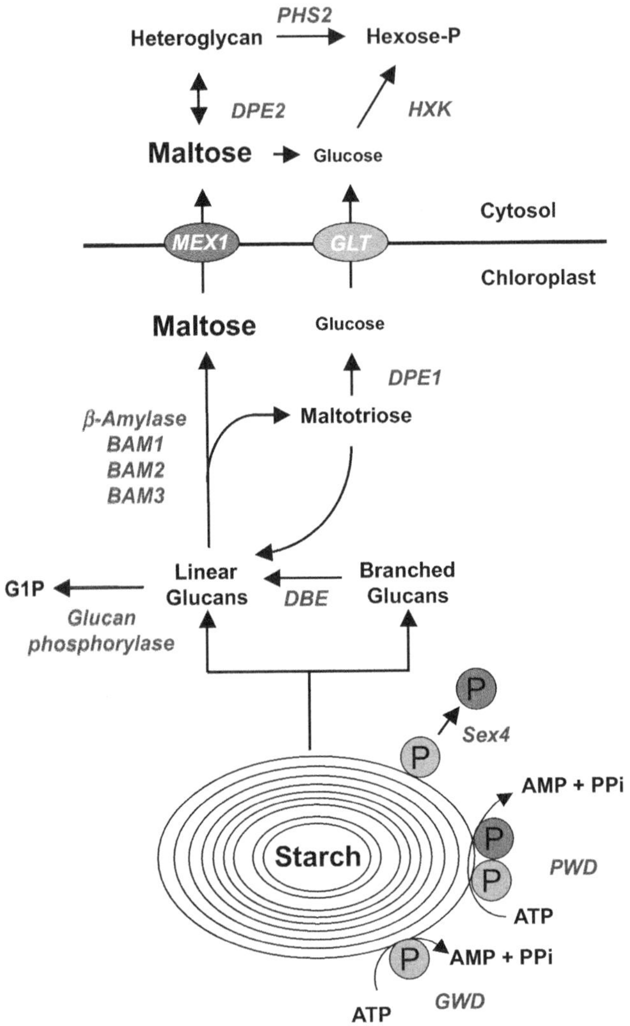

Figure 1.2 *Starch degradation in leaves. The genes encoding for the major enzymes in this pathway in* Arabidopsis *are indicated. Abbreviations:* GWD, *glucan-water dikinase;* PWD, *phosphoglucan-water dikinase;* Sex4, *phosphoglucan phosphatase;* DBE, *debranching enzyme;* DPE1, *plastidic disproportionating enzyme;* MEX1, *maltose transporter;* GLT, *glucose transporter;* DPE2, *cytosolic disproportionating enzyme;* PHS2, *cytosolic glucan phosphorylase;* HXK, *hexokinase; G1P, glucose-1-phosphate*

knock-out mutants (Baunsgaard *et al.*, 2005; Kötting *et al.*, 2005). This enzyme uses a phosphorylated glucan as a substrate and thus requires the prior action of GWD. Recent evidence suggests that phosphorylation disrupts the crystalline structure of amylopectin and thus mediates its transition into a soluble state (Hejazi *et al.*, 2008). This is required for downstream enzymes to further degrade the glucan polymer. Several independent studies identified a glucan-binding phosphatase in *Arabidopsis* leaves (Kerk *et al.*, 2006; Niittylä *et al.*, 2006; Sokolov *et al.*, 2006). A loss-of-function mutant (*sex4*) has decreased rates of starch degradation and accumulates soluble phospho-oligosaccharides during the night (Kötting *et al.*, 2009). Recombinant glucan-binding phosphatase can dephosphorylate semi-crystalline amylopectin, acting on phosphate groups either on C3 or C6 positions (Hejazi *et al.*, 2010). Taken together, this suggests that removal of the phosphate groups added by GWD and PWD is also necessary for complete starch degradation.

Further hydrolysis of the glucan chains is catalysed primarily by β-amylases that act as exo-amylases, releasing maltose from exposed non-reducing ends of chains. β-amylases cannot hydrolyse $\alpha(1{\rightarrow}6)$ linkages or act close to them. Thus, the complete degradation of amylopectin requires DBEs that release linear malto-oligosaccharides (MOSs). These are probably further metabolised by β-amylases to yield maltose and maltotriose, with the latter being too short to act as a substrate for further degradation by β-amylases. It appears that maltotriose is further metabolised by the disproportionating enzyme (DPE1), which preferentially transfers a maltosyl residue from maltotriose to an acceptor glucan, generating glucose and longer glucan that can be further degraded by β-amylases (Zeeman *et al.*, 2007). The *Arabidopsis* genome encodes for nine β-amylases (*BAM1–9*), four of which have been shown to be plastid localised (BAMs 1–4) (Fulton *et al.*, 2008). BAM1 and BAM3 are active enzymes and appear to have overlapping functions as *bam1* and *bam3* mutants have a wild-type and a mild *sex* phenotype, respectively, while the *bam1/bam3* double mutant has a strong *sex* phenotype (Fulton *et al.*, 2008). Interestingly, BAM4 has no apparent catalytic activity but the *bam4* mutant has a sex phenotype. Thus, a regulatory role for BAM4 in starch degradation has been proposed (Fulton *et al.*, 2008).

Alternatively to being broken down hydrolytically, linear glucans can potentially also be metabolised by the chloroplast-localised α-glucan phosphorylase that liberates G1P (Beck and Ziegler, 1989). The precise role and importance of phosphorolysis in starch breakdown is yet not clear. Potato plants with reduced plastid localised phosphorylase activity have a wild-type phenotype (Sonnewald *et al.*, 1995).

There is now strong evidence that maltose is the major metabolite exported from chloroplasts during the night. *Arabidopsis* lacking a plastidic maltose transporter (mex1) accumulate high levels of maltose in their leaves, are severely impaired in growth and display a starch excess phenotype (Niittylä *et al.*, 2004). In addition to the maltose produced by β-amylases, glucose accumulates to a lesser extent during starch breakdown, produced from the disproportionation of maltotriose by DPE1. A specific transporter that could mediate glucose export from the plastid into the cytosol has been identified in several plant species (Weber *et al.*, 2000), although the importance of this protein for starch breakdown has not yet been established.

In the cytosol, maltose and glucose undergo a series of reactions that finally lead to the formation of sucrose, substrates for cellular respiration, or building blocks for biosyntheses. In *Arabidopsis*, maltose is further metabolised by transglucosidase named DPE2 (Chia *et al.*, 2004). DPE2 catalyses the transfer of one of the glucosyl moieties of maltose to a soluble heteroglycan and releases the other as glucose (Fettke *et al.*, 2009). *Arabidopsis dpe2* mutants accumulate high levels of maltose and show impaired starch breakdown, suggesting that metabolism via DPE2 is the major route for maltose metabolisation in the cytosol (Chia *et al.*, 2004). While it is clear that glucose released from maltose is subsequently channelled into the hexose-P pool via the action of hexokinase, metabolism of heteroglycan is much less well understood. Recent evidence from work with transgenic potato indicates that the cytosolic isoform of starch-phosphorylase (Pho2/PHS2) catalyses the reversible glucosyltransfer of glucosyl residues to the heteroglycans. Hence, the enzyme can utilise orthophosphate as an acceptor and the heteroglycan as donor, yielding the formation of G1P (Fettke *et al.*, 2009).

1.5 Sucrose to Starch Conversion in Storage Organs

The pathway of starch synthesis in developing storage organs is relatively well understood (Figure 1.3). In all organs apart from cereal endosperms, sucrose entering the storage parenchyma is converted to G6P in the cytosol. In the case of the potato tuber, sucrose delivered by the phloem from source tissue can be metabolised in different ways. It can either be hydrolysed by apoplastic or cytosolic invertase, respectively, resulting in glucose and fructose, or converted into UDP-Glc and fructose by sucrose synthase (SuSy). The prevailing route of sucrose cleavage depends upon the developmental stage of the tuber. At the onset of tuberisation when cell division takes place, hydrolytic degradation of sucrose

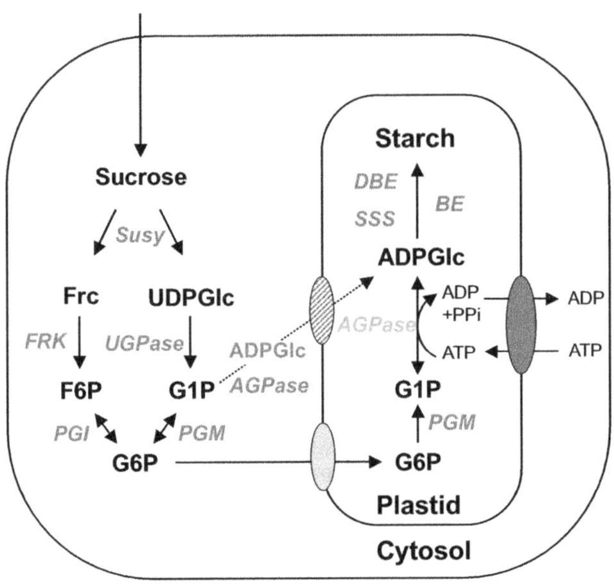

Figure 1.3 Principle pathway leading to the formation of starch in storage organs. The alternative route of ADPGlc via generation within the cytosol and subsequent uptake into the amyloplast, as it occurs in the endosperm cells of graminaceous species, is shown by dotted arrows. Susy, sucrose synthase; UDPGlc, UDP-glucose; F6P, fructose-6-phosphate; FRK, fructokinase; UGPase, UDP-glucose pyrophosphorylase; PGI, phosphoglucose isomerase; PGM, phosphoglucose mutase; G1P; glucose-6-phosphate; AGPase, ADP-glucose pyrophosphorylase; ADPGlc, ADP-glucose; SSS, starch synthase; BE, branching enzyme; DBE, debranching enzyme

by invertases dominates. Later at the beginning of the storage phase of the developing tuber, a switch to SuSy mediated sucrose degradation occurs (Appeldoorn *et al.*, 1997). The products of sucrose cleavage enter metabolism by the concerted action of UGPase and fructokinase or hexokinase in the case of the SuSy or invertase pathways, respectively. As a result, the end products of sucrose degradation will enter the hexose-P pool, which consists of an equilibrium mixture of F6P, G6P and G1P, because of readily reversible reactions catalysed by the cytosolic isoforms of PGI and PGM (Fernie *et al.*, 2002).

The crucial role for Susy in sucrose breakdown during the storage phase has been established from transgenic potato plants expressing a Susy antisense construct driven by the CaMV 35S promoter (Zrenner *et al.*, 1995). In transgenic tubers a reduction in Susy activity of up to 95% resulted in a reduction in starch and storage protein content of mature tubers, but surprisingly the sucrose levels remained unchanged. There was, however, a significant increase in the levels of hexoses which was paralleled by a 40-fold increase in invertase activity. The fact that the induction of invertase activity could not compensate for the loss in Susy activity argues for metabolic channelling of sucrose via the Susy-dominated pathway into starch, and thus indicates that Susy plays a predominant role in determining potato tuber sink strength (Zrenner *et al.*, 1995). Accordingly, expression of a yeast-derived invertase either in the apoplast or in the cytosol of transgenic tubers did not result in increased sink strength in these tubers, despite their higher sucrolytic capacity (Sonnewald *et al.*, 1997).

Since starch synthesis in potato tubers is confined to amyloplasts, it entirely relies on the translocation of metabolites from the cytosol through the amyloplast envelope. It now seems clear that carbon for starch synthesis enters the amyloplast in the form of G6P (Tauberger *et al.*, 2000). Uptake is mediated by specialised glucose-6-phosphate/phosphate translocator (GPT) residing in the inner envelope membrane of amyloplasts (Kammerer *et al.*, 1998). Following uptake of carbon into the tuber amyloplast, starch synthesis proceeds via the concerted action of plastidial PGM, AGPase and the polymerising reactions already described for the synthesis of transitory starch in chloroplasts (see above). The important role of AGPase for starch synthesis in potato tubers has been proven by antisense studies in which a reduction of AGPase activity led to a dramatic reduction in the level of starch (Müller-Röber *et al.*, 1992). Concerning its regulatory properties, potato tuber AGPase resembles the leaf enzyme (Tiessen *et al.*, 2002).

Whereas in chloroplasts the ATP necessary for starch synthesis can be readily provided through photosynthesis, potato tuber amyloplasts have to import ATP from the cytosol via an ATP/ADP transport protein (NTT) located on the inner-envelope membrane (Neuhaus and Emes, 2000). Tjaden *et al.* (1998) showed that a relatively small decrease in ATP/ADP transporter activity leads to a reduced level of total starch content and a lower amylose to amylopectin ratio. In contrast, increased transporter activity correlated with higher starch contents and a higher amylose to amylopectin ratio. In total, these observations indicated that the rate of ATP import exerts considerable control on the rate of starch synthesis and affects the molecular composition of starch in potato tubers.

Although the later polymerising reactions of starch synthesis catalysed by the various isoforms of SSs, SBEs and DBEs do not appear to play a role in the control of starch accumulation, they are crucial in determining the structure of starch (Kossmann and Lloyd, 2000; Tetlow *et al.*, 2004).

A recent comparative analysis of gene expression in potato leaves and tubers revealed that transient and storage starch biosynthesis are strikingly similar in respect to the expression of particular isoforms of starch biosynthetic genes (Ferreira *et al.*, 2010).

Starch synthesis in endosperms of graminaceous species, including cereals, differs from that in potato tubers in that the synthesis of ADP-Glc occurs largely in the cytosol (Figure 1.3), via a cytosolic isoform of AGPase. ADP-Glc is imported into the plastid by a specific sugar nucleotide transporter (Tomlinson and Denyer, 2003).

1.6 Metabolic Engineering of Carbohydrate Metabolism

Photosynthetic primary metabolism is a classical field of plant biochemical research, and many fundamental insights into plant function have been obtained in this field over the past decades. Due to the fact that photosynthesis and post-photosynthetic processes are supposed to be the main drivers of crop yield, there is continued interest in further unravelling the underlying regulatory mechanisms and networks. It is hoped that this will lead to novel knowledge-based strategies to engineer carbohydrate metabolism towards altered crop composition using biotechnology. Starch and sugar accumulation have been targeted by metabolic engineering in transgenic plants since the advent of plant biotechnology (Lytovchenko *et al.*, 2007; Smith, 2008), and examples from this area of metabolic engineering are highlighted below.

1.6.1 Increasing Starch Content

Besides its importance as a staple in human and animal diets, starch is also used as a renewable raw material for a wide range of industrial purposes (Jobling, 2004). The source for industrial starch is mainly corn, but significant amounts are also extracted from a range of other species, including rice, wheat, cassava and potato. Starches from different species differ in their polymer composition (e.g. amylose to amylopectin ratio) and structure. These characteristics determine the functional properties and thus the range of applications for which a given starch is used. The modification of starch metabolism in crops could be beneficial to increase starch accumulation in harvestable organs, to prevent or increase starch degradation, or to modify starch structure to enhance or diversify its functionality for industrial uses.

To increase the efficiency of the pathway and thus to increase starch accumulation in crop plants, molecular strategies have initially concentrated on AGPase, the enzyme assumed to catalyse the rate-limiting step of starch synthesis. In an early attempt to increase the activity of the starch biosynthetic pathway in potato tubers, Stark *et al.* (1992) overexpressed a deregulated bacterial AGPase in the potato variety Russet Burbank. Overall, the transformed lines were reported to have an average of 35% more tuber starch than the controls. However, this effect was lost upon transformation of a different potato cultivar (Sweetlove *et al.*, 1996). In the latter case starch degradation was up-regulated in addition to starch synthesis, resulting in no net change in starch accumulation.

Attempts to increase starch contents through manipulation of AGPase in cereal seeds have made use of a variant of the maize AGPase gene (*shrunken2*) whose gene product is less sensitive to inhibition by phosphate when compared with the wild-type protein.

Smidansky and colleagues (Smidansky *et al.*, 2002, 2003, 2007) have shown that maize, rice and wheat plants expressing this AGPase allele in the endosperm and grown under controlled conditions display an increase in individual seed weight as well as in seed yield per plant. However, in field trials, transgenic wheat plants only showed a yield enhancement under conditions of minimal inter-plant competition and optimal water supply (Meyer *et al.*, 2007).

A second general approach to increase starch accumulation in storage organs is to increase sucrose catabolism, thereby diverting more carbon from sucrose to starch. Expression of yeast invertase within the apoplast of transgenic tubers led to a dramatic reduction in tuber sucrose contents and a corresponding large increase in glucose content. Although an increase in size was observed in the transgenic tubers, an increase in yield was largely compensated for by a simultaneous reduction in the number of tubers per plant (Sonnewald *et al.*, 1997). To further stimulate hexose utilisation in invertase-expressing plants, a bacterial glucokinase was co-expressed along with invertase. Despite the massive accumulation of hexose-phosphates the transgenic tubers showed no increase in starch synthesis, but were rather characterised by an induction of glycolysis and a massive partitioning of carbon into respiration (Trethewey *et al.*, 1998). The reason for this dramatic change in partitioning remains mysterious, but it clearly demonstrates the flexibility of plant metabolism and highlights the necessity for a detailed understanding of the underlying factors that regulate it.

Starch content in potato tubers is very sensitive to manipulation of the plastidial adenylate transporter providing the ATP necessary for the AGPase reaction. Overexpression of an adenylate transporter from *Arabidopsis* in potato tubers resulted in 16–36% more starch per gram fresh weight, indicating that ATP supply to the plastid limits starch synthesis (Tjaden *et al.*, 1998; Geigenberger *et al.*, 2001). Recently, a further increase in potato tuber starch content was achieved by the simultaneous overexpression of a GPT from pea and an *Arabidopsis* adenylate translocator (NTT). Double transformants exhibited an increase in tuber yield of up to 19% in addition to an increase in starch content of 28%, when compared with control plants (Zhang *et al.*, 2008). Both effects taken together led to a calculated increase in potato tuber starch of up to 44%. The authors concluded that starch synthesis in potato tubers is co-limited by the ATP supply as well as by the import of carbon skeletons into the amyloplast (Zhang *et al.*, 2008). Further evidence for an energy limitation of starch synthesis in potato tubers comes from transgenic plants with reduced expression of plastidial adenylate kinase (ADK) (Regierer *et al.*, 2002). In this study a strong negative influence of ADK activity on starch accumulation was found, suggesting that ADK normally competes with starch synthesis for plastidial ATP.

Taken together, successful attempts to increase starch content through metabolic engineering are scarce. The analyses so far suggest that in potato tubers considerable control of starch synthesis lies outside of the linear pathway, as both the adenylate transporter as well as the plastidial ADK appear to exert higher control over the pathway than AGPase, the enzyme widely believed to be rate-limiting (Geigenberger *et al.*, 2004).

1.6.2 Altering Starch Quality

As stated above, one of the major factors influencing starch quality is the amylose to amylopectin ratio. Due to their contrasting physico-chemical properties, it is advantageous if starches for industrial applications are composed mainly of either amylose or amyolopectin.

The synthesis of amylose is accomplished through the activity of a particular isoform of starch synthase, the GBSS, and antisense inhibition of this gene has led to amylose-free potato starch (Visser *et al.*, 1991). No penalties in starch content have been observed and thus these plants are seemingly suited for commercialisation. Amylose-free potato starch has improved paste clarity and stability and can be expected to find application in both the food industry and in paper manufacture. Although produced almost 20 years ago, it was not before the year 2010 that a transgenic amylose-free potato variety received approval for commercial cultivation in Europe. Meanwhile, a non-functional mutant GBSS from a non-transgenic amylose-free mutant potato line, originally identified in a diploid variety (Muth *et al.*, 2008), was bred into a commercial cultivar, and amylose-free starch is produced from that novel variety on a commercial scale.

Also high-amylose starches have been of great interest although they are much more difficult to obtain (Jobling, 2004). Recently, in an innovative approach to increase amylose content by the inhibition of starch branching enzyme A (SBE A) activity, the enzyme responsible for introducing $\alpha(1\rightarrow6)$ linkages into amylopectin has been reported (Jobling *et al.*, 2003). The authors of this study expressed a single-chain antibody targeted against the active centre of SBE A, thereby neutralising its activity. They found that immunomodulation increased the amylose content of starch granules from about 20% in wild-type tubers up to 74% in the best transgenic line, exceeding the levels of amylose achieved by conventional antisense strategies (Jobling *et al.*, 2003).

1.7 Engineering Soluble Sugars

For several crop species, soluble sugar content is much more important than that of starch. This is either because soluble sugars such as sucrose are major reserve carbohydrates (e.g. in sugar cane and sugar beet), or, as in fruit-bearing species, because sugar is an important component of taste. An increase in sugar content in strawberry has been achieved through fruit-specific antisense repression of AGPase. Transgenic strawberry fruits showed a decrease in starch content of approximately 50% and an increase in total soluble solids of up to 37% (Park *et al.*, 2006). In general, relatively little is known at the biochemical or genetic level about the factors that control the rate of sucrose storage in sugar beet taproots or sugar cane nodes. This reflects the intractability of both crops using genetics and the difficulties in assessing storage metabolism at the biochemical level. Factors controlling sucrose accumulation in storage tissues are photoassimilate partitioning on the whole plant level, but also the control of phloem unloading, the nature and kinetics of sugar transporters in storage tissues, and the control of futile cycling of sucrose in the cytosol of storage cells (Smith, 2008). One of the rare examples of a successful increase in soluble sugar content was recently reported for sugar cane. Wu and Birch (2007) introduced a bacterial sucrose isomerase (SI) gene tailored for vacuolar compartmentation into transgenic sugar cane. SI activity converts sucrose into its non-metabolisable isomer isomaltulose (IM) and transgenic SI-expressing lines accumulated substantial amounts of IM in their culm. Remarkably, this was not at the expense of sucrose levels, resulting in a total sugar concentration of up to double in harvested juice. The reason for this boost in sugar concentration is not understood, but it has been hypothesised that IM accumulation in the culm leads to enhanced sink strength which fosters import of additional carbon from source tissues (Wu and Birch,

2007). It remains to be shown whether this strategy allows increasing total sugar content in other sucrose-storing crops such as sugar beet.

1.8 Production of Novel Carbohydrates in Transgenic Plants

In addition to attempts aiming at manipulating the contents and properties of endogenous carbohydrates, there have been several successful approaches for the production of novel carbohydrates in transgenic plants. By expressing enzymes that act on sucrose or sucrose biosynthetic intermediates, novel compounds can theoretically accumulate to high levels.

Fructans, or polyfructosylsucroses, are alternative storage carbohydrates that are highly soluble and are stored within the vacuole as opposed to the plastid: they are present in approximately 15% of all flowering plants (Hellwege *et al.*, 2000). Fructan synthesis is initiated by sucrose:sucrose 1-fructosyltransferase (SST) which catalyses the fructosyltransfer from one sucrose molecule to another, resulting in the trisaccharide 1-kestose. In subsequent steps fructosyltransferase (FFT) catalyses the reversible transfer of fructosyl residues from one fructan to another, producing a mixture of fructans with different chain lengths (Ritsema and Smeekens, 2003). One of the simplest fructans is inulin, which consists of $\beta(1\rightarrow2)$-linked fructose residues, while fructans of the levan type are $\beta(2\rightarrow6)$-linked fructose polymers.

From a biotechnological viewpoint, interest in fructans has continued to increase as they have been recognised as beneficial food ingredients. As part of the human diet, they are considered to be prebiotics as they selectively promote the growth of beneficial intestinal bacteria. Furthermore, fructans are assumed to have anti-cancer activity, promote mineral absorption, decrease cholesterol levels and decrease insulin levels. Fructans are normally isolated from plants with low agronomic value, such as the Jerusalem artichoke (*Helianthus tuberosus*) and chicory. Thus, attempts have been made to produce transgenic plants with higher fructan yield or making fructans with specific properties. Transformation of sugar beet with an SST gene from Jerusalem artichoke resulted in the conversion of 90% of the vacuolar sucrose into fructan (Sevenier *et al.*, 1998); since the sugar beet accumulates to concentrations approaching 600 mM sucrose, this represents a massive fructan yield. Weyens *et al.* (2004) introduced a pair of FFTs, namely sucrose:sucrose 1-fructosyltransferase (1-SST) and fructan:fructan 6-G-fructosyltransferase (6G-FFT) from onion into transgenic sugar beet. Expression of these two enzymes resulted in high-level accumulation of onion-type fructans in transgenic taproots without affecting storage capacity. Potato, as another crop naturally not accumulating fructans, was used to express plant fructosyltransferases. The SST and FFT enzymes from globe artichoke were engineered into potato and led to the accumulation of the full range of fructans found in globe artichoke itself (Hellwege *et al.*, 2000).

The production of isomaltulose (IM) in transgenic sugar cane expressing a bacterial sucrose isomerase (SI) has already been introduced (see section above). IM is an excellent sucrose substitute in foods as it shares many physico-chemical properties with sucrose, but is non-cariogenic and has a low calorific value. IM is produced on an industrial scale from sucrose by an enzymatic rearrangement using immobilised bacterial cells expressing a SI. Expression of a SI gene from *Erwinia rhapontici* within the apoplast of transgenic potato tubers led to a nearly total conversion of sucrose into IM (Börnke *et al.*, 2002). Despite the soluble carbohydrates having been altered within the tuber, growth of SI-expressing transgenic potato plants was indistinguishable from wild-type plants. This example, together with the study on SI-expressing sugar cane (Wu and Birch, 2007), demonstrates that transgenic plants provide a valid platform for high-level IM production in storage tissues.

1.9 Network Analysis of Carbohydrate Metabolism

Traditionally, metabolism has been divided into discrete pathways. However, it has become increasingly apparent that metabolism operates as a highly integrated network (Sweetlove *et al.*, 2008). Synthesis of a metabolite usually requires the operation of many pathways, and metabolites are not synthesised in isolation from each other; rather, large sets of metabolites must often be synthesised simultaneously. Analysis of metabolic networks at a systems level depends upon the integration of data obtained from more than one level of molecular entity (that is DNA, RNA, proteins, metabolites, organelles, cell types, organs, etc.). The expanding development of high-throughput data generation technologies (so-called "omics" such as genomics, transcriptomics, proteomics, metabolomics, etc.) made it possible to measure (profile) most or even all components of one class (e.g. transcripts, proteins, etc.) in a highly parallel way. These high-throughput profiling technologies provide a rich source of quantitative biological information that allows researchers to move beyond a reductionist approach by both integrating and understanding interactions between multiple components in cells and organisms (Yuan *et al.*, 2008; Fukushima *et al.*, 2009; Stitt *et al.*, 2010). At the simplest level, correlation networks can be used to identify which components might be functionally related based on the 'guilt by association' principle. Gene-to-metabolite networks, for instance, define the interactions amongst genes and metabolites and are typically constructed using multivariate analysis, i.e. a statistical approach used to analyse more than one variable at a time, or data mining of gene expression profiling and metabolite profiling data under different conditions or in different genetic backgrounds, respectively. The outputs from statistical analyses are often visualised based upon the distance calculated among genes and metabolites according to their profiling patterns. If a gene is determined as being 'close' to a metabolite, it might be implicated in the synthesis or regulation of the latter (Yuan *et al.*, 2008). A recent example of the correlative approaches was provided by Sulpice *et al.* (2009) who used the combination of molecular phenotyping techniques, including metabolite- and transcriptional profiling, to reveal correlations between molecular markers and biomass in 92 *Arabidopsis* accessions. By using a multivariate regression approach, the authors found that starch content at the end of the light period and the protein content negatively correlate with leaf biomass. This analysis prompted the hypothesis that fast-growing accessions utilise their starch more efficiently for growth, and that decreased synthesis of protein is one of the factors that contributes to this increase in the use of carbon. In addition, from the results of an integrative approach with transcript profiling and detailed genotyping of leaf samples, two candidate genes associated with increased biomass production were identified which might represent candidate lead genes with the potential to increase biomass production (Sulpice *et al.*, 2009).

Given the increasing number of publications dealing with correlative approaches (Fukushima *et al.*, 2009), it appears highly likely that these kinds of studies will greatly enhance our understanding of the connectivity that underpins the regulatory network of carbon metabolism and the relative importance of varying environments or different genetic backgrounds, respectively.

Acknowledgements

Research in our laboratories is supported by grants from the Bundesministerium für Bildung und Forschung (BMBF) and by the Deutsche Forschungsgemeinschaft (DFG).

References

Appeldoorn, N.J.G., de Bruijn, S.M., Koot-Gronsveld, E.A.M., Visser, R.G.F., Vreugdenhil, D. and van der Plas, L.H.W. (1997) Developmental changes of enzymes involved in conversion of sucrose to hexose-phosphate during early tuberisation of potato. *Planta*, **202**, 220–226.

Baunsgaard, L., Lutken, H., Mikkelsen, R., Glaring, M.A., Pham, T.T. and Blennow, A. (2005) A novel isoform of glucan, water dikinase phosphorylates pre-phosphorylated alpha-glucans and is involved in starch degradation in Arabidopsis. *Plant J.*, **41**, 595–605.

Beck, E. and Ziegler, P. (1989) Biosynthesis and degradation of starch in higher plants. *Annu. Rev. Plant Phys.*, **40**, 95–117.

Börnke, F., Hajirezaei, M. and Sonnewald, U. (2002) Potato tubers as bioreactors for palatinose production. *J. Biotechnol.*, **96**, 119–124.

Castleden, C.K., Aoki, N., Gillespie, V.J., MacRae, E.A., Quick, W.P., Buchner, P., Foyer, C.H., Furbank, R.T. and Lunn, J.E. (2004) Evolution and function of the sucrose-phosphate synthase gene families in wheat and other grasses. *Plant Physiol.*, **135**, 1753–1764.

Chen, S., Hajirezaei, M. and Börnke, F. (2005a) Differential expression of sucrose-phosphate synthase isoenzymes in tobacco reflects their functional specialization during dark-governed starch mobilization in source leaves. *Plant Physiol.*, **139**, 1163–1174.

Chen, S., Hajirezaei, M., Peisker, M., Tschiersch, H., Sonnewald, U. and Börnke, F. (2005b) Decreased sucrose-6-phosphate phosphatase level in transgenic tobacco inhibits photosynthesis, alters carbohydrate partitioning, and reduces growth. *Planta*, **221**, 479–492.

Chia, T., Thorneycroft, D., Chapple, A., Messerli, G., Chen, J., Zeeman, S.C., Smith, S.M. and Smith, A.M. (2004) A cytosolic glucosyltransferase is required for conversion of starch to sucrose in *Arabidopsis* leaves at night. *Plant J.*, **37**, 853–863.

Daie, J. (1993) Cytosolic fructose-1,6-bisphosphatase: a key enzyme in the sucrose biosynthetic pathway. *Photosynth Res.*, **38**, 5–14.

Draborg, H., Villadsen, D. and Nielsen, T.H. (2001) Transgenic *Arabidopsis* plants with decreased activity of fructose-6-phosphate,2-kinase/fructose-2,6-bisphosphatase have altered carbon partitioning. *Plant Physiol.*, **126**, 750–758.

Fernie, A.R., Willmitzer, L. and Trethewey, R.N. (2002) Sucrose to starch: a transition in molecular plant physiology. *Trends Plant Sci.*, **7**, 35–41.

Ferreira, S.J., Senning, M., Sonnewald, S., Kessling, P.M., Goldstein, R. and Sonnewald, U. (2010) Comparative transcriptome analysis coupled to X-ray CT reveals sucrose supply and growth velocity as major determinants of potato tuber starch biosynthesis. *BMC Genomics*, **11**, 93.

Fettke, J., Hejazi, M., Smirnova, J., Höchel, E., Stage, M. and Steup, M. (2009) Eukaryotic starch degradation: integration of plastidial and cytosolic pathways. *J. Exp. Bot.*, **60**, 2907–2922.

Flügge, U.-I. (1998) Metabolite transporters in plastids. *Curr. Opin. Plant Biol.*, **1**, 201–206.

Fu, Y.B., Ballicora, M.A., Leykam, J.F. and Preiss, J. (1998) Mechanism of reductive activation of potato tuber ADP-glucose pyrophosphorylase. *J. Biol. Chem.*, **273**, 25045–25052.

Fukushima, A., Kusano, M., Redestig, H., Arita, M. and Saito, K. (2009) Integrated omics approaches in plant systems biology. *Curr. Opin. Chem. Biol.*, **13**, 532–538.

Fulton, D.C., Stettler, M., Mettler, T., Vaughan, C.K., Li, J., Francisco, P., Gil, M., Reinhold, H., Eicke, S., Messerli, G., Dorken, G., Halliday, K., Smith, A.M., Smith, S.M. and Zeeman, S.C. (2008) β-AMYLASE4, a noncatalytic protein required for starch breakdown, acts upstream of three active beta-amylases in *Arabidopsis* chloroplasts. *Plant Cell*, **20**, 1040–1058.

Geigenberger, P., Stamme, C., Tjaden, J., Schulz, A., Quick, P.W., Betsche, T., Kersting, H.J. and Neuhaus, H.E. (2001) Tuber physiology and properties of starch from tubers of transgenic potato plants with altered plastidic adenylate transporter activity. *Plant Physiol.*, **125**, 1667–1678.

Geigenberger, P., Stitt, M. and Fernie, A.R. (2004) Metabolic control analysis and regulation of the conversion of sucrose to starch in growing potato tubers. *Plant Cell Environ.*, **27**, 655–673.

Gibon, Y., Bläsing, O.E., Palacios-Rojas, N., Pankovic, D., Hendriks, J.H., Fisahn, J., Hohne, M., Günther, M. and Stitt, M. (2004) Adjustment of diurnal starch turnover to short days: depletion of sugar during the night leads to a temporary inhibition of carbohydrate utilization, accumulation of sugars and post-translational activation of ADP-glucose pyrophosphorylase in the following light period. *Plant J.*, **39**, 847–862.

Häusler, R.E., Schlieben, N.H., Schulz, B. and Flügge, U.I. (1998) Compensation of decreased triose phosphate/phosphate translocator activity by accelerated starch turnover and glucose transport in transgenic tobacco. *Planta*, **204**, 366–376.

Hejazi, M., Fettke, J., Haebel, S., Edner, C., Paris, O., Frohberg, C., Steup, M. and Ritte, G. (2008) Glucan, water dikinase phosphorylates crystalline maltodextrins and thereby initiates solubilization. *Plant J.*, **55**, 323–334.

Hejazi, M., Fettke, J., Kotting, O., Zeeman, S.C. and Steup, M. (2010) The Laforin-like dual-specificity phosphatase SEX4 from Arabidopsis hydrolyzes both C6- and C3-phosphate esters introduced by starch-related dikinases and thereby affects phase transition of alpha-glucans. *Plant Physiol.*, **152**, 711–722.

Hellwege, E.M., Czapla, S., Jahnke, A., Willmitzer, L. and Heyer, A.G. (2000) Transgenic potato (Solanum tuberosum) tubers synthesize the full spectrum of inulin molecules naturally occurring in globe artichoke (*Cynara scolymus*) roots. *Proc. Natl. Acad. Sci. USA*, **97**, 8699–8704.

Hendriks, J.H.M., Kolbe, A., Gibon, Y., Stitt, M. and Geigenberger, P. (2003) ADP-glucose pyrophosphorylase is activated by posttranslational redox-modification in response to light and to sugars in leaves of Arabidopsis and other plant species. *Plant Physiol.*, **133**, 838–849.

Huber, S.C. and Huber, J.L. (1996) Role and regulation of sucrose-phosphate synthase in higher plants. *Ann. Rev Plant Physiol.*, **47**, 431–444.

Jobling, S. (2004) Improving starch for food and industrial applications. *Curr. Opin. Plant Biol.*, **7**, 210–218.

Jobling, S.A., Jarman, C., Teh, M.M., Holmberg, N., Blake, C. and Verhoeyen, M.E. (2003) Immunomodulation of enzyme function in plants by single-domain antibody fragments. *Nature Biotechnol.*, **21**, 77–80.

Kammerer, B., Fischer, K., Hilpert, B., Schubert, S., Gutensohn, M., Weber, A. and Flügge, U.I. (1998) Molecular characterization of a carbon transporter in plastids from heterotrophic tissues: The glucose 6-phosphate phosphate antiporter. *Plant Cell*, **10**, 105–117.

Kerk, D., Conley, T.R., Rodriguez, F.A., Tran, H.T., Nimick, M., Muench, D.G. and Moorhead, G.B. (2006) A chloroplast-localized dual-specificity protein phosphatase in *Arabidopsis* contains a phylogenetically dispersed and ancient carbohydrate-binding domain, which binds the polysaccharide starch. *Plant J.*, **46**, 400–413.

Kolbe, A., Tiessen, A., Schluepmann, H., Paul, M., Ulrich, S. and Geigenberger, P. (2005) Trehalose 6-phosphate regulates starch synthesis via posttranslational redox activation of ADP-glucose pyrophosphorylase. *Proc. Natl. Acad. Sci. USA*, **102**, 11118–11123.

Kossmann, J. and Lloyd, J. (2000) Understanding and influencing starch biochemistry. *Crit. Rev. Biochem. Mol.*, **35**, 141–196.

Kötting, O., Pusch, K., Tiessen, A., Geigenberger, P., Steup, M. and Ritte, G. (2005) Identification of a novel enzyme required for starch metabolism in *Arabidopsis* leaves. The phosphoglucan, water dikinase. *Plant Physiol.*, **137**, 242–252.

Kötting, O., Santelia, D., Edner, C., Eicke, S., Marthaler, T., Gentry, M.S., Comparot-Moss, S., Chen, J., Smith, A.M., Steup, M., Ritte, G. and Zeeman, S.C. (2009) STARCH-EXCESS4 is a laforin-like phosphoglucan phosphatase required for starch degradation in *Arabidopsis thaliana*. *Plant Cell*, **21**, 334–346.

Kötting, O., Kossmann, J., Zeeman, S.C. and Lloyd, J.R. (2010) Regulation of starch metabolism: the age of enlightenment? *Curr. Opin. Plant Biol.*, **13**, 321–329.

Lee, S.K., Jeon, J.S., Börnke, F., Voll, L., Cho, J.I., Goh, C.H., Jeong, S.W., Park, Y.I., Kim, S.J., Choi, S.B., Miyao, A., Hirochika, H., An, G., Cho, M.H., Bhoo, S.H., Sonnewald, U. and Hahn, T.R. (2008) Loss of cytosolic fructose-1,6-bisphosphatase limits photosynthetic sucrose synthesis and causes severe growth retardations in rice (*Oryza sativa*). *Plant Cell Environ.*, **31**, 1851–1863.

Lorberth, R., Ritte, G., Willmitzer, L. and Kossmann, J. (1998) Inhibition of a starch-granule-bound protein leads to modified starch and repression of cold sweetening. *Nature Biotechnol.*, **16**, 473–477.

Lunn, J.E. and MacRae, E. (2003) New complexities in the synthesis of sucrose. *Curr. Opin. Plant Biol.*, **6**, 208–214.

Lunn, J.E., Feil, R., Hendriks, J.H., Gibon, Y., Morcuende, R., Osuna, D., Scheible, W.R., Carillo, P., Hajirezaei, M.R. and Stitt, M. (2006) Sugar-induced increases in trehalose 6-phosphate are correlated with redox activation of ADPglucose pyrophosphorylase and higher rates of starch synthesis in *Arabidopsis thaliana*. *Biochem. J.*, **397**, 139–148.

Lytovchenko, A., Sonnewald, U. and Fernie, A.R. (2007) The complex network of non-cellulosic carbohydrate metabolism. *Curr. Opin. Plant Biol.*, **10**, 227–235.

Meyer, F.D., Talbert, L.E., Martin, J.M., Lanning, S.P., Greene, T.W. and Giroux, M.J. (2007) Field evaluation of transgenic wheat expressing a modified ADP-glucose pyrophosphorylase large subunit. *Crop Sci.*, **47**, 336–342.

Michalska, J., Zauber, H., Buchanan, B.B., Cejudo, F.J. and Geigenberger, P. (2009) NTRC links built-in thioredoxin to light and sucrose in regulating starch synthesis in chloroplasts and amyloplasts. *Proc. Natl. Acad. Sci. USA*, **106**, 9908–9913.

Mikkelsen, R., Mutenda, K.E., Mant, A., Schürmann, P. and Blennow, A. (2005) Alpha-glucan, water dikinase (GWD): a plastidic enzyme with redox-regulated and coordinated catalytic activity and binding affinity. *Proc. Natl. Acad. Sci. USA*, **102**, 1785–1790.

Müller-Röber, B., Sonnewald, U. and Willmitzer, L. (1992) Inhibition of the ADP-glucose pyrophosphorylase in transgenic potatoes leads to sugar-storing tubers and influences tuber formation and expression of tuber storage protein genes. *EMBO J.*, **11**, 1229–1238.

Muth, J., Hartje, S., Twyman, R.M., Hofferbert, H.R., Tacke, E. and Prüfer, D. (2008) Precision breeding for novel starch variants in potato. *Plant Biotechnol. J.*, **6**, 576–584.

Neuhaus, H.E. and Emes, M.J. (2000) Nonphotosynthetic metabolism in plastids. *Ann. Rev. Plant Physiol.*, **51**, 111–140.

Nielsen, T.H., Rung, J.H. and Villadsen, D. (2004) Fructose-2,6-bisphosphate: a traffic signal in plant metabolism. *Trends Plant Sci.*, **9**, 556–563.

Niittylä, T., Messerli, G., Trevisan, M., Chen, J., Smith, A.M. and Zeeman, S.C. (2004) A previously unknown maltose transporter essential for starch degradation in leaves. *Science*, **303**, 87–89.

Niittylä, T., Comparot-Moss, S., Lue, W.L., Messerli, G., Trevisan, M., Seymour, M.D., Gatehouse, J.A., Villadsen, D., Smith, S.M., Chen, J., Zeeman, S.C. and Smith, A.M. (2006) Similar protein phosphatases control starch metabolism in plants and glycogen metabolism in mammals. *J. Biol. Chem.*, **281**, 11815–11818.

Okita, T.W., Nakata, P.A., Anderson, J.M., Sowokinos, J., Morell, M. and Preiss, J. (1990) The subunit structure of potato-tuber ADPglucose pyrophosphorylase. *Plant Physiol.*, **93**, 785–790.

Park, J.-I., Lee, Y.-K., Chung, W.-I., Lee, I.-H., Choi, J.-H., Lee, W.-M., Ezura, H., Lee, S.-P. and Kim, I.-J. (2006) Modification of sugar composition in strawberry fruit by antisense suppression of an ADP-glucose pyrophosphorylase. *Mol. Breeding*, **17**, 269–279.

Preiss, J. (1988) Biosynthesis of starch and its degradation, in *The Biochemistry of Plants* (eds M.D. Hatch and N.K. Boardman), Academic Press, New York, pp. 181–254.

Regierer, B., Fernie, A.R., Springer, F., Perez-Melis, A., Leisse, A., Koehl, K., Willmitzer, L., Geigenberger, P. and Kossmann, J. (2002) Starch content and yield increase as a result of altering adenylate pools in transgenic plants. *Nature Biotechnol.*, **20**, 1256–1260.

Riesmeier, J.W., Flügge, U.I., Schulz, B., Heineke, D., Heldt, H.W., Willmitzer, L. and Frommer, W.B. (1993) Antisense repression of the chloroplast triose phosphate translocator affects carbon partitioning in transgenic potato plants. *Proc. Natl. Acad. Sci. USA*, **90**, 6160–6164.

Ritsema, T. and Smeekens, S.C.M. (2003) Engineering fructan metabolism in plants. *Journal Plant Physiol.*, **160**, 811–820.

Ritte, G., Lloyd, J.R., Eckermann, N., Rottmann, A., Kossmann, J. and Steup, M. (2002) The starch-related R1 protein is an alpha-glucan, water dikinase. *Proc. Natl. Acad. Sci. USA*, **99**, 7166–7171.

Roldán, I., Wattebled, F., Mercedes Lucas, M., Delvallé, D., Planchot, V., Jimenez, S., Perez, R., Ball, S., D'Hulst, C. and Merida, A. (2007) The phenotype of soluble starch synthase IV defective mutants of *Arabidopsis thaliana* suggests a novel function of elongation enzymes in the control of starch granule formation. *Plant J.*, **49**, 492–504.

Rung, J.H., Draborg, H.H., Jorgensen, K. and Nielsen, T.H. (2004) Carbon partitioning in leaves and tubers of transgenic potato plants with reduced activity of fructose-6-phosphate,2-kinase/fructose-2,6-bisphosphatase. *Physiol. Plantarum*, **121**, 204–214.

Schneider, A., Häusler, R.E., Kolukisaoglu, U., Kunze, R., van der Graaff, E., Schwacke, R., Catoni, E., Desimone, M. and Flügge, U.I. (2002) An *Arabidopsis thaliana* knock-out mutant of the chloroplast triose phosphate/phosphate translocator is severely compromised only when starch synthesis, but not starch mobilisation is abolished. *Plant J.*, **32**, 685–699.

Serrato, A.J., Perez-Ruiz, J.M., Spinola, M.C. and Cejudo, F.J. (2004) A novel NADPH thioredoxin reductase, localized in the chloroplast, which deficiency causes hypersensitivity to abiotic stress in *Arabidopsis thaliana*. *J. Biol. Chem.*, **279**, 43821–43827.

Serrato, A.J., de Dios Barajas-Lopez, J., Chueca, A. and Sahrawy, M. (2009) Changing sugar partitioning in FBPase-manipulated plants. *J. Exp. Bot.*, **60**, 2923–2931.

Sevenier, R., Hall, R.D., van der Meer, I.M., Hakkert, H.J., van Tunen, A.J. and Koops, A.J. (1998) High level fructan accumulation in a transgenic sugar beet. *Naure Biotechnol.*, **16**, 843–846.

Smidansky, E.D., Clancy, M., Meyer, F.D., Lanning, S.P., Blake, N.K., Talbert, L.E. and Giroux, M.J. (2002) Enhanced ADP-glucose pyrophosphorylase activity in wheat endosperm increases seed yield. *Proc. Natl. Acad. Sci. USA*, **99**, 1724–1729.

Smidansky, E.D., Martin, J.M., Hannah, L.C., Fischer, A.M. and Giroux, M.J. (2003) Seed yield and plant biomass increases in rice are conferred by deregulation of endosperm ADP-glucose pyrophosphorylase. *Planta*, **216**, 656–664.

Smidansky, E.D., Meyer, F.D., Blakeslee, B., Weglarz, T.E., Greene, T.W. and Giroux, M.J. (2007) Expression of a modified ADP-glucose pyrophosphorylase large subunit in wheat seeds stimulates photosynthesis and carbon metabolism. *Planta*, **225**, 965–976.

Smith, A.M. (2008) Prospects for increasing starch and sucrose yields for bioethanol production. *Plant J*, **54**, 546–558.

Smith, A.M. and Stitt, M. (2007) Coordination of carbon supply and plant growth. *Plant Cell Environ.*, **30**, 1126–1149.

Smith, A.M., Zeeman, S.C. and Smith, S.M. (2005) Starch degradation. *Ann. Rev. Plant Biol.*, **56**, 73–98.

Sokolov, L.N., Dominguez-Solis, J.R., Allary, A.L., Buchanan, B.B. and Luan, S. (2006) A redox-regulated chloroplast protein phosphatase binds to starch diurnally and functions in its accumulation. *Proc. Natl. Acad. Sci. USA*, **103**, 9732–9737.

Sonnewald, U., Basner, A., Greve, B. and Steup, M. (1995) A second L-type isozyme of potato glucan phosphorylase: cloning, antisense inhibition and expression analysis. *Plant Mol. Biol.*, **27**, 567–576.

Sonnewald, U., Hajirezaei, M.R., Kossmann, J., Heyer, A., Trethewey, R.N. and Willmitzer, L. (1997) Increased potato tuber size resulting from apoplastic expression of a yeast invertase. *Nature Biotechnol.*, **15**, 794–797.

Stark, D.M., Timmerman, K.P., Barry, G.F., Preiss, J. and Kishore, G.M. (1992) Regulation of the amount of starch in plant-tissues by Adp glucose pyrophosphorylase. *Science*, **258**, 287–292.

Stitt, M., Huber, S.C. and Kerr, P.S. (1987) Control of photosynthetic sucrose formation, in *The Biochemistry of Plants* (eds M.D. Hatch and N.K. Boardman), Academic Press, New York, pp. 327–409.

Stitt, M., Sulpice, R. and Keurentjes, J. (2010) Metabolic networks: how to identify key components in the regulation of metabolism and growth. *Plant Physiol.*, **152**, 428–444.

Strand, A., Zrenner, R., Trevanion, S., Stitt, M., Gustafsson, P. and Gardestrom, P. (2000) Decreased expression of two key enzymes in the sucrose biosynthesis pathway, cytosolic fructose-1,6-bisphosphatase and sucrose phosphate synthase, has remarkably different consequences for photosynthetic carbon metabolism in transgenic Arabidopsis thaliana. *Plant J.*, **23**, 759–770.

Sulpice, R., Pyl, E.T., Ishihara, H., Trenkamp, S., Steinfath, M., Witucka-Wall, H., Gibon, Y., Usadel, B., Poree, F., Piques, M.C., Von Korff, M., Steinhauser, M.C., Keurentjes, J.J., Guenther, M., Hoehne, M., Selbig, J., Fernie, A.R., Altmann, T. and Stitt, M. (2009) Starch as a major integrator in the regulation of plant growth. *Proc. Natl. Acad. Sci. USA*, **106**, 10348–10353.

Sweetlove, L.J., Burrell, M.M. and ap Rees, T. (1996) Characterization of transgenic potato (*Solanum tuberosum*) tubers with increased ADPglucose pyrophosphorylase. *Biochem. J.*, **320**, 487–492.

Sweetlove, L.J., Fell, D. and Fernie, A.R. (2008) Getting to grips with the plant metabolic network. *Biochem. J.*, **409**, 27–41.

Tauberger, E., Fernie, A.R., Emmermann, M., Renz, A., Kossmann, J., Willmitzer, L. and Trethewey, R.N. (2000) Antisense inhibition of plastidial phosphoglucomutase provides compelling evidence that potato tuber amyloplasts import carbon from the cytosol in the form of glucose-6-phosphate. *Plant J.*, **23**, 43–53.

Tetlow, I.J., Morell, M.K. and Emes, M.J. (2004) Recent developments in understanding the regulation of starch metabolism in higher plants. *J. Exp. Bot.*, **55**, 2131–2145.

Tiessen, A., Hendriks, J.H.M., Stitt, M., Branscheid, A., Gibon, Y., Farré, E.M. and Geigenberger, P. (2002) Starch synthesis in potato tubers is regulated by post-translational redox modification of ADP-glucose pyrophosphorylase: A novel regulatory mechanism linking starch synthesis to the sucrose supply. *Plant Cell*, **14**, 2191–2213.

Tjaden, J., Möhlmann, T., Kampfenkel, K., Henrichs, G. and Neuhaus, H.E. (1998) Altered plastidic ATP/ADP-transporter activity influences potato (*Solanum tuberosum* L.) tuber morphology, yield and composition of tuber starch. *Plant J.*, **16**, 531–540.

Tomlinson, K. and Denyer, K. (2003) Starch synthesis in cereal grains. *Adv. Bot. Res.*, **40**, 1–61.

Toroser, D. and Huber, S.C. (1997) Protein phosphorylation as a mechanism for osmotic-stress activation of sucrose-phosphate synthase in spinach leaves. *Plant Physiol.*, **114**, 947–955.

Toroser, D., Athwal, G.S. and Huber, S.C. (1998) Site-specific regulatory interaction between spinach leaf sucrose-phosphate synthase and 14-3-3 proteins. *FEBS Lett.*, **435**, 110–114.

Trethewey, R.N., Geigenberger, P., Riedel, K., Hajirezaei, M.R., Sonnewald, U., Stitt, M., Riesmeier, J.W. and Willmitzer, L. (1998) Combined expression of glucokinase and invertase in potato tubers leads to a dramatic reduction in starch accumulation and a stimulation of glycolysis. *Plant J.*, **15**, 109–118.

Visser, R.G.F., Somhorst, I., Kuipers, G.J., *et al.* (1991) Inhibition of the expression of the gene for granule-bound starch synthase in potato by antisense constructs. *Molecular and General Genetics MGG*, **225**, 289–296.

Weber, A., Servaites, J.C., Geiger, D.R., Kofler, H., Hille, D., Groner, F., Hebbeker, U. and Flügge, U.I. (2000) Identification, purification, and molecular cloning of a putative plastidic glucose translocator. *Plant Cell*, **12**, 787–802.

Weyens, G., Ritsema, T., Van Dun, K., Meyer, D., Lommel, M., Lathouwers, J., Rosquin, I., Denys, P., Tossens, A., Nijs, M., Turk, S., Gerrits, N., Bink, S., Walraven, B., Lefebvre, M. and Smeekens, S. (2004) Production of tailor-made fructans in sugar beet by expression of onion fructosyltransferase genes. *Plant Biotechnol. J.*, **2**, 321–327.

Winter, H. and Huber, S.C. (2000) Regulation of sucrose metabolism in higher plants: localization and regulation of activity of key enzymes. *Crit. Rev. Plant Sci.*, **19**, 31–67.

Wu, L. and Birch, R.G. (2007) Doubled sugar content in sugarcane plants modified to produce a sucrose isomer. *Plant Biotechnol J.*, **5**, 109–117.

Yu, T.S., Kofler, H., Häusler, R.E., Hille, D., Flügge, U.I., Zeeman, S.C., Smith, A.M., Kossmann, J., Lloyd, J., Ritte, G., Steup, M., Lue, W.L., Chen, J.C. and Weber, A. (2001) The *Arabidopsis sex1* mutant is defective in the R1 protein, a general regulator of starch degradation in plants, and not in the chloroplast hexose transporter. *Plant Cell*, **13**, 1907–1918.

Yuan, J.S., Galbraith, D.W., Dai, S.Y., Griffin, P. and Stewart, C.N., Jr. (2008) Plant systems biology comes of age. *Trends Plant Sci.*, **13**, 165–171.

Zeeman, S.C. and ap Rees, T. (1999) Changes in carbohydrate metabolism and assimilate partitioning in starch-excess mutants of *Arabidopsis*. *Plant Cell Environ.*, **22**, 1445–1453.

Zeeman, S.C., Smith, S.M. and Smith, A.M. (2007) The diurnal metabolism of leaf starch. *Biochem J.*, **401**, 13–28.

Zhang, L., Häusler, R.E., Greiten, C., Hajirezaei, M.R., Haferkamp, I., Neuhaus, H.E., Flügge, U.I. and Ludewig, F. (2008) Overriding the co-limiting import of carbon and energy into tuber amyloplasts increases the starch content and yield of transgenic potato plants. *Plant Biotechnol. J.*, **6**, 453–464.

Zrenner, R., Salanoubat, M., Willmitzer, L. and Sonnewald, U. (1995) Evidence of the crucial role of sucrose synthase for sink strength using transgenic potato plants (*Solanum tuberosum* L.). *Plant J.*, **7**, 97–107.

Zrenner, R., Krause, K.P., Apel, P. and Sonnewald, U. (1996) Reduction of the cytosolic fructose-1,6-bisphosphatase in transgenic potato plants limits photosynthetic sucrose biosynthesis with no impact on plant growth and tuber yield. *Plant J.*, **9**, 671–681.

2

Lipid Biosynthesis

David Hildebrand

2.1 Introduction

The term lipid can refer to almost any molecule hydrophobic enough such that it would partition into an organic solvent from an aqueous solution. Major lipids in plants include the glycerolipids and other fatty acid derivatives, isoprenoids and some phenylpropanoids. In the narrow sense 'lipids' often only includes 'glycerolipids' and some other fatty acid derivatives. Glycerolipids are composed of one, two or three fatty acids esterified to glycerol. Free fatty acids rarely accumulate in healthy living tissues. Cell membranes are mainly composed of diacylglycerides with a polar head group, usually phospholipids or glycolipids. Most storage lipids are composed of triacylglycerides (TAGs). Woody plant tissues are dominated by secondary cell walls and, therefore, are mostly composed of cellulose, lignin and other cell wall polymers. The surface of some xerophytic plant tissues have thick wax deposits. Some storage tissues such as seeds can be up to 75% TAG and some fruits can have TAG at almost 90% of dry weight (e.g. oil palm fruit)! Most leaves highly active in photosynthetic function are about 6% lipid, of which ~56–57% is glycerolipid and about a quarter chlorophyll.

The major fatty acids of plants (and most other eukaryotic organisms) have a chain length of 16 or 18 carbons and contain from zero to three cis-double bonds. Five fatty acids (18:1, 18:2, 18:3, 16:0 and in some species 16:3) make up over 90% of acyl chains of structural glycerolipids of almost all plant membranes (Ohlrogge and Browse, 1995; Somerville *et al.*, 2000) (Table 2.1).

Most plant seeds accumulate storage products during seed development to provide nutrients and energy for seedlings to grow competitively for light and nutrients, especially

Plant Metabolism and Biotechnology, First Edition. Edited by Hiroshi Ashihara, Alan Crozier, and Atsushi Komamine.

Table 2.1 *Fatty acid composition of some important plant oils. Data from Maguire* et al. *(2004) and Reeves and Weihrauch (1979)*

	Fatty acid (% oil)								
Commodity	14:0	16:0	18:0	16:1	18:1	18:2	18:3	Other	Sats
Nutmeg butter (*Myristica fragrans*)	87	5	0	0	5	0	0	3	95
Coconut (*Cocos nucifera*)	18	9	3	0	6	2	0	62	92
Palm kernel (*Elaeis guineensis*)	17	9	3	0	12	2	0	57	86
Cocoa butter (*Theobroma cacao*)	0	27	35	0	35	3	0	0	62
Palm (*Elaeis guineensis*)	1	46	4	0	38	10	0	1	51
Sheanut (*Butyrospermum paradoxum*)	0	5	41	0	46	5	0	2	48
Hazelnut (*Corylus avellana*)	0	5	2	0	82	11	0	0	8
Olive (*Olea europaea*)	0	11	2	1	76	8	1	1	14
Avocado (*Persea americana mill.*)	0	11	1	3	71	13	1	0	12
Almond (*Prunus dulcis*)	0	7	2	1	73	18	0	0	9
Canola (*Brassica napus*)	0	4	2	0	59	21	10	3	7
Peanut (*Arachis hypogea*)	0	10	2	0	48	34	0	5	17
Sesame (*Sesamum spp*)	0	9	5	0	41	43	0	0	14
Macadamia (*Macadamia integrifolia*)	0	9	8	25	54	1	0	3	17
Corn (*Zea mays*)	0	11	2	0	25	61	1	0	13
Soybean (*Glycine max*)	0	11	4	0	24	54	7	0	15
Sunflower (*Helianthus annuus*)	0	7	4	0	21	69	0	0	11
Walnut (*Juglans regia*)	0	7	2	0	23	56	11	0	9
Cotton seed (*Gossypium spp.*)	1	24	2	1	18	54	0	0	37
Grape seed (*Vitis vinifera*)	0	7	3	0	17	73	0	0	10
Safflower (*Carthamus tinctorius*)	0	7	2	0	12	78	0	0	9

*14:0, myristic; 16:0, palmitic; 18:0, stearic; 16:1, palmitoleic; 18:1, oleic; 18:2, linoleic; 18:3, linolenic acid; Sats, total saturated fatty acid.

nitrogen. Most seeds either accumulate starch or TAG as an energy store for seedling establishment. Many seed crops such as corn, wheat, rice, peas and common beans (*Phaseolus vulgaris*) accumulate starch as the main form of energy storage in the seeds, although the embryo or germ portion is often high in oil. Oilseeds such as soybean (*Glycine max*), canola (*Brassica napus*), sunflower (*Helianthus annus*), cotton seed (*Gossypium spp.*), peanuts (*Arachis hypogea*) and many other oilseeds accumulate oil instead of starch. Like many tiny seeds, *Arabidopsis thaliana* and tobacco (*Nicotiana tabacum*) seeds accumulate oil as an energy store, with *Arabidopsis* seeds usually being ~42% oil (O'Neill *et al.*, 2003; Zhang *et al.*, 2005).

The genomics of lipid biosynthesis in *Arabidopsis* has been studied rather well by Ohlrogge and colleagues (Beisson *et al.*, 2003; Ruuska *et al.*, 2004; Wallis and Browse, 2010). There is an *Arabidopsis* lipid gene database, http://www.plantbiology.msu.edu/lipids/genesurvey/index.htm, which is being added to The *Arabidopsis* Book (TAB), http://www.aspb.org/publications/arabidopsis/. To date they have identified more than 620 genes in *Arabidopsis* involved in acyl-lipid metabolism. This is about 2.4% of the total number of predicted genes in the *Arabidopsis* genome. They are classified into eight groups plus a miscellaneous class. Interestingly, the largest group in terms of numbers are lipid signalling genes. Sequencing of the soybean genome was recently completed (Schmutz *et al.*, 2010). The groups of greatest importance in the synthesis of seed oil, and the emphasis in this chapter, are the synthesis of plastid fatty acids, endomembrane lipid synthesis and TAG synthesis and storage. Interestingly, and in agreement with the reported multiplicity of genes in the soybean genome, more lipid gene isozymes are present in soybean than in *Arabidopsis* (Schmutz *et al.*, 2010) (Table 2.2). The increased number of isozymes is probably a result of duplication in the polyploid soybean genome. The soybean genome sequence available at soybase was screened to identify sequences likely to encode proteins with significant similarities to known fatty acid biosynthetic proteins from *Arabidopsis* (Table 2.1). Putative orthologous sequences from soybean identified for these *Arabidopsis* genes are listed in Table 2.2.

Relevant and excellent reviews on this subject include Browse and Somerville (1991), Thelen and Ohlrogge (2002), Lung and Weselake (2006) and Wallis and Browse (2010). Additionally, Chapter 10 on lipids in the ASPB Plant Biochemistry reference (Somerville *et al.*, 2000) is an excellent starting place for good background material. This chapter will focus on fatty acid synthesis and desaturation, membrane and storage lipid biosynthesis, genetic engineering of plant oils, and plant oils as a renewable resource. Lipid catabolism is not covered. Readers interested in this latter subject can find a thorough review on storage lipid mobilisation by Graham (2008).

2.2 Fatty Acid Synthesis

In sink tissue, sucrose is converted into hexose phosphates and then to fructose-1,6-bisphosphate which is cleaved into triosphosphates. Triosphosphates such as dihydroxyacetone phosphate can be reduced to glycerol-3-phosphate which provides the glycerol backbone for membrane lipids and TAGs (Figures 2.1 and 2.3). It can also be oxidised to 3-phosphoglycerate, which can be isomerised to phosphoenol pyruvate (PEP) and then to pyruvate. Pyruvate and possibly PEP enter the plastids and the pyruvate can be converted to acetyl-CoA by pyruvate dehydrogenase. Pyruvate synthesis and/or transport to plastids might be a limiting step in fatty acid biosynthesis and accumulation in oil (Murase, 1999; Lonien and Schwender, 2009; Maeo *et al.*, 2009). Acetyl-CoA is a precursor to many molecules in plants and other organisms in multiple organelles. The first committed step of fatty acid biosynthesis is the conversion of acetyl-CoA into malonyl-CoA by acetyl-CoA carboxylase and then to malonyl-ACP (acyl carrier protein) by a transacylase. In most tissues of most eukaryotic organisms including plants, malonyl-ACP is elongated in eight cycles, two carbon units at a time, via the fatty acid synthase complex, to palmitoyl (16:0)-ACP (or -CoA). The fatty acid synthase complex involves four different enzymatic reactions with each cycle starting with a condensation, followed by a reduction, a dehydration and a second reduction (Ohlrogge and Jaworski, 1997). The condensation reactions are catalyzed by enzymes known as 3-ketoacyl-ACP synthases or KASs. The first condensation reaction going from acetyl-CoA to 3-ketobutyrate is catalyzed by KAS III, the reaction from butyryl-ACP (C4) to palmitoyl-ACP (C16) by KAS I, and from palmitoyl-ACP to stearoyl-ACP (C18) by KAS II. The reaction stops at C16 and C18 fatty acids not only by virtue of the specificity of the KAS enzymes, but also by the action of thioesterases

Table 2.2 *Lipid biosynthetic genes*

Name	Description	Arabidopsis accessions	Soybean, *G. max*, accessions
PDH	Pyruvate dehydrogenase	At1g01090	Glyma02g03080, Glyma03g42190, Glyma07g05550, Glyma08g40380, Glyma16g02090, Glyma19g44930.
ACCase	Acetyl CoA carboxylase	At2g38040, At5g 16390	Glyma18g42280, Glyma18g42300, Glyma18g42310.
	ACP S-malonyl transferase	At2g30200	Glyma11g29490, Glyma18g06500.
ACP	Acyl carrier protein	At1g54580, At1g54630, At2g44620, At3g05020, At4g25050, At5g27200	Glyma03g40250, Glyma05g38490, Glyma07g04800, Glyma08g01180, Glyma09g38260, Glyma09g38400, Glyma10g30000, Glyma13g28590, Glyma15g10520, Glyma16g01380, Glyma18g47950, Glyma18g48110, Glyma19g42860.
KAS	3-ketoacyl ACP synthase KAS I KAS II KAS III	 At5g46290 At1g74960 At1g62640	Glyma05g25970, Glyma05g36690, Glyma08g02850, Glyma08g08910, Glyma10g04680, Glyma17g05200, Glyma13g17290, Glyma13g19010, Glyma09g41380, Glyma15g00550, Glyma18g44350.
	3-ketoacyl ACP reductase	At1g24360	Glyma08g10760, Glyma16g04630, Glyma18g01280.
	3-hydroxyacyl ACP dehydratase	At2g22230, At5g10160	Glyma05g24650, Glyma15g05800.
MOD1	Enoyl ACP reductase	At2g05990	Glyma08g45990, Glyma11g10770, Glyma12g03060, Glyma18g31780.
SACPD	Stearoyl-ACP-desaturase	At1g43800, At2g43710, At3g02610, At3g02620, At3g02630, At5g16230, At5g16240.	Glyma02g15600, Glyma07g32850, Glyma14g27990.
ACT1	Glycerol-3-phosphate acyl transferase (plastidal)	At1g32200	Glma01g01800, Glma09g34110.
GPAT	Glycerol-3-phosphate acyl transferase	At1g06520, At1g02390, At1g01610, At2g38110, At2g38110, At4g01950, At4g00400, At5g06090, At5g60620.	Glyma02g45600*, Glyma07g17720*, Glyma14g03210*, Glyma18g42580*, Glyma01g27900, Glyma02g01400, Glyma02g41660, Glyma03g01070, Glyma03g37970, Glyma07g07580, Glyma08g42210, Glyma10g01420, Glyma14g07290, Glyma18g12750, Glyma19g40590, Glyma20g16980.
FAT	Acyl ACP thioesterase FATA1 FATA2 FATB	 At3g25110 At4g13050 At1g08510	Glyma04g21910, Glyma04g37420, Glyma05g08060, Glyma06g23560, Glyma08g46360, Glyma10g41420, Glyma17g12940, Glyma18g36130.

Table 2.2 *(Continued)*

Name	Description	Arabidopsis accessions	Soybean, *G. max*, accessions
FAD	Fatty acid desaturase FAD2-18:2 synthesising FAD3-18:3 synthesising FAD6-16:2/18:2 synthesising (plastidal) FAD7-16:3/18:3 synthesising (plastidal) FAD8-16:3/18:3 synthesising (B100+B11plastidal)	At3g12120 At2g29980 At4g30950 At3g11170 At5g05580	Glyma01g29630, Glyma01g05470, Glyma02g11820, Glyma02g36460, Glyma02g39230, Glyma03g07570, Glyma03g30070, Glyma07g18350, Glyma08g41120, Glyma09g17170, Glyma10g42470, Glyma11g27190, Glyma14g10890, Glyma14g37350, Glyma17g34640, Glyma18g06950, Glyma18g15620, Glyma18g43210, Glyma19g32930, Glyma19g32940, Glyma20g24530.
DGAT	Acyl-CoA:diacylglycerol-acyltransferase		Glyma01g36010, Glyma03g40350, Glyma09g32790, Glyma11g09410, Glyma09g32790, Glyma11g09410, Glyma13g16560, Glyma16g21960,
	DGAT1	At2g19450	Glyma13g16560, Glyma16g21960, Glyma16g21970, Glyma17g06120.
	DGAT2	At3g51520	Glyma16g21970, Glyma17g06120, Glyma13g16790, Glyma07g04080.
PDAT	Phospholipid:diacylglycerol-acyltransferase	At5g13640, At3g44830, At3g03310**, At1g04010**	Glyma13g16790, Glyma07g04080, Glyma16g00790, Glyma12g08920, Glyma16g00790, Glyma12g08920, Glyma11g19570, Glyma17g05910.
		Atg19860**	Glyma11g19570, Glyma17g05910.

*Annotated as G3P acyltransferases, all others annotated simply as acyltransferases.
**Putative PDATs.

(TEs) which hydrolyse the acyl-S-ACP thioester bonds. Some plants such as coconuts have unusual TEs, known as medium chain TEs, which stop the reaction at C8, C10, C12 or C14 fatty acid chain lengths, and these plants can accumulate medium chain fatty acids in their seed oil (Voelker *et al.*, 1992, 1996; Yuan *et al.*, 1995; Budziszewski *et al.*, 1996).

Oil is synthesized during the second stage of seed maturation (Harwood and Page, 1994; Le *et al.*, 2007), at which time the relevant biosynthetic enzymes are highly expressed. The major fatty acids of plants (and most other eukaryotic organisms) have a chain length of 16 or 18 carbons and contain from zero to three *cis*-double bonds. The nature of the acyl composition of the TAG is dependent upon the availability of fatty acids from the acyl-CoA substrate pool as well as the selectivity of the acyltransferases of the Kennedy pathway (Harwood, 1998) and possibly transacylases. These same five fatty acids are the main fatty acids present in most plant oils (Table 2.1).

Reactions of fatty acid synthesis are terminated by hydrolysis or transfer of the acyl chains from the ACP by ACP-hydrolase or an acyltransferase consecutively. The 'competition' for substrate is thus a competition between the termination of synthesis, a function of thioesterase and transferase activity, and extension, a function of KAS I and KAS II isoforms (Voelker *et al.*, 1992; Budziszewski *et al.*, 1996). ACP-thioesterases are one of two main types (Klaus *et al.*, 2004). One thioesterase is relatively specific for 18:1 ACP, encoded by Fat A, and a second more specific for saturated acyl-ACPs encoded by Fat B. Fat A molecules formed in the chloroplast stroma are released from ACPs by thioesterases and cross the membrane. As Fat A molecules cross the membrane, they are converted to acyl-CoA esters through the activity of an acyl-CoA synthetase (ACS) located on the outer membrane. Plants have multiple ACSs that participate in lipid metabolism (Schnurr *et al.*, 2002; Shockey *et al.*, 2002). ACS enzymes encoded by different genes can have differential specificities for particular fatty acids (McKeon *et al.*, 2006).

2.3 Fatty Acid Desaturases

In most plant tissues, over 75% of the fatty acids are unsaturated. Two types of desaturases have been identified, one soluble and the other membrane bound, that have different consensus motifs. Database searching for these motifs reveals that these enzymes belong to two distinct multifunctional classes, each of which includes desaturases, acetylenases, hydrolases and epoxygenases that act on fatty acids or other substrates (Lee *et al.*, 1998; Shanklin and Cahoon, 1998). Free fatty acids are not thought to be desaturated *in vivo*; rather they are esterified to either acyl carrier protein (ACP) for the soluble plastid desaturase, to coenzyme-A (CoA) or to phospholipids for integral membrane desaturases.

2.3.1 Δ-9 Desaturases

The first double bond in unsaturated FAs in plants is introduced by the soluble enzyme stearoyl-ACP desaturase. This fatty acid desaturase is unusual in that only a few other known desaturases are soluble. Soluble Δ-9 stearoyl-ACP desaturases are found in all plant cells and are essential for the biosynthesis of unsaturated membrane lipids (Shanklin and Somerville, 1991; Cahoon *et al.*, 1994, 1997; Kaup *et al.*, 2002). Desaturases that convert saturated fatty acids to mono-unsaturated fatty acids share several common characteristics. They perform stereospecific Δ-9 desaturation of an 18:0/16:0 substrate with the removal of the 9-Δ and 10-Δ hydrogens (Bloomfield and Bloch, 1960; Mudd and Stumpf, 1961). Protein crystallographic studies on the purified desaturase from castor bean have shown that it contains a diiron cluster (Fox *et al.*, 1993). The protein is active as a homodimer and consists of a single domain of 11 helices. This diiron centre is the active site of the desaturase (Lindqvist *et al.*, 1996).

Expression and regulation of Δ-9 desaturase in plants have been extensively studied (Fawcett *et al.*, 1994; Slocombe *et al.*, 1994). The expression of the promoter of the *Brassica napus* stearoyl desaturase gene in tobacco was found to be temporally regulated in developing seed tissues. However, the promoter is also particularly active in other oleogenic tissues such as tapetum and pollen grains, raising the interesting question of whether seed-expressed lipid synthesis genes are regulated by separate tissue-specific determinants or by a single factor common to all oleogenic tissues (Slocombe *et al.*, 1994). In *Saccharomyces cerevisiae*, addition of saturated fatty acids induces Δ-9 fatty acid desaturase mRNA (Ole1 mRNA) by 1.6-fold, whereas a large family of unsaturated fatty acids represses Ole1 transcription by 60-fold. A 111 bp fatty acid regulation region (FAR) approximately 580 bp upstream of the start codon that is essential for the transcription activation and unsaturated fatty acid repression was identified (Quittnat *et al.*, 2004). In addition to the transcriptional regulation, unsaturated fatty acids mediate changes in the half-life of the Ole1 mRNA (Gonzalez and Martin,1996).

Currently, industries that manufacture shortening, margarine, and confectionery products use considerable amounts of stearate (18:0) produced mainly from partially hydrogenated plant oils (Facciotti *et al.*, 1999). Hydrogenation not only adds extra cost, but also generates significant amounts of *trans*-fatty acids which have been associated with an elevated risk of heart disease (Katan *et al.*, 1995; Nelson, 1998; Facciotti *et al.*, 1999). Industries manufacturing shortenings and confectionery products could benefit from oil crops that accumulate high levels of stearate. However, stearate (18:0) does not naturally accumulate to abundant levels in most cultivated oil crops including soybeans, and the production of a high-stearate phenotype has only had modest success so far through conventional breeding and mutagenesis techniques (Facciotti *et al.*, 1999). Although stearic acid (18:0) is one of the major saturated fatty acids in most seed oils, its percentages vary among the different oilseed crops from 1.0% in rape seed oil to 3.6% in sesame and corn seed oils and 4.0% in soybean oil, with a range from 2.2% to 7.2% for the soybean genotypes available in the world germplasm collection (Hymowitz *et al.*, 1972; Downey and McGregor, 1975; Rahman *et al.*, 1997). The fatty acid composition of soybeans has been improved by using selective breeding techniques utilising natural variants or induced mutagenesis (Ladd and Knowles, 1970; Graef *et al.*, 1985a,b). Hammond and Fehr (1983) were able to increase the amount of stearate (C18:0) produced in the soybean oil to levels up to about 28.1% of the total fatty acid content using mutagenesis. Rahman *et al.* (2003) reported a novel soybean germplasm with high stearic levels. This novel soybean was obtained as a consequence of the combination of the loci of high palmitic and stearic acids to determine the effects of altered contents of these two fatty acids on other fatty acids. As a result, two lines (M25 and HPS) with a five-fold increase in stearic acid (from 34 to 181 and 171 g kg^{-1}) were developed. This increase in stearic acid was also found to be associated with a change in oleic and linoleic acid contents. Furthermore, these authors reported that when the palmitic and stearic acid levels in the oil of HPS were combined, this line had a saturated fatty acid content of >380 g kg^{-1}. Thus, such oil might have the potential to increase the utility and also to improve the quality of soybean oil for specific purposes (Rahman *et al.*, 2003; Clemente and Cahoon, 2009). Aghoram *et al.* (2006) showed that the *fap*$_2$ locus which mediates an elevated seed palmitate phenotype in soybeans is due to a point mutation resulting in a premature stop codon.

Vegetable oils rich in mono-unsaturated fatty acids (MUFA) are not only important in human nutrition but can also be used as renewable sources of industrial chemicals (Cahoon *et al.*, 1997). Palmitoleic acid, cis-9-hexadecenoic acid, (16:1Δ9) has the nutritional and industrial chemical advantages of the much more common longer chain MUFA, oleic acid (18:1Δ9), but with much better cold flow properties. Sea buckthorn (*Hippophae rhamnoides*) and cat's claw, *Macfadyena unguis-cati* (formerly *Doxantha unguis-cati* L.), have high levels of 16:1Δ9 accumulating 40% and 80%, respectively (Chisholm and Hopkins, 1965; Cahoon *et al.*, 1998; Yang and Kallio, 2001). Macadamia ((*Macadamia integrifolia*) nut oil is the main commercially available source of palmitoleic acid. Its oil is unique among edible sources in that mono-unsaturated fatty acids are the predominant component (about 80%) and a considerable portion (17–21%) of this is palmitoleic acid (a component not present in substantial amounts in olive oil) (Curb *et al.*, 2000). Grayburn and Hildebrand (1995) and Wang *et al.* (1996) reported large increases in palmitoleic acid (16:1 Δ-9) after expressing a mammalian or yeast Δ-9 desaturase gene in both tobacco and tomato. Since soybeans are an important oil source that is high in linoleic and saturated fatty acids (mostly linoleic and palmitic acid: about 55% and 15%), conversion of all or part of these saturated fatty acids into palmitoleic acid would be a great benefit for health, while converting much of the remaining PUFAs (polyunsaturated fatty acids) into palmitoleic acid could have industrial value. Liu *et al.* (1996) reported converting about half of the palmitic acid of soybean somatic embryos into palmitoleic acid with good expression of a Δ9-CoA desaturase. The transformed embryos had 16:1 levels from 0% to over 10% of total fatty acids, while the levels of 16:0 dropped from 25% to approximately 5% of total fatty acids.

A number of studies have demonstrated apparently beneficial effects of diets based on high MUFA content primarily derived from olive oil (Kris-Etherton *et al.*, 1988; Spiller *et al.*, 1992; Hegsted *et al.*, 1993; Curb *et al.*, 2000). The health implications of palmitoleic acid were first addressed by Yamori *et al.* (1986) and Abraham *et al.* (1989). Curb *et al.* (2000) compared the effects of (i) a typical American diet high in saturated fat – "37% energy from fat", (ii) the American Heart Association 'step 1' diet – "30% energy from fat" (half the saturated fatty acids, normal amounts of MUFAs and PUFAs, and high levels of carbohydrates), and (iii) a macadamia nut-based mono-unsaturated fat diet (37% energy from fat). When compared with the typical diet, the 'step 1' and macadamia nut diets both had potentially beneficial effects on cholesterol and LDL cholesterol levels. These results are consistent with previously reported lipid-altering benefits of MUFA-rich diets, particularly those involving macadamia nut oil (Ako and Okuda, 1995; Griel *et al.*, 2008). Palmitoleic acid has also been reported to protect rats from stroke (Yamori *et al.*, 1986), apparently by increasing cell membrane fluidity, clearing lipids from the blood, and altering the activity of important cell membrane transport systems particularly through inhibition of the Na^+,$K(^+)$-ATPase activity within a narrow range (Swarts *et al.*, 1990). In men and women, elevated blood/tissue levels of palmitoleic acid were found to be correlated with protection from ventricular arrhythmias (Abraham *et al.*, 1989) and negatively correlated with markers of artherosclerosis (Theret *et al.*, 1993). Palmitoleic acid was also found to reportedly inhibit mutagenesis in animals (Hayatsu *et al.*, 1988) and to be negatively correlated with breast cancer incidence in women (Simonsen *et al.*, 1998).

Many vegetable oils are partially hydrogenated to increase the stability of cooking oils, and hydrogenated further for use as margarines and shortenings. To reduce or eliminate the need for hydrogenation of vegetable oils used for margarines and shortenings, a goal of plant geneticists has been to develop high stearate oils. As described in the section above on high stearate oils, many groups have been successful in achieving this goal with a variety of vegetable oils using different approaches, including genetic engineering of soybeans to a 53% stearic acid content of oil (Knutzon *et al.*, 1992; Kridl, 2002; Martinez-Force *et al.*, 2002). There have been several strategies to increase stearic acid levels in oilseed crops, and the increases achieved have usually been at the expense of oleic (18:1) and linoleic (18:2) acids. Among other strategies, both anti-sense suppression and co-suppression to reduce or knock out the activity of stearoyl-ACP desaturase, which is responsible for converting stearoyl-ACP (saturated) to oleoyl-ACP (unsaturated) (Budziszewski *et al.*, 1996), have been used routinely. Also the stearoyl-ACP thioesterase is another possible metabolic target. Up-regulation of this enzyme by sense-oriented reintroduction of the stearoyl-ACP thioesterase has been found to result in an increase of free stearate release. Kridl (2002)

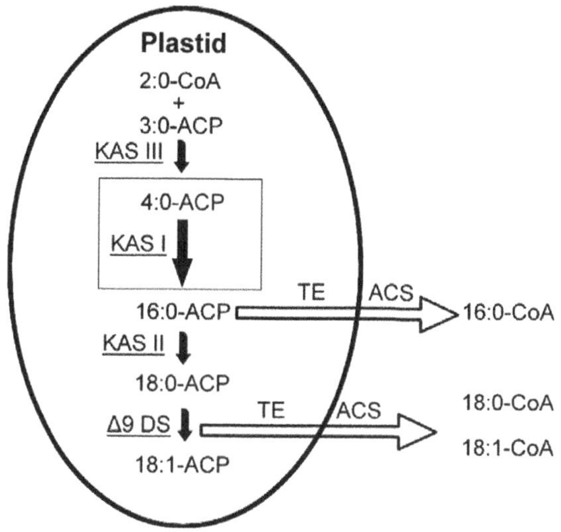

Figure 2.1 *Synthesis of the main saturated and mono-unsaturated fatty acids and acyl-CoAs in plant cells. The fatty acid acyl groups are named by numbers of carbons:number of double bonds. 2:0 = acetyl; 3:0 = malonyl; 4:0 = butryl; 16:0 = palmitoyl (hexadecanoate); 18:0 = stearyl (octadecanoate); 18:1 = oleoyl (octadecenoate); ACP = acyl carrier protein; ACS = acyl CoA-synthetase; KAS = 3-ketoacyl-ACP synthase; TE = thioesterase*

reported transgenic soybeans with stearate levels as high as *ca.* 53% compared with *ca.* 4% in non-transformed control plants. This line is low in linoleic and linolenic acids and high in oleic acid and stearic acid. It is important that these large increases in stearate are seed-specific and more so in triacylglycerol than in membrane lipids, because high stearate in membranes can reduce membrane fluidity and result in relatively poor germination rates (Voelker and Kinney, 2001; Kaup *et al.*, 2002).

Changes in palmitate levels of soybean oil has been another long-time goal of soybean breeders and geneticists, and development of genotypes with levels of <4% and >40% has been achieved (Stoltzfus *et al.*, 2000). Because saturated fatty acids, especially mid-chain saturated fatty acids such as 16:0, are dietary health-risk factors particularly for cardiovascular health (www.americanheart.org), reduced palmitate has been the main aim. Alleles for altered palmitate in soybean oil are known as *fap* alleles. A low palmitate mutant, A22 with 6.8% 16:0, has a single recessive *fap3* allele (Schnebly *et al.*, 1994). A mutant soybean line with elevated 16:0 containing the *fap2* allele has a single base-pair substitution in codon 299 of the *GmKASIIA* gene with TGG → TAG converting a trp to a premature stop codon (Aghoram *et al.*, 2006). *KASII* encodes the keto-acyl-ACP synthase that catalyzes the condensation reaction of the fatty acid complex involved in elongation of 16:0-ACP to 18:0-ACP (Figure 2.1).

2.3.2 Δ-12 Desaturases

Plant Δ-12 desaturases are plastid membrane- or ER membrane-bound enzymes. *Arabidopsis* plastid Δ-12 desaturases have been isolated using degenerate oligonucleotides, based on amino acid sequences conserved between plant and cyanobacterial desaturases, to screen cDNA libraries (Falcone *et al.*, 1994). The *Arabidopsis* Δ-12 desaturase was also used to screen rape and soybean cDNA libraries and the homologous sequences were isolated (Falcone *et al.*, 1994). These plant chloroplast Δ-12 desaturases all show a high degree of similarity (around 50%) with cyanobacterial Δ-12 desaturases, but less with cyanobacterial and plant ω-3 desaturases.

The Δ-12 desaturase is particularly active in microsomal preparations from developing seed cotyledons of some oilseed species where it is associated with the biosynthesis of triacylglycerols (Stymne and Stobart, 1986; Griffiths *et al.*, 1988). The microsomal Δ-12 desaturase requires NAD(P)H as reductant and molecular oxygen, and is inhibited by cyanide but not carbon monoxide, suggesting that cytochrome P450 is not involved in the electron transport chain (Griffiths *et al.*, 1985).

More than ten plant microsomal and a similar number of plastid Δ-12 cDNAs and genes have been isolated to date. *Arabidopsis* mutants lacking both microsomal Δ-12 (*fad*2) and ω-3 desaturases (*fad*3) have been isolated (Browse *et al.*, 1986, 1993). Mutants at the *Fad*2 locus of *Arabidopsis* that are deficient in Δ-12 desaturase of the eukaryotic pathway have been isolated and characterised. It was shown that the *Arabidopsis fad*2 mutants had similar growth characteristics to wild type at 22°C, but at 12°C their growth was greatly impaired, and at 6°C the mutants died (Miquel *et al.*, 1993). This experiment showed that *Arabidopsis* requires polyunsaturated fatty acids for low temperature survival (Tocher *et al.*, 1998). Subsequently, Okuley *et al.* (1994) isolated the entire *Arabidopsis fad*2 cDNA sequence with T-DNA tagged line with a higher 18:1 content in seeds, roots and leaves than the wild-type line.

After screening soybean libraries with the *Arabidopsis fad*2 cDNA, two different Δ-12 desaturase cDNAs, *FAD*2-1 and *FAD*2-2 were isolated (Heppard *et al.*, 1996). *FAD*2-1 is mainly expressed in developing seeds, whereas *FAD*2-2 is expressed in all or almost all tissues (leaves, roots and stems) in addition to developing seeds. *FAD*2-1 is mainly responsible for polyunsaturated fatty acid biosynthesis in soybean seed oil. Both soybean *FAD*2-1 and *FAD*2-2 have two gene members (Schlueter *et al.*, 2007; Bachlava *et al.*, 2008) consistent with soybean being a polyploid (Schmutz *et al.*, 2010).

Comparison of available sequence information reveals that there is a high degree of similarity between the same class of membrane-bound desaturases in different plant species, but much less similarity between different classes of desaturases, even in the same species (Murphy and Piffanelli, 1998). Membrane-bound enzymes most likely contain similar diiron complexes (Fox *et al.*, 1993). The most strictly conserved feature is the presence of eight histidines in three separate clusters. These clusters are held in position by a different ligation sphere, which may involve the three histidine boxes (Shanklin *et al.*, 1994) that are characteristic for this group of enzymes (HX3-4H, HX2-3HH, (H/Q)X2HH). This motif was also found in the Δ-12 oleate hydroxylases from castor bean and *Lesquerella fendleri* (van de Loo *et al.*, 1995; Broun *et al.*, 1997), epoxygenase from *Vernonia galamensis* (Hatanaka *et al.*, 2004; Hitz, 1998), acetylenase and epoxygenase from *Crepis* spp. (Lee *et al.*, 1998), and the Δ-6 linoleate desaturase from borage (Beremand *et al.*, 1997).

Another major goal of plant breeding has been to develop oils with high oxidative stability, without the need for hydrogenation, that are liquid at room temperature. Oils high in 18:1 are one way to achieve this. Mutant alleles affecting oleate levels in soybean are given the *ol* designation. An oleate content of >70% has been achieved by conventional breeding/mutagenesis (Alt *et al.*, 2005b). The high oleate mutant, M23, has a deletion in *FAD2-1a* (Alt *et al.*, 2005a). Using sense-mediated PTGS (co-suppression) targeting the Δ12-desaturase that converts oleic acid to linoleic acid, Toni Kinney and colleagues at DuPont Co. (Heppard *et al.*, 1996; Kinney, 1998a,b), by suppressing the *FAD*2-1 gene, succeeded in producing a soybean containing oil with high oxidative stability, as a consequence of a total polyunsaturated content of less than 5% and an oleic acid content of 85%. Normal soybeans contain about 20% oleic acid (18:1) (Warner *et al.*, 1997; Knowlton, 1999). This increase in oleate levels was accompanied by a reduction in 18:2 levels from 55% to less than 1%, and saturated fatty acids down to 10% (Heppard *et al.*, 1996; Beisson *et al.*, 2003). An oleate content of >90% by seed-specific suppression of *FAD*2-1 has been reported by (Buhr *et al.*, 2002).

Fatty acid desaturases in all organisms are subject to several different types of regulation, depending on their localisation and function. Those desaturases involved in membrane lipid biosynthesis have important 'housekeeping' functions and are, therefore, constitutively regulated (Murphy and Piffanelli, 1998). A cold-inducible plastidial ω-3 desaturase gene has been isolated from *Arabidopsis* (Gibson *et al.*, 1994), and there are several other reports that are consistent with the presence of cold-inducible ω-3 and Δ-12 desaturase genes in soybean (Rennie and Tanner 1989; Kinney 1994). However, there are other reports of the isolation of *Arabidopsis* and soybean Δ-12 desaturase genes that are not regulated by low temperature (Okuley *et al.*, 1994; Heppard *et al.*, 1996). Since multigene families encode many desaturases, it is possible that some plant species may have both cold-inducible and non-cold-inducible forms of the same class of desaturase enzyme and/or gene (Murphy and Piffanelli, 1998).

2.3.3 ω-3 Desaturases

The Δ-12 and ω-3 desaturases introduce the second and the third double bonds in the biosynthesis of 18:2 and 18:3 fatty acids which are important constituents of plant membranes. In most species, the fatty acids present in the galactolipids of the chloroplast membrane are ~70–80% trienoic fatty acids. In leaf tissue, there are two distinct pathways for polyunsaturated fatty acid biosynthesis, one located in the microsomes and the other located in the plastid membranes. In non-green tissues and developing seeds, the microsomal pathway predominates. Cytosolic and plastid ω-3 desaturation that result in the production of triene fatty acids are controlled by the *FAD*3, *FAD*7 and *FAD*8 loci in *Arabidopsis* (Browse *et al.*, 1986; Lemieux *et al.*, 1990; Arondel *et al.*, 1992; Yadav *et al.*, 1993; McConn *et al.*, 1994).

Microsomal ω-3 desaturases are responsible for the production of extraplastidal 18:3. This enzyme accounts for over 80% of the 18:3 in *Arabidopsis* root tissues. *Arabidopsis FAD*3 mutants are characterised by reduced levels of 18:3 and a concomitant increase in 18:2 levels. However, studies with the *Arabidopsis FAD*3 mutants revealed that exchange of lipid between chloroplast and ER allows the chloroplast desaturase to provide highly unsaturated lipid to the extrachloroplast membranes of leaf cells (Browse *et al.*, 1993). Changing 18:3 levels of soybean seed oil has long been a goal of plant breeders, and a number of low 18:3 mutants have been generated. Soybean genotypes A5 and A23 have reduced linolenic acid contents when compared with current cultivators. Byrum *et al.* (1997) reported that the reduced linolenic acid concentration in A5 was at least in some part the result of partial or full deletion of a microsomal ω-3 desaturase gene. Alleles for reduced 18:3 in soybeans are designated *fan* alleles with the allele for reduced linolenate in A5 controlled by the *fan1* allele (Byrum *et al.*, 1997). The soybean genome has at least three *FAD3* ω-3 desaturase genes designated *GmFAD3A*, *GmFAD3B* and *GmFAD3C* (Bilyeu *et al.*, 2005). Combining mutations *GmFAD3A*, *GmFAD3B* and *GmFAD3C* into single soybean lines (e.g. A29) can result in linolenate levels of *ca.* 1% (Sarmiento *et al.*, 1997; Anai *et al.*, 2005; Bilyeu *et al.*, 2006). Experimental soybean lines with >50% 18:3 have been reported as a result of increased expression of a *FAD3* gene in transgenic soybean (Cahoon, 2003).

2.4 Lipid Signals

As mentioned in Section 2.1, the largest group of lipid-related genes in *Arabidopsis* is the lipid signalling group (Beisson *et al.*, 2003; Ruuska *et al.*, 2004; Wallis and Browse, 2010). Increasing evidence also implicates fatty acids and their derivatives as signalling molecules, modulating normal as well as disease-related metabolism in animals, plants and single-celled organisms. For example, 18:1 and linoleic acid (18:2) induce protein kinase C-mediated activation of NADPH oxidase, resulting in the production of reactive oxygen species (Cury-Boaventura and Curi, 2005). The T-cell response to infection is modulated by eicosapentanoic acid, which induces anti-inflammatory effects (Denys *et al.*, 2001). Free fatty acids also serve as alarm molecules to repel phylogenetically-related and unrelated species of insects (Rollo *et al.*, 1994). Unsaturated fatty acids and their derivatives regulate sporulation, sexual structure development and host seed colonisation in mycotoxic *Aspergillus spp.* (Calvo *et al.*, 1999; Wilson *et al.*, 2004). In plants, fatty acids modulate a variety of responses to both biotic and abiotic stresses. For instance, polyunsaturated fatty acid levels in chloroplastic membranes affect membrane lipid fluidity and regulate acclimatisation to temperature stress (Routaboul *et al.*, 2000; Iba, 2002). Linolenic acid (18:3) is involved in protein modifications in heat-stressed plants (Yamauchi *et al.*, 2008). Fatty acids also regulate salt, drought and heavy metal tolerance, as well as wounding-induced responses and defence against insect/herbivore feeding in plants (Thoma *et al.*, 2003; Tumlinson and Engelberth, 2008; Upchurch, 2008). The fatty acid-derived phytohormone, jasmonic acid (JA), is particularly well known for its role in wound responses and plant defence against insects and pathogens (Creelman and Mulpuri, 2002; Browse, 2009). Other fatty acid oxidation products collectively known as oxylipins are also involved in plant defence reactions (Hildebrand, 1989; Halitschke and Baldwin, 2004). Work in Kachroo's laboratory identified a novel defence signalling pathway in plants induced in response to reductions in 18:1 levels (Kachroo *et al.*, 2003, 2007; Kachroo and Kachroo, 2009). The pathway is conserved amongst diverse plants and it interfaces with resistance protein-derived signalling, and in fact recruits components of resistance-derived pathways (Kachroo *et al.*, 2003). Additionally, a role for fatty acid biosynthetic genes in systemic acquired resistance has recently been demonstrated (Xia *et al.*, 2009).

2.5 Algae

Nine divisions of eukaryotic algae are known, with the classification mainly based on pigmentation patterns. The groups with the most known species are the Chlorophyceae (green algae), Phaeophyceae (brown algae), Pyrrophyceae (dinoflagellates), Chrysophyceae (golden-brown algae), Bacillariophyceae (diatoms) and Rhodophyceae (red algae). Eukaryotic algae are particularly diverse in fatty acid composition and many contain unusually high amounts of very polyunsaturated fatty acids (Harwood and Guschina, 2009). Many marine algae contain high concentrations of the very long chain polyunsaturated fatty acids, arachidonic, eicosapentaenoic (EPA) and docosahexanoic (DHA) acids, which are of keen interest in the health of humans and other animals. EPA and DHA are ω-3 fatty acids which are very important and often deficient in human diets. Fish that accumulate high levels of EPA and DHA ω-3 fatty acids do not actually synthesize these molecules but accumulate them from the algae or other plants that produce them at the base of the food chain. Since brain tissue is particularly high in DHA, and humans have a very high brain to total body mass ratio, we benefit from high DHA levels in our diets, especially as infants. Human breast milk is much higher in DHA than common milk sources such as cow's milk. Some marine algae, most notably the Thraustochytrids, accumulate unusually high levels of DHA, and biotechnology companies have been producing DHA from such algae for fortification of infant formula and other foods. Due to the high cost of DHA production in algae, oilseeds are being genetically engineered for production of this valuable fatty acid (Napier, 2007). The main mono-unsaturated fatty acid in algae is palmitoleic acid (16:1 Δ9), rather than oleic acid (18:1 Δ9) which is the main mono-unsaturated fatty acid in higher plants. Algae have also been used successfully for the commercial production of high-value lipids such as arachidonic acid, which is also sometimes added to infant formula.

2.6 Membrane Synthesis

The main storage lipid is TAG and the main polar lipids are phospholipids and glycolipids. The distributions of these different lipids in *Arabidopsis* leaves, roots and seeds are shown in Table 2.3. Seeds are 95% TAG with cytosolic phospholipids making up most of the remainder. Leaves and roots normally have little or no detectable TAG. Cytosolic phospholipids dominate lipids of non-photosynthetic tissues such as roots whereas plastid lipids, particularly galactolipids (galactose containing glycolipids), are most abundant in photosynthetically active leaves (Table 2.3). Plastid glycolipids and phospholipids, mainly in thylakoids, make up 76% of leaf lipids. The main phospholipids in plants are phosphatidylcholine (PC), phosphatidylethanolamine (PE), phosphatidylglycerol (PG), phosphatidylinositol (PI), phosphatidylserine (PS) and cardiolipin (CL). The relative abundance

Table 2.3 *Lipid distributions of* Arabidopsis *seed, leaf and root tissues (Somerville* et al., *2000)*

Lipid type	Seed	Leaf	Root
Plastid galactolipids	1%	60%	5%
Plastid phospholipids	1%	16%	12%
Cytosolic phospholipids	3%	24%	83%
Triaglycerols	95%	0%	0%

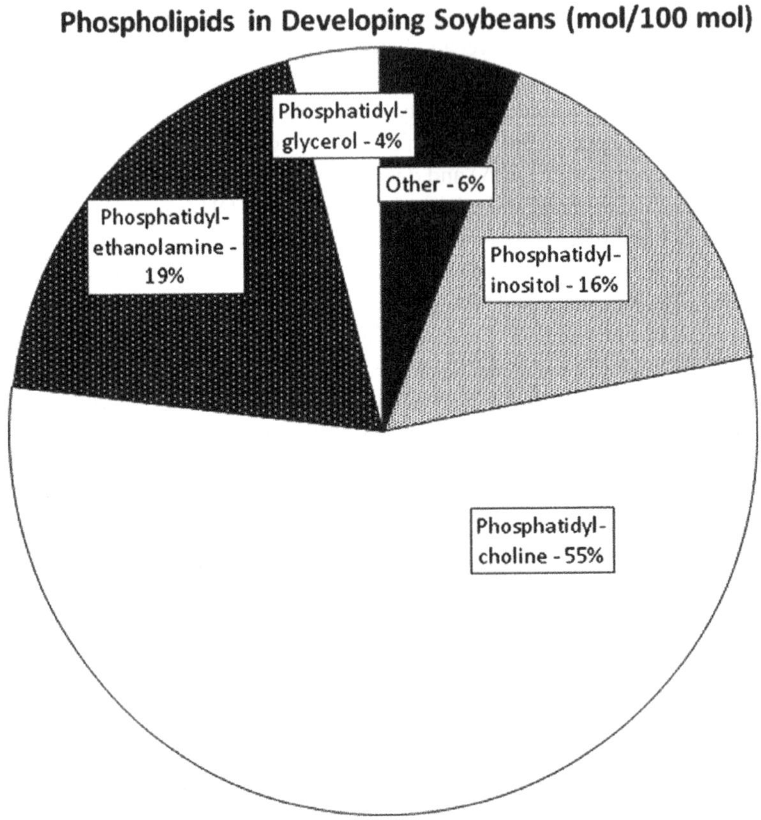

Figure 2.2 *Relative amounts of individual phospholipids in developing soybean seed. Reproduced with permission from Slack* et al. *1978 Copyright (1978) the Biochemical Society*

of different phospholipids in developing soybeans is typical of non-photosynthetic plant tissues (Slack *et al.*, 1978). These are dominated by PC at 50-60% followed by PE and PI (Figure 2.2). The principal glycolipids in order of abundance in chloroplasts are monogalactosyldiacylglycerol (MGD), digalactosyldiacylglycerol (DGD) and sulfoquinovosyldiacylglycerol (SQD) (Table 2.4). Plants have higher glyco and sulfo lipid levels than animals, and this can provide a mechanism for membrane synthesis with limited phosphate (Benning 2009). 18:3 is the most abundant fatty acid in all chloroplast lipids especially MGD and DGD. DGD and especially MGD are also relatively high in 16:3 in 16:3 plants such as Arabidopsis. MGD fatty acids are nearly all polyunsaturated and DGD only has small amounts of 16:0. SQD has a relatively high amount of the saturated fatty acid 16:0. PG is unusual in that it has a high level of trans-Δ3-16:1 (Table 2.4). This can be used as a diagnostic feature of PG in lipid separation studies.

Higher plants can be classified as 16:3 or 18:3 plants (Mongrand *et al.*, 1998; Somerville *et al.*, 2000). Eukaryotic plants are thought to have evolved from a symbiotic relationship with a photosynthetic bacterium, and the plastid endosymbiont originally made all its own

Table 2.4 Major Arabidopsis chloroplast lipids and fatty acid composition of these lipid classes (Somerville et al., 2000)

Fatty acid↓ Lipid type→	MGD	DGD	SQD	PG
16.1t	0%	0%	0%	3%
18.3	32%	23%	8%	4%
18.2	<1%	1%	2%	≈1%
18.1	<1%	<1%	0%	<1%
16.3	10%	3%	0%	<1%
16	0%	1%	7%	2%
Total	43%	29%	17%	11%

lipids (Gould *et al.*, 2008). All higher plants still have prokaryotic and eukaryotic lipid biosynthesis, but it is thought that during evolution many angiosperms have lost much of the endogenous plastid prokaryotic lipid biosynthesis and must rely on transfer of much of their plastid lipids from the ER (Benning, 2009). Plastids of 16:3 plants still can synthesize MGD, DGD, SQD and PG, but plastids of 18:3 plants can only synthesize PG (Wallis and Browse, 2010). Some 16:3 plants contain significant amounts (5–10%) of cis-δ-7,10, 13-hexadecatrienoic acid (16:3) in leaf lipids mainly in MGD (where it is >20% of MGD fatty acids) and also in DGD (Table 2.4). 18:3 plants do not contain detectable levels of 16:3. 16:3 is not known to accumulate in seed oil even in 16:3 plants such as *Arabidopsis* or *Brassica* spp. (Table 2.4). Mongrand *et al.* (1998) evaluated the fatty acid composition of 468 plant species in 141 families and found that only 12% of angiosperm species are 16:3 plants. 16:3 is found to be ubiquitous among lower eukaryotic land plant species examined of the bryophytes, pteridophytes, gymnosperms, prespermatophytes, chlamydosperms, Gnetaceae and Ephedraceae. Common high 16:3 crop plant families include the Brassicaceae and Solanaceae. Chenopodiaceae, such as spinach, are also 16:3 plants. The most important crop plant families, Poaceae (Graminaceae) and Fabaceae (Leguminaceae), are 18:3 plants, but all nine species of the very important fruit family, Rosaceae, and two out of three Rutaceae examined have small amounts of 16:3 (Mongrand *et al.*, 1998).

Both membrane and TAG synthesis begins with the acylation of *sn*-glycerol-3-phosphate producing lysophosphatidic acid, catalyzed by glycerol-3-phosphate acyltransferase (GPAT). A second acylation of lysophosphatidic acid catalyzed by lysophatidic acid acyl transferase (LPAT) produces phosphatidic acid (PA) (Figure 2.3). LPAT is known to display specificity for substrates with certain fatty acids and can be important in determining final TAG composition (Franzosi *et al.*, 1998). The PA formed can be subsequently dephosphorylated to diacylglycerol (DAG). The DAG then serves as a precursor for membrane lipids and TAG. The fatty acid transferred by LPAT is predominately 16:0 in the prokaryotic pathway and an unsaturated fatty acid, usually 18:1, in the eukaryotic pathway (Somerville *et al.*, 2000; Wallis and Browse, 2010). This difference can be used as a measure of the relative contributions of the prokaryotic and eukaryotic pathways in the formation of particular lipids. The PA is either converted to diacylglycerol by PA phosphatase or reacts with cytidine 5′-triphosphate (CTP) forming the nucleotide-activated form of DAG, CDP-DAG. This nucleotide activation provides the energy for polar headgroup attachment. CDP-DAG reacts with myo-inositol, serine and G3P to form PI, PS and phosphatidyl

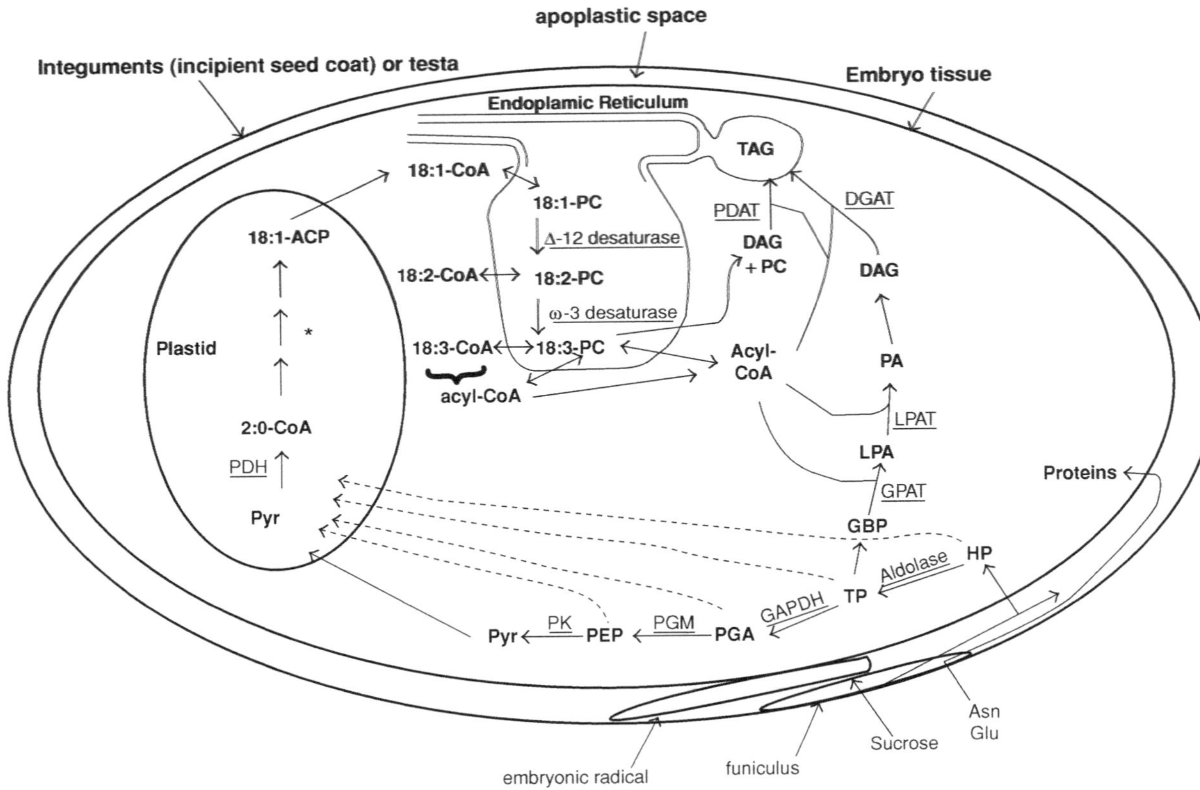

Figure 2.3 *Overall scheme from seed assimilates to final seed oil (and protein) accumulation. See Figure 4.1. Enzymes shown are underlined. Abbreviations: DGAT, acyl-CoA: diacylglycerol acyltransferase; GAPDH, glyceraldehyde 3-phosphate dehydrogenase; G3P, glycerol-3 phosphate; GPAT, glycerol-3 phosphate acyltransferase; HP, hexose phosphate; LPA, lysophosphatidic acid; LPAT, lysophosphatidic acid acyltransferase; PDAT, phosphatidylcholine: diacylglycerol acyltransferase; PDH, Pyruvate dehydrogenase; PEP, phosphoenolpyruvate; PGA, 3-phosphoglycerate; PGM, phosphoglycerate mutase; PK, pyruvate kinase; Pyr, pyruvate; TAG, triacylglycerol; TP, triose phosphate. 18:2 = linolyl (octadecadienoate); 18:3 = linolenyl (octadecatrienoate)*

glycerol (PG). In the synthesis of the other membrane lipids, an activated head group reacts with DAG. CDP-choline, CDP-ethanolamine, UDP-galactose and UDP-sulfoquinovose react with DAG, forming PC, PE, MGD and SL. DGD is synthesized from MGD. In oil seeds phosphatidyl choline (PC) has been identified as a key intermediate in oil biosynthesis and plays a central role in the production of polyunsaturated fatty acids by serving as a substrate for Δ-6, Δ-9, Δ-12 and Δ-15 desaturases (Jackson *et al.*, 1998; Lu *et al.*, 2009). Acyl CoA:lysoPC acyltransferase (LPCAT) provides a mechanism of enriching the acyl CoA pool with polyunsaturated fatty acids synthesized on PC (Wallis and Browse, 2010). Similarly phosphatidylcholine:diacylglycerol cholinephosphotransferase (PDCT) can provide polyunsaturated DAG for synthesis of other membrane lipids and TAG (Lu *et al.*, 2009).

2.7 TAG Biosynthesis

The hydrocarbon fuelling TAG accumulation in most oilseeds is sucrose. For many Rosaceae seeds such as apple seeds, sorbitol is the main assimilate translocated from leaves (Nosarzewski and Archbold, 2007). The main steps have been described, from delivery of assimilates from leaves to developing seeds to final seed oil, TAG (and protein) accumulation (Bewley and Black, 1994; Ohlrogge and Browse, 1995; Bates *et al.*, 2009; Lonien and Schwender, 2009; Weselake *et al.*, 2009). Assimilates move through the vascular connection to the mother plant through the funiculus and move through vascular connections in the integuments and then into developing embryo tissue across the apoplastic space. The amino acids for seed protein synthesis, including storage proteins, arrive from leaves mainly in the form of the N-rich amino acids asparagine and glutamine. In nitrogen-fixing tropical legume oilseeds such as soybeans and peanuts, fixed nitrogen is exported from nodules as ureides which are reconverted into amino acids in leaves, and ureides do not directly enter developing seeds. Sucrose is cleaved into hexose sugar monomers and hexose phosphates can be cleaved into triose phosphates. Triose phosphates can be reduced to glycerol-3 phosphate, the backbone for glycerolipids, or oxidised to 3-phosphoglycerate. 3-phosphoglycerate can be rearranged to phospho*enol*pyruvate which can be dephosphorylated to pyruvate. Intermediates from hexose phosphates to pyruvate and pyruvate itself can be translocated into plastids of developing seeds, and pyruvate converted (decarboxylated) into acetyl-CoA (2:0-CoA). The acetyl-CoA then provides hydrocarbon for fatty acid biosynthesis to 18:1 in plastids, but acetyl-CoA itself is not translocated into plastids from other compartments. Acetyl-CoA is a precursor to many molecules in plants and other organisms in multiple organelles. The first committed step of fatty acid biosynthesis is the conversion of acetyl-CoA into malonyl-CoA as described in the section on fatty acid biosynthesis above.

Oil is biosynthesized during the second main stage of seed maturation (Goldberg *et al.*, 1989; Harwood and Page, 1994; Le *et al.*, 2007), at which time the relevant biosynthetic enzymes are highly expressed. The nature of the acyl composition of the TAG is dependent upon the availability of the fatty acids from the acyl-CoA substrate pool, as well as the selectivity of the acyltransferases of the Kennedy pathway (Harwood and Page, 1994; Harwood, 1997) and possibly transacylases. These same five fatty acids are the main fatty acids present in plant oils in proportions generally quite different from that of membrane lipids.

Acyl-CoA: diacylglycerol (DAG) acyltransferase (DGAT; EC 2.3.1.20) activity has long been detected in various animal and plant tissues active in TAG synthesis. A DGAT was purified to apparent homogeneity from lipid body fractions of an oleaginous fungi, *Mortierella ramanniana* (Kamisaka *et al.*, 1997), now *Umbelopsis ramanniana* (Lardizabal *et al.*, 2008). The purified DGAT utilised a broad range of molecular species of both DAG and acylcoenzyme-A as substrates (Gavilano *et al.*, 2006) and higher plants (Vogel and Browse, 1996). The first plant DGAT was cloned recently from *Arabidopsis* (Hobbs *et al.*, 1999); the amino acid sequence shared 38% identity and 59% similarity with the mouse DGAT. Amino acid sequence analysis revealed nine membrane-spanning helices and also a 14 kD hydrophilic domain at the *N*-terminus. It had no significant sequence homology with plant GPAT and LPAT genes. Studies on expression of the homologue in *Brassica napus* showed that the DGAT mRNA was present in the highest concentrations in developing embryos, petals of flowers, and developing flower buds, but in very low amounts in leaf and stem tissues. DGAT catalyzes the reaction:

$$\text{Acyl}-\text{CoA}+\text{DAG}\rightarrow\text{TAG}+\text{CoASH}$$

As expected this enzyme is membrane bound or associated, and difficult to work with biochemically. As such the first DGAT gene was cloned from mice (Cases *et al.*, 1998) and subsequently from *Arabidopsis* (Hobbs *et al.*, 1999; Zou *et al.*, 1999; Hobbs and Hills, 2000). The cloning of a second class of DGAT, DGAT2, from the oleagineous fungus *U. ramanniana* established that it has no homology to the earlier identified DGAT sequences now known as DGAT1s (Lardizabal *et al.*, 2001). Cases *et al.* (2001) also cloned a mammalian DGAT2, and it is now known that humans have seven DGAT2s (Turkish *et al.*, 2005). Only a single DGAT1 gene (At2g19450) and a single DGAT2 gene (At3g51520) are present in the *Arabidopsis* genome (Beisson *et al.*, 2003; Mhaske *et al.*, 2005). Soybeans have at least two DGAT1s (Hildebrand *et al.*, 2008). A draft of the soybean genome has recently been reported (Schmutz *et al.*, 2010). A third soluble DGAT in peanuts was described by Saha *et al.* (2006).

A second mechanism for biosynthesis of TAG in yeast and plants was discovered and reported by Dahlqvist *et al.* (2000) and Oelkers *et al.* (2000) who showed homology to lecithin cholesterol acyltransferases (LCATs). This is catalyzed by an enzyme known as phospholipid: diacylglycerol acyltransferase (PDAT, EC 2.3.1.158) which transfers an acyl group (fatty acid) from a phospholipid (PL) to DAG, forming TAG and a lysophospholipid (LPL):

$$\text{PL}+\text{DAG}\rightarrow\text{TAG}+\text{LPL}$$

Arabidopsis has six PDAT/LCAT homologues (Ståhl *et al.*, 2004) of which At5g13640 is most closely related to the PDAT identified in yeast. Ståhl *et al.* (2004) demonstrated that this gene is expressed widely in different *Arabidopsis* tissues and has PDAT activity. In humans most TAG is synthesized by DGAT1 and DGAT2 and the only human gene similar to PDAT has phospholipaseA_2 and phospholipid:ceramide transacylase activities (Hiraoka *et al.*, 2002). Mhaske *et al.* (2005) generated a knockout for At5g13640, and their studies plus those of Stahl *et al.* (2004) suggested no role for this gene in TAG synthesis in *Arabidopsis* seeds. However, when Zhang *et al.* (2009) made a double knockout of this PDAT1 gene together with a knockout of DGAT1 in *Arabidopsis*, they found that most of the seed oil is made by DGAT1 with the remainder being synthesized by this PDAT1. A second *Arabidopsis* PDAT/LCAT homologue most related to (At5g13640) (57% homology) is At5g44830. This PDAT/LCAT-like gene was found to be expressed mainly in developing seeds by Ståhl *et al.* (2004), who speculated that it might have a role in seed oil biosynthesis. However the data of Zhang *et al.* (2009) indicate that AtDGAT1 and AtPDAT1 account for all *Arabidopsis* seed oil; no other AtPDAT nor AtDGAT2 appear to contribute to seed TAG in *Arabidopsis*. In contrast, DGAT2 does appear to have a role in oil biosynthesis in seeds that accumulate unusual fatty acids (see Section 2.8).

A third TAG biosynthetic activity involving a DAG/DAG transacylase (DGTA) has been reported in animals (Lehner and Kuksis, 1993) and plants (Stobart *et al.*, 1997), with Ståhl *et al.* (2004) detecting such activity in *Arabidopsis*. DGTA catalyzes the reaction:

$$\text{DAG} + \text{DAG} \rightarrow \text{TAG} + \text{MAG(monoacylglycerol)}$$

To date no DGTA has been biochemically characterised or the corresponding genes cloned.

Another reaction that appears to be involved in TAG accumulation is the reversible conversion of PC into DAG in the presence of CDP choline transferase. Slack *et al.* (1985) gave indirect evidence for the reversibility of PC by labelling studies *in vivo* with linseed (*Linum usitatissimum*) cotyledons and *in vitro* with safflower (*Carthamus tinctorius*) cotyledons. When sunflower microsomes were incubated with radiolabelled PC, the radioactivity was progressively incorporated into DAG. When the concentration of the microsomal protein was increased the CDP choline transferase activity also increased, indicating the reversible reaction of choline transferase (Triki *et al.*, 1998). A soybean cDNA encoding an aminoalcoholphosphotransferase (AAPTase) that demonstrates high levels of CDP-choline:sn-1,2-diacylglycerol cholinephosphotransferase activity was isolated by Dewey *et al.* (1994) by complementation of a yeast strain deficient in this function. AAPTases utilise diacylglycerols and cytidine diphosphate (CDP)-amino-alcohols as substrates in the synthesis of the main membrane lipids phosphatidylcholine and phosphatidylethanolamine, and can possibly affect DAG pools for TAG synthesis. Lu *et al.* (2009) described a new enzyme, phosphatidylcholine:diacylglycerol cholinephosphotransferase (PDCT), that is important in enriching the polyunsaturated fatty acid levels of the DGAT pool from PC. In animals at least TAG can be synthesized by two major pathways, the glycerol 3-phosphate pathway and the monoacylglycerol pathway (Hiramine *et al.*, 2010).

A number of mutants with reduced seed oil contents have been reported in *Arabidopsis* and they have been found to be due to defects in DGAT1 (Katavic *et al.*, 1995; Focks and Benning, 1998; Routaboul *et al.*, 1999; Zou *et al.*, 1999; Lu and Hills, 2002) or are impaired in transfer of carbon from sucrose and glucose to TAG, possibly due to impaired hexokinase and pyrophosphate-dependent phosphofructokinase (Focks and Benning, 1998). *Arabidopsis* DGAT1 mutants have 25–50% reductions in seed oil contents. DGAT1 is reported to be maximally expressed in developing seeds at a stage of high oil synthesis (Lu *et al.*, 2003). Silencing of DGAT1 in tobacco has also been reported to reduce seed oil content (Zhang *et al.*, 2005). Preliminary data of Hildebrand *et al.* (2008) indicate a role of DGAT1(s) in soybean oil synthesis, but this has not yet been addressed directly. The role of DGAT2 in oil accumulation in common oilseeds, such as soybean, has not yet been investigated, although it is known that, in contrast to developing castor and Vernonia seeds, there is no increase in DGAT2 transcript levels in developing soybean and *Arabidopsis* seed at a stage of high TAG biosynthesis (Li *et al.*, 2010b). It has been shown that DGAT2 does not influence seed oil levels in *Arabidopsis* and there is no other obvious phenotypic effect of a DGAT2 knockout (Zhang *et al.*, 2009).

Several reports indicate a role for DGAT in oil accumulation in developing soybean seeds (Kwanyuen and Wilson, 1986, 1990; Kwanyuen *et al.*, 1988; Settlage *et al.*, 1998). Currently there are no soybean mutants with large changes in oil levels or defects in DGAT. It is unclear what contributions of DGAT1, DGAT2 and possible other DGATs play in soybean oil biosynthesis. Transcripts for DGAT1, DGAT2 and PDAT have been detected in soybean tissues including developing seeds (Hildebrand *et al.*, 2008; Hildebrand, unpublished data). Developing soybean seeds accumulate TAG after most cell division has ceased and cotyledons have been formed and cell expansion initiated (Dahmer *et al.*, 1991; Le *et al.*, 2007). As with most green tissues, linolenate (18:3) is the most abundant fatty acid of soybean oil early in seed development. The 18:3 levels of soybean oil continues to decline throughout seed development, with linoleate (18:2) and oleate (18:1) becoming the predominate fatty acids of soybean oil as seeds mature (Dahmer *et al.*, 1991). DGAT levels correlate with oil accumulation. Li *et al.* (2010a) have found that expression of a DGAT in leaves can cause accumulation of TAG in leaf tissue. Baud and Lepiniec (2009) and Baud *et al.* (2009) report that expression of the B3 family domain transcription factor, LEC2, can cause accumulation of TAG in *Arabidopsis* leaves. LEC2 induces other transcription factors, including FUS3, that can increase lipid biosynthesis leading to TAG accumulation in plant tissues including leaves (Wang *et al.*, 2007a; Vyacheslav *et al.*, 2009). Over-expression of the related transcription factor LEC1 (Baud and Lepiniec, 2009) also causes lipid accumulation in *Arabidopsis* (Mu *et al.*, 2008) and other plants (Shen *et al.*, 2010). Another transcription factor, WR1, can also increase oil accumulation in plants (Cernac and Benning, 2004). Soybean *Dof*-type transcription factor genes are reported to increase lipid levels in transgenic *Arabidopsis* seeds (Wang *et al.*, 2007b). High expression of ZmLEC1 increases maize seed oil by up to 48% but reduces seedling and whole plant growth. High expression of the downstream regulatory element ZmWRI1 results in a similar enhancement of oil levels without impacting upon plant growth or yield (Shen *et al.*, 2010). WRI1 is an AP2-type transcription factor that binds to the AW-box in many genes involved in fatty acid biosynthesis, including genes for a subunit of pyruvate kinase (Pl-PKb1), acetyl-CoA carboxylase (BCCP2), acyl carrier protein (ACP1), and ketoacyl-acyl carrier protein synthase (KAS1) (Maeo *et al.*, 2009). Over-expression of WRI1 does not increase DGAT1 expression.

Seeds accumulate TAG in special organelles known as oil bodies. There is strong evidence that oil bodies form with the accumulation of TAG inside the phospholipid bilayer in specialised regions of the ER ballooning out from the accumulating TAG and the remaining phospholipid forming a monolayer surrounding the growing lipid body. Concurrent with this, the oil body-specific protein, oleosin, is co-translationally inserted into the phospholipid monolayer of the oil bodies (Tzen *et al.*, 1990; Kalinski *et al.*, 1991; Loer and Herman, 1993; Sarmiento *et al.*, 1997; Siloto *et al.*, 2006).

2.8 Genetic Engineering of Oilseed for Industrial Uses

Plant oil TAGs are known to accumulate more than 300 unusual fatty acids (UFAs) in addition to the five fatty acids common in almost plant tissues and the main components of most commodity oils (van de Loo *et al.*, 1993). However, in most cases sources of these UFAs cannot be produced economically on a commercial scale. Hydroxy, epoxy, conjugated, acetylenic, very long chain and branched chain fatty acids and liquid waxes are among the industrial targets of greatest interest. Oils high in hydroxy fatty acids can be produced from castor and *Lesquerella*, but they could be produced more economically on a large scale currently with canola or soybeans engineered with genes for hydroxy fatty acid accumulation. This is due principally to the value of the meal co-product and better developed agronomic properties providing higher yields and greater ease of planting and harvest. Genes for most of these UFAs have been cloned and good reviews have been written on this subject, including that of Napier (2007). It is easy to produce UFAs in transgenic oilseeds with the genes encoding enzymes for UFA biosynthesis, but it has been very difficult to achieve accumulation of UFAs to more than 10% of the total lipids. This is in contrast to the accumulation of as much of 95% of the seed oil TAG being composed of a single UFA such as the hydroxy fatty acid ricinoleate in castor oil, the epoxy fatty acid vernoleate in *Bernardia pulchella* oil, and the short-chain fatty acid caproate in *Cuphea koehneana* oil. In the cases where the details of the biosynthesis of the UFA are known they are made from phosphatidyl choline (PC) in the ER, but then selectively accumulate in seed oil TAG. They do not accumulate in membrane lipids such as the starting PC in the plants that accumulate high levels in TAG, but do not show such selective distribution in transgenic oilseeds with UFA biosynthetic genes alone. This has led to studies on whether TAG biosynthetic enzymes might have selectivity for such fatty acids that accumulate in the TAG.

In order to induce large changes in oil composition, LPAT has been considered an important target enzyme because of its selective discrimination ability (Franzosi *et al.*, 1998). Rapeseed (*Brassica napus*) and meadowfoam (*Limnanthes*) have 60% and 90% erucic acid in their TAGs. In meadowfoam, erucic acid is present in the *sn*2 position of TAGs, whereas it is excluded in rapeseed. This difference was attributed to the substrate specificity of LPAT in the two species (Cao *et al.*, 1990). To alter the stereochemical composition of rapeseed oil, a cDNA encoding *Limnanthes* seed-specific LPAT was expressed in *Brassica napus* plants using a napin expression cassette. In the transgenic plants, 22.3% erucic acid was present at the *sn*2 position leading to the production of trierucin. However, alteration of erucic acid at the *sn*2 position did not affect the total erucic acid content. It may be that the meadowfoam LPAT does not increase the erucic acid content of rapeseed (Lassner *et al.*, 1995) because of the limited pool size of the 22:1 coenzyme A in the maturing embryos of *B. napus*. The metabolism of laurate was found to be different in transgenic *Brassica napus* lines (transformed with a California bay lauroyl-acyl carrier protein thiosesterase cDNA driven by napin promoter) and the natural laurate accumulators coconut, oil palm and *Cuphea wrightii*. When tested at the mid-stage of embryo development, the PC had up to 26 mol% of laurate in the transgenic rapeseed high laurate line, whereas in other species it ranged between 1 to 4 mol%. The laurate in the *Brassica* TAG was almost totally confined to the outer *sn*1 and *sn*3 positions, whereas the laurate in coconut and *Cuphea* was highest in the *sn*2 position. Very low amounts of laurate were found in the *sn*2 position in DAG and PC of the rapeseed lipids, indicating that no arrangement of laurate between the outer and *sn*2 positions occurred in any of the lipids. There was an enhanced activity of lauroyl-PC metabolising enzymes in the laurate-producing rapeseed when embryos were fed with ^{14}C-lauroyl-PC and ^{14}C-palmitoyl-PC. The data indicated that DAG was preferentially utilised from natural laurate accumulators like oil palm, coconut and *Cuphea* (Wiberg *et al.*, 1997). Transgenic rapeseed oil expressing California bay thioesterase produced 60% saturated FA, with laurate alone comprising 48%. In these plants laurate was present only at the *sn*1 and *sn*3 positions. When these plants were crossed with transgenic lines expressing coconut LPAT laurate was present at the *sn*2 position along with *sn*1 and *sn*3 positions in the resulting hybrids. An overall increase in the oil content was also observed.

When the yeast LPAT genes SLC1 and SLC1-1 (mutant form of yeast LPAT) were expressed in *Brassica napus* and *Arabidopsis* under the CaMV35S promoter, the TAG and VLCFA contents were increased by 56% and 80%, respectively (Zou *et al.*, 1997). In the transgenic plants, seed weight increased indicating at least a partial contribution from enhanced oil content. In the total oil content, 60–75% consisted of VLCFAs and up to 40% that of non-VLCFAs such as palmitate, oleate, linoleate and linolenate. No increase in total oil content was reported in coconut or meadowfoam LPAT-transformed rapeseed. This could be due to different regulatory properties of the plant and yeast LPAT enzymes. The plant LPAT genes have 62% amino acid identity among themselves, whereas the yeast genes have 24% homology. In transgenic plants the high expression of SLC1-1 gene did not correlate with an increased oil content, indicating that even small levels of expression were sufficient to overcome the PA limitations during TAG biosynthesis. Although SLC1-1 levels were stronger in leaves than in seeds, no significant changes were observed in the fatty acid composition of leaves, indicating that the pools of available LPA and/or acyl-CoAs may be more tightly regulated in leaves (source) than in seeds (sink). There are preliminary data indicating a 1–2% increase in oil content of soybeans expressing the yeast SLC1 gene (Rao and Hildebrand, 2009).

In studies on the expression profiles of genes encoding TAG biosynthetic enzymes, it was found that DGAT1, unlike DGAT2 or PDAT, has an expression profile in different tissues of soybeans and *Arabidopsis* consistent with a role in seed oil synthesis. DGAT1 and DGAT2, in contrast, display expression consistent with a role in seed oil synthesis in high epoxy and hydroxy fatty acid accumulating plants, implicating DGATs, and particularly DGAT2, as playing an important role in the selective accumulation of UFA in TAG (Li *et al.*, 2005, 2010a). As mentioned above, Zhang *et al.* (2009) have established that DGAT2 has no role in TAG biosynthesis in *Arabidopsis*. However, there is good evidence that DGAT2 from tung trees, *Vernicia fordii*, has specificity for the conjugated fatty acid that accumulates in tung oil, eleostearic acid (Shockey *et al.*, 2006a,b). Coexpression of hydroxylase and DGAT2 from castor and epoxygenase + DGAT2 from *Bernardia* and *Vernonia* can increase the accumulation of hydroxy and epoxy fatty acids in seed oil up to five-fold over expression of the hydroxylase and epoxygenase genes alone (Burgal *et al.*, 2008; Zhou *et al.*, 2008; Li *et al.*, 2010a). Li *et al.* (2010a) were the first to demonstrate this in a commercial oilseed going from about 5% epoxy fatty acid in seed lipids of soybeans expressing an epoxygenase, increasing to >10% in soybeans expressing an epoxygenase + a DGAT1 from a high epoxy fatty acid accumulating plant, and to >30% in soybeans expressing an epoxygenase + a DGAT2 from the same high epoxy fatty acid accumulating plant (unpublished data).

2.9 Plant Oils as a Renewable Resource

Global plant oil production is dominated by four main oil crops: palm, soybeans, rapeseed or canola, and sunflowers (Wilson and Hildebrand, 2010) (Figure 2.4). Plant oil production

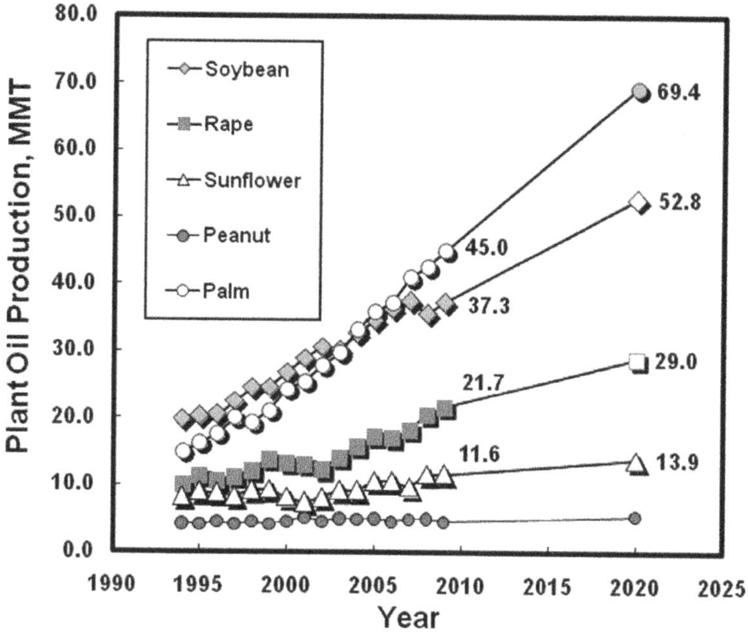

Figure 2.4 *Trends in global production of major plant oils*

in 2009 exceeded 120 million metric tons. Worldwide palm and soybean production has increased rapidly and rapeseed has also shown steady increases, and this trend is expected to continue with a projected global oil production of over 170 million metric tons in 2020 (Figure 2.4). Palm oil production has been dominated by Malaysia and Indonesia, although expansion in South America and Africa is feasible. The vast majority of soybeans are produced in the US, Brazil, China and Argentina, with further expansion possible in southern Africa and South America. Global oilseed production is dominated by soybeans followed by rapeseed as a distant second (Wilson and Hildebrand, 2010) (Figure 2.5). This is because among oilseeds, soybeans are low in oil and high in protein, making soybeans the dominant global protein source. Over 200 million metric tons of soybean seeds have been produced annually in recent years. Soybeans average ~20% oil and 40% protein on a dry weight basis, whereas rapeseed is ~50% oil, and palm fruit close to 90% oil. Oil palm fruit contains seeds from which palm kernel oil is derived. Palm fruit and kernel oils are quite different in fatty acid composition, with palm kernel oil being dominated by medium chain fatty acids similar to coconut oil. Palm fruit cannot be stored in large quantities or transported long distances, unlike most oilseeds.

Breeding for increased oilseed yield per unit land area continues to progress, with steady soybean yield increases being a good example (Egli, 2008a,b). This is often with little or no increased inputs, making renewable oil production from plants less expensive over time and progressively more competitive with petroleum as an industrial chemical feedstock. Further, it is becoming increasingly possible to alter hydrocarbon flux into oil in some oilseeds, increasing the oil yield per unit land area independent of yield increases. This is particularly true for low-oil oilseeds such as soybeans. A number of studies indicate that oil content

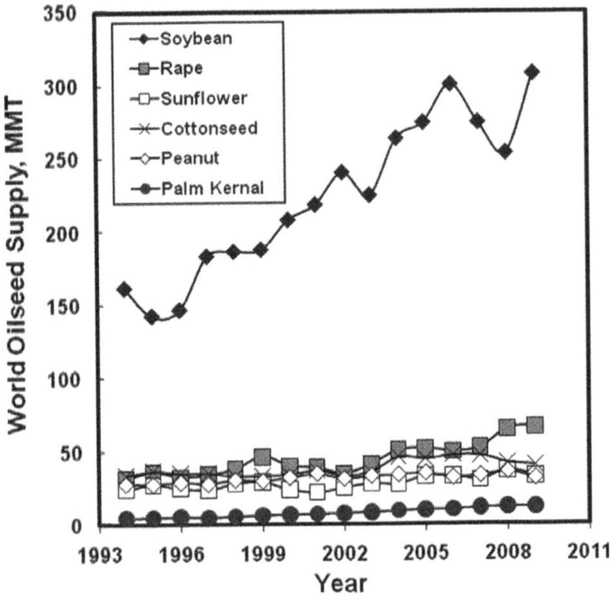

Figure 2.5 *Global production (supply) of major oilseeds*

increases with higher expression of TAG biosynthetic genes (Lardizabal *et al.*, 2008; Rao and Hildebrand, 2009; Taylor *et al.*, 2009; Vyacheslav *et al.*, 2009). Increased expression of regulatory genes that up-regulate multiple enzymes for fatty acid biosynthesis can also result in higher oil levels, as mentioned above (Vyacheslav *et al.*, 2009; Shen *et al.*, 2010).

Most plant oil produced is used for food purposes, but an increasing proportion is being utilised for industry, and the proportion of industrial versus food usage has increased from

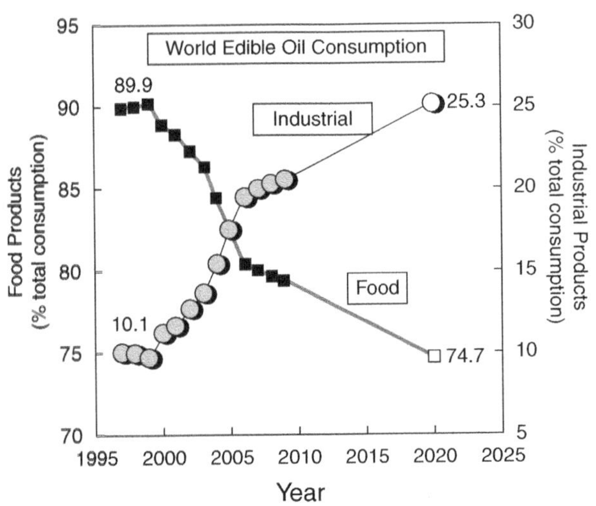

Figure 2.6 *Food vs industrial utilisation (consumption) of plant oils*

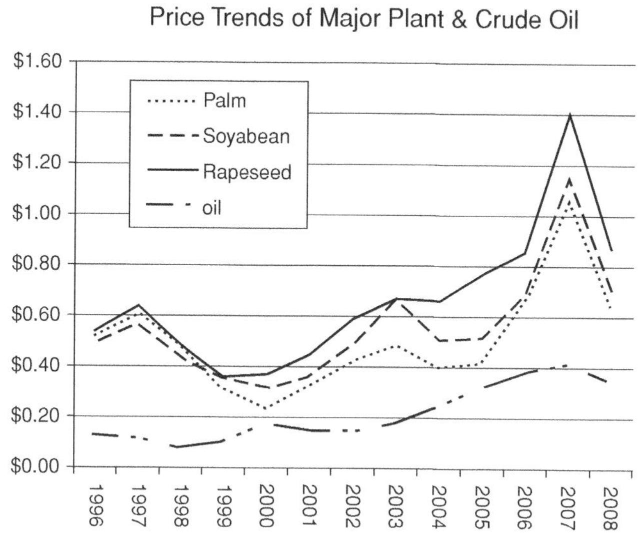

Figure 2.7 *Price trends of the major plant oils vs crude oil (petroleum) in the past 15 years in US $/kg*

ca. 10% to around 20% in the last ten years (Figure 2.6). With the overall oil production showing a large increase, still much more oil is available for food purposes and this trend is projected to continue (Figures 2.4 and 2.6). A large proportion of the increased non-food usage has been for fuel, most notably biodiesel, but with advances in genetic engineering for specific properties for industrial chemical usages, as outlined above, and advances in chemical engineering of vegetable oils for certain end uses, plant oils will become increasingly attractive as industrial chemical feedstocks. Also, although historically more expensive than petroleum, the cost of producing plant oils is decreasing while petroleum production is becoming more expensive, and reserves are finite and expected to be largely depleted in the next 50 years. Although plant oils are still more expensive than crude petroleum, the price difference is decreasing (USDA, 2009; Anon., 2010) (Figure 2.7). In the mid-late 1990s, plant oils sold for between four and six times that of crude oil, whereas they are currently around double the price. It should be more efficient to use biomass for electrical generation and for increasing numbers of electrical passenger vehicles in the near future, while larger commercial vehicles are projected to run on biodiesel in the foreseeable future (Ohlrogge *et al.*, 2009). In time, specially tailored plant oils for certain industrial chemical applications are likely to become more economical than petroleum-derived sources.

Acknowledgements

Thanks to the United Soybean Board (USB), the Consortium for Plant Biotechnology Research (CPBR) and the Kentucky Agricultural Experiment Station for support of our lipid biosynthesis research. Helpful input on this chapter from Tom Zimmerman, Nazia Nahid and Hirotada Fukushige is much appreciated.

References

Abraham, R., Riemersma, R., Wood, D., Elton, R. and Oliver, M. (1989) Adipose fatty acid composition and the risk of serious arrhythmias in acute myocardial infarction. *Am. J. Cardiol.*, **63**, 269–272.

Aghoram, K., Wilson, R.F., Burton, J.W. and Dewey, R.E. (2006) A mutation in a 3-keto-acyl-ACP synthase II gene is associated with elevated palmitic acid levels in soybean seeds. *Crop Sci.*, **46**, 2453–2459.

Ako, H. and Okuda, D. (1995) Healthful new oil from macadamia nuts. *Nutr.*, **11**, 286–288.

Alt, J.L., Fehr, W.R., Welke, G.A. and Sandhu, D. (2005a) Phenotypic and molecular analysis of oleate content in the mutant soybean line M23. *Crop Sci.*, **45**, 1997–2000.

Alt, J.L., Fehr, W.R., Welke, G.A. and Shannon, J.G. (2005b) Transgressive segregation for oleate content in three soybean populations. *Crop Sci.*, **45**, 2005–2007.

Anai, T., Yamada, T., Kinoshita, T., Rahman, S.M. and Takagi, Y. (2005) Identification of corresponding genes for three low-[α]-linolenic acid mutants and elucidation of their contribution to fatty acid biosynthesis in soybean seed. *Plant Sci.*, **168**, 1615–1623.

Anon. (2010) *Historical oil prices*, http://www.inflationdata.com/inflation/Inflation_Rate/Historical_Oil_Prices_Table.asp [accessed 19 November 2010].

Arondel, V., Lemieux, B., Hwang, I., Gibson, S., Goodman, H. and Somerville, C. (1992) Map-based cloning of a gene controlling omega-3 fatty acid desaturation in *Arabidopsis*. *Science*, **258**, 1353–1355.

Bachlava, E., Dewey, R.E., Auclair, J., Wang, S., Burton, J.W. and Cardinal, A.J. (2008) Mapping genes encoding microsomal {omega}-6 desaturase enzymes and their cosegregation with QTL affecting oleate content in soybean. *Crop Sci.*, **48** 640–650.

Bates, P.D., Durrett, T.P., Ohlrogge, J.B. and Pollard, M. (2009) Analysis of acyl fluxes through multiple pathways of triacylglycerol synthesis in developing soybean embryos. *Plant Physiol.*, **150**, 55–72.

Baud, S. and Lepiniec, L. (2009) Regulation of *de novo* fatty acid synthesis in maturing oilseeds of *Arabidopsis*. *Plant Physiol. Biochem.*, **47**, 448–455.

Baud, S., Wuilleme, S., To, A., Rochat, C. and Lepiniec, L. (2009) Role of WRINKLED1 in the transcriptional regulation of glycolytic and fatty acid biosynthetic genes in *Arabidopsis*. *Plant J.* **60**, 933–947.

Beisson, F., Koo, A.J., Ruuska, S., Schwender, J., Pollard, M., Thelen, J.J., Paddock, T., Salas, J.J., Savage, L., Milcamps, A., Mhaske, V.B., Cho, Y., Ohlrogge, J.B. (2003) Arabidopsis genes involved in acyl lipid metabolism. A 2003 census of the candidates, a study of the distribution of expressed sequence tags in organs, and a web-based database. *Plant Physiol.* **132**, 681–697.

Benning, C. (2009) Mechanisms of lipid transport involved in organelle biogenesis in plant cells. *Annu. Rev. Cell Develop. Biol.*, **25**, 71–91.

Beremand, P.D., Nunberg, A.N., Reddy, A.S. and Thomas, T. (1997) Production of γ-linolenic acid by transgenic plants expressing cyanobacterial or plant □6-desaturase genes, in *Physiology, Biochemistry and Molecular Biology of Plant Lipids* (eds J.P. Williams, M.U. Khan and N.W. Lem), Kluwer Academic Publishers, Dordrecht, pp. 351–353.

Bewley, J. and Black, M. (1994) *Seeds: Physiology of Development and Germination*, Plenum Press, New York.

Bilyeu, K., Palavalli, L., Sleper, D. and Beuselinck, P. (2005) Mutations in soybean microsomal omega-3 fatty acid desaturase genes reduce linolenic acid concentration in soybean seeds. *Crop Sci.*, **45**, 1830–1836.

Bilyeu, K., Palavalli, L., Sleper, D.A. and Beuselinck, P. (2006) Molecular genetic resources for development of 1% linolenic acid soybeans. *Crop Sci.*, **46**, 1913–1918.

Bloomfield, D.K. and Bloch, K. (1960) Formation of D9-unsaturated fatty acids. *J. Biol. Chem.*, **235**, 337–345.

Broun, P., Hauker, N. and Somerville, C.R. (1997) Expression of castor and *Lesquerella fendleri* oleate 12-hydroxylases in transgenic plants, in *Physiology, Biochemistry and Molecular Biology of Plant Lipids* (eds J.P. Williams, M.U. Khan and N.W. Lem), Kluwer Academic Publishers, Dordrecht, pp. 342–344.

Browse, J. (2009) Jasmonate passes muster: a receptor and targets for the defense hormone. *Ann. Rev. Plant Biol.* **60**, 183–205.

Browse, J., Warwick, N., Somerville, C. and Slack, C. (1986) Fluxes through the prokaryotic and eukaryotic pathways of lipid synthesis in the '16:3' plant *Arabidopsis thaliana. Biochem. J.*, **235**, 25–31.

Browse, J. and Somerville, C. (1991) Glycerolipid synthesis: biochemistry and regulation. *Ann. Rev. Plant Physiol. Plant Mol. Biol.*, **42**, 467–506.

Browse, J., McConn, M., James, D.J. and Miquel, M. (1993) Mutants of *Arabidopsis* deficient in the synthesis of α-linolenate. Biochemical and genetic characterization of the endoplasmic reticulum linoleoyl desaturase. *J. Biol. Chem.*, **268**, 16345–16351.

Budziszewski, G., Croft, K.P.C. and Hildebrand, D.F. (1996) Uses of biotechnology in modifying plant lipids. *Lipids*, **31**, 557–569.

Buhr, T., Sato, S., Ebrahim, F., Xing, A., Zhou, Y., Mathiesen, M., Schweiger, B., Kinney, A., Staswick, P. and Clemente, T. (2002) Ribozyme termination of RNA transcripts down-regulate seed fatty acid genes in transgenic soybean. *Plant J.*, **30**, 155–163.

Burgal, J., Shockey, J., Lu, C.F., Dyer, J., Larson, T., Graham, I. and Browse, J. (2008) Metabolic engineering of hydroxy fatty acid production in plants: RcDGAT2 drives dramatic increases in ricinoleate levels in seed oil. *Plant Biotech. J.* **6**, 819–831.

Byrum, J.R., Kinney, A.J., Stecca, K.L., Grace, D.J. and Diers, B.W. (1997) Alteration of the omega-3 fatty acid desaturase gene is associated with reduced linolenic acid in the A5 soybean genotype. *Theor. Appl. Genet.* **94**, 356–359.

Cahoon, E.B. (2003) Genetic enhancement of soybean oil for industrial uses: Prospects and challenges. *AgBioForm* **6**, 11–13.

Cahoon, E., Becker, C., Shanklin, J. and Ohlrogge, J. (1994) cDNAs for isoforms of the delta 9-stearoyl-acyl carrier protein desaturase from *Thunbergia alata* endosperm. *Plant Physiol.* **106**, 807–808.

Cahoon, E., Lindqvist, Y., Schneider, G. and Shanklin, J. (1997) Redesign of soluble fatty acid desaturases from plants for altered substrate specificity and double bond position. *Proc. Natl. Acad. Sci. USA*, **94**, 4872–4877.

Cahoon, E.B., Shah, S., Shanklin, J. and Browse, J. (1998) A determinant of substrate specificity predicted from the acyl-acyl carrier protein desaturase of developing cat's claw seed. *Plant Physiol.*, **117**, 593–598.

Calvo, A., Hinze, L., Gardner, H. and Keller, N. (1999) Sporogenic effect of polyunsaturated fatty acids on development of *Aspergillus* spp. *Appl. Environ. Microbiol.*, **65**, 3668–3673.

Cao, Y., Oo, K.C. and Huang, A.H.C. (1990) Lysophosphatidate acyltransferase in the microsomal fraction from maturing seeds of meadowfoam (*Limnanthes alba*). *Plant Physiol.*, **9**, 1199–1206.

Cases, S., Smith, S.J., Zheng, Y., Myers, H.M., Lear, S.R., Sande, E., Novak, S., Collins, C., Welch, C.B., Lusis, A.J., Erickson, S.K., Farese, R.V. (1998) Identification of a gene encoding an acyl CoA:diacylglycerol acyltransferase, a key enzyme in triacylglycerol synthesis. *Proc. Natl. Acad. Sci. USA*, **95**, 13018–13023.

Cases, S., Stone, S.J., Zhou, P., Yen, E., Tow, B., Lardizabal, K.D., Voelker, T. and Farese, R.V., Jr. (2001) Cloning of DGAT2, a second mammalian diacylglycerol acyltransferase, and related family members. *J. Biol. Chem.* **276**, 38870–38876.

Cernac, A. and C. Benning. (2004) WRINKLED1 encodes an AP2/EREB domain protein involved in the control of storage compound biosynthesis in *Arabidopsis*. *Plant J.*, **40**, 575–585.

Chisholm, M. and Hopkins, C. (1965) Fatty acids of doxantha seed oil. *J. Am. Oil Chem. Soc.*, **42**, 49–50.

Clemente, T.E. and Cahoon, E.B. (2009) Soybean oil: genetic approaches for modification of functionality and total content. *Plant Physiol.*, **151**, 1030–1040.

Creelman, R.A. and Mulpuri, R. (2002) The oxylipin pathway in *Arabidopsis*, in *The Arabidopsis Book* (eds C.R. Somerville and E.M. Meyerowitz), American Society of Plant Biologists, Rockville, MD, pp. 1–24.

Curb, J., Wergowske, G., Dobbs, J., Abbott, R. and Huang, B. (2000) Serum lipid effects of a high-monounsaturated fat diet based on macadamia nuts. *Arch. Intern. Med.*, **160**, 1154–1158.

Cury-Boaventura, M. and Curi, R. (2005) Regulation of reactive oxygen species (ROS) production by C18 fatty acids in Jurkat and Raji cells. *Clin. Sci. (Lond.)*, **108**, 245–253.

Dahlqvist, A., Stahl, U., Lenman, M., Banas, A., Lee, M., Sandager, L., Ronne, H. and Stymne, S. (2000) Phospholipid:diacylglycerol acyltransferase: an enzyme that catalyzes the acyl-CoA-independent formation of triacylglycerol in yeast and plants. *Proc. Natl. Acad. Sci. USA*, **97**, 6487–6492.

Dahmer, M.L., Collins, G.B. and Hildebrand, D.F. (1991) Lipid concentration and composition of soybean zygotic embryos maturing in vitro and in planta. *Crop Sci.*, **31**, 735–740.

Denys, A., Hichami, A. and Khan, N. (2001) Eicosapentaenoic acid and docosahexaenoic acid modulate MAP kinase (ERK1/ERK2) signaling in human T cells. *J. Lipid Res.*, **42**, 2015–2020.

Dewey, R.E., Wilson, R.F., Novitzky, W.P. and Goode, J.H. (1994) The AAPT1 gene of soybean complements a cholinephosphotransferase-deficient mutant of yeast. *Plant Cell*, **6**, 1495–1507.

Downey, R.K. and McGregor, D.I. (1975) Breeding for modified fatty acid composition. *Curr. Adv. Plant Sci.*, **12**, 151–167.

Egli, D.B. (2008a) Comparison of corn and soybean yields in the United States: Historical trends and future prospects. *Agron. J.*, **100**, S79–S80.

Egli, D.B. (2008b) Soybean yield trends from 1972 to 2003 in mid-western USA. *Field Crops Res.*, **106**, 53–59.

Facciotti, M.T., Bertain, P.B. and Yuan, L. (1999) Improved stearate phenotype in transgenic canola expressing a modified acyl-acyl carrier protein thioesterase. *Nature Biotechnol.*, **17**, 593–597.

Falcone, D., Gibson, S., Lemieux, B. and Somerville, C. (1994) Identification of a gene that complements an *Arabidopsis* mutant deficient in chloroplast omega 6 desaturase activity. *Plant Physiol.*, **106**, 1453–1459.

Fawcett, T., Simon, W.J., Swinhoe, R., Shanklin, J., Nishida, I., Christie, W.W. and Slabas, A.R. (1994) Expression of mRNA and steady-state levels of protein isoforms of enoyl-ACP reductase from *Brassica napus*. *Plant Mol. Biol.*, **26**, 155–163.

Focks, N. and Benning, C. (1998) wrinkled1: A novel, low-seed-oil mutant of arabidopsis with a deficiency in the seed-specific regulation of carbohydrate metabolism. *Plant Physiol.*, **118**, 91–101.

Fox, B., Shanklin, J., Somerville, C. and Munck, E. (1993) Stearoyl-acyl carrier protein delta 9 desaturase from *Ricinus communis* is a diiron-oxo protein. *Proc. Natl. Acad. Sci. USA*, **90**, 2486–2490.

Franzosi, G., Battistel, E., Santoro, M. and Iannacone, R. (1998) LPAAT and DAGAT activity and specificity in rapeseed (*Brassica napus* L. var. canola) and sunflower (*Helianthus annuus*) developing seeds, in *Advances in Plant Lipid Research* (eds J. Sánchez, E. Cerdá-Olmedo and E. Martínez-Force), Universidad De Sevilla, Secretariado de Publicaciones, Sevilla, pp. 679–682.

Gavilano, L.B., Coleman, N.P., Burnley, L.E., Bowman, M.L., Kalengamaliro, N.E., Hayes, A., Bush, L. and Siminszky, B. (2006) Genetic engineering of *Nicotiana tabacum* for reduced nornicotine content. *J. Agric. Food Chem.*, **54**, 9071–9078.

Gibson, S., Arondel, V., Iba, K. and Somerville, C. (1994) Cloning of a temperature-regulated gene encoding a chloroplast omega-3 desaturase from *Arabidopsis thaliana*. *Plant Physiol.*, **106**, 1615–1621.

Goldberg, R.B., Barker, S.J. and Perez-Grau, L. (1989) Regulation of gene expression during plant embryogenesis. *Cell*, **56**, 149–160.

Gonzalez, C.I. and Martin, C.E. (1996) Fatty acid-responsive control of mRNA stability. Unsaturated fatty acid-induced degradation of the saccharomyces ole1 transcript. *J. Biol. Chem.*, **271**, 25801–25809.

Gould, S.B., Waller, R.F. and McFadden, G.I. (2008) Plastid evolution. *Ann. Review Plant Biol.*, **59**, 491–517.

Graef, G.L., Fehr, W.R. and Hammond, E.G. (1985a) Inheritance of three stearic acid mutants of soybean. *Crop Sci.*, **25**, 1076–1079.

Graef, G.L., Miller, L.A., Fehr, W.R. and Hammond., E.G. (1985b) Fatty acid development in a soybean mutant with high stearic acid. *J. Am. Oil Chem. Soc.*, **62**, 773–775.

Graham, I.A. (2008) Seed storage oil mobilization. *Ann. Review Plant Biol.*, **59**, 115–142.

Grayburn, W.S. and Hildebrand, D.F. (1995) Progeny analysis of tobacco that express a mammalian $\Delta 9$ desaturase. *J. Am. Oil Chem. Soc.*, **72**, 317–321.

Griel, A.E., Cao, Y.M., Bagshaw, D.D., Cifelli, A.M., Holub, B. and Kris-Etherton, P.M. (2008) A macadamia nut-rich diet reduces total and LDL-cholesterol in mildly hypercholesterolemic men and women. *J. Nutr.*, **138**, 761–767.

Griffiths, G., Stobart, A. and Stymne, S. (1985) The acylation of sn-glycerol 3-phosphate and the metabolism of phosphatidate in microsomal preparations from the developing cotyledons of safflower (*Carthamus tinctorius* L.) seed. *Biochem. J.*, **230**, 379–388.

Griffiths, G., Stobart, A.K. and Stymne, S. (1988) Δ6- and Δ12-desaturase activities and phosphatidic acid formation in microsomal preparations from the developing cotyledons of common borage (*Borago officinalis*). *Biochem. J.*, **252**, 641–647.

Halitschke, R. and Baldwin, I.T. (2004) Jasmonates and related compounds in plant-insect interactions. *J. Plant Growth Regul.* **23**, 238–245.

Hammond, E.G. and Fehr, W.R. (1983) Registration of A5 germplasm line of soybean. *Crop Sci.*, **23**, 192–193.

Harwood, J.L. (1997) Plant lipid metabolism, in *Plant Biochemistry* (eds P.M. Dey and J.B. Harborne), Academic Press, London, pp. 237–272.

Harwood, J.L (1998) What's so special about plant lipids?, in *Plant Lipid Biosynthesis Fundamentals and Agricultural Applications* (ed. J.L. Harwood), Cambridge University Press, New York, pp. 1–26.

Harwood, J.L. and Page, R.A. (1994) Biochemistry of oil synthesis. In *Designer Oil Crops* (ed. D.J. Murphy), VCH, Weinheim, pp. 165–194.

Harwood, J.L. and Guschina, I.A. (2009) The versatility of algae and their lipid metabolism. *Biochimie*, **91**, 679–684.

Hatanaka, T., Shimizu, R. and Hildebrand, D. (2004) Expression of a *Stokesia laevis* epoxygenase gene. *Phytochemistry*, **65**, 2189–2196.

Hayatsu, H., Arimoto, S. and Negishi, T. (1988) Dietary inhibitors of mutagenesis and carcinogenesis. *Mutat Res.*, **202**, 429–446.

Hegsted, D.M., Ausman, L.M., Johnson, J.A. and Dallal, G.E. (1993) Dietary fat and serum lipids: an evaluation of experimental data. *Am. J. Clin. Nutr.*, **57**, 875–883.

Heppard, E.P., Kinney, A.J., Stecca, K.L. and Miao, G.H. (1996) Developmental and growth temperature regulation of two different microsomal omega-6 desaturase genes in soybeans. *Plant Physiol.*, **110**, 311–319.

Hildebrand, D.F. (1989) Lipoxygenases. *Physiol. Plant.*, **76**, 249–253.

Hildebrand, D.F., Li, R. and Hatanaka, T. (2008) Genomics of soybean oil traits, in *Genetics and Genomics of Soybean* (ed. G. Stacey), Springer, New York, pp. 185–210.

Hiramine, Y., Emoto, H., Takasuga, S. and Hiramatsu, R. (2010) Novel acyl-coenzyme A:monoacylglycerol acyltransferase (MGAT) plays an important role in hepatic triacylglycerol secretion. *J. Lipid Res.*, **51**, 1424–1431

Hiraoka, M., Abe, A. and Shayman, J.A. (2002) Cloning and characterization of a lysosomal phospholipase A2, 1-O-acylceramide synthase. *J. Biol. Chem.*, **277**, 10090–10099.

Hitz, W.D. (1998) Fatty acid modifying enzymes from developing seeds of *Vernonia galamensis*. US Patent no. 5,846,784. DuPont, United States.

Hobbs, D.H., Lu, C. and Hills, M.J. (1999) Cloning of a cDNA encoding diacylglycerol acyltransferase from Arabidopsis thaliana and its functional expression. *FEBS Lett.*, **452**, 145–149.

Hobbs, D.H. and Hills, M.J. (2000) Expression and characterization of diacylglycerol acyltransferase from Arabidopsis thaliana in insect cell cultures. *Biochem. Soc. Trans.*, **28**, 687–689.

Hymowitz, T., Palmer, R.G. and Hadley, H.H. (1972) Seed weight, protein, oil, and fatty acid relationships within the genus *Glycine*. *Trop. Agr.*, **49**, 245–250.

Iba, K. (2002) Acclimative response to temperature stress in higher plants: approaches of gene engineering for temperature tolerance. *Ann. Rev. Plant Biol.*, **53**, 225–245.

Jackson, F.M., Michaelson, L., Fraser, T.C.M., Stobart, A.K. and Griffiths, G. (1998) Biosynthesis of triacylglycerol in the filamentous fungus *Mucor circinelloides*. *Microbiology*, **144**, 2639–2645.

Kachroo, A., Lapchyk, L., Fukushige, H., Hildebrand, D., Klessig, D. and Kachroo, P. (2003) Plastidial fatty acid signaling modulates salicylic acid- and jasmonic acid-mediated defense pathways in the Arabidopsis ssi2 mutant. *Plant Cell*, **15**, 2952–2965.

Kachroo, A., Shanklin, J., Whittle, E., Lapchyk, L., Hildebrand, D. and Kachroo, P. (2007) The Arabidopsis stearoyl-acyl carrier protein-desaturase family and the contribution of leaf isoforms to oleic acid synthesis. *Plant Mol. Biol.*, **63**, 257–271.

Kachroo, A. and Kachroo, P. (2009) Fatty acid derived signals in plant defense. *Ann. Rev. Phytopathol.*, **47**, 153–176.

Kalinski, A., Loer, D.S., Weisemann, J.M., Matthews, B.F. and Herman, E.M. (1991) Isoforms of soybean seed oil body membrane protein 24 kDa oleosin are encoded by closely related cDNAs. *Plant Mol. Biol.*, **17**, 1095–1098.

Kamisaka, Y., Mishra, S. and Nakahara, T. (1997) Purification and characterization of diacylglycerol acyltransferase from the lipid body fraction of an oleaginous fungus. *J Biochem* **121**, 1107–1114.

Katan, M.B., Zook, P.L. and Mensik, R.P. (1995) Trans-fatty acids and their effects on lipoproteins in humans. *Ann. Rev. Nutr.* **15**, 473–493.

Katavic, V., Reed, D.W., Taylor, D.C., Giblin, E.M., Barton, D.L., Zou, J., Mackenzie, S.L., Covello, P.S. and Kunst, L. (1995) Alteration of seed fatty acid composition by an ethyl methanesulfonate-induced mutation in *Arabidopsis thaliana* affecting diacylglycerol acyltransferase activity. *Plant Physiol.*, **108**, 399–409.

Kaup, M.T., Froese, C.D. and Thompson, J.E. (2002) A role for diacylglycerol acyltransferase during leaf senescence. *Plant Physiol.*, **129**, 1616–1626.

Kinney, A.J. (1994) Genetic modification of the storage lipids of plants. *Curr. Opin. Biotechnol.*, **5**, 144–151.

Kinney, A.J. (1998a) Plants as industrial chemical factories-new oils from genetically engineered soybeans. *Fett/Lipid*, **100**, 173–176.

Kinney, A.J. (1998b) Production of specialised oils for industry, in *Plant Lipid Biosynthesis; Fundamentals and Agricultural Applications* (ed. J.L. Harwood), Cambridge University Press, Cambridge, pp. 273–285.

Klaus, D., Ohlrogge, J.B., Neuhaus, H.E. and Dormann, P. (2004) Increased fatty acid production in potato by engineering of acetyl-CoA carboxylase. *Planta* **219**, 389–396.

Knowlton, S. (1999) Soybean oil having high oxidative stability. US Patent no. 5,981,781.

Knutzon, D.S., Thompson, G.A., Radke, S.E., Johnson, W.B., Knauf, V.C. and C, K.J. (1992) Modification of *Brassica* seed oil by antisense expression of a stearoyl-acyl carrier protein desaturase gene. *Proc. Nat. Acad. Sci. USA*, **89**, 2624–2628.

Kridl, J. (2002) Method for increasing stearate content in soybean oil. US Patent no. 6,380,462.

Kris-Etherton, P.M., Krummel, D., Russell, M.E., Dreon, D., Mackey, S., Borchers, J. and Wood, P.D. (1988) The effect of diet on plasma lipids, lipoproteins, and coronary heart disease. *J. Am. Dietet. Assoc.*, **88**, 1373–1400.

Kwanyuen, P. and Wilson, R.F. (1986) Isolation and purification of diacylglycerol acyltransferase from germinating soybean cotyledons. *Biochim. Biophys. Acta.*, **877**, 238–245.

Kwanyuen, P., Wilson, R.F. and Burton, J.W. (1988) Substrate specificity of diacylglycerol acyltransferase purified from soybean, in *Proceeding: World Conference on Biotechnology for the Fats and Oils Industry* (ed. T.H. Applewhite), American Oil Chemists' Society, Champaign, Ill. pp. 294–297.

Kwanyuen, P. and Wilson, R.F. (1990) Subunit and amino acid composition of diacylglycerol acyltransferase from germinating soybean cotyledons. *Biochim. Biophys. Acta.* **1039**, 67–72.

Ladd, S.L. and Knowles, P.F. (1970) Inheritance of stearic acid in the seed oil of safflower (*Carthamus tinctorius* L.). *Crop Sci.*, **10**, 525–527.

Lardizabal, K., Effertz, R., Levering, C., Mai, J., Pedroso, M.C., Jury, T., Aasen, E., Gruys, K. and Bennett, K. (2008) Expression of *Umbelopsis ramanniana DGAT2A* in seed increases oil in soybean. *Plant Physiol.*, **148**, 89–96.

Lardizabal, K.D., Mai, J.T., Wagner, N.W., Wyrick, A., Voelker, T. and Hawkins, D.J. (2001) DGAT2 is a new diacylglycerol acyltransferase gene family. Purification, cloning, and expression in insect cells of two polypeptides from *Mortierella ramanniana* with diacylglycerol acyltransferase activity. *J. Biol. Chem.*, **276**, 38862–38869.

Lassner, M.W., Levering, C.K., Davies, H.M. and Knutzon, D.S. (1995) Lysophosphatidic acid acyltransferase from meadowfoam mediates insertion of erucic acid at the sn-2 position of triacylglycerol in transgenic rapeseed oil. *Plant Physiol.*, **109**, 1389–1394.

Le, B.H., Wagmaister, J.A., Kawashima, T., Bui, A.Q., Harada, J.J. and Goldberg, R.B. (2007) Using genomics to study legume seed development. *Plant Physiol.*, **144**, 562–574.

Lee, M., Lenman, M., Bana, A., Bafor, M., Singh, S., Schweizer, M., Nilsson, R., Liljenberg, C., Dahlqvist, A., Gummeson, P.O., Sjödahl, S., Green, A., Stymne, S. (1998) Identification of non-heme diiron proteins that catalyze triple bond and epoxy group formation. *Science*, **280**, 915–918.

Lehner, R. and Kuksis, A. (1993) Triacylglycerol synthesis by an sn-1,2(2,3)-diacylglycerol transacylase from rat intestinal microsomes. *J. Biol. Chem.*, **268**, 8781–8786.

Lemieux, B.M.M., Somerville, C. and Browse, J. (1990) Mutants of Arabidopsis with alterations in seed lipid fatty acid composition. *Theor. Appl. Genet.* **80**, 234–240.

Li, R., Yu, K., Hatanaka, T. and Hildebrand, D.F. (2005) Characterization of cDNAs involved in seed oil synthesis in some high epoxy and hydroxy fatty acid accumulators. In *Plant & Animal Genomes XIII Conference*, San Diego, CA. (abstract p. 251, poster # P725).

Li, R., Yu, K., Hatanaka, T. and Hildebrand, D.F. (2010a) *Vernonia* DGATs increase accumulation of epoxy fatty acids in oil. *Plant Biotechnol. J.*, **8**, 184–195.

Li, R., Yu, K. and Hildebrand, D. (2010b) DGAT1, DGAT2 and PDAT expression in seeds and other tissues of epoxy and hydroxy fatty acid accumulating plants. *Lipids*, **45**, 145–157.

Lindqvist, Y., Huang, W., Schneider, G. and Shanklin, J. (1996) Crystal structure of delta9 stearoyl-acyl carrier protein desaturase from castor seed and its relationship to other di-iron proteins. *EMBO J.*, **15**, 4081–4092.

Liu, W., Torisky, R.S., McAllister, K.P., Avdiushko, S., Hildebrand, D. and Collins, G.B. (1996) A mammalian desaturase gene lowers saturated fatty acid levels in transgenic soybean embryos. *Plant Cell Tiss. Organ Cult.*, **47**, 33–42.

Loer, D.S. and Herman, E.M. (1993) Cotranslational integration of soybean (*Glycine max*) oil body membrane protein oleosin into microsomal membranes. *Plant Physiol.*, **101**, 993–998.

Lonien, J. and Schwender, J. (2009) Analysis of metabolic flux phenotypes for two *Arabidopsis* mutants with severe impairment in seed storage lipid synthesis. *Plant Physiol.* **151**, 1617–1634.

Lu, C. and Hills, M.J. (2002) *Arabidopsis* mutants deficient in diacylglycerol acyltransferase display increased sensitivity to abscisic acid, sugars, and osmotic stress during germination and seedling development. *Plant Physiol.*, **129**, 1352–1358.

Lu, C.L., de Noyer, S.B., Hobbs, D.H., Kang, J., Wen, Y., Krachtus, D. and Hills, M.J. (2003) Expression pattern of diacylglycerol acyltransferase-1, an enzyme involved in triacylglycerol biosynthesis in *Arabidopsis thaliana. Plant Mol. Biol.* **52**, 31–41.

Lu, C., Xin, Z., Ren, Z., Miquel, M. and Browse, J. (2009) An enzyme regulating triacylglycerol composition is encoded by the ROD1 gene of *Arabidopsis. Proc. Nat. Acad. Sci. USA*, **106**, 18837–18842.

Lung, S.-C. and Weselake, R. (2006) Diacylglycerol acyltransferase: a key mediator of plant triacylglycerol synthesis. *Lipids*, **41**, 1073–1088.

Maeo, K., Tokuda, T., Ayame, A., Mitsui, N., Kawai, T., Tsukagoshi, H., Ishiguro, S. and Nakamura, K. (2009) An AP2-type transcription factor, WRINKLED1, of *Arabidopsis thaliana* binds to the AW-box sequence conserved among proximal upstream regions of genes involved in fatty acid synthesis. *Plant J.*, **60**, 476–487.

Maguire, L.S., Sullivan, S.M., Galvin, K., Connor, T.P. and Brien, N.M. (2004) Fatty acid profile, tocopherol, squalene and phytosterol content of walnuts, almonds, peanuts, hazelnuts and the macadamia nut. *Int. J. Food Sci. Nutr.*, **55**, 171–178.

Martinez-Force, E., Fernandez-Moya, V. and Garces, R. (2002) Plant, seeds and oil with saturated triacylglycerol content and oil having a high stearic content. US Patents 6,410,831 and 6,486,336.

McConn, M., Hugly, S., Browse, J. and Somerville, C. (1994) A mutation at the *fad8* locus of *Arabidopsis* identifies a second chloroplast ω–3 desaturase. *Plant Physiol.*, **106**, 1609–1614.

McKeon, T., Kang, S.-T., Turner, C., He, X., Chen, G. and Lin, J.-T. (2006) Enzymatic synthesis of intermediates in castor oil biosynthesis. *97th J. Am. Oil Chem. Soc. Annual Meeting & Expo*, p. 15.

Mhaske, V., Beldjilali, K., Ohlrogge, J. and Pollard, M. (2005) Isolation and characterization of an *Arabidopsis thaliana* knockout line for phospholipid: diacylglycerol transacylase gene (At5g13640). *Plant Physiol. Biochem.* **43**, 413–417.

Miquel, M., James, D., Dooner, H. and Browse, J. (1993) Arabidopsis requires polyunsaturated lipids for low-temperature survival. *Proc. Natl. Acad. Sci. USA*, **90**, 6208–6212.

Mongrand, S., Bessoule, J.-J., Cabantous, F. and Cassagne, C. (1998) The C16:3\C18:3 fatty acid balance in photosynthetic tissues from 468 plant species. *Phytochemistry*, **49**, 1049–1064.

Mu, J.Y., Tan, H.L., Zheng, Q., Fu, F.Y., Liang, Y., Zhang, J.A., Yang, X.H., Wang, T., Chong, K., Wang, X.J. and Zuo, J.R. (2008) LEAFY COTYLEDON1 is a key regulator of fatty acid biosynthesis in *Arabidopsis*. *Plant Physiol.*, **148**, 1042–1054.

Mudd, J.B. and Stumpf, P.K. (1961) Fat metabolism in plants. IXV. Factors affecting the synthesis of oleic acid by particulate preparations from avocado mesocarp. *J. Biol. Chem.*, **236**, 2602–2609.

Murase, M. (1999) *Method for increasing storage lipid content in plant seed.* Mitsubishi Co., US.

Murphy, D.J. and Piffanelli, P. (1998) Fatty acid desaturases: structure, mechanism and regulation, in *Plant Lipid Biosynthesis Fundamentals and Agricultural Applications* (ed. J.L. Harwood), Cambridge University Press, New York, pp. 1–26.

Napier, J.A. (2007) The production of unusual fatty acids in transgenic plants. *Ann. Rev. Plant Biol.*, **58**, 295–319.

Nelson, G.J. (1998) Dietary fat, trans-fatty acids, and risk of coronary heart disease. *Nutr. Rev.*, **56**, 250–252.

Nosarzewski, M. and Archbold, D.D. (2007) Tissue-specific expression of sorbitol dehydrogenase in apple fruit during early development. *J. Exp. Bot.*, **58**, 1863–1872.

Oelkers, P., Tinkelenberg, A., Erdeniz, N., Cromley, D., Billheimer, J.T. and Sturley, S.L. (2000) A lecithin cholesterol acyltransferase-like gene mediates diacylglycerol esterification in yeast. *J. Biol. Chem.*, **275**, 15609–15612.

Ohlrogge, J.B. and Browse, J. (1995) Lipid biosynthesis. *Plant Cell*, **7**, 957–970.

Ohlrogge, J.B. and Jaworski, J.G. (1997) Regulation of fatty acid synthesis. *Ann. Rev. Plant Physiol. Plant Molec. Biol.*, **48**, 109–136.

Ohlrogge, J., Allen, D., Berguson, B., DellaPenna, D., Shachar-Hill, Y. and Stymne, S. (2009) Driving on biomass. *Science*, **324**, 1019–1020.

Okuley, J., Lightner, J., Feldmann, K., Yadav, N., Lark, E. and Browse, J. (1994) Arabidopsis *FAD2* gene encodes the enzyme that is essential for polyunsaturated lipid synthesis. *Plant Cell*, **6**, 146-158.

O'Neill, C.M., Gill, S., Hobbs, D., Morgan, C. and Bancroft, I. (2003) Natural variation for seed oil composition in *Arabidopsis thaliana*. *Phytochemistry*, **64**, 1077–1090.

Quittnat, F., Nishikawa, Y., Stedman, T.T., Voelker, D.R., Choi, J.Y., Zahn, M.M., Murphy, R.C., Barkley, R.M., Pypaert, M., Joiner, K.A. and Coppens, I. (2004) On the biogenesis of lipid bodies in ancient eukaryotes: synthesis of triacylglycerols by a *Toxoplasma* DGAT1-related enzyme. *Mol. Biochem. Parasitol.*, **138**, 107–122.

Rahman, S.M., Takagi, Y. and Kinshita, T. (1997) Genetic control of high stearic content in seed oil of two soybean mutants. *Ther. Appl. Genet.*, **95**, 772–776.

Rahman, S., Anai, T., Kinoshita, T. and Takagi, Y. (2003) A novel soybean germplasm with elevated saturated fatty acids. *Crop Sci.*, **43**, 527–531.

Rao, S. and Hildebrand, D. (2009) Changes in oil content of transgenic soybeans expressing the yeast SLC1 gene. *Lipids*, **44**, 945–951.

Reeves, J.I. and Weihrauch, J. (1979) *Composition of Foods: Fats and Oils.* United States Department of Agriculture, Human Nutrition Information Service, Washington, DC.

Rennie, B.D. and Tanner, J.W. (1989) Fatty acid composition of oil from soybean seeds grown at extreme temperatures. *J. Am. Oil Chem. Soc.*, **86**, 1622–1624.

Rollo, C., Czyzewska, E. and Borden, J. (1994) Fatty acid necromones for cockroaches. *Naturwissenschaften*, **81**, 409–410.

Routaboul, J.-M., Benning, C., Bechtold, N., Caboche, M. and Lepiniec, L. (1999) The TAG1 locus of *Arabidopsis* encodes for a diacylglycerol acyltransferase. *Plant Physiol. Biochem.*, **37**, 831–840.

Routaboul, J.-M., Fischer, S. and Browse, J. (2000) Trienoic fatty acids are required to maintain chloroplast function at low temperatures. *Plant Physiol.* **124**, 1697–1705.

Ruuska, S.A., Schwender, J. and Ohlrogge, J.B. (2004) The capacity of green oilseeds to utilize photosynthesis to drive biosynthetic processes. *Plant Physiol.*, **136**, 2700–2709.

Saha, S., Enugutti, B., Rajakumari, S. and Rajasekharan, R. (2006) Cytosolic triacylglycerol biosynthetic pathway in oilseeds. Molecular cloning and expression of peanut cytosolic iacylglycerol acyltransferase. *Plant Physiol.*, **141**, 1533–1543.

Sarmiento, C., Ross, J.H., Herman, E. and Murphy, D.J. (1997) Expression and subcellular targeting of a soybean oleosin in transgenic rapeseed. Implications for the mechanism of oil-body formation in seeds. *Plant J.*, **11**, 783–796.

Schlueter, J.A., Vasylenko-Sanders, I.F., Deshpande, S., Yi, J., Siegfried, M., Roe, B.A., Schlueter, S.D., Scheffler, B.E. and Shoemaker, R.C. (2007) The FAD2 gene family of soybean: insights into the structural and functional divergence of a paleopolyploid genome. *Crop Sci.*, **47**, S14–S26.

Schmutz, J., Cannon, S.B., Schlueter, J., Ma, J., Mitros, T., *et al.* (2010) Genome sequence of the palaeopolyploid soybean. *Nature*, **463**, 178–183.

Schnebly, S.R., Fehr, W.R., Welke, G.A., Hammond, E.G. and Duvick, D.N. (1994) Inheritance of reduced and elevated palmitate in mutant lines of soybean. *Crop Sci.*, **34**, 829–833.

Schnurr, J.A., Shockey, J.M., de Boer, G.-J. and Browse, J.A. (2002) Fatty acid export from the chloroplast. Molecular characterization of a major plastidial acyl-coenzyme A synthetase from *Arabidopsis*. *Plant Physiol.*, **129**, 1700–1709.

Settlage, S.B., Kwanyuen, P. and Wilson, R.F. (1998) Relation between diacylglycerol acyltransferase activity and oil concentration in soybean. *J. Am. Oil Chem. Soc.*, **75**, 775–781.

Shanklin, J. and Somerville, C. (1991) Stearoyl-acyl-carrier-protein desaturase from higher plants is structurally unrelated to the animal and fungal homologs. *Proc. Natl. Acad. Sci. USA*, **88**, 2510–2514.

Shanklin, J., Whittle, E. and Fox, B.G. (1994) Eight histidine residues are catalytically essential in a membrane-associated iron enzyme, stearoyl-CoA desaturase, and are conserved in alkane hydroxylase and xylene monooxygenase. *Biochemistry*, **33**, 12787–12794.

Shanklin, J. and Cahoon, B.E. (1998) Desaturation and related modifications of fatty acids. *Mol. Biol.*, **49**, 109–136.

Shen, B., Allen, W.B., Zheng, P., Li, C., Glassman, K., Ranch, J., Nubel, D. and Tarczynski, M.C. (2010) Expression of *ZmLEC1* and *ZmWRI1* increases seed oil production in maize. *Plant Physiol.*, **153**, 980–987.

Shockey, J.M., Fulda, M.S. and Browse, J.A. (2002) Arabidopsis contains nine long-chain acyl-coenzyme A synthetase genes that participate in fatty acid and glycerolipid metabolism. *Plant Physiol.*, **129**, 1710–1722.

Shockey, J., Gidda, S., Burgal, J., Chapital, D., Kuan, J.-C., Rothstein, S., Mullen, R., Browse, J. and Dyer, J. (2006a) A new magic bullet? Type-2 diacylglycerol acyltransferases are key components to novel fatty acid accumulation in transgenic systems. In *17th International Symposium on Plant Lipids*, East Lansing, MI (abstract).

Shockey, J.M., Gidda, S.K., Chapital, D.C., Kuan, J.-C., Dhanoa, P.K., Bland, J.M., Rothstein, S.J., Mullen, R.T. and Dyer, J.M. (2006b) Tung tree DGAT1 and DGAT2 have nonredundant functions in triacylglycerol biosynthesis and are localized to different subdomains of the endoplasmic reticulum. *Plant Cell*, **18**, 2294–2313.

Siloto, R.M.P., Findlay, K., Lopez-Villalobos, A., Yeung, E.C., Nykiforuk, C.L. and Moloney, M.M. (2006) The accumulation of oleosins determines the size of seed oil bodies in *Arabidopsis*. *Plant Cell*, **18**, 1961–1974.

Simonsen, N.R., Navajas, J.F.-C., Martin-Moreno, J.M., Strain, J.J., Huttunen, J.K., Martin, B.C., Thamm, M., Kardinaal, A.F., Veer, P.v.t., Kok, F.J. and Kohlmeier, L. (1998) Tissue stores of individual monounsaturated fatty acids and breast cancer: the EURAMIC study. *Am. J. Clin. Nutr.*, **68**, 134–141.

Slack, C.R., Roughan, P.G. and Balasingham, N. (1978) Labelling of glycerolipids in the cotyledons of developing oilseeds by [1-^{14}C]acetate and [2-^{3}H]glycerol. *Biochem. J.*, **170**, 421–433.

Slack, C.R., Roughan, P.G., Browse, J.A. and Gardiner, S.E. (1985) Some properties of choline phosphotransferase from developing safflower cotyledons. *Biochim. Biophys. Acta.*, **833**, 438–448.

Slocombe, S.P., Piffanelli, P., Fairbairn, D., Bowra, S., Hatzopoulos, P., Tsiantis, M. and Murphy, D.J. (1994) Temporal and tissue-specific regulation of a *Brassica napus* stearoyl-acyl carrier protein desaturase gene. *Plant Physiol.*, **104**, 1167–1176.

Somerville, C., Browse, J., Jaworski, J.G. and Ohlrogge, J.B. (2000) Lipids, in *Biochemistry and Molecular Biology of Plants* (eds B.B. Buchanan, W. Gruissem and R.L. Jones), American Society of Plant Physiologists, Rockville, Maryland, pp. 456–527.

Spiller, G.A., Jenkins, D.J., Cragen, L.N., Gates, J.E., Bossello, O., Berra, K., Rudd, C., Stevenson, M. and Superko, R. (1992) Effects of a diet high in monounsaturated fat from almonds on plasma cholesterol and lipoproteins. *Am. J. Clin. Nutr.*, **11**, 126–130.

Ståhl, U., Carlsson, A., Lenman, M., Dahlqvist, A., Huang, B., Bana, W., Bana, A. and Stymne, S. (2004) Cloning and functional characterization of a phospholipid:diacylglycerol acyltransferase from *Arabidopsis*. *Plant Physiol.*, **135**, 1324–1335.

Stobart, K., Mancha, M., Lenman, M., Dahlqvist, A. and S, S. (1997) Triacylglycerols are synthesised and utilized by transacylation reactions in microsomal preparations of developing safflower (*Carthamus tinctorius* L.) seeds. *Planta*, **203**, 58–66.

Stoltzfus, D.L., Fehr, W.R. and Welke, G.A. (2000) Relationship of elevated palmitate to soybean seed traits. *Crop Sci.*, **40**, 52–54.

Stymne, S. and Stobart, A.K. (1986) Biosynthesis of γ-linolenic acid in cotyledons and microsomal preparations of the developing seeds of common borage (*Borago officinalis*). *Biochem. J.*, **240**, 385–393.

Swarts, H., Schuurmans, S.F. and De Pont, J. (1990) Binding of unsaturated fatty acids to N+, K(+)-ATPase leading to inhibition and inactivation. *Biochim. Biophys. Acta*, **1024**, 32–40.

Taylor, D.C., Yan, Z., Kumar, A., Francis, T., Giblin, E.M., *et al.* (2009) Molecular modification of triacylglycerol accumulation by over-expression of DGAT1 to produce canola with increased seed oil content under field conditions. *Botany*, **87**, 533–543.

Thelen, J.J. and Ohlrogge, J.B. (2002) Metabolic engineering of fatty acid biosynthesis in plants. *Metabol. Eng.* **4**, 12–21.

Theret, N., Bard, J., Nuttens, M., Lecerf, J., Delbart, C., Romon, M., Salomez, J. and Fruchart, J. (1993) The relationship between the phospholipids fatty acid composition of red blood cells, plasma lipids, and apolipoproteins. *Metabolism*, **42**, 526–528.

Thoma, I., Loeffler, C., Sinha, A., Gupta, M., Steffan, B., Krischke, M., Roitsch, T. and Mueller, M. (2003) Cyclopentenone isoprostanes induced by reactive oxygen species trigger defense gene activation and phytoalexin accumulation in plants. *Plant J.*, **34**, 363–375.

Tocher, D.R., Leaver, M.J. and Hodgson, P.A. (1998) Recent advances in the biochemistry and molecular biology of fatty acyl desaturases. *Prog. Lipid Res.*, **37**, 73–117.

Triki, S., Ben Hamida, J. and Mazliak, P. (1998) About the reversibility of the cholinephosphotransferase in developing sunflower seed microsomes, in *Advances in Plant Lipid Research* (eds J. Sanchez, E. Cerda-Olmedo, and E. Martinze-Force), Univ. Sevilla, pp. 236–239.

Tumlinson, J. and Engelberth, J. (2008) Fatty acid derived signals that induce or regulate plant defenses against herbivory, in *Induced Plant Resistance to Herbivory* (ed. A. Schaller), Springer, Amsterdam, pp. 389–407.

Turkish, A.R., Henneberry, A.L., Cromley, D., Padamsee, M., Oelkers, P., Bazzi, H., Christiano, A.M., Billheimer, J.T. and Sturley, S.L. (2005) Identification of two novel human acyl-CoA wax alcohol acyltransferases: members of the diacylglycerol acyltranserase 2 (DGAT2) gene superfamily. *J. Biol. Chem.*, **280**, 14755–14764.

Tzen, J.T.C., Lai, Y.K., Chan, K.L. and Huang, A.H.C. (1990) Oleosin isoforms of high and low molecular weights are present in the oil bodies of diverse seed species. *Plant Physiol.*, **94**, 1282–1289.

Upchurch, R. (2008) Fatty acid unsaturation, mobilization, and regulation in the response of plants to stress. *Biotechnol. Lett.*, **30**, 967–977.

USDA (2009) *Oilseeds: World Markets and Trade, FOP 10-09* (ed. F.A. Service), Washington, DC.

van de Loo, F.J., Fox, B.G. and Somerville, C. (1993) Unusual fatty acids, in *Lipid Metabolism in Plants* (ed. J.T.S. Moore), CRC Press, Boca Raton, pp. 91–126.

van de Loo, F.J., Broun, P., Turner, S. and Somerville, C. (1995) An oleate 12-hydroxylase from *Ricinus communis* L. is a fatty acyl desaturase homolog. *Proc. Natl. Acad. Sci. USA*, **92**, 6743–6747.

Voelker, T.A., Worell, A.C., Anderson, L., Bleiaum, J., Fan, C., Hawkins, D.J., Radke, S.E. and Davies, H.M. (1992) Fatty acid biosynthesis redirected to medium chains in transgenic oilseed plants. *Science*, **257**, 72–74.

Voelker, T.A., Hayes, T.R., Cranmer, A.M., Turner, J.C. and Davies, H.M. (1996) Genetic engineering of a quantitative trait: metabolic and genetic parameters influencing the accumulation of laurate in rapeseed. *Plant J.*, **9**, 229–241.

Voelker, T. and Kinney, A. (2001) Variations in the biosynthesis of seed-storage lipids. *Ann. Rev. Plant Physiol. Plant Mol. Biol.*, **52**, 335–361.

Vogel, G. and Browse, J. (1996) Cholinephosphotransferase and diacylglycerol acyltransferase: substrate specificities at a key branch point in seed lipid metabolism. *Plant Physiol.*, **110**, 923–931.

Vyacheslav, A., Nikolai, B., Natalia, P., Anita, B., Joseph, D., *et al.* (2009) Tobacco as a production platform for biofuel: overexpression of *Arabidopsis DGAT* and *LEC2* genes increases accumulation and shifts the composition of lipids in green biomass. *Plant Biotechnol. J.*, **8**, 1–11.

Wallis, J.G. and Browse, J. (2010) Lipid biochemists salute the genome. *Plant J.*, **61**, 1092–1106.

Wang, C., Chin, C.-K., Ho, C.-T., Hwang, C.-F., Polashoc, J.J. and Martin, C.E. (1996) Changes of fatty acids and fatty acid-derived flavor compounds by expressing the yeast Δ-9 desaturase gene in tomato. *J. Agric. Food Chem.*, **44**, 3399–3402.

Wang, H.W., Zhang, B., Hao, Y.J., Huang, J., Tian, A.G., Liao, Y., Zhang, J.S. and Chen, S.Y. (2007a) The soybean Dof-type transcription factor genes, *GmDof4* and *GmDof11*, enhance lipid content in the seeds of transgenic *Arabidopsis* plants. *Plant J.*, **52**, 716–729.

Wang, H.Y., Guo, J.H., Lambert, K.N. and Lin, Y. (2007b) Developmental control of *Arabidopsis* seed oil biosynthesis. *Planta*, **226**, 773–783.

Warner, K., Orr, P. and Glynn, M. (1997) Effect of fatty acid composition of oils on flavor and stability of fried foods. *J. Am. Oil Chem. Soc.*, **74**, 347–356.

Weselake, R.J., Taylor, D.C., Rahman, M.H., Shah, S., Laroche, A., McVetty, P.B.E. and Harwood, J.L. (2009) Increasing the flow of carbon into seed oil. *Biotechnol. Adv.*, **27**, 866–878.

Wiberg, E., Banas, A. and Stymne, S. (1997) Fatty acid distribution and lipid metabolism in developing seeds of laurate-producing rape (*Brassica napus* L.). *Planta*, **203**, 341–348.

Wilson, R., Calvo, A., Chang, P. and Keller, N. (2004) Characterization of the *Aspergillus parasiticus* Δ12-desaturase gene: a role for lipid metabolism in the *Aspergillus*-seed interaction. *Microbiology*, **150**, 2881–2888.

Wilson, R.F. and Hildebrand, D. (2010) Engineering status, challenges and advantages of oil crops. In *Plant Biotechnology for Sustainable Production of Energy and Coproducts* (eds P. Mascia, J. Scheffran, S. Thomas and J.M. Widholm), Springer, New York (in press).

Xia, Y., Gao, Q.-M., Yu, K., Lapchyk, L., Navarre, D., Hildebrand, D., Kachroo, A. and Kachroo, P. (2009) An intact cuticle in distal tissues is essential for the induction of systemic acquired resistance in plants. *Cell Host Microbe*, **5**, 151–165.

Yadav, N.S., Wierzbicki, A., Aegerter, M., Caster, C.S., Perez-Grau, L., Kinney, A.J., Hitz, W.D., Booth, J.R.J., Schweiger, B. and Stecca, K.L. (1993) Cloning of higher plant omega-3 fatty acid desaturases. *Plant Physiol.*, **103**, 467–476.

Yamauchi, Y., Furutera, A., Seki, K., Toyoda, Y., Tanaka, K. and Sugimoto, Y. (2008) Malondialdehyde generated from peroxidized linolenic acid causes protein modification in heat-stressed plants. *Plant Physiol. Biochem.*, **46**, 786–793.

Yamori, Y., Nara Y., Tsubouchi T., Sogawa Y, Ikeda K, Horie R. (1986) Dietary prevention of stroke and its mechanism in stoke-prone spontaneously hypertensive rats – preventive effect of dietary fibre and palmitoleic acid. *J. Hypertens.*, **4**, S449–452.

Yang, B. and Kallio, H.P. (2001) Fatty acid composition of lipids in sea buckthorn (*Hippophaë rhamnoides* L.) berries of different origins. *J. Agric. Food Chem.*, **49**, 1939–1947.

Yuan, L., Voelker, T.A. and Hawkins, D.J. (1995) Modification of the substrate specificity of an acyl-acyl carrier protein thioesterase by protein engineering. *Proc. Nat. Acad. Sci. USA*, **92**, 10639–10643.

Zhang, F.-Y., Yang, M.-F. and Xu, Y.-N. (2005) Silencing of *DGAT1* in tobacco causes a reduction in seed oil content. *Plant Sci.*, **169**, 689–694.

Zhang, M., Fan, J., Taylor, D.C. and Ohlrogge, J.B. (2009) DGAT1 and PDAT1 Acyltransferases have overlapping functions in *Arabidopsis* triacylglycerol biosynthesis and are essential for normal pollen and seed development. *Plant Cell*, **21**, 3885–3901.

Zhou, X.-R., Singh, S. and Green, A. (2008) Increased accumulation of epoxy fatty acids in Arabidopsis by transgenic expression of TAG assembly genes from *Bernardia pulchella*. In *18th International Symposium on Plant Lipids*, Bordeaux, France (abstract).

Zou, J.T., Katavic, V., Giblin, E.M., Barton, D.L., MacKenzie, S.L., Keller, W.A., Hu, X. and Taylor, D.C. (1997) Modification of seed oil content and acyl composition in the *Brassicaceae* by expression of a yeast *sn*-2 acyltransferase gene. *Plant Cell*, **9**, 909–923.

Zou, J., Wei, Y. and Taylor, D.C. (1999) The *Arabidopsis thaliana* TAG1 mutant has a mutation in a diacylglycerol acyl transferase gene. *Plant J.* **19**, 645–654.

3

Symbiotic Nitrogen Fixation

Hiroshi Kouchi

3.1 Nitrogen Fixing Organisms and the Nitrogenase System

3.1.1 Nitrogen Fixing Microorganisms

Nitrogen for all living organisms ultimately comes from atmospheric dinitrogen (N_2). Biological nitrogen fixation is a reaction whereby N_2 is reduced to ammonium (NH_3) by a special enzymatic system, of which the primary enzyme is the nitrogenase complex. The nitrogenase system is found in various prokaryotes (diazotrophic bacteria or diazotrophs). They can be grouped into two categories: one is the group of free-living nitrogen fixers which inhabit the soil, fresh water and seawater, and the other is a group of bacteria that exhibits a highly efficient nitrogen fixation system in endosymbiotic association with plants. Symbiotic nitrogen fixation in legumes is the prominent representative of the second group. Legume plants form root nodules by symbiosis with a group of soil bacteria collectively termed *Rhizobium*, in which endosymbiotic rhizobia are capable of fixing N_2. Nitrogen fixation by the legume-*Rhizobium* symbiosis accounts for more than 50% of biological nitrogen fixation and has a critical importance in agriculture. Some kinds of symbiotic nitrogen fixers are able to fix nitrogen in their free-living state, and such free-living nitrogen fixing bacteria can reside (not intracellularly) inside the tissues of higher plants. For example, some kinds of endophytic bacteria, which inhabit the intercellular space and/or the vascular cylinders of plants, fix N_2, and are thought to contribute nitrogen nutrition to these plants. In addition, various free-living nitrogen-fixing bacteria have been identified from the rhizosphere of many plants, and make a loose symbiotic association with those plants. Representatives of nitrogen-fixing microorganisms are shown in Table 3.1.

Plant Metabolism and Biotechnology, First Edition. Edited by Hiroshi Ashihara, Alan Crozier, and Atsushi Komamine.

Table 3.1 Representatives of nitrogen-fixing microorganisms

Free-living diazotrophs	
Group	Representative genus
Archaebacteria	*Methanosarcia, Methanococcus*
Obligate anaerobes	*Clostridium, Desulphovibrio*
Facultative anaerobes	*Klebsiella, Erwinia, Enterobacter*
Microaerobes	*Azospirillum, Arthrobacter*
Aerobic bacteria	*Azotobacter, Derxia, Acetobacter*
Photosynthetic bacteria	*Rhodospirillum, Chromatium*
Cyanobacteria	*Nostoc, Anabaena, Calothrix*
Symbiotic nitrogen-fixers	
Group	Representative genus
Rhizobia and plants (legumes)	*Rhizobium, Sinorhizobium, Bradyrhizobium, Azorhizobium, Mesorhizobium*
Actinomyces and plants	*Frankia*
Cyanobacteria and plants	*Anabaena, Nostoc*

3.1.2 The Nitrogenase Complex

The nitrogenase complex catalyzes the following reactions;

$$\begin{array}{lllllllll} \text{I.} & N_2 + & 12ATP + & 6e^- + & 6H^+ = & 2NH_3 + & & 12ADP + & 12Pi \\ \text{II.} & & 4ATP + & 2e^- + & 2H^+ = & & H_2 + & 4ADP + & 4Pi \\ \hline & N_2 + & 16ATP + & 8e^- + & 8H^+ = & 2NH_3 + & H_2 + & 16ADP + & 16Pi \end{array}$$

Nitrogenase reactions consume large amounts of ATP to cleave the highly stable triple covalent bond between two nitrogen atoms. The nitrogenase reaction is accompanied by formation of H_2 by proton reduction (reaction II); at least 25% of the reducing potential is consumed to produce H_2. The exact energy cost of N_2 reduction *in vivo* cannot be easily estimated, since the stoichiometry of reactions I and II varies according to cellular conditions such as cytoplasmic pH, and ATP/ADP ratio.

The nitrogenase complex has two components, an MoFe protein (component I, dinitrogenase) and an Fe protein (component II, dinitrogenase reductase). The MoFe protein is a tetramer ($\alpha_2\beta_2$) of MW = 220,000 Da, and the Fe protein is a homodimer of MW = 70,000 Da. Reduction of N_2 to NH_3 occurs on molybdenum (Mo) in the iron-molybdenum cofactor (FeMo-cofactor, MW = *ca.* 5000 Da), the active site binding to α subunits of the MoFe protein, although the exact reaction mechanism from N_2 to NH_3 on the FeMo-cofactor has not been completely determined (see Rees and Howard, 2000). Molybdenum in the active site is known to be substituted by vanadium (V) or iron (Fe) in some kinds of *Azotobacter* species (Hales, 1990). In many diazotrophs, electron transfer through ferredoxin and/or flavodoxin reduces dinitrogenase reductase (Fe protein) which in turn reduces

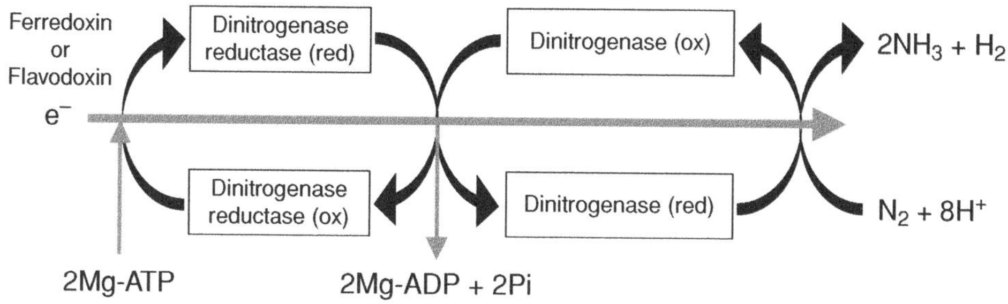

Figure 3.1 *Scheme of the reaction of the nitrogenase complex*

dinitrogenase (MoFe protein), and ATP hydrolysis is coupled to reduction of dinitrogenase by dinitrogenase reductase (Figure 3.1).

The structure and function of the two nitrogenase components are highly conserved among the diazotrophs. The nitrogenase system is thought to have been established in a very early era of the evolution of life, when the Earth was anaerobic, and as a consequence the nitrogenase proteins are highly O_2-sensitive (i.e. inhibited irreversibly in aerobic conditions). Induction of the genes coding the nitrogenase complex also requires anaerobic conditions. Nitrogenase can reduce and/or bind various kinds of double and/or triple bonds other than N_2. Among them, the reduction of acetylene (C_2H_2) to ethylene (C_2H_4) is of practical importance as a simple detection method for nitrogenase activity. It should be noted, however, that C_2H_2 reduction activity does not necessarily represent N_2 reduction activity *in vivo*.

3.1.3 *Nif* Genes and Regulation

The *Nif* genes which encode the nitrogenase complex and related enzymes are structurally well conserved in the diazotrophs. The structure and function of *Nif* gene clusters have been well-studied in the free-living nitrogen-fixing bacteria, *Klebsiella pneumoniae* (Figure 3.2) and *Azotobacter vinelandii* (Arnold *et al.*, 1988; Jacobson *et al.*, 1989). *NifH*, *D* and *K* are the nitrogenase structural genes encoding an Fe protein subunit, and the α and β subunits

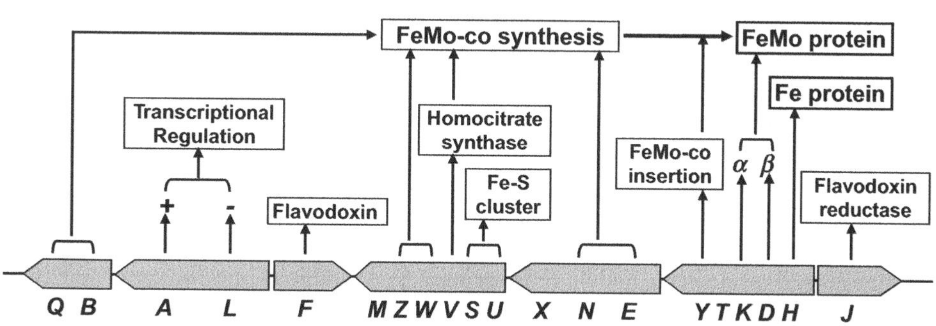

Figure 3.2 *Structure and functions of the* Nif *regulon of* Klebsiella pneumoniae

of the MoFe protein, respectively. *NifE, Y, N, B* and *V* are involved in biosynthesis of the FeMo-cofactor. *NifS* and *U* are involved in the formation of the Fe-S clusters required for electron transfer between Fe and MoFe proteins. These *Nif* genes are clustered in a few operons, which are regulated positively by NifA and negatively by NifL. The NifA protein is a transcriptional activator of other *Nif* operons, and the NifL protein modulates NifA activity in response to the presence of combined nitrogen sources and molecular oxygen (Schmitz *et al.*, 2002).

Although *Nif* genes essential for assembly of the nitrogenase complex have highly consensus sequences among different diazotrophs, regulation mechanisms of the *Nif* genes differ between organisms (see Dixon and Kahn, 2004). In some kinds of aerobic diazotrophs, including *Rhizobium* bacteria, *Fix* genes are also involved in nitrogen fixation and its regulation. Most notably, a two-component system composed of FixL, a low O_2 concentration sensor localised in the envelopes of bacterial cells, and FixJ, a regulator protein in the cytoplasm, plays a key role in the activation of NifA and FixK, which in turn activate transcription of the downstream *Nif* and *Fix* gene operons (Fischer, 1994; Miyatake *et al.*, 1999). However, details of regulatory cascades of *Nif* gene induction and the exact roles of individual *Fix* genes in *Rhizobium* bacteria remain to be solved, because most rhizobia species exhibit nitrogenase activity only under symbiotic conditions with their corresponding host legumes, and the mechanisms underlying symbiotic nitrogen fixation are not yet fully understood.

In obligate aerobes, coincidental nitrogen fixation and aerobic respiration appears to be paradoxical. In *Azotobacter* species, nitrogen fixation and aerobic respiration is thought to be circumvented by a so-called 'respiratory protection'. This 'uncoupled' respiration ensures rapid consumption of ambient O_2 at the cell surface to keep intracellular free O_2 concentration very low, and hence protects the nitrogenase system from irreversible O_2 damage (Poole and Hill, 1997). Rhizobia circumvent oxygen damage to the nitrogenase system through symbiosis with legumes, where they inhabit a specialised symbiotic organ, the root nodule (see later sections).

3.2 Symbiotic Nodule Formation in Legume Plants

3.2.1 Infection and the Nodulation Process

Interactions between legume plants and *Rhizobium* bacteria are generally initiated from the colonisation of rhizobia on growing root hairs, where colonies of rhizobia induce characteristic morphological changes in root hairs called root hair curling. Rhizobia invade the root hair cells by inducing the production of an infection thread, which is a tubular structure that fills with rapidly proliferating rhizobia. Infection threads grow and successively penetrate and ramify through root cortical cells. Concomitantly with infection thread initiation, cell division is induced in the root cortex to form a nodule primordium, which develops into a symbiosis-specific organ, the root nodule. Rhizobia are released by endocytosis from infection threads and reside in the nodule cells enclosed by a symbiosis-specific membrane (the peribacteroid membrane, PBM), which is derived from the plant cell plasma membrane. The structure of a PBM-enclosed *Rhizobium* is similar to a cell organelle, and is termed 'symbiosome', and the PBM is thus termed the 'symbiosome membrane' (SM). Rhizobia in

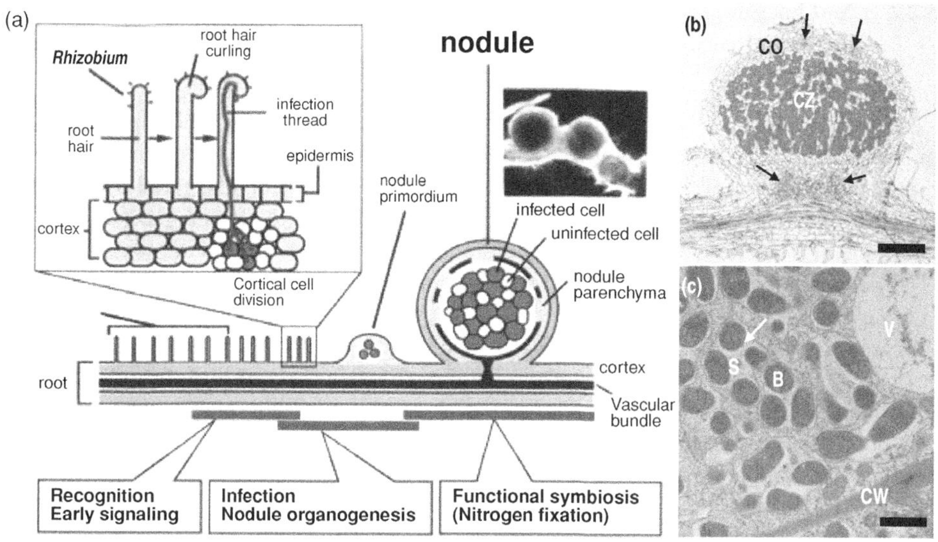

Figure 3.3 The nodulation process and the structure of the mature nodule of Lotus japonicus. (a) A schematic illustration of rhizobial infection and the nodulation process (Modified with permission from Kawaguchi et al. (2002) Copyright 2002 The American Phytopathological Society). (b) Light micrograph of a nodule section. CZ, infected zone; CO, cortex. Vascular bundles are indicated by arrows. Bar = 200μm. (c) Electron micrograph of infected cells. B, bacteroid; S, symbiosome; CW, cell wall; V, vacuole. Peribacteroid membrane is indicated by an arrow. Bar = 2μm

the symbiosomes differentiate into 'bacteroids', which are distinct from free-living rhizobia in both morphological and physiological characteristics, and finally N_2 fixation takes place (Figure 3.3).

Nodule ontogeny has evolutionarily diverged into two types among legume species. In most temperate legumes, such as alfalfa, pea and vetch, cell division that forms nodule primordia is initiated at the inner cortex near the endodermis, while in most tropical legumes, such as soybean and the common bean, nodule primordia are initiated at the outer cortex just beneath the epidermis. The former develops indeterminate nodules, which have a permanent meristem at the tip of an elongated structure, whereas the latter forms determinate nodules, which have a spherical shape and where meristematic activity is restricted to the early stages of nodule formation. Despite these differences in nodule ontogeny and morphology, the structural organisation inside the nodule is essentially the same between these two types; a central zone contains bacterial infected cells and is surrounded by a nodule cortex in which vascular bundles develop from the root vascular cylinders. Light and electron micrographs of *Lotus japonicus* nodule sections are shown in Figure 3.3. With only a few exceptions, in most legume species the nodule central zone contains both bacterial infected cells and uninfected cells, and there exists a differentiation of metabolic function between these two cell types.

3.2.2 Nod Factors, Early Symbiotic Signalling and the Rhizobial Infection Process

Interactions between legume plants and *Rhizobium* bacteria are controlled genetically by a strict species–species specificity with few exceptions. Individual rhizobial species infect only a limited range of host legumes. This specificity (host specificity) is primarily determined by lipochitin oligosaccharide signal molecules (Nod factors) secreted by rhizobia in response to plant signals (mainly flavonoids/isoflavonoids). Nod factors have a common backbone of *N*-acetyl glucosamine tetramer or pentamer, of which the non-reducing end is *N*-acylated with a $C_{16\text{-}18}$ fatty acid and the reducing end is decorated by various molecules (Lerouge *et al.*, 1990; Spaink *et al.*, 1991). These modifications and a number of *N*-acetyl glucosamine residues in the backbone determine host specificity (see Cohn *et al.*, 1998, for review). In the decade after the first identification of Nod factor structures from *Sinorhizobium meliloti*, a microsymbiont of *Medicago* (Lerouge *et al.*, 1990), biochemical functions of a number of rhizobial nodulation (*Nod*) genes have been demonstrated (Figure 3.4, see Spaink, 1995). Recent studies are, however, revealing complex regulatory mechanisms of bacterial *Nod* genes, as well as the presence of bacterial factors involved in symbiosis, other than Nod factors, including exo- and/or lipo-polysaccharides, in particular regarding suppression of plant defence responses leading to the establishment of successful infection (Gibson *et al.*, 2008; Deakin and Broughton, 2009).

Nod factors elicit both rhizobial infection and initiation of nodule primordia, as has been well demonstrated for various legume species using purified and/or chemically synthesized

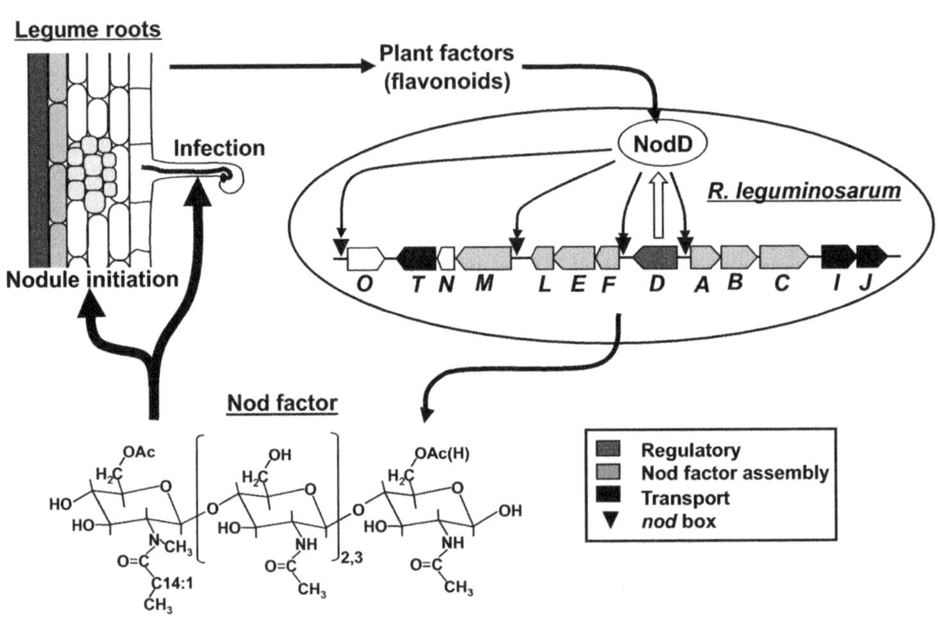

Figure 3.4 Molecular basis of the mutual recognition and intial interactions between legume host and Rhizobium. *The* Nod *genes and structure of Nod factor are shown for* R. leguminosarum *bv.* viciae

Nod factors (Stokkermans and Peters, 1994; Minami *et al.*, 1996; Niwa *et al.*, 2001). The host legume mechanisms involved in Nod factor perception and subsequent symbiotic signal transduction have been mostly investigated by means of molecular genetics using two model legume species, *Lotus japonicus* and *Medicago truncatula*. On the basis of the resources established for genome research in these model legumes, a number of host legume genes essential for symbiosis have been identified.

Lysin motif receptor kinases (LysM-RKs), NFR1 and NFR5 in *L. japonicus*, and LYK3 and NFP1 from *M. truncatula*, have all been identified as Nod factor receptors (Radutoiu *et al.*, 2003; Madsen *et al.*, 2003; Limpens *et al.*, 2003; Arrighi *et al.*, 2006). The LysM receptor domain is known to be responsible for chitin binding. In *L. japonicus*, NFR1 and NFR5 are supposed to make a receptor complex responsible to specific recognition of Nod factors secreted from *Mesorhizobium loti*, since their co-transformation has been shown to be able to extend the host range of *M. loti* to the heterologous plant species (Radutoiu *et al.*, 2007). The earliest and most prominent physiological response of legume plants to Nod factors is Ca^{2+} spiking, which is an oscillation of Ca^{2+} concentration in the perinuclear region of root hair cells (Ehrhardt *et al.*, 1996). Several genes downstream of the Nod factor receptors have been shown to be required for generating Ca^{2+} spiking, and are also essential for symbiosis. SYMRK/DMI2 is an LRR (leucine-rich repeat) type receptor kinase (Stracke *et al.*, 2002; Endre *et al.*, 2002), and CASTOR and POLLUX/DMI1 are potassium channel proteins (Ane *et al.*, 2004; Imaizumi-Anraku *et al.*, 2005; Charpentier *et al.*, 2008). NUP133 and NUP85 are components of the nucleoporin complex (Kanamori *et al.*, 2006; Saito *et al.*, 2007). Exact biochemical functions of these proteins in generating Ca^{2+} spiking are still to be elucidated. CCaMK/DMI3 is a Ca^{2+}- and calmodulin (CaM)-dependent protein kinase present in the nucleus, and is a putative decoder of Ca^{2+} signals. Its loss-of-function results in a defect in nodulation, while its gain-of-function mutation (resulting in a Ca^{2+} independently active form) induces spontaneous nodulation in the absence of rhizobia, suggesting a pivotal role of CCaMK in early symbiotic signalling cascade(s) leading to nodulation (Levy *et al.*, 2004; Tirichine *et al.*, 2006; Gleason *et al.*, 2006). CYCLOPS/IPD3 is a putative transcription factor and shown to interact directly with CCaMK *in planta* and is phosphorylated by CCaMK *in vitro* (Messinese *et al.*, 2007; Yano *et al.*, 2008). It is worth noting that these proteins, except the LysM-RKs, are also required for plant-fungus endosymbiosis (i.e. mycorrhization), which makes benefits for the host plants to be supplied efficiently with nutrients such as phosphorus from the soil. Thus, these genes are termed 'common symbiosis genes' and the signalling pathway mediated by these genes is termed the 'common symbiosis pathway' (CSP). Since the mycorrhizal symbiosis is widely distributed in the plant kingdom and has its evolutionary origin far preceding the legume-*Rhizobium* symbiosis, it is very likely that legumes acquired the ability of symbiosis with rhizobia on the basis of pre-existing mechanisms for mycorrhizal symbiosis. Indeed, these gene sets are also present in non-legume plants that have mycorrhizal symbiosis, and have been shown to be essential for symbiosis with mycorrhizae. In addition, at least some of them have been demonstrated to be functionally conserved between legume and non-legume plant species (Godfroy *et al.*, 2006; Chen *et al.*, 2007; Markmann *et al.*, 2008; Banba *et al.*, 2008).

Downstream of the above-mentioned genes, a few genes have been isolated which are required for rhizobial infection and/or nodulation processes. NIN, NSP1 and NSP2 have been shown to be required for rhizobial infection and also for the induction of cortical cell division leading to nodule primordium formation (Schauser *et al.*, 1999; Heckmann *et al.*,

2006; Murakami *et al.*, 2006). From their predicted sequences, these proteins are thought to function as transcription factors. The rhizobial infection process and nodule organogenesis are governed by genetically distinct programs, even though their coordinated progression is essential for the development of effective symbiosis. This has been shown by gain-of-function mutations of CCaMK or LHK1 (a cytokinin receptor kinase: Tirichine *et al.*, 2007) which lead to spontaneous nodule organogenesis in the absence of rhizobia, while a loss-of-function of LHK1 leads to formation of excessive infection threads penetrating into the root cortex without inducing timely cortical cell division after rhizobial inoculation (Murray *et al.*, 2007). In contrast, the loss-of-function mutants of CCaMK exhibit defects in initiation of both infection threads and nodule primordia when inoculated with rhizobia (Tirichine *et al.*, 2006), indicating that CCaMK, different from LHK1, is essential for both infection and nodule organogenesis. Genetic analyses of *Lotus* and *Medicago* symbiotic mutants strongly suggest that an additional signalling pathway(s) is derived from LysM-RKs, separately from the pathway involving Ca^{2+} spiking (CSP), and is required for successful rhizobial infection (Figure 3.5) (Radutoiu *et al.*, 2003; Geurts *et al.*, 2005; Smit *et al.*, 2007). This additional pathway has been suggested to involve Ca^{2+} influx into root hair cells, which is observed prior to Ca^{2+} spiking within a few minutes after Nod factor

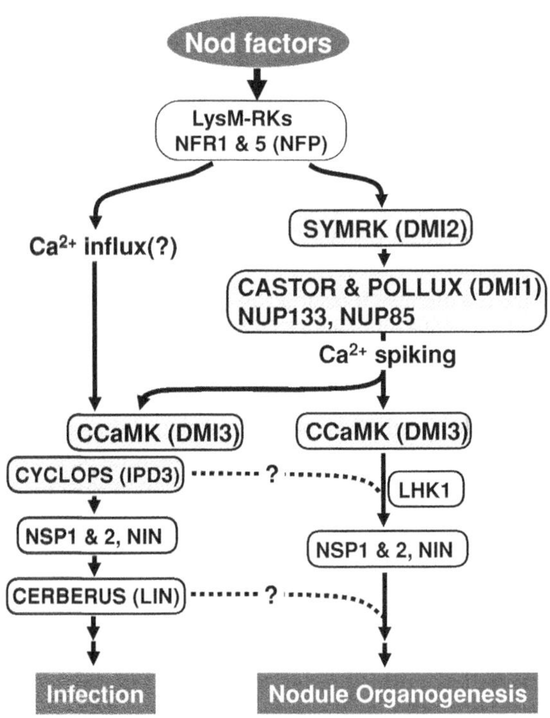

Figure 3.5 *Proposed model for early symbiotic signalling. Light-shaded genes are required for both rhizobial and mycorrhizal symbiosis. Gene names are from* L. japonicus, *and their orthologues in* M. truncatula *are shown in parentheses. Modified with permission from Hayashi* et al. *(2010) Copyright 2010 John Wiley & Sons, Ltd*

application (Miwa *et al.*, 2006). Although no gene has been attributed to this pathway so far, the assumption is supported by the fact that plants with a mutation in the genes on CSP exhibit more or less significant root hair swelling and/or deformation, which precede and are prerequisite for infection thread initiation, whereas a defect in LysM-RKs shows no such response at all. Recent evidence through a genetic approach also strongly suggests the operation of a symbiotic signalling pathway distinct from CSP with regard to successful rhizobial infection (Hayashi *et al.*, 2010; Madsen *et al.*, 2010). One of the functions of CCaMK (or more probably the complex with other proteins including CYCLOPS) could be the integration of these two pathways. It should be noted that rhizobial infection events occur in the root epidermis, while nodule organogenesis is initiated in the root cortex. Thus spatio-temporally regulated activation of CCaMK may play a key role in the coordination of the infection process and nodule primordium initiation.

CERBERUS recently isolated from *L. japonicus* (Yano *et al.*, 2009) and its orthologue in *M. truncatula*, *LIN* (Kiss *et al.*, 2009), as well as *CYCLOPS*, are identified as the genes essential for the infection process, that is, the initiation and/or growth of infection threads. CERBERUS/LIN is postulated to be an E3 ubiquitin ligase because of the presence of a U-box domain which is known to be involved in interaction with the E2 ubiquitin-conjugating enzyme. The defects in these genes result in not only a defect in infection thread formation but also the arrest of nodule organogenesis at a stage of small bumps without endosymbiotic rhizobia. However, when these mutants were transformed with a gain-of-function CCaMK, they were able to form genuine nodule structures spontaneously, even at very low frequencies, indicating that these genes play primary roles in infection thread formation and/or its growth but not for the nodule organogenesis program itself. Incomplete nodule organogenesis in these mutants is thought to be an indirect consequence of the failure in rhizobial infection (Yano *et al.*, 2009). However, despite the defect in infection thread formation in the *cerberus* mutant, the frequency of nodule primordium formation in the mutants transformed by a gain-of-function CCaMK is significantly reinforced by inoculation with *M. loti*, suggesting possible involvement of infection signals through the additional pathway(s) other than CSP, as described above, in nodule initiation (Yano *et al.*, 2008, 2009). Thus the interdependency of rhizobial infection process and nodule organogenesis is still an open question with regard to early symbiotic signalling cascades triggered by Nod factors.

NIN is induced exclusively in response to Nod factors, and has been shown to be required for both nodulation and infection: the *nin* mutant displays excess root hair deformation in response to rhizobial inoculation or Nod factor application, but neither infection thread formation nor nodule primordium formation were induced (Schauser *et al.*, 1999). In contrast, mutants defective in *NSP2* exhibit no obvious root hair responses to Nod factors, even though NSP2 is located downstream of Ca^{2+} spiking (Murakami *et al.*, 2006). Based on phenotypic analyses of the mutants including symbiosis-specific gene induction, together with genetic analyses, *CERBERUS/LIN* is positioned downstream of *NSP1*, *NSP2* and *NIN*, and they are all downstream of *CYCLOPS*, but their exact epistatic relationships and possible interactions with each other are still obscure. A current working model of early symbiotic signalling is illustrated in Figure 3.5. The host mechanisms underlying infection thread formation are still unknown, but possibly CERBERUS/LIN functions in the very early steps of infection thread initiation (Yano *et al.*, 2009). In addition, rearrangement of microtubules and actin filaments in root hair cells has also been shown to be prerequisite for infection thread formation. Mutations in *NAP1* or *PIR1*, which are both involved in actin cytoskeleton organisation, have demonstrated that infection thread formation and its growth are dependent on actin cytoskeleton rearrangement in root hair cells (Yokota *et al.*, 2009). It has also been shown that rearrangement of microtubule dynamics occurs quickly in response to Nod factor application (Vassileva *et al.*, 2005).

Although recent progress in molecular genetic studies of model legumes has allowed us to identify a number of host genes involved in Nod factor perception and the immediate downstream symbiotic signalling pathways, our understanding remains, as a whole, at the level of identification of the individual host components. Major study topics in the near future will be epistatic relationships and/or possible interactions of these components, together with their exact biochemical functions, for reaching an understanding of symbiotic signalling cascades as well as the interrelationships between the rhizobial infection process and nodule organogenesis. More details of early steps in symbiotic signalling events can be found in recent reviews by Oldroyd and Downie (2008), Herder and Parniske (2009), Markmann and Parniske (2009), and Kouchi *et al.* (2010).

3.3 Mutual Interactions between Host Cells and Bacteroids in Legume Nodules

3.3.1 Differentiation of Rhizobia into Bacteroids

Induction of nitrogen-fixing activity of rhizobia is achieved by symbiotic association with their compatible legume host. In general, rhizobia do not undergo nitrogen fixation when they inhabit the soil in a free-living state. Rhizobia are released from infection threads into the host cells by endocytosis, and they are then present in the host cell cytoplasm enclosed by the PBM (symbiosome membrane), differentiate to a symbiosis-specific form, the bacteroid, and finally carry out efficient nitrogen-fixing activity (Figure 3.6). The PBM forms a physical and functional interface between the host cell cytoplasm and the microsymbiont. Bacteroids at full maturation cease cell division and PBM-enclosed bacteroids act like an intracellular organelle, which is ultimately fuelled by plant photosynthates and in turn provides NH_4^+ as a nitrogen nutrient for the host plant. The differentiation of rhizobia into bacteroids is under the control of complex interactions between host legume cells and intracellular bacteria. However, the mechanisms underlying differentiation of endosymbiotic rhizobia in symbiosomes to the bacteroid form are still largely unknown. Gene expression profiles, including specific induction of *Nif* and *Fix* genes, are quite different between bacteroids and free-living rhizobia, as demonstrated by transcriptome analysis (Uchiumi *et al.*, 2004).

Arrest of proliferation of endosymbiotic bacteria must be a prerequisite for their persistent symbiosis in the root nodules, otherwise rhizobia would spread throughout the nodule cells and tissues, and become parasitic. In those legumes forming indeterminate nodules, recent evidence indicates that rhizobial proliferation is terminated by DNA endoreduplication (polyploidy) triggered by host plant factors suggested as nodule-infected cell-specific cysteine-rich (NCR) peptides (Mergaert *et al.*, 2006; Van de Velde *et al.*, 2010). The NCR peptides are very similar to defensin-type antimicrobial proteins, and have been suggested to be targeted at the secretory pathway. *DNF1*, which encodes a component of the signal peptidase complex (Wang *et al.*, 2010), has been suggested to be involved in transport of

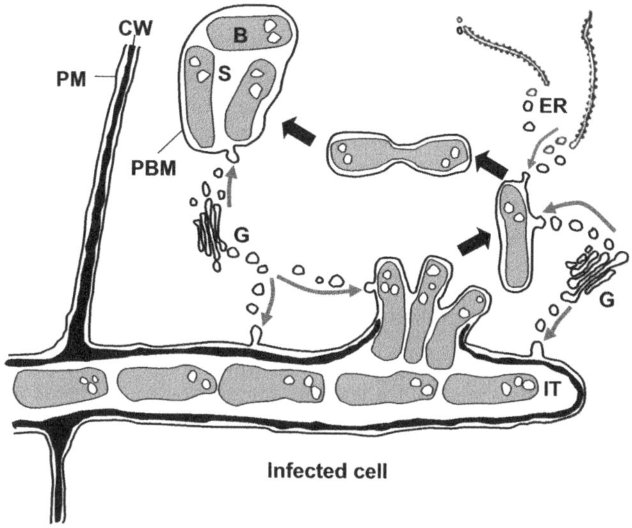

Figure 3.6 *Rhizobial release from infection threads into nodule infected cells and symbiosomes (Modified from Bergersen (1982) Copyright Research Studies Press/John Wiley & Sons, Ltd). IT, infection thread; S, symbiosome; B, bacteroid; CW, cell wall; PM, plasma membrane; G, Golgi apparatus; ER, endoplasmic reticulum; PBM, peribacteroid (symbiosome) membrane*

the NCR peptides into symbiosomes and then into bacteroids (Van de Velde *et al.*, 2010). Thus, it is likely that in indeterminate nodules, the host plants control the differentiation of rhizobia into bacteroids by targeting NCR peptides through the nodule-specific protein secretory system. From the bacterial side, the *BacA* gene identified from *S. meliloti* is essential for bacteroid development (Glazebrook *et al.*, 1993), and it has been proposed that it is involved in the uptake of host legume-derived peptides (Marlow *et al.*, 2009). These findings may provide an important clue to understanding the control of bacteroid differentiation by host-derived factors. In determinate nodules, however, the host mechanism controlling rhizobial proliferation is totally unknown. In *M. loti*, the microsymbiont of *Lotus* plants, *BacA* is also present in its genome, but its disruption does not affect bacteroid differentiation or symbiotic nitrogen fixation (Maruya and Saeki, 2010; see Mergaert *et al.*, 2006). In addition, a nodule-specific cysteine-rich protein family is not found in legumes forming determinate nodules.

A few host legume genes or loci have been suggested to be involved in bacteroid differentiation and/or maintenance as speculated from the phenotypes of Fix^- (forming ineffective nodules; see later sections) mutants of *L. japonicus*, but their exact functions are still obscure (Suganuma *et al.*, 2003; Hossain *et al.*, 2006; Kumagai *et al.*, 2007). In indeterminate nodules, a symbiosome generally contains a single bacteroid, while in determinate nodules it contains multiple bacteroids (sometimes a few tens of bacteroids in a single symbiosome). In *L. japonicus*, most of symbiosomes contain a single bacteroid in young nodules, but the number of bacteroids inside a symbiosome increases in parallel with the development of nodules. Thus it appears that bacterial proliferation continues, even if strictly controlled by interactions with the host cell, during determinate nodule growth, which is mainly accomplished by infected cell enlargement rather than new cells from a persistent meristem. Furthermore, terminally differentiated bacteroids in indeterminate nodules are functional but not viable, while the bacteroids in determinate nodules have been shown to survive reversibly, that is, reverting to the free-living state when released in the soil after senescence and subsequent disruption of nodules. Therefore, the mechanisms of bacteroid differentiation and the situation of bacteroids inside nodule cells may be considerably different between indeterminate and determinate nodules.

3.3.2 Protection of the Nitrogenase System from Oxygen in Root Nodules

As mentioned above, expression of the rhizobial *Nif* and *Fix* genes requires anaerobic conditions, and nitrogenase proteins are irreversibly inactivated by the presence of molecular oxygen. *Rhizobium* bacteria are obligate aerobes, and hence their highly energy-consuming nitrogenase activity depends on active aerobic respiration. Thus, one of the reasons rhizobia do not exhibit nitrogen fixation in the free-living state in soil is suppression due to the presence of oxygen. In legume nodules, this is overcome by the presence of O_2-binding nodule-specific proteins, leghaemoglobins (Lbs). Lbs are composed of multiple species of heme proteins closely related to each other, which more resemble mammal myoglobin rather than haemoglobin, and are found abundantly in nodule infected cells. Lbs have a high affinity for O_2 and maintain the free O_2 concentration in the infected cells at a very low level (<20 nM), while at the same time ensuring a large influx of oxygen to the bacteroids. Indeed, the knockdown of Lb gene expression by RNA interference (RNAi) in *L. japonicus* results in severe inhibition of nitrogen fixation (Ott *et al.*, 2005, 2009).

Another mechanism involved in the regulation of free O_2 in nodules is the 'oxygen diffusion barrier' presumably present in nodule parenchyma surrounding the infected zone. This was proposed on the basis of the measurement of the free-O_2 gradient in nodules by microelectrode as well as of gas diffusion properties (see Hunt and Layzell, 1993, for review). This diffusion barrier is shown to be variable physiologically, adapting to ambient O_2 concentrations and nitrogen fixation activity in the nodules. However, the 'diffusion barrier' hypothesis still lacks substantial evidence, although the apoplastic aqueous layer in the nodule parenchyma has been proposed to control O_2 diffusion into the central infected zone. Since Lbs oxygenated in the nodules under normal ambient O_2 conditions are estimated to be only around 20–30% of total Lb, surplus concentration of Lbs in the central infection zone may be an alternative interpretation of the variable function of the 'oxygen diffusion barrier', even though the rate of O_2 diffusion into the infected zone itself is an important factor limiting nitrogen-fixing activity.

Haemoglobins (Hbs) are now known to be widely present in the plant kingdom, not just in legume nodules, and they are divided into two major classes: class 1 Hbs and class 2 Hbs (Arredondo-Peter *et al.*, 1998; Hunt *et al.*, 2002). Representatives of the class 2 Hbs are leghaemoglobins in legume nodules, while class 1 Hbs have been identified in a wide variety of plants including non-legumes and legumes. Although those non-symbiotic Hbs have a high affinity to O_2, it is unlikely that they participate in oxygen transport because of their low dissociation constant with O_2 (Arredondo-Peter *et al.*, 1997). Rather, it has been proposed that they play roles in sensing low O_2 tension and/or scavenging nitric oxide (NO) generated when the plants are affected by stress conditions, including pathogen attack. It has been shown recently that class 1 Hbs are induced in the early steps of legume-*Rhizobium* interactions, and play a role in preventing legume plants from invoking defence reactions by themselves against rhizobial infection and invasion, by scavenging nitric oxide (NO) (Shimoda *et al.*, 2005, 2009). NO is also present in nodule infected cells and severely inhibits nitrogenase activity. It is worth noting that over-expression of non-symbiotic (class 1) Hbs in *L. japonicus* results in a significant increase in both nodule numbers and their nitrogen-fixing activity, suggesting a possible genetic improvement strategy to increase nitrogen fixation in legume nodules (Shimoda *et al.*, 2009).

3.3.3 Metabolite Exchange between the Plant Cell Cytosol and Bacteroids

Bacteroids are fuelled by plant photosynthates as their carbon and energy source, and in exchange supply NH_4^+ formed by fixation of atmospheric nitrogen to the host plants. For these mutual benefits, specialised metabolic machineries and metabolite exchange systems have evolved in both symbiotic partners. Carbon and nitrogen metabolism in nodules and their exchange between plant cell cytosol and bacteroids have been well-documented in many comprehensive reviews in the past decade (Tajima and Kouchi, 1996; Udvardi and Day, 1997; Lodwig and Poole, 2003; White *et al*; 2007; Prell and Poole, 2008). Therefore, this review will briefly summarise the basic concepts, with special emphasis on nutrient exchange between the two symbiotic partners across the symbiosome membrane. Carbon and nitrogen metabolism and their compartmentation in nodule host cells will also not be described in this review. For these aspects, refer to the review articles described above, together with Tajima *et al.* (2004) for ureide metabolism in nodules.

3.3.3.1 Dicarboxylates as Substrates for Bacteroid Respiration

One of the most characteristic features of metabolite exchange between the plant cell cytoplasm and bacteroids is that bacteroids utilise mainly C_4-dicarboxylates, such as malate and succinate, as their carbon source, but cannot utilise sugars; this is different from free-living rhizobia. This was first shown by the pioneering work of Tujimura and Meguro (1960), who observed that respiration of bacterial fractions prepared from soybean nodules is significantly enhanced by malate and succinate, but not by glucose and sucrose. Conclusive evidence has been given by the finding that bacterial mutants with no ability for dicarboxylate transport and/or their catabolism form ineffective (no nitrogen-fixation) nodules with no exception (Ronson *et al.*, 1981; Finan *et al.*, 1983). In contrast, rhizobial mutants with defects in sugar catabolism have been shown to form effective nodules (Ronson and Primrose, 1979; Lafontaine *et al.*, 1989). Structure and regulation of the genes for rhizobial dicarboxylate transporter (Dct) systems have been extensively studied (Ronson *et al.*, 1987; Jording *et al.*, 1993). Malate and/or succinate enter into the TCA cycle in bacteroids and acetyl-CoA is produced by NAD-dependent malic enzyme to form pyruvate, thus enabling continuous operation of the TCA cycle in bacteroids (Driscoll and Finan, 1993; Chen *et al.*, 1998).

3.3.3.2 Ammonium Export and Amino Acid Cycling

Nitrogen fixed by nitrogenase is excreted from bacteroids into the plant cell cytoplasm as ammonium ions (NH_4^+). The product of nitrogenase activity, NH_3, has been thought to diffuse passively across the bacteroid membranes, and then convert to NH_4^+ in the symbiosome space (the space between the symbiosome membrane and bacteroid) where it is relatively acidic (Udvardi and Day, 1997). Primary ammonium assimilation pathways (GS/GOGAT) as well as the ammonium uptake system appear to be almost shut down in bacteroids (Taté *et al.*, 1998).

However, the nutritional status for nitrogen in bacteroids is more complex. Amino acid cycling between the plant cell and bacteroid was first proposed by Kahn *et al.* (1985), who proposed the operation of a malate-aspartate shuttle, analogous to mitochondria. This model was based on consideration of the question of how bacteroids are able to continue nitrogen fixation even under the combined nitrogen-rich environment in the plant cytoplasm, despite the fact that many diazotrophs exhibit nitrogen fixation only when the available exogenous nitrogen is limited. When bacteroids are fuelled by N-containing compounds from the plant cell cytosol, of which biosynthesis is dependent on NH_4^+ exported from the N_2-fixing bacteroids, continuing N_2 fixation and exporting NH_4^+ could meet bacterial needs to increase carbon and energy supplied to it by the host plants. Following this influential suggestion by Kahn *et al.* (1985), the participation of a malate-aspartate shuttle has been suggested experimentally for pea bacteroids by enzymological studies (Appels and Haaker, 1991). In addition, bacteroids isolated anaerobically from soybean nodules are able to utilise glutamate as their carbon source to support respiration and nitrogenase activity, and when bacteroids are incubated with malate and glutamate, large amounts of aspartate and alanine are excreted in the incubation medium (Kouchi *et al.*, 1991). Since these early experiments were done *in vitro* using anaerobically-isolated bacteroids, it is still uncertain whether the malate-aspartate shuttle really operates between bacteroids and the host cell cytoplasm in exactly the same way as in mitochondria. It is, however, very likely that the supply of amino acids from plants to bacteroids is essential for nitrogen fixation, instead of a shutdown of primary ammonium assimilation systems in bacteroids. Indeed, aspartate aminotransferase in *S. meliloti* is shown to be required for nitrogen fixation in alfalfa nodules, implicating aspartate catabolism in bacteroids as essential for nitrogen-fixing symbiosis (Rastogi and Watson, 1991). More recently, by using a *Rhizobium leguminosarum* strain with mutated *aap* and *bra* genes, which both encode the ABC-type transporter with broad specificity to various amino acids, Lodwig *et al.* (2003) reported that the amino-acid cycling is essential for symbiotic nitrogen fixation, and proposed that it thus prevents the symbiosis being dominated by the host plants. Very recently, however, the same group showed that these rhizobial ABC-type transporters are responsible for the supply of branched-chain amino acids (LIV) to bacteroids, in which the biosynthetic pathways of LIV are suppressed, and thus essential for nitrogen fixation by the endosymbionts (Prell *et al.*, 2009).

3.3.3.3 Symbiosome and Symbiosome Membrane

The symbiosome membrane serves both as a physical interface and as a mediator of metabolite exchange between plant cell and endosymbiont. The symbiosome membrane is derived from the plant plasma membrane, and the massive synthesis of symbiosome membrane depends upon protein delivery through the endoplasmic reticulum and Golgi apparatus (Figure 3.6). Proteomic analyses have shown that the symbiosome membrane has a highly specific composition compared with either plasma or tonoplast membranes (Saalbach *et al.*, 2002; Wienkoop and Saalbach, 2003; Catalano *et al.*, 2004). Although the protein-targeting mechanisms to symbiosome membranes are still largely unknown, a symbiosome membrane-specific Syntaxin (t-SNARE; soluble *N*-ethylmaleimide-sensitive factor adaptor protein receptor), MtSYP132, has recently been identified from *M. truncatula* (Catalano *et al.*, 2007). Syntaxin is known to facilitate vesicle docking and fusion to membrane systems. MtSYP132 has also been shown to localise to the plasma membrane surrounding the infection thread, thus suggesting that it functions in infection thread development as well as in the early stages of symbiosome formation. Along with bacteroid proliferation, symbiosomes are believed to propagate to fill the nodule infected cell. However, at least in determinate nodules which contain multiple bacteroids in single symbiosomes, fusion of individual symbiosomes also appears to contribute to the development of mature symbiosomes (Fedorova *et al.*, 1999).

There is no doubt that the symbiosome membrane plays a critical role in nutrient exchange between bacteroids and the cell cytoplasm, and is thus responsible for bacteroid differentiation. However, biochemical evidence of the exact function of the symbiosome membrane in carbon and nitrogen exchange between the two symbiotic partners is still very limited. The presence of systems involving C_4-dicarboxylate import, amino acids import/export, NH_4^+ export, and some ion transport through the symbiosome membrane has been reported based on a number of experiments using isolated symbiosomes or symbiosome membrane vesicles (Udvardi and Day, 1997; Mouritzen and Rosendahl, 1997; Roberts and Tyerman, 2002). However, only a few transporter proteins have been isolated so far.

The symbiosome membrane is decorated by a number of nodule-specific proteins (nodulins). Among them, Nodulin-26 has been intensively studied (Weaver *et al.*, 1994). Nodulin-26 is one of the most abundant proteins in the symbiosome membrane in soybean nodules, and was originally shown to be a water channel that participates in regulating osmotic pressure inside the symbiosome. Recent evidence has indicated that Nodulin-26 belongs to a functional subclass of a voltage-dependent MIP (major intrinsic protein) channel family and is multifunctional, involved in osmoregulation as well as the transport of small organic molecules (Wallace *et al.*, 2006). Because of the high abundance of Nodulin-26 proteins in the symbiosome membrane, it may play a central role(s) in nutrient exchange. However, it is unlikely that Nodulin-26 participates in C_4-dicarboxylate transport, because of its low affinity to anionic ions and molecules. A cDNA coding for a dicarboxylate transporter (AgDCAT1) involved in symbiotic nitrogen fixation has been isolated from *Alnus* plants, which form nitrogen-fixing nodules in symbiosis with actinomycetes (termed collectively *Frankia*) (Jeong *et al.*, 2004). Although no obvious legume counterpart of AgDCAT1 has yet been assigned, nodule-specific genes annotated as peptide transporters by transcriptome analyses (Colebatch *et al.*, 2004; Kouchi *et al.*, 2004) could be possible candidates because AgDCAT1 structurally belongs to a peptide transporter family.

Another example of a well-defined symbiosome membrane-specific transporter is a sulfate transporter, SST1, which was identified as the causal gene of an *L. japonicus* symbiotic mutant with a defect in effective nitrogen fixation (Krusell *et al.*, 2005). SST1 is shown to function as a high-affinity sulfate transporter by its ability to complement a *Saccharomyces cerevisiae* sulfate transporter mutant and has been demonstrated to be a component of the symbiosome membrane (Wienkoop and Saalbach, 2003). Sulfur is an important element of many biological processes, but in bacteroids it has special importance as a component of metal-sulfur clusters within the nitrogenase complex as well as within the related electron transfer proteins. Therefore, SST1 could meet the high demand for sulfur by bacteroids inside symbiosomes. Nodule-specific transporters, GmN70 from soybean and LjN70 from *L. japonicus*, have also been found in the symbiosome membrane, and are postulated to act as broad-specificity inorganic anion transporters with enhanced preference to nitrate (Szczyglowski *et al.*, 1998; Vincill *et al.*, 2005). Although the exact function of these transporters in the symbiosis have not yet been fully elucidated, they are proposed to be involved in the regulation of ion and membrane potential homeostasis, in particular in the regulation of nitrate concentrations in the symbiosome space, because nodulation and nitrogen fixation activity are both highly sensitive to nitrate. Fe transporter proteins have been identified from symbiosome membranes isolated from soybean root nodules (see Day *et al.*, 2001, for review; Kaiser *et al.*, 2003). A zinc (Zn) transporter (GmZip1) has also been isolated as a component of the symbiosome membrane of soybean nodules and characterised in great detail (Moreau *et al.*, 2002). As for some other metal ions, such as Mo and Cu, which have a special importance in bacteroid nitrogenase activity, their specific transporter systems are predicted to be present in the symbiosome membrane, but there is so far no substantial evidence.

Nutrient transport across the symbiosome membrane is thought to be dependent upon its membrane potential energisation by proton motive force, which is generated by an H^+-pumping ATPase. The major H^+-ATPase activity found in symbiosome membranes has been shown to be of the plasma membrane type ATPase (P-ATPase). Immunological studies also support the prominence of P-ATPase in symbiosome membranes (Fedorova *et al.*, 1999). However, vacuolar type ATPase (V-ATPase) has also been found in symbiosome membranes of *L. japonicus* by proteomic analysis (Wienkoop and Saalbach, 2003). Because of the presence of an H^+-pumping ATPase which forms a membrane potential with the positive inside the symbiosome, dicarboxylate transporters, sulfate transporters and other anion transporters are likely to have a uniport mechanism (Udvardi and Day, 1997; Krusell *et al.*, 2005). H^+-ATPase activity also makes the symbiosome space acidic, so that NH_3 that diffuses from the bacteroid is converted immediately to NH_4^+ in the symbiosome space and then exported to the plant cell cytoplasm through voltage-dependent channel proteins, which has been demonstrated by the patch-clamp method using symbiosome membranes isolated from *L. japonicus* nodules (Roberts and Tyerman, 2002). A detailed biochemical analysis of a channel responsible for NH_4^+ currents across the symbiosome membrane has been described, but its molecular identification is still to be solved (Obermeyer and Tyerman, 2005). The overall nutrient exchange between bacteroids and the plant cell cytoplasm across the symbiosome membrane is outlined in Figure 3.7.

A number of transporter activities have been attributed to the symbiosome membrane biochemically during the past decade. However, molecular identification of those transporters is still very limited. This is mainly due to the inherent difficulties in preparation of symbiosomes or symbiosome membranes with sufficient purity, uniformity and intactness. Recent development of infrastructures for genome research for the genetically tractable legumes, such as *L. japonicus* and *M. truncatula*, has provided a powerful tool for the identification of the constituents of the symbiosome membrane. For instance, transcriptome and/or proteome analyses have allowed us to identify a large number of nodule-specific genes and/or proteins of putative transporter function (Wienkoop and Saalbach, 2003; Colebatch *et al.*, 2004; Kouchi *et al.*, 2004; Benedito *et al.*, 2008; Høgslund *et al.*, 2009). Because of the overwhelming abundance of symbiosome membrane in the nodule membrane system, a

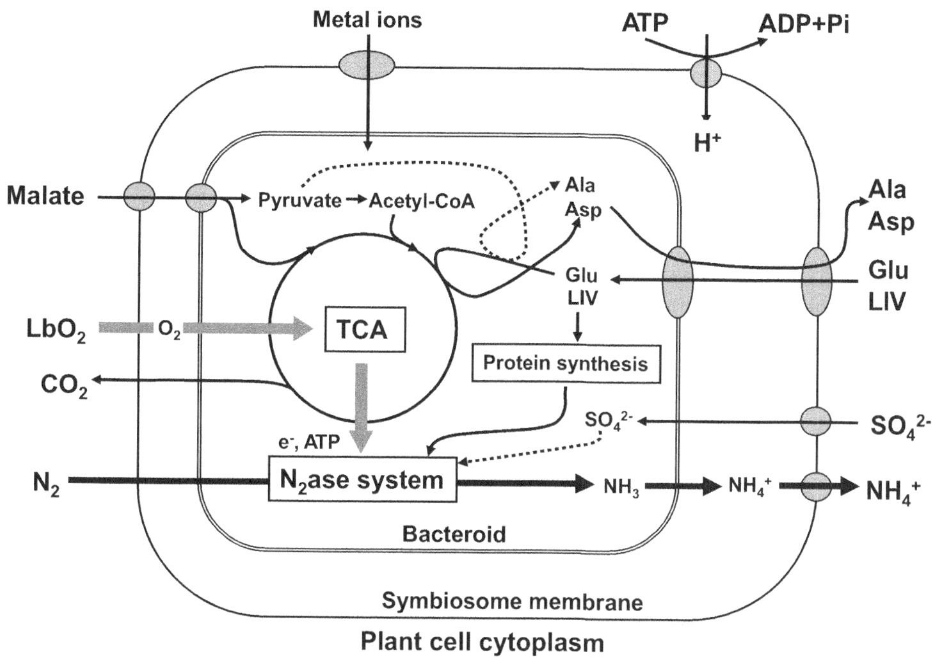

Figure 3.7 A scheme showing metabolic interrelationships between bacteroid and host cell cytoplasm across the symbiosome membrane

majority of these nodule-specific and/or nodule-enhanced putative transporters could be attributed to the symbiosome membrane. Analysis of their *in situ* localisation, as well as their functions by means such as RNA interference technology, or identification and analysis of their defective mutants, will uncover more details of symbiosome membrane function as well as its biogenesis.

3.4 Molecular Genetic Approaches to the Host Regulation of Nitrogen Fixation

3.4.1 The Fix$^-$ Mutants: Host Legume-Determined Ineffective Nodules

Symbiotic mutants of legumes have been generated by various means, such as chemical mutagenesis (mainly EMS), T-DNA insertion or transposon tagging, somatic mutation, fast neutron or heavy ion beam radiation, and so on. Based on recent developments in genome sequencing, construction of genetic maps, and accumulation of ESTs for the model legumes, a number of genes essential for symbiosis have been isolated. In *Lotus japonicus*, these symbiotic mutants were classified into three categories according to the stages of symbiosis development to which the defects were attributed (Kawaguchi *et al.*, 2002; Sandal *et al.*, 2006). Non-nodulating (Nod$^-$) mutants are attributed to defects in Nod factor perception and/or immediate downstream signalling pathways, and thus neither infection nor nodulation occur. Cooperative histogenesis-defective (Hist$^-$) mutants are characterised by defects in infection thread formation and are accompanied by incomplete nodule organogenesis. Representatives of genes of these two categories are described in Section 3.2.2. In addition, the genes involved in negative control of nodulation are also positioned in early symbiotic signalling. This is known as 'autoregulation', which is mediated through long-distance signal transport between roots and shoots in response to perception of Nod factors (Oka-Kira and Kawaguchi, 2006, Kouchi *et al.*, 2010). Defects of the genes involved in this process result in the formation of an excess number of nodules (Nod^{++} phenotypes). An LRR receptor kinase, HAR1, plays a central role in the autoregulation (Nishimura *et al.*, 2002; Krusell *et al.*, 2002). Recently, the root-derived signals have been identified as small peptides belonging to the CLE family, which putatively interact with HAR1 in the shoots (Okamoto *et al.*, 2008). In turn, HAR1-dependent shoot-derived signals have been shown to be transmitted to roots (Yamaya and Arima, 2010; Lin *et al.*, 2010). Their molecular identification is, however, still elusive.

The third category of *Lotus* symbiotic mutants includes the ineffective nitrogen fixation (Fix$^-$) mutants, which develop morphologically normal nodules with endosymbiotic bacteria but display very low or no nitrogen fixation activity, and thus the plants cannot grow depending solely on symbiotic nitrogen fixation for their nitrogen source. Hist$^-$ mutants are sometimes included in the Fix$^-$ mutant category because they apparently form nodule structures even though their development is incomplete. However, they should be distinguished because the former is characterised primarily by defects in the rhizobial infection process, and nodule structures formed (small bumps in most cases) do not contain endosymbiotic rhizobia, although some of them form nodules with rhizobia inside on very rare occasions. In contrast, Fix$^-$ mutants form nodules normally developed and filled with endosymbiotic rhizobia, thus indicating that in these mutants, bacterial invasion and nodule organogenesis are both comparable to the wild-type nitrogen-fixing nodules. In another model legume, *M. truncatula*, Fix$^-$ mutants have been further grouped into three subcategories based on the expression profiles of both plant and rhizobial genes in the nodules (Starker *et al.*, 2006).

Up to now, a total of six plant proteins have been demonstrated to be essential for nitrogen-fixation in nodules. Among them, leghaemoglobin was shown by RNAi knock-down of the corresponding gene(s) in *L. japonicus* (Ott *et al.*, 2005), and nodule-specific phosphoenolpyruvate carboxylase was also shown to be essential for nitrogen fixation by antisense technology (Nomura *et al.*, 2006). Sucrose synthase was identified from a pea mutant line *rug4* (Craig *et al.*, 1999; Gordon *et al.*, 1999). Sucrose synthase plays a key role in breakdown of sucrose translocated from plant shoots to nodules, and is thus essential for the supply of reduced carbon and energy for bacteroid respiration and nitrogenase activity. The other three, SST1, IGN1 and FEN1, were isolated from *L. japonicus* Fix$^-$ mutants. In addition to these identified Fix$^-$ genes, a number of Fix$^-$ loci have been found from *Lotus* mutant libraries, as summarised in Table 3.2 (Sandal *et al.*, 2006).

The genes identified from these Fix$^-$ mutants are attributed to the late stages of symbiotic nodule development. Even after formation of mature nodules with endosymbionts, it is obvious that specific recognition and interactions, which would be different from those mediated by Nod factors, operate between legume host and the endosymbiotic bacteria. For example, *Rhizobium etli* strain CE3, which is the microsymbiont of bean (*Phaseolus vulgaris*), produces Nod factors with the same structures as those from *M. loti*, a symbiotic partner of *Lotus* plants. Indeed, *R. etli* CE3 can form functional nodules on *L. japonicus*,

Table 3.2 *Representatives of* L. japonicus *Fix$^-$ mutants. In the column of N_2 fixation, −/+ means extremely low; +/− means low but significant C_2H_2 reduction activity*

Locus	Mutant phenotypes: N2 fixation	Mutant phenotypes: Early senescence	Gene cloning status	References
Fen1	−/+	+	Homocitrate synthase	Hakoyama *et al.* (2009)
Sst1	+/−	+	Sulfate transporter	Krusell *et al.* (2005)
Sen1	−	+	Cloned*	Suganuma *et al.* (2003)
Ign1	−/+	+++	Ankyrin membrane protein	Kumagai *et al.* (2007)
Sym102	−/+	+	Cloned*	H.Yamaya and Y. Umehara**
Sym104	+/−	+	Cloned*	Y. Umehara**
Sym103	−/+	n.d	Under way	Y. Umehara**
Sym105	+/−	+++	Cloned*	Hossain *et al.* (2006)
Sym89	+/−	++	Cloned*	Y. Umehara**

* Gene cloned but not published.
** These mutant phenotypes have not yet been described elsewhere.

but the nodules senesce prematurely with disintegration of the infected cell cytoplasm, and thus fail to develop further nitrogen-fixing symbiosis (Banba *et al.*, 2000). This observation indicates the presence and significance of host and *Rhizobium* species-specific interactions after full maturation of symbiotic nodules. However, our knowledge of these interactions in nodule function and of both the host and bacterial genes involved in such interactions is still very limited.

Typical phenotypes of the growth and nitrogen fixation of three representative Fix$^-$ mutants of *L. japonicus* are shown in Figure 3.8. Wild-type *Lotus* nodules are red because of the abundant accumulation of leghaemoglobins in the central infected zone, while Fix$^-$ nodules are pale pink in colour or almost white. Such Fix$^-$ plants display severe nitrogen deficiency symptoms when cultured under symbiotic conditions (i.e. without a combined nitrogen source). Nodule sizes are smaller than the wild-type, and in many cases the number of nodules formed is increased compared with wild-type plants. Fix$^-$ mutants isolated so far exhibit very low (but not zero) nitrogenase activity with only one exception, *sen1*, which completely lacks nitrogenase activity throughout all nodule developmental stages. Nitrogenase proteins in bacteroids of *sen1* nodules are difficult to detect, and based on observations by scanning electron microscopy, bacteroid differentiation has been suggested to be incomplete in *sen1* nodules (Suganuma *et al.*, 2003) The *Sen1* gene has been cloned recently and shown to code for a putative transporter protein most likely localised on the symbiosome membrane, but its exact function is still to be elucidated (N. Suganuma and T. Hakoyama, unpublished results). In contrast, *fen1* and *ign1* both exhibit low nitrogenase activity at least for a short period just after full maturation of nodule structures and the onset of nitrogen fixation. The *sst1* mutants show relatively high nitrogenase activity (30–40% of that of wild-type nodules) through the nodule growth period (Krusell *et al.*, 2005). This

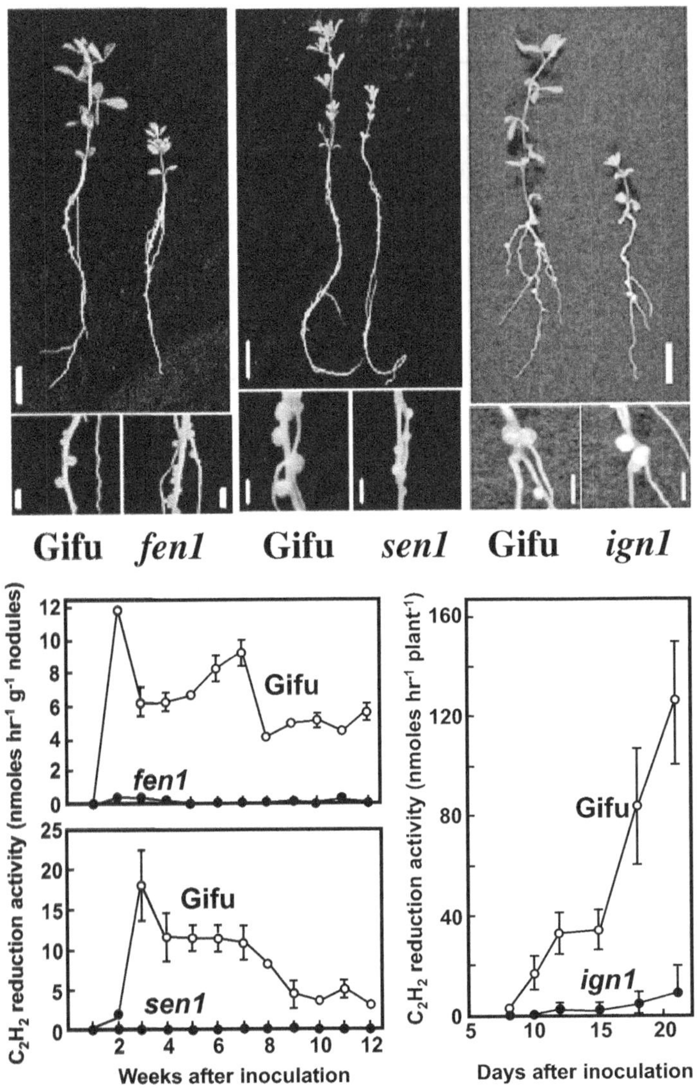

Figure 3.8 Growth and nitrogen fixation phenotypes of three L. japonicus *Fix- mutants,* fen1, sen1 *and* ign1 *(Reprinted with permission from Hakoyama* et al., *2009; Suganuma* et al., *2003; Kumagai* et al., *2007 Copyright 2009 Macmillan Publishers Ltd., 2003 Springer Science+Business Media., and 2007 American Society of Plant Biologists)*

is also the case for the *Ljsym105* mutants isolated from *L. japonicus* accession MG-20 by somatic mutation (Hossain *et al.*, 2006). The *Ljsym105* nodules exhibit around 50% of nitrogenase activity compared with wild-type MG-20 nodules on a nodule fresh weight basis. Nevertheless, these mutants both suffer severe nitrogen starvation under symbiotic conditions. One possible reason for this inconsistency could be due to the fact that estimation of nitrogenase activity of nodules has always been carried out by the C_2H_2 reduction assay, which although an easy and simple way to measure nitrogenase activity, represents only the maximal potential of nitrogenase activity in nodules, and does not always reflect actual N_2 fixation activity. *In vivo* estimation of N_2 fixation activity by means of the $^{15}N_2$ feeding method and chasing the metabolic fate of fixed ^{15}N in nodules will be necessary to better understand the genetic defects in these mutant nodules.

Alternatively, if the mutant has, for instance, a defect in assimilation and/or export of fixed nitrogen, apparent nitrogenase activity in the bacteroids may not be affected so greatly, at least in the initial stage of nitrogen fixation. This suggestion has been made for the case of the ineffective nitrogen fixation of pea plants by inoculation with *R. leguminosarum* with *aap*/*bra* double mutations with regard to significance of amino acid cycling and/or utilisation by bacteroids (Lodwig *et al.*, 2003; Prell *et al.*, 2009). In this case, the nitrogenase activity of the nodules, estimated by both C_2H_2 reduction and $^{15}N_2$ fixation assays, accounted for 30–50% of the wild-type rates, but the plants suffered severe nitrogen starvation. In *Ljsym105*, however, nitrogenase activity per plant remains quite low despite the relatively high nitrogenase activity of individual nodules, and this is possibly due to extreme retardation of nodule infected cell enlargement in the nodules of this mutant (Hossain *et al.*, 2006). The causal gene of *Ljsym105* has been cloned recently and its functional analysis is underway in our laboratory (Y. Umehara, unpublished results).

3.4.2 Metabolic Partnerships Unveiled by Fix$^-$ Mutants

The interactions in mature nodules represented by Fix$^-$ mutants are most probably attributed to: (1) the differentiation and/or persistence of bacteroids including induction of nitrogenase activity; (2) organisation of metabolic functions in the host cells required for nitrogen fixation; and (3) transport functions relating to metabolite exchange between bacteroids and the host cell cytoplasm, in particular in symbiosome membranes. Sucrose synthase belongs to the second, and SST1 (see Section 3.3.3) is in the third category. As such, identification of genes for these Fix$^-$ mutants would provide important clues for elucidating how the legume plant genes facilitate and/or regulate rhizobial nitrogen fixation in the symbiosis.

The *fen1* mutant was isolated and characterised by a systematic effort to screen symbiotic mutants of *L. japonicus* accession B-129 ‘Gifu’ (Imaizumi-Anraku *et al.*, 1997; Kawaguchi *et al.*, 2002). It forms pale pink, small nodules with very low nitrogen-fixing activity and thus cannot grow normally under symbiotic conditions (Figure 3.8). However, when supplied with sufficient amounts of combined nitrogen source, the mutant growth was perfectly comparable to the wild-type Gifu plants, indicating that the defect in the mutant is only in symbiotic nitrogen fixation.

The causal gene *Fen1* was cloned by map-based cloning, and its function has been clarified (Hakoyama *et al.*, 2009). By complementation experiments and *in vivo* assays of enzyme activity with *Saccharomyces cerevisiae* mutants, the FEN1 protein was demonstrated to be homocitrate synthase (HCS) which catalyzes a reaction to form homocitrate by condensation of 2-oxoglutarate and acetyl-CoA (Figure 3.9). The expression of *Fen1* is strictly nodule-specific, and in fact the wild-type *Lotus* nodules contain a large amount of homocitrate, which is quite an unusual compound in higher plants. In contrast, homocitrate is barely detectable in nodules formed on the *fen1* mutants.

Homocitrate is well-known as a component of the FeMo-cofactor of nitrogenase, on which nitrogen fixation is thought to occur (Hoover *et al.*, 1987, 1989; see also

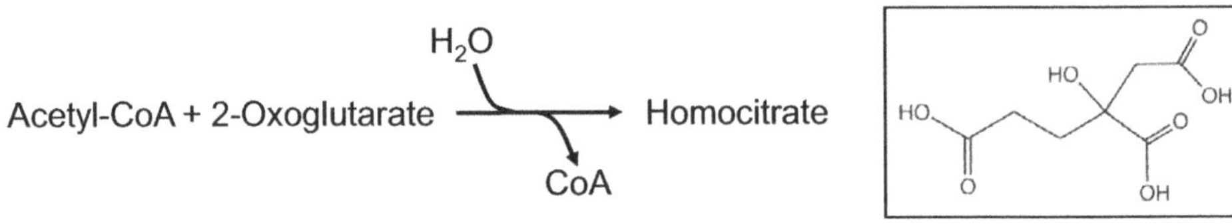

Figure 3.9 *Reaction catalyzed by homocitrate synthase (HCS)*

Section 3.1.2). *NifV*, which encodes HCS, has been identified from many diazotrophs and shown to be essential for their nitrogenase activity (Hoover *et al.*, 1988; Zheng *et al.*, 1997). However, the *NifV* orthologues are not found in most of the *Rhizobium* species, which exert highly efficient nitrogen fixation only in symbiotic association with compatible host legumes. This raised the hypothesis that nodule-specific HCS encoded in the host legume genome could compensate for the lack of *NifV* in endosymbiotic rhizobia. This hypothesis was confirmed by the fact that *M. loti* transformed with *Fen1* under the control of a bacterial *NifH* promoter could complement ineffectiveness of nodules formed on the *fen1* mutants, and consequently the plant growth recovered to a level comparable to that of wild-type plants (Figure 3.10). The same experiment using the authentic *NifV* gene (*AvNifV*)

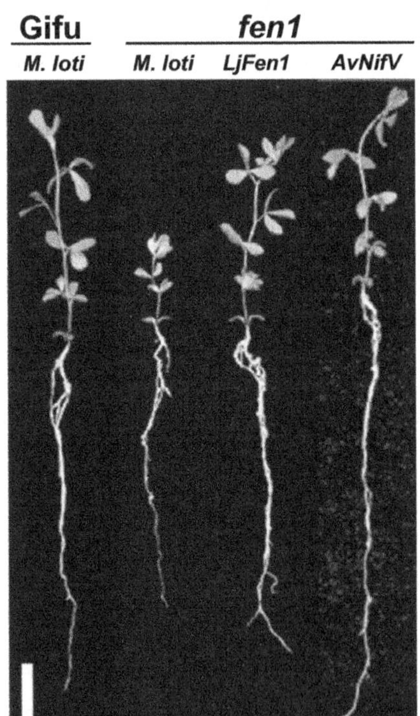

Figure 3.10 *Complementation of the* fen1 *mutant phenotype with inoculation of* M. loti *transformed with* pNifH::Fen1 *or* pNifH::AvNifV. *Bar = 1 cm. Reprinted by permission from Macmillan Publishers Ltd, Hakoyama* et al. *(2009), copyright 2009*

from *Azotobacter vinelandii* gave exactly the same results. This was the first example of the complementation of a genetic defect in host legume plants by genetic manipulation of the microsymbionts. A soybean *Fen1* orthologue, *GmN56*, which had been isolated previously as one of the nodulin genes (Kouchi and Hata, 1995), was shown to function as HCS by cross-species complementation, and it is thus likely that nodule-specific HCS is widely distributed in legume plants. In addition, the exogenous supply of homocitrate can rescue the ineffectiveness of the *fen1* mutant nodules through *de novo* synthesis of the nitrogenase proteins.

The discovery of the *Fen1* gene and its function in symbiosis indicates that the host genome compensates for the lack of a bacterial gene, which is essential for assembly of the active nitrogenase complex, by supplying homocitrate to bacteroids from the plant cell cytoplasm. This demonstrates at the molecular level the complementary and indispensable partnership between legumes and rhizobia in symbiotic nitrogen fixation (Figure 3.11). Furthermore, this raises an important issue in exploring the co-evolution of legume plants and *Rhizobium* bacteria. The FEN1 protein shows a high similarity (about 70% amino acid identity) to 2-isopropylmalate synthase (IPMS) identified from various plant species (De Kraker *et al.*, 2007). IPMS catalyzes the condensation of acetyl-CoA and 2-oxoisovalerate (3-methyl-2-oxobutanoate) to form 2-isopropylmalate, which is a precursor of leucine biosynthesis. Thus IPMS and HCS somewhat resemble each other in that they transfer an acyl group from acetyl-CoA to a 2-oxo acid to generate the alkyl group on the 2-oxo acid.

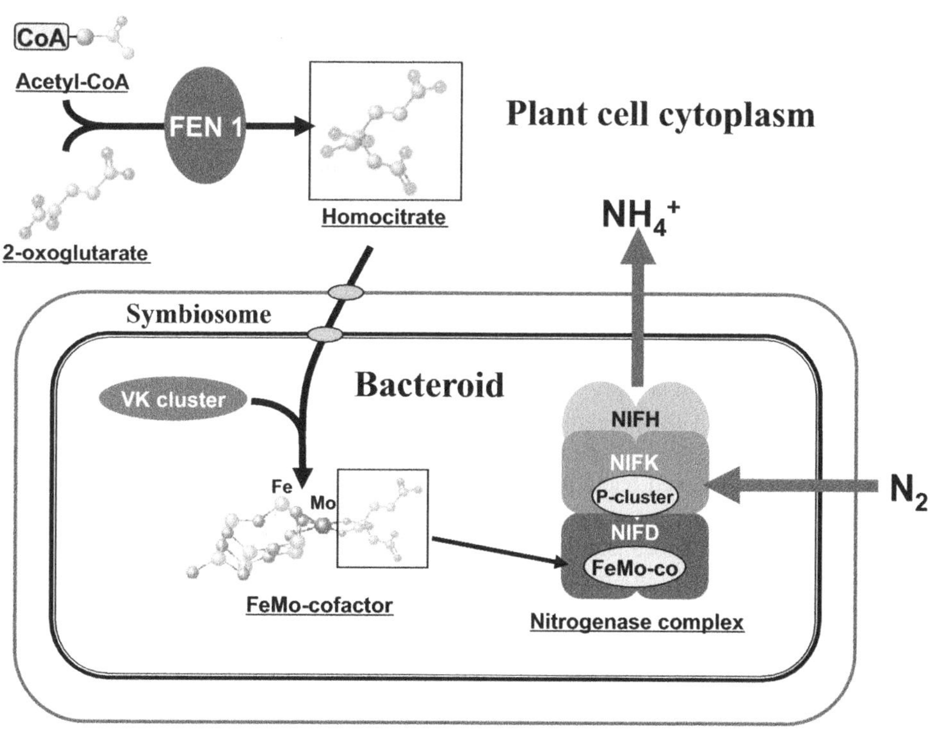

Figure 3.11 *A schematic model for the function of FEN1 in symbiotic nitrogen fixation*

Indeed, IPMS has been shown to be capable of using various 2-oxo acids as substrates. However, FEN1 has no IPMS activity but instead has HCS activity, which is not involved in plant metabolism *per se*. Furthermore, genes with highly similar sequences to *Fen1* are located at positions very close to *Fen1* in the *L. japonicus* genome, and the protein coded in one of these genes has been found to confer IPMS activity (N. Suganuma and T. Hakoyama, unpublished data). Therefore, it is very likely that the *Fen1* gene has evolved by gene duplication from a pre-existing house-keeping *IPMS* gene during evolution of the symbiosis. Tracing the evolution from IPMS to HCS at the molecular level will further support the idea that the acquirement of HCS by the nodule specific *Fen1* gene in legumes constituted one of the key genetic events for the establishment of a highly efficient nitrogen-fixing symbiosis by legumes and rhizobia.

3.4.3 Premature Senescence in Fix^- Nodules and Symbiosome Organisation

Nodules formed on Fix^- mutants commonly show early (premature) senescence, which is characterised by unusually excessive vacuolation of infected cells, and distorted enlargement and disintegration of symbiosomes, followed by disruption of cytoplasmic integrity of the infected cells. These phenotypes have been reported in more or less most of the Fix^- mutants isolated from various legume species. It has been shown that the premature senescence of nodules is not simply due to nitrogen deficiency in the host legumes caused by ineffective nitrogen fixation (Banba *et al.*, 2000). Rather, it is likely that it reflects the activation of a kind of defence response in the host legumes to exclude inefficient endosymbionts which confer no benefit to the host legume, since nodules formed by nitrogenase-defective *Rhizobium* mutant strains also show symptoms of premature senescence. However, some Fix^- mutants exhibit very rapid and drastic disruption of symbiosomes and then of the whole infected cell cytoplasm; the genes attributed to those Fix^- mutants are proposed to be intimately connected to the mutualistic nature of the symbiosis.

The temporal patterns and severity of premature senescence are considerably different between Fix^- mutants. An *L. japonicus* Fix^- mutant, *ign1* (Figure 3.8), which was isolated by somatic mutagenesis, is a typical example (Kumagai *et al.*, 2007). The *ign1* mutants exhibit low but significant nitrogenase activity at initial stages of nodule development, but fail to develop further nitrogen-fixing symbiosis. The most striking feature of this mutant is its extremely rapid premature senescence; irregularly enlarged symbiosomes with multiple bacteroids were observed at early stages (7–9 days post-inoculation) of nodule formation, followed by disruption of the symbiosomes and disintegration of nodule infected cell cytoplasm with aggregation of the bacteroids (Figure 3.12). Sometimes it appeared that symbiosomes and/or naked bacteroids are incorporated into lytic vacuoles. The cloned *Ign1* gene was predicted to encode a novel ankyrin-repeat protein with a transmembrane domain, and shown to be localised in the plasma membrane rather than in the symbiosome membrane. Interestingly, *Ign1* expression is not specific to nodules, but is also detected in all organs of *L. japonicus* plants at low levels. Nevertheless, the mutant phenotype appeared only in the symbiotic defect, suggesting a role of IGN1 in surveillance or control of responses to biotic infection. In this regard, it is noteworthy that the IGN1 protein has the same domain structure as the *Arabidopsis* ACD6, which is also localised in the plasma membrane and is postulated to be a positive regulator of defence responses against virulent bacteria and of salicylic acid-dependent cell death (Lu *et al.*, 2003, 2005), although the

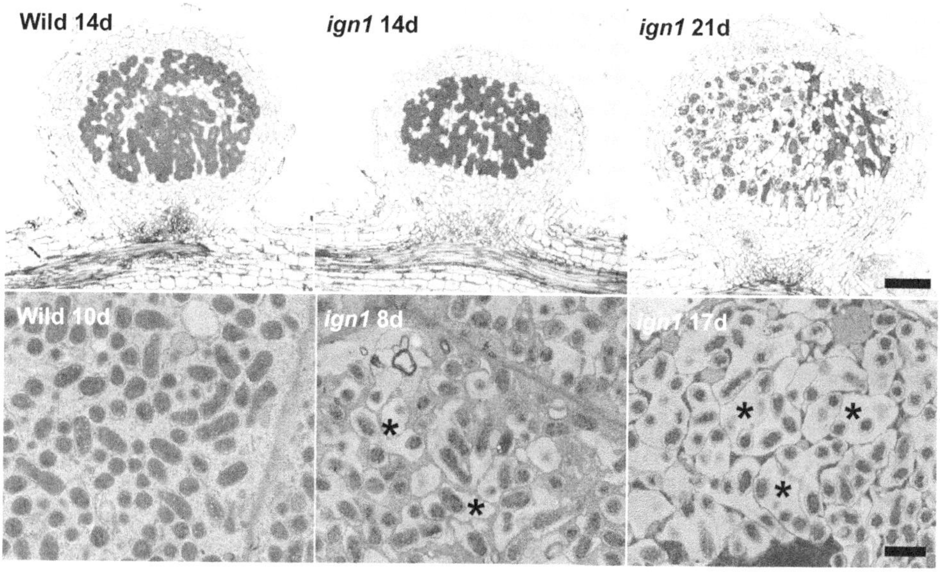

Figure 3.12 *Nodule and infected cell structures of the* ign1 *mutant. d, days after* M. loti *inoculation. Distorted enlargement of symbiosomes is indicated by asterisks. Bars = 200 μm (upper panels) and 2 μm (lower panels). Reprinted with permission from Kumagai* et al. *(2007). Copyright 2007 American Society of Plant Biologists*

amino-acid sequence similarity between IGN1 and ACD6 is quite low. Based on these findings, it has been hypothesised that IGN1 is required for preventing the host plant cells from inappropriately invoking a defence system against compatible microsymbionts, thus being essential for differentiation and/or persistence of bacteroids and symbiosomes.

Besides *ign1*, *Ljsym105* has been also characterised by very rapid deterioration of symbiosomes, despite the fact that it retains a relatively high rate of nitrogenase activity (Hossain *et al.*, 2006). In contrast, premature senescence in *sen1* mutants appears to be not so rapid or drastic compared with the two Fix^- mutants, but it has no nitrogenase activity at all (Suganuma *et al.*, 2003). Therefore, the extent of premature senescence represented by distorted enlargement and disorganisation of symbiosomes is not apparently correlated with nitrogenase activities at various Fix^- mutant loci.

Since symbiotic nitrogen fixation is supposed to be regulated by complex interactions between both host legume and bacterial genes, the number of identified independent Fix^- loci in host legumes is still increasing. In addition, even from the bacterial side, bacteroid differentiation has not been well studied (see Section 3.3.1). In a sense, premature senescence of the nodule, as a dominant and common phenotype in many Fix^- symbiotic mutants, makes it somewhat difficult to assign possible functions of individual Fix^- loci in the establishment and persistence of the nitrogen-fixing symbiosis from analyses of their apparent phenotypes. Nevertheless, more detailed analyses of symbiosomes and successive bacteroid disruptions in these Fix^- mutants will be required for understanding the mechanisms underlying organisation of symbiosomes and bacteroid differentiation, which both constitute the central events in establishing functional symbiosis. Systematic and comparative studies of many individual Fix$^-$ mutants and wild-type nodules with regard to gene-expression profiles (transcriptome analysis: see Høgslund *et al.*, 2009), and proteome and metabolome analyses of both bacteria and host plants, will be very useful for ordering the Fix$^-$ loci within the context of a pathway leading to the establishment of a fully functional symbiosis.

Acknowledgements

The author thanks Drs Norio Suganuma, Tsuneo Hakoyama, Yosuke Umehara and Haruko Imaizumi-Anraku for helpful discussion and permission to include their unpublished data in this review, and Prof. Robert W. Ridge for critical reading of and comments on the manuscript.

References

Ane, J.M., Kiss, G.B., Riely, B.K., Penmetsa, R.V., Oldroyd, G.E.D., Ayax, C., Levy, J., Debelle, F., Baek, J.M., Kalo, P., Rosenberg, C., Roe, B.A., Long, S.R., Denarie, J. and Cook, D.R. (2004) *Medicago truncatula* DMI1 required for bacterial and fungal symbioses in legumes. *Science*, **303**, 1364–1367.

Appels, M.A. and Haaker, H. (1991) Glutamate oxaloacetate transaminase in pea root nodules: participation in a malate/aspartate shuttle between plant and bacteroid. *Plant Physiol.*, **95**, 740–747.

Arnold, W., Rump, A., Klipp, W., Priefer, U.B. and Pühler, A. (1988) Nucleotide sequence of a 24,206-base-pair DNA fragment carrying the entire nitrogen fixation gene cluster of *Klebsiella pneumoniae*. *J. Mol. Biol.*, **203**, 715–738.

Arredondo-Peter, R., Hargrove, M.S., Sarath, G., Moran, J.F., Lohrman, J., Olson, J.S. and Klucas, R.V. (1997) Rice hemoglobins. Gene cloning, analysis, and O_2-binding kinetics of a recombinant protein synthesized in *Escherichia coli*. *Plant Physiol.*, **115**, 1259–1266.

Arredondo-Peter, R., Hargrove, M.S., Moran, J.F., Sarath, G. and Klucas, R.V. (1998) Plant hemoglobins. *Plant Physiol.*, **118**, 1121–1125.

Arrighi, J.F., Barre, A., Amor, B.B., Bersoult, A., Soriano, L.C., Mirabella, R., Carvalho-Niebel, J. F., Journet, E.P., Gherardi, M. Huguet, T., Geurts, R., Denarie, J., Rouge, P. and Gough, C. (2006) The *Medicago truncatula* lysine motif-receptor-like kinase gene family includes NFP and new nodule-expressed genes. *Plant Physiol.*, **142**, 265–279.

Banba, M., Siddique, A.-B. M., Kouchi, H., Izui. K. and Hata, S. (2000) *Lotus japonicus* forms early senescent root nodules with *Rhizobium etli*. *Mol. Plant-Microbe Interact.*, **14**, 173–180.

Banba, M., Gutjahr, C., Miyao, A., Hirochika, H., Paszkowski, U., Kouchi, H. and Imaizumi-Anraku, H. (2008) Divergence of evolutionary ways among common sym genes: *CASTOR* and *CCaMK* show functional conservation between two symbiosis systems and constitute the root of a common signaling pathway. *Plant Cell Physiol.*, **49**, 1659–1671.

Benedito, V.A., Torres-Jerez, I., Murray, J.D., Andriankaja, A., Allen, S., Kakar, K., Wandrey, M., Verdier, J., Zuber, H., Ott, T., Moreau, S., Niebel, A., Frickey, T., Weiller, G.,

He, J., Dai, X., Zhao, P.X., Tang, Y. and Udvardi, M.K. (2008) A gene expression atlas of the model legume *Medicago truncatula*. *Plant J.*, **55**, 504–513.

Bergersen, F.J. (1982) *Root Nodules of Legumes: Structure and Functions*, John Wiley & Sons Ltd, Chichester, UK.

Catalano, C.M, Lane, W.S. and Sherrier, D.J. (2004) Biochemical characterization of symbiosome membrane proteins from *Medicago truncatula* root nodules. *Electrophoresis*, **25**, 519–531.

Catalano, C.M., Czymmek, K.J., Gann, J.G. and Sherrier, D.J. (2007) *Medicago truncatula* syntaxin SYP132 defines the symbiosome membrane and infection droplet membrane in root nodules. *Planta*, **225**, 541–550.

Charpentier, M., Bredemeier, R., Wanner, G., Takeda, N., Schleiff, E. and Parniske, M. (2008) *Lotus japonicus* CASTOR and POLLUX are ion channels essential for perinuclear calcium spiking in legume root endosymbiosis. *Plant Cell*, **20**, 3467–3479.

Chen, F., Okabe, Y., Osano, K. and Tajima, S. (1998) Purification and characterization of an NAD-malic enzyme from *Bradyrhizobium japonicum* A1017. *Appl. Environ. Microbiol.*, **64**, 4073–4075.

Chen, C., Gao, M., Liu, J. and Zhu, H. (2007) Fungal symbiosis in rice requires an ortholog of a legume common symbiosis gene encoding a Ca^{2+}/calmodulin-dependent protein kinase. *Plant Physiol.*, **145**, 1619–1628.

Cohn, J., Day, R.B. and Stacey, G. (1998) Legume nodule organogenesis. *Trends in Plant Sci.*, **3**, 105–110.

Colebatch, G., Desbrosses, G., Ott, T., Krusell, L., Montanari, O., Kloska, S., Kopka, J. and Udvardi, M.K. (2004) Global changes in transcription orchestrate metabolic differentiation during symbiotic nitrogen fixation in *Lotus japonicus*. *Plant J.*, **39**, 487–512.

Craig, J., Barratt, P., Tatge, H., Dejardin, A., Handley, L., Gardner, C.D., Barber, L., Wang, T., Hedley, C., Martin, C. and Smith, A.M. (1999) Mutations at the rug4 locus alter the carbon and nitrogen metabolism of pea plants through an effect on sucrose synthase. *Plant J.*, **17**, 353–362.

Day, D.A., Kaiser, B.N., Thomson, R., Udvardi, M.K., Moreau, S. and Puppo, A. (2001) Nutrient transport across symbiotic membranes from legume nodules. *Aust. J. Plant Physiol.*, **28**, 667-674.

Deakin, W.J. and Broughton, W.J. (2009) Symbiotic use of pathogenic strategies: rhizobial protein secretion systems. *Nature Rev. Microbiol.*, **7**, 312–320.

De Kraker, J.W., Luck, K., Textor, S., Tokuhisa, J.G. and Gershenzon, J. (2007) Two Arabidopsis genes (*IPMS1* and *IPMS2*) encode isopropylmalate synthase, the branchpoint step in the biosynthesis of leucine. *Plant Physiol.*, **143**, 970–986.

Dixon, R. and Kahn, D. (2004) Genetic regulation of biological nitrogen fixation. *Nature Rev. Microbiol.*, **2**, 621–631.

Driscoll, B.T. and Finan, T.M. (1993) NAD^+-dependent malic enzyme of *Rhizobium meliloti* is required for symbiotic nitrogen fixation. *Mol. Microbiol.*, **7**, 865–873.

Ehrhardt, D.W., Wais, R. and Long, S.R. (1996) Calcium spiking in plant root hairs responding to *Rhizobium* nodulation signals. *Cell*, **85**, 673–681.

Endre, G., Kereszt, A., Kevei, Z., Mihacea, S., Kalo, P. and Kiss, G.B. (2002) A receptor kinase gene regulating symbiotic nodule development. *Nature*, **417**, 962–966.

Fedorova, E., Thomson, R., Whitehead, L.F., Maudoux, O., Udvardi, M.K. and Day, D.A. (1999) Localization of H(+)-ATPases in soybean root nodules. *Planta*, **209**, 25–32.

Finan, T.M., Wood, J.M. and Jordan, D.C. (1983) Symbiotic properties of C4-dicarboxylic acid transport mutants of *Rhizobium leguminosarum*. *J. Bacteriol.*, **154**, 1403–1413.

Fischer, H.M. (1994) Genetic regulation of nitrogen fixation in rhizobia. *Microbiol. Rev.*, **58**, 352–386.

Geurts, R., Fedorova, E. and Bisseling, T. (2005). Nod factor signaling genes and their function in the early stages of *Rhizobium* infection. *Curr. Opin. Plant Biol.*, **8**, 346–352.

Gibson, K.E., Kobayashi, H. and Walker, G.C. (2008) Molecular determinants of a symbiotic chronic infection. *Ann. Rev. Genet.*, **42**, 413–441.

Glazebrook, J., Ichige, A. and Walker, G.C. (1993) A *Rhizobium meliloti* homolog of the *Escherichia coli* peptide-antibiotic transport protein SbmA is essential for bacteroid development. *Genes Dev.*, **7**, 1485–1497.

Gleason, C., Chaudhuri, S., Yang, T., Munoz, A., Poovaiah, B.W. and Oldroyd, G.E. (2006). Nodulation independent of rhizobia induced by a calcium-activated kinase lacking autoinhibition. *Nature*, **441**, 1149–1152.

Godfroy, O., Debelle, F., Timmers, T. and Rosenberg, C. (2006) A rice calcium- and calmodulin-dependent protein kinase restores nodulation to a legume mutant. *Mol. Plant-Microbe Interact.*, **19**, 495–501.

Gordon, A.J., Minchin, F.R., James, C.L. and Komina, O. (1999) Sucrose synthase in legume nodules is essential for nitrogen fixation. *Plant Physiol.*, **120**, 867–878.

Hakoyama, T., Niimi, K., Watanabe, H., Tabata, R., Matsubara, J., Sato, S., Nakamura, Y., Tabata, S., Jichun, L., Matsumoto, T., Tatsumi, K., Nomura, M., Tajima, S., Ishizaka, M., Yano, K., Imaizumi-Anraku, H., Kawaguchi, M., Kouchi, H. and Suganuma, N. (2009) Host plant genome overcomes the lack of a bacterial gene for symbiotic nitrogen fixation. *Nature*, **462**, 514–517.

Hales, B.J. (1990) Alternative nitrogenase. *Adv. Inorg. Biochem.*, **8**, 165–198.

Hayashi, T., Banba, M., Shimoda, Y., Kouchi, H., Hayashi, M. and Imaizumi-Anraku, H. (2010) A dominant function of CCaMK in intracellular accommodation of bacterial and fungal endosymbionts. *Plant J.*, published online: 16 April 2010.

Heckmann, A.B., Lombardo, F., Miwa, H., Perry, J.A., Bunnewell, S., Parniske, M., Wang, T.L. and Downie, J.A. (2006). *Lotus japonicus* nodulation requires two GRAS domain regulators, one of which is functionally conserved in a non-legume. *Plant Physiol.*, **142**, 1739–1750.

Herder, G.D. and Parniske, M. (2009) The unbearable naivety of legumes in symbiosis. *Curr. Opin. Plant Biol.*, **12**, 491–499.

Høgslund, N., Radutoiu, S., Krusell, L., Voroshilova, V., Hannah, M.A., Goffard, N., Sanchez, D.H., Lippold, F., Ott, T., Sato, S., Tabata, S., Liboriussen, P., Lohmann, G.V., Schauser, L., Weiller, G.F., Udvardi, M.K. and Stougaard, J. (2009) Dissection of symbiosis and organ development by integrated transcriptome analysis of Lotus japonicus mutant and wild-type plants. *PLoS One*, **4**, e6556.

Hoover, T.R., Robertson, A.D., Cerny, R.L., Hayes, R.N., Imperial, J., Shah, V.K. and Ludden, P.W. (1987) Identification of the V factor needed for synthesis of the iron-molybdenum cofactor of nitrogenase as homocitrate. *Nature*, **329**, 855–857.

Hoover, T.R., Imperial, J., Ludden, P.W. and Shah, V.K. (1988) Homocitrate cures the NifV- phenotype in *Klebsiella pneumoniae*. *J. Bacteriol.*, **170**, 1978–1979.

Hoover, T.R., Imperial, J., Ludden, P.W. and Shah, V.K. (1989) Homocitrate is a component of the iron-molybdenum cofactor of nitrogenase. *Biochemistry*, **28**, 2768–2771.

Hossain, Md.S., Umehara, Y. and Kouchi, H. (2006) A novel Fix$^-$ symbiotic mutant of *Lotus japonicus*, Ljsym105, shows impaired development and premature deterioration of nodule infected cells and symbiosomes. *Mol. Plant-Microbe Interact.*, **19**, 780–788.

Hunt, S. and Layzell, D.B. (1993) Gas exchange of legume nodules and the regulation of nitrogenase activity. *Ann. Rev. Plant Physiol. Plant Mol. Biol.*, **44**, 483–511.

Hunt, P.W., Klok, E.J., Trevaskis, B., Watts, R.A., Ellis, M.H., Peacock, W.J. and Dennis, E.S. (2002) Increased level of hemoglobin 1 enhances survival of hypoxic stress and promotes early growth in Arabidopsis thaliana. *Proc. Natl. Acad. Sci. USA*, **99**, 17197–17202.

Imaizumi-Anraku, H., Kawaguchi, M., Koiwa, H., Akao, S. and Syono, K. (1997) Two ineffective-nodulating mutants of *Lotus japonicus* – different phenotypes caused by the blockage of endocytotic bacterial release and nodule maturation. *Plant Cell Physiol.*, **38**, 871–881.

Imaizumi-Anraku, H., Takeda, N., Charpentier, M., Perry, J., Miwa, H., Umehara, Y., Kouchi, H., Murakami, Y., Mulder, L., Vickers, K., Pike, J., Downie, A., Wang, T., Sato, S., Asamizu, E., Tabata, S., Yoshikawa, M., Murooka, Y., Wu, G.-J., Kawaguchi, M., Kawasaki, S., Parniske, M. and Hayashi, M. (2005) Plastid proteins crucial for symbiotic fungal and bacterial entry into plant roots. *Nature*, **433**, 527–530.

Jacobson, M.R., Brigle, K.E., Bennett, L.T., Setterquist, R.A., Wilson, M.S., Cash, V.L., Beynon, L., Newton, W.E. and Dean, D.R. (1989) Physical and genetic map of the major *Nif* gene cluster from *Azotobacter vinelandii*. *J. Bacteriol.*, **171**, 1017–1027.

Jeong, J., Suh, S., Guan, C., Tsay, Y.F., Moran, N., Oh, C.J., An, C.S., Demchenko, K.N., Pawlowski, K. and Lee, Y. (2004) A nodule-specific dicarboxylate transporter from alder is a member of the peptide transporter family. *Plant Physiol.*, **134**, 969–978.

Jording, D., Sharma, P.K., Schmidt, R., Engelke, T., Uhde, C. and Puhler, A. (1993) Regulatory aspects of the C_4-dicarboxylate transport in *Rhizobium meliloti* – Transcriptional activation and dependence on effective symbiosis. *J. Plant Physiol.*, **141**, 18–27.

Kahn, M.L., Kraus, J. and Sommerville, J.E. (1985) A model of nutrient exchange in the Rhizobium-legume symbiosis, in *Nitrogen Fixation Research Progress* (eds H.J. Evans, P.J. Bottomley and W.E. Newton), Martinus Nijhoff, Dordrecht, pp. 193–199.

Kaiser, B.N., Moreau, S., Castelli, J., Thomson, R., Lambert, A., Bogliolo, S., Puppo, A. and Day, D.A. (2003) The soybean NRAMP homologue, GmDMT1, is a symbiotic divalent metal transporter capable of ferrous iron transport. *Plant J.*, **35**, 295–304.

Kanamori, N., Madsen L.H., Radutoiu, S., Frantescu, M., Quistgaard, E.M.H., Miwa, H., Downie, J.A., James, E.K., Felle, H.H., Haaning L.L., Jensen T.H., Sato, S., Nakamura, Y., Tabata, S., Sandal, N. and Stougaard, J. (2006) A nucleoporin is required for induction of Ca^{2+} spiking in legume nodule development and essential for rhizobial and fungal symbiosis. *Proc. Natl. Acad. Sci. USA*, **103**, 359–364.

Kawaguchi, M., Imaizumi-Anraku, H., Koiwa, H., Niwa, S., Ikuta, A., Syono, K. and Akao, S. (2002) Root, root hair, and symbiotic mutants of the model legume *Lotus japonicus*. *Mol. Plant-Microbe Interact.*, **15**, 17–26.

Kiss, E., Olah, B., Kalo, P., Morales, M., Heckmann, A.B., Borbola, A., Lozsa, A., Kontar, K., Middleton, P., Downie, J.A., Oldroyd, G.E. and Endre, G. (2009). LIN, a novel type of U-box/WD40 protein, controls early infection by rhizobia in legumes. *Plant Physiol.*, **151**, 1239–1249.

Kouchi, H., Fukai, K. and Kihara, A. (1991) Metabolism of glutamate and aspartate in bacteroids isolated from soybean root nodules. *J. Gen. Microbiol.*, **137**, 2901–2910.

Kouchi, H. and Hata, S. (1995) GmN56, a novel nodule-specific cDNA from soybean root nodules encodes a protein homologous to isopropylmalate synthase and homocitrate synthase. *Mol. Plant-Microbe Interact.*, **8**, 172–176.

Kouchi, H., Shimomura, K., Hata, S., Hirota, A., Wu, G.-J., Kumagai, H., Tajima, S., Suganuma, N., Suzuki, A., Aoki, T., Hayashi, M., Yokoyama, T., Ohyama, T., Asamizu, E., Kuwata, C., Shibata, D. and Tabata, S. (2004) Large-scale analysis of gene expression profiles during early stages of root nodule formation in a model legume, *Lotus japonicus*. *DNA Res.*, **11**, 263–274.

Kouchi, H., Imaizumi-Anraku, H., Hayashi, M., Hakoyama, T., Nakagawa, T., Umehara, Y., Suganuma, N. and Kawaguchi, M. (2010) How many peas in a pod? Legume genes responsible for mutualistic symbioses underground. *Plant Cell Physiol.*, **51**, 1381–1397.

Krusell, L., Madsen, L.H., Sato, S., Aubert, G., Genua, A., Szczyglowski, K., Duc, G., Kaneko, T., Tabata, S., de Bruijn, F., Pajuelo, E., Sandal, N. and Stougaard, J. (2002) Shoot control of root development and nodulation is mediated by a receptor-like kinase. *Nature*, **420**, 422–426.

Krusell, L., Krause, K., Ott, T., Desbrosses, G., Kramer, U., Sato, S., Nakamura, Y., Tabata, S., James, E.K., Sandal, N., Stougaard, J., Kawaguchi, M., Miyamoto, A., Suganuma, N. and Udvardi, M.K. (2005) The sulfate transporter SST1 is crucial for symbiotic nitrogen fixation in *Lotus japonicus* root nodules. *Plant Cell*, **17**, 1625–1636.

Kumagai, H., Hakoyama, T., Umehara, Y., Sato, S., Kaneko, T., Tabata, S. and Kouchi, H. (2007) A novel ankyrin-repeat membrane protein IGN1 is required for persistence of nitrogen-fixing symbiosis in root nodules of *Lotus japonicus*. *Plant Physiol.*, **143**, 1293–1305.

Lafontaine, P.J., Lafrenière, C. and Antoun, H. (1989) Some properties of carbohydrate and C4-dicarboxylic acid utilization negative mutants of *Rhizobium leguminosarum* biovar *phaseoli* strain P121. *Plant Soil*, **120**, 195–201.

Lerouge, P., Roche, P., Faucher, C., Maillet, F., Truchet, G., Prome, J.C. and Denarie, J. (1990) Symbiotic host-specificity of *Rhizobium meliloti* is determined by a sulphated and acylated glucosamine oligosaccharide signal. *Nature*, **344**, 781–784.

Levy, J., Bres, C., Geurts, R., Chalhoub, B., Kulikova, O., GerardDuc, Journet, E.-P., Ane, J.-M., Lauber, E., Bisseling, T., Denarie, J., Rosenberg, C. and Debelle, F. (2004) A putative Ca^{2+} and calmodulin dependent protein kinase required for bacterial and fungal symbioses. *Science*, **303**, 1361–1364.

Limpens, E., Franken, C., Smit, P., Willemse, J., Bisseling, T. and Geurts, R. (2003) LysM domain receptor kinases regulating rhizobial Nod factor-induced infection. *Science*, **302**, 630–633.

Lin, Y.H., Ferguson, B.J., Kereszt, A. and Gresshoff, P.M. (2010) Suppression of hypernodulation in soybean by a leaf-extracted, NARK- and Nod factor-dependent, low molecular mass fraction. *New Phytol.*, **185**, 1074–1086.

Lodwig, E.M., Hosie, A.H.F., Bordes, A., Findlay, K., Allaway, D., Karunakaran, R., Downie, J.A. and Poole, P.S. (2003) Amino-acid cycling drives nitrogen fixation in the legume-*Rhizobium* symbiosis. *Nature*, **422**, 722–726.

Lodwig, E. and Poole, P. (2003) Metabolism of *Rhizobium* bacteroids. *Crit. Rev. Plant Sci.*, **33**, 37–78.

Lu, H., Rate, D.N., Song, J.T. and Greenberg, J.T. (2003) ACD6, a novel ankyrin protein, is a regulator and an effector of salicylic acid signaling in the *Arabidopsis* defense response. *Plant Cell*, **15**, 2408–2420.

Lu, H., Liu, Y. and Greenberg, J.T. (2005) Structure-function analysis of the plasma membrane-localized *Arabidopsis* defense component ACD6. *Plant J.*, **44**, 798–809.

Madsen, E.B., Madsen, L.H., Radutoiu, S., Olbryt, M., Rakwalska, M., Szczyglowski, K., Sato, S., Kaneko, T., Tabata, S., Sandal, N., Stougaard, J. (2003) A receptor kinase gene of the LysM type is involved in legume perception of rhizobial signals. *Nature*, **425**, 637–640.

Madsen, L.H., Tirichine, L., Jurkiewicz, A., Sullivan, J.T., Heckmann, A.B., Bek, A.S., Ronson, C.W., James, E.K. and Stougaard, J. (2010) The molecular network governing nodule organogenesis and infection in the model legume *Lotus japonicus*. *Nature Commun.*, **1**, 10. DOI: 10.1038

Markmann, K., Giczey, G. and Parniske, M. (2008) Functional adaptation of a plant receptor-kinase paved the way for the evolution of intracellular root symbioses with bacteria. *PLoS Biol.*, **6**, e68.

Markmann, K. and Parniske, M. (2009) Evolution of root endosymbiosis with bacteria: how novel are nodules? *Trends Plant Sci.*, **14**, 77–86.

Marlow, V.L., Haag, A.F., Kobayashi, H., Fletcher, V., Scocchi, M., Walker, G.C. and Ferguson, G.P. (2009) Essential role for the BacA protein in the uptake of a truncated eukaryotic peptide in *Sinorhizobium meliloti*. *J. Bacteriol.*, **191**, 1519–1527.

Maruya, J. and Saeki, K. (2010) The *bacA* gene homologue, mlr7400, in *Mesorhizobium loti* MAFF303099 is dispensable for symbiosis with *Lotus japonicus* but partially capable to support symbiotic function of the *bacA* in *Sinorhizobium meliloti*. *Plant Cell Physiol.*, **51**, 1443–1452.

Mergaert, P., Uchiumi, T., Alunni, B., Evanno, G., Cheron, A., Catrice, O., Mausset, A.E., Barloy-Hubler, F., Galibert, F., Kondorosi, A. and Kondorosi, E. (2006) Eukaryotic control on bacterial cell cycle and differentiation in the *Rhizobium*-legume symbiosis. *Proc. Natl. Acad. Sci. USA*, **103**, 5230–5235.

Messinese, E., Mun, J.H., Yeun, L.H., Jayaraman, D., Rouge, P., Barre, A., Lougnon, G., Schornack, S., Bono, J.J., Cook, D.R. and Ane, J.M. (2007) A novel nuclear protein interacts with the symbiotic DMI3 calcium- and calmodulin-dependent protein kinase of *Medicago truncatula*. *Mol. Plant–Microbe Interact.*, **20**, 912–921.

Minami, E., Kouchi, H., Cohn, J.R., Ogawa, T. and Stacey, G. (1996) Expression of the early nodulin, ENOD40, in soybean roots in response to various lipo-chitin signal molecules. *Plant J.*, **10**, 23–32.

Miwa, H., Sun, J., Oldroyd, G.E. and Downie, J.A. (2006) Analysis of Nod-factor-induced calcium signaling in root hairs of symbiotically defective mutants of *Lotus japonicus*. *Mol. Plant-Microbe Interact.*, **19**, 914–923.

Miyatake, H., Mukai, M., Adachi, S., Nakamura, H., Tamura, K., Iizuka, T., Shiro, Y., Strange, R.W. and Hasnain, S.S. (1999) Iron coordination structures of oxygen sensor FixL characterized by Fe K-edge extended X-ray absorption fine structure and resonance raman spectroscopy. *J. Biol. Chem.*, **274**, 23176–23184.

Moreau, S., Thomson, R.M., Kaiser, B.N., Trevaskis, B., Guerinot, M.L. Udvardi, M.K., Puppo, A. and Day, D.A. (2002) *GmZIP1* encodes a symbiosis-specific zinc transporter in soybean. *J. Biol. Chem.*, **277**, 4738–4746.

Mouritzen, P. and Rosendahl, L. (1997) Identification of a transport mechanism for NH_4^+ in the symbiosome membrane of pea root nodules. *Plant Physiol.*, **115**, 519–526.

Murakami, Y., Miwa, H., Imaizumi-Anraku, H., Kouchi, H., Downie, J.A., Kawaguchi, M. and Kawasaki, S. (2006) Positional cloning identifies *Lotus japonicus* NSP2, a putative transcription factor of the GRAS family, required for NIN and ENOD40 gene expression in nodule initiation. *DNA Res.*, **13**, 255–265.

Murray, J.D., Karas, B.J., Sato, S., Tabata, S., Amyot, L. and Szczyglowski, K. (2007) A cytokinin perception mutant colonized by *Rhizobium* in the absence of nodule organogenesis. *Science*, **315**, 101–104.

Nishimura, R., Hayashi, M., Wu, G.-J., Kouchi, H., Imaizumi-Anraku, H., Murakami, Y., Kawasaki, S., Akao, S., Ohmori, M., Nagasawa, M., Harada, K. and Kawaguchi, M. (2002) HAR1 mediates systemic regulation of symbiotic organ development. *Nature*, **420**, 426–429.

Niwa, S., Kawaguchi, M., Imaizumi-Anraku, H., Chechetka, S.A., Ishizaka, M., Ikuta, A. and Kouchi, H. (2001) Responses of a model legume *Lotus japonicus* to lipochitin oligosaccharide nodulation factors purified from *Mesorhizobium loti* JRL501. *Mol. Plant-Microbe Interact.*, **14**, 848–856.

Nomura, M., Mai, H.T., Fujii, M., Hata, S., Izui, K. and Tajima, S. (2006) Phosphoenolpyruvate carboxylase plays a crucial role in limiting nitrogen fixation in *Lotus japonicus* nodules. *Plant Cell Physiol.*, **47**, 613–621.

Obermeyer, G. and Tyerman, S.D. (2005) NH_4^+ currents across the peribacteroid membrane of soybean. Macroscopic and microscopic properties, inhibition by Mg^{2+}, and temperature dependence indicate a SubpicoSiemens channel finely regulated by divalent cations. *Plant Physiol.*, **139**, 1015–1029.

Oka-Kira, E. and Kawaguchi, M. (2006) Long-distance signaling to control root nodule number. *Curr. Opin. Plant Biol.*, **9**, 496–502.

Okamoto, S., Ohnishi, E., Sato, S., Takahashi, H., Nakazono, M., Tabata, S. and Kawaguchi, M. (2009) Nod factor/nitrate-induced CLE genes that drive HAR1-mediated systemic regulation of nodulation. *Plant Cell Physiol.*, **50**, 67–77.

Oldroyd, G.E. and Downie, J.A. (2008) Coordinating nodule morphogenesis with rhizobial infection in legumes. *Ann. Rev. Plant Biol.*, **59**, 519–546.

Ott, T., van Dongen, J.T., Gunther, C., Krusell, L., Desbrosses, G., Vigeolas, H., Bock, V., Czechowski, T., Geigenberger, P. and Udvardi, M.K. (2005) Symbiotic leghemoglobins are crucial for nitrogen fixation in legume root nodules but not for general plant growth and development. *Curr. Biol.*, **15**, 531–535.

Ott, T., Sullivan, J., James, E.K., Flemetakis, E., Gunther, C., Gibon, Y., Ronson, C. and Udvardi, M. (2009) Absence of symbiotic leghemoglobins alters bacteroid and plant cell differentiation during development of *Lotus japonicus* root nodules. *Mol. Plant-Microbe Interact.*, **22**, 800–808.

Poole, R.K. and Hill, S. (1997) Respiratory protection of nitrogenase activity in *Azotobacter vinelandii* – roles of the terminal oxidases. *Biosci. Rep.*, **17**, 303–317.

Prell, J. and Poole, P. (2008) Metabolic changes of rhizobia in legume nodules. *Trends Microbiol.*, **14**, 161–168.

Prell, J., White, J.P., Bourdes, A., Bunnewell, S., Bongaerts, R.J. and Poole, P.S. (2009) Legumes regulate *Rhizobium* bacteroid development and persistence by the supply of branched-chain amino acids. *Proc. Natl. Acad. Sci. USA*, **106**, 12477–12482.

Radutoiu, S., Madsen, L.H., Madsen, E.B., Felle, H.H., Umehara, Y., Gronlund, M., Sato, S., Nakamura, Y., Tabata, S., Sandal, N. and Stougaard, J. (2003) Plant recognition of symbiotic bacteria requires two LysM receptor-like kinases. *Nature*, **425**, 585–592.

Radutoiu, S., Madsen, L.H., Madsen, E.B., Jurkiewicz, A., Fukai, E., Quistgaard, E.M., Albrektsen, A.S., James, E.K., Thirup, S. and Stougaard, J. (2007) LysM domains mediate lipochitin-oligosaccharide recognition and *Nfr* genes extend the symbiotic host range. *EMBO J.*, **26**, 3923–3935.

Rastogi, V.K. and Watson, R.J. (1991) Aspartate aminotransferase activity is required for aspartate catabolism and symbiotic nitrogen fixation in *Rhizobium meliloti*. *J. Bacteriol.*, **173**, 2879–2887.

Rees, D.C. and Howard, J.B. (2000) Nitrogenase: standing at the crossroads. *Curr. Opin. Chem. Biol.*, **4**, 559–566.

Roberts, D.M. and Tyerman, S.D. (2002) Voltage-dependent cation channels permeable to NH_4^+, K^+, and Ca^{2+} in the symbiosome membrane of the model legume *Lotus japonicus*. *Plant Physiol.*, **128**, 370–378.

Ronson, C.W. and Primrose, S.B. (1979) Carbohydrate metabolism in Rhizobium trifolii: identification and symbiotic properties of mutants. *J. Gen. Microbiol.*, **112**, 77–88.

Ronson, C.W., Lyttleton, P. and Robertson, J.G. (1981) C_4-dicarboxylate transport mutants of *Rhizobium trifolii* form ineffective nodules on *Trifolium repens*. *Proc. Natl Acad. Sci. USA*, **78**, 4284–4288.

Ronson, C.W., Astwood, P.M., Nixon, B.T. and Ausubel, F.M. (1987) Deduced products of C_4-dicarboxylate transport regulatory genes of *Rhizobium leguminosarum* are homologous to nitrogen regulatory gene products. *Nucleic Acids Res.*, **15**, 7921–7934.

Saalbach, G., Erik, P. and Wienkoop, S. (2002) Characterisation by proteomics of peribacteroid space and peribacteroid membrane preparations from pea (*Pisum sativum*) symbiosomes. *Proteomics*, **2**, 325–337.

Saito, K., Yoshikawa, M., Yano, K., Miwa, H., Uchida, H., Asamizu, E., Sato, S., Tabata, S., Imaizumi-Anraku, H., Umehara, Y., Kouchi, H., Murooka, Y., Szczyglowski, K., Downie, J.A., Parniske, M., Hayashi, M. and Kawaguchi, M. (2007) NUCLEOPORIN85 is required for calcium spiking, fungal and bacterial symbioses, and seed production in *Lotus japonicus*. *Plant Cell*, **19**, 610–624.

Sandal, N., Petersen, T.R., Murray, J., Umehara, Y., Karas, B., Yano, K., Kumagai, H., Yoshikawa, M., Saito, K., Hayashi, M., Murakami, Y., Wang, X.-W., Hakoyama, T., Imaizumi-Anraku, H., Sato, S., Kato, T., Chen, W.-L., Shakhawat-Hossain, M., Shibata, S., Wang,, T., Yokota, K., Larsen, K., Kanamori, N., Madsen, E., Radutoiu, S., Madsen, L.H., Radu, T.G., Krusell, L., Ooki, Y., Banba, M., Betti, M., Rispail, N., Skot, L., Tuck, E., Perry, J., Yoshida, S., Vickers, K., Pike, J., Mulder, L., Charpentier, M., Muller, J., Ohtomo, R., Kojima, T., Ando, S., Marquez, A.J., Gresshoff, P.M., Harada, K., Webb, J., Hata, S., Suganuma, N., Kouchi, H., Kawasaki, S., Tabata, S., Hayashi, M., Parniske, M., Szczyglowski, K., Kawaguchi, M. and Stougaard, J. (2006) Genetics of symbiosis in *Lotus japonicus*: Recombinant inbred lines, comparative genetic maps and map position of 35 symbiotic loci. *Mol. Plant-Microbe Interact.*, **19**, 80–91.

Schauser, L., Roussis, A., Stiller, J. and Stougaard, J. (1999) A plant regulator controlling development of symbiotic root nodules. *Nature*, **402**, 191–195.

Schmitz, R.A., Klopprogge, K. and Grabbe, R. (2002) Regulation of nitrogen fixation in *Klebsiella pneumoniae* and *Azotobacter vinelandii*: NifL, transducing two environmental signals to the nif transcriptional activator NifA. *J. Mol. Microbiol. Biotechnol.*, **4**, 235–242.

Shimoda, Y., Nagata, M., Suzuki, A., Abe, M., Sato, S., Kato, T., Tabata, S., Higashi, S. and Uchiumi, T. (2005) Symbiotic rhizobium and nitric oxide induce gene expression of non-symbiotic hemoglobin in *Lotus japonicus*. *Plant Cell Physiol.*, **46**, 99–107.

Shimoda, Y., Shimoda-Sasakura, F., Kucho, K., Kanamori, N., Nagata, M., Suzuki, A., Abe, M., Higashi, S. and Uchiumi, T. (2009) Overexpression of class 1 plant hemoglobin genes enhances symbiotic nitrogen fixation activity between *Mesorhizobium loti* and *Lotus japonicus*. *Plant J.*, **57**, 254–263.

Smit, P., Limpens, E., Geurts, R., Fedorova, E., Dolgikh, E., Gough, C. and Bisseling, T. (2007) Medicago LYK3, an entry receptor in rhizobial nodulation factor signaling. *Plant Physiol.*, **145**, 183–191.

Spaink, H.P. (1995) The molecular basis of infection and nodulation by rhizobia: the ins and outs of sympathogenesis. *Ann. Rev. Phytopathol.*, **33**, 345–368.

Spaink, H.P., Sheeley, D.M., Van Brussel, A.A.N., Glushka, J., York, W.S., Tak, T., Geiger, O., Kennedy, E.P., Reinhold, V.N. and Lugtenberg, B.J.J. (1991) A novel highly unsaturated fatty acid moiety of lipo-oligosaccharide signals determines host specificity of *Rhizobium*. *Nature*, **354**, 125–130.

Starker, C.G., Parra-Colmenares, A.L., Smith, L., Mitra, R.M. and Long, S.R. (2006) Nitrogen fixation mutants of *Medicago truncatula* fail to support plant and bacterial symbiotic gene expression. *Plant Physiol.*, **140**, 671–680.

Stokkermans, T.J. and Peters, N.K. (1994). *Bradyrhizobium elkanii* lipo-oligosaccharide signals induce complete nodule structures on *Glycine soja* Siebold et Zucc. *Planta*, **193**, 413–420.

Stracke, S., Kistner, C., Yoshida, S., Mulder, L.. Sato, S., Kaneko, T., Tabata, S., Sandal, N., Stougaard, J., Szczyglowski, K. and Parniske, M. (2002) A plant receptor-like kinase required for both bacterial and fungal symbiosis. *Nature*, **417**, 959–962.

Suganuma, N., Nakamura, Y., Yamamoto, M., Ohta, T., Koiwa, H., Akao, S. and Kawaguchi, M. (2003) The *Lotus japonicus* Sen1 gene controls rhizobial differentiation into nitrogen-fixing bacteroids in nodules. *Mol. Genet. Genom.*, **269**, 312–320.

Szczyglowski, K., Kapranov, P., Hamburger, D. and deBruijn, F.J. (1998) The *Lotus japonicus LjNOD70* nodulin gene encodes a protein with similarities to transporters. *Plant Mol. Biol.*, **37**, 651–661.

Taté, R., Riccio, A., Merrick, M., and Patriarca, E.J. (1998) The *Rhizobium etli amtB* gene coding for an NH_4^+ transporter is down-regulated early during bacteroid differentiation. *Mol. Plant-Microbe Interact.*, **11**, 188–198.

Tajima, S. and Kouchi, H. (1996) Metabolism and compartmentation of carbon and nitrogen in legume nodules, in *Plant-Microbe Interactions*, Vol. **2** (eds G. Stacey and N. Keen), Chapman & Hall, New York, pp. 27–60.

Tajima, S., Nomura, M. and Kouchi, H. (2004) Ureide biosynthesis in legume nodules. *Front. Biosci.*, **9**, 1374–1381.

Tirichine, L., Imaizumi-Anraku, H., Yoshida, S., Murakami, Y., Madsen, L.H., Miwa, H., Nakagawa, T., Sandal, N., Albrektsen, A.S., Kawaguchi, M., Downie, A., Sato, S., Tabata, S., Kouchi, H., Parniske, M., Kawasaki, S. and Stougaard, J. (2006) Deregulation of a Ca^{2+}/calmodulin-dependent kinase leads to spontaneous nodule development. *Nature*, **441**, 1153–1156.

Tirichine, L., Sandal, N., Madsen, L.H., Radutoiu, S., Albrektsen, A.S., Sato, S., Asamizu, E., Tabata, S. and Stougaard, J. (2007) A gain-of-function mutation in a cytokinin receptor triggers spontaneous root nodule organogenesis. *Science*, **315**, 104–107.

Tujimura, K. and Meguro, H. (1960) Respiration substrate of *Rhizobium* in the nodules. *J. Biochem.*, **47**, 391–397.

Uchiumi, T., Ohwada, T., Itakura, M., *et al.* (2004) Expression islands clustered on the symbiosis island of the *Mesorhizobium loti* genome. *J. Bacteriol.*, **186**, 2439–2448.

Udvardi, M.K. and Day, D.A. (1997) Metabolite transport across symbiotic membranes of legume nodules. *Ann. Rev. Plant Physiol. Plant Mol. Biol.*, **48**, 493–523.

Van de Velde, W., Zehirov, G., Szatmari, A., Debreczeny, M., Ishihara, H., Kevei, Z., Farkas, A., Mikulass, K., Nagy, A., Tiricz, H., Satiat-Jeunemaitre, B., Alunni, B., Bourge, M., Kucho, K., Abe, M., Kereszt, A., Maroti, G., Uchiumi, T., Kondorosi, E. and Mergaert, P. (2010) Plant peptides govern terminal differentiation of bacteria in symbiosis. *Science*, **327**, 1122–1126.

Vassileva, V., Kouchi, H. and Ridge, R.W. (2005) Microtubule dynamics in living root hairs: transient slowing by lipochitin oligosaccharide nodulation signals. *Plant Cell*, **17**, 1777–1787.

Vincill, E.D., Szczyglowski, K. and Roberts, D.M. (2005) GmN70 and LjN70. Anion transporters of the symbiosome membrane of nodules with a transport preference for nitrate. *Plant Physiol.*, **137**, 1435–1444.

Wallace, I.S., Choi, W.G. and Roberts, D.M. (2006) The structure, function and regulation of the nodulin 26-like intrinsic protein family of plant aquaglyceroporins. *Biochim. Biophys. Acta*, **1758**, 1165–1175.

Wang, D., Griffiths, J., Starker, C., Fedorova, E., Limpens, E., Ivanov, S., Bisseling, T. and Long, S. (2010) A nodule-specific protein secretory pathway required for nitrogen-fixing symbiosis. *Science*, **327**, 1126–1129.

Weaver, C.D., Nirah, H. Shomerg, N.H., Louis, C.F. and Roberts, D.M. (1994) Nodulin 26, a nodule-specific symbiosome membrane protein from soybean, is an ion channel. *J. Biol. Chem.*, **269**, 17858–17862.

White, J., Prell, J., James, E.K. and Poole, P. (2007) Nutrient sharing between symbionts. *Plant Physiol.*, **144**, 604–614.

Wienkoop, S. and Saalbach, G. (2003) Proteome analysis. Novel proteins identified at the peribacteroid membrane from *Lotus japonicus* root nodules. *Plant Physiol.*, **131**, 1080–1090.

Yamaya, H. and Arima, Y. (2010) Evidence that a shoot-derived substance is involved in regulation of the super-nodulation trait in soybean. *Soil Sci. Plant Nutr.*, **56**, 115–122.

Yano, K., Yoshida, S., Müller, J., Singh, S., Banba, M., Vickers, K., Markmann, K., White, C., Schuller, B., Sato, S., Asamizu, E., Tabata, S., Murooka, Y., Perry, J., Wang, T.L., Kawaguchi, M., Imaizumi-Anraku, H., Hayashi, M. and Parniske, M. (2008) CYCLOPS, a mediator of symbiotic intracellular accommodation. *Proc. Natl. Acad. Sci. USA*, **105**, 20540–20545.

Yano, K., Shibata, S., Chen, W.-L., Sato, S., Kaneko, T., Jurkiewicz, A., Sandal, N., Banba, M., Imaizumi-Anraku, H., Kojima, T., Ohtomo, R., Szczyglowski, K., Stougaard, J., Tabata, S., Hayashi, M., Kouchi, H. and Umehara, Y. (2009) *CERBERUS*, a novel U-box protein containing WD-40 repeats is required for infection thread formation and nodule development in the legume-*Rhizobium* symbiosis. *Plant J.*, **60**, 168–180.

Yokota, K., Fukai, E., Madsen, L.H., Jurkiewicz, A., Rueda, P., Radutoiu, S., Held, M., Hossain, M. S., Szczyglowski, K., Morieri, G., Oldroyd, G.E., Downie, J.A., Nielsen, M.W., Rusek, A.M., Sato, S., Tabata, S., James, E.K., Oyaizu, H., Sandal, N. and Stougaard, J. (2009) Rearrangement of actin cytoskeleton mediates invasion of *Lotus japonicus* roots by *Mesorhizobium loti*. *Plant Cell*, **21**, 267–284.

Zheng, L., White, R.H. and Dean, D.R. (1997) Purification of the *Azotobacter vinelandii* nifV-encoded homocitrate synthase. *J. Bacteriol.* **179**, 5963–5966.

4

Sulfur Metabolism

Hideki Takahashi

4.1 Introduction

Sulfur is an essential element and classified as one of the six macroelements that significantly affect plant growth (Marschner, 1995). Sulfur is found in both inorganic and organic compounds in nature. The sulfur cycle is globally balanced between assimilatory and dissimilatory functions of plants and microorganisms (Crawford *et al.*, 2000; Leustek *et al.*, 2000; Saito 2000, 2004; Kopriva, 2006). Plants and microorganisms assimilate inorganic sulfur to produce organic sulfur, and animals will utilise the organic forms of sulfur as nutrient sources. Plants normally utilise sulfate ($SO_4{}^{2-}$), the most oxidised form of sulfur (+VI redox state), to synthesize organic sulfur-compounds (Leustek *et al.*, 2000; Saito, 2000, 2004). Inorganic sulfur-compounds in reduced states such as hydrogen sulfide (H_2S) and sulfur dioxide (SO_2) are erupted from volcanoes and hot springs, although they are not the sulfur sources primarily utilised by higher plant species. Organic sulfur is present in soil as sulfated esters and sulfonates. Microorganisms in soil degrade these organic sulfur-compounds and release sulfate. In addition, wastes and remains from plants and animals enter the degradation pathways in soil microbes. Catabolic degradation of organic sulfur to inorganic sulfate is an essential step for recycling sulfur source for plants.

Sulfur is present in both major cellular constituents and specialised compounds. Cysteine and methionine are the sulfur-containing essential amino acids. For folding of proteins, the thiol residue of cysteine serves as a moiety to form S-S bonds. It also serves as active sites for electron transfer in enzyme reactions. Glutathione, ferredoxins and thioredoxins are peptides and proteins which contain cysteine to function in redox regulation (Buchanan and Balmer, 2005). Sulfur can be also found in sulfolipid which is an essential component

Plant Metabolism and Biotechnology, First Edition. Edited by Hiroshi Ashihara, Alan Crozier, and Atsushi Komamine.

of chloroplast membranes (Benning, 1998). Other than being cellular constituents like amino acids, proteins and lipids, sulfur is present in secondary sulfur-metabolites having specific biological functions. Vitamins and cofactors such as thiamine, biotin and coenzyme A contain sulfur. Some secondary sulfur-metabolites are important as being beneficial for human. Glucosinolates in *Brassicaceae* plants and *S*-alkylcysteine sulfoxides in *Alliaceae* plants are known to induce detoxifying enzymes and prevent tumour formation (Talalay and Fahey, 2001; Bianchini and Vainio, 2001). They also have special odours and pungency, and are repellants against insects and microorganisms (Jones *et al.*, 2004; Halkier and Gerschenzon, 2006; Grubb and Abel, 2006). Sulfated forms of *N*-acylated chitooligosaccharides are known as Nod factors of symbiotic nitrogen-fixing rhizobacteria promoting nodule formation in legumes (Fisher and Long, 1992).

As mentioned here, plants synthesize a wide variety of sulfur-compounds that serve for maintenance of cell viability. Plants are autotrophic for sulfur and able to assimilate inorganic sulfate to synthesize all these metabolites. In other words, sulfur in all metabolites present in plants derives from sulfate (Figure 4.1). The input of sulfate therefore influences the remaining metabolic pathways. This review will focus on how sulfate is transported and metabolised to cysteine and methionine, and how sulfur metabolisms are controlled under regulatory components that optimise sulfur use efficiencies, responding to environmental sulfur availabilities and intrinsic signals.

4.2 Sulfate Transport

4.2.1 Sulfate Transport Mechanisms

Uptake of sulfate is the entry step of sulfur metabolism (Figure 4.1). Plants have eukaryotic-type membrane-bound sulfate transport proteins that facilitate uptake of sulfate from extracellular space to cytoplasm across plasma membranes. At a whole-plant level, sulfate moves from cell to cell and transfers from roots to above-ground organs through xylems to fulfill the requirements of sulfur in photosynthetic tissues. Multiple transport systems eventually mediate the intercellular and interorgan transport processes. Following uptake to the cell, sulfate accumulates in vacuoles for storage or enters chloroplast/plastids.

A plasma membrane-bound protein that facilitates the uptake of sulfate to the cell requires a driving force to import sulfate, a negatively charged ion, against the gradients of membrane potential and sulfate concentration (Figure 4.2). Under these circumstances, influx of sulfate can be mediated by a secondary active co-transport system that utilises proton gradients as motive force or by an anion exchange system that needs to be coupled with an outward movement of another anion. Current understanding indicates that a proton/sulfate co-transport system probably mediates the influx of sulfate in plant cells (Lass and Ullrich-Eberius, 1984; Hawkesford *et al.*, 1993). Proton pumping by H^+-ATPase is suggested to be coupled with the influx of sulfate, but at the same time the outside positive potential can facilitate the efflux of sulfate through an unknown passive transport mechanism. This mechanism applies to accumulation of sulfate in vacuoles where H^+-ATPase and H^+-pyrophosphatase at tonoplast form positive potentials on vacuolar lumen side (Martinoia *et al.*, 2000, 2007). It is known that both saturable and linear components may work as facilitators for influx of sulfate to vacuoles (Kaiser *et al.*, 1989). Efflux of sulfate from

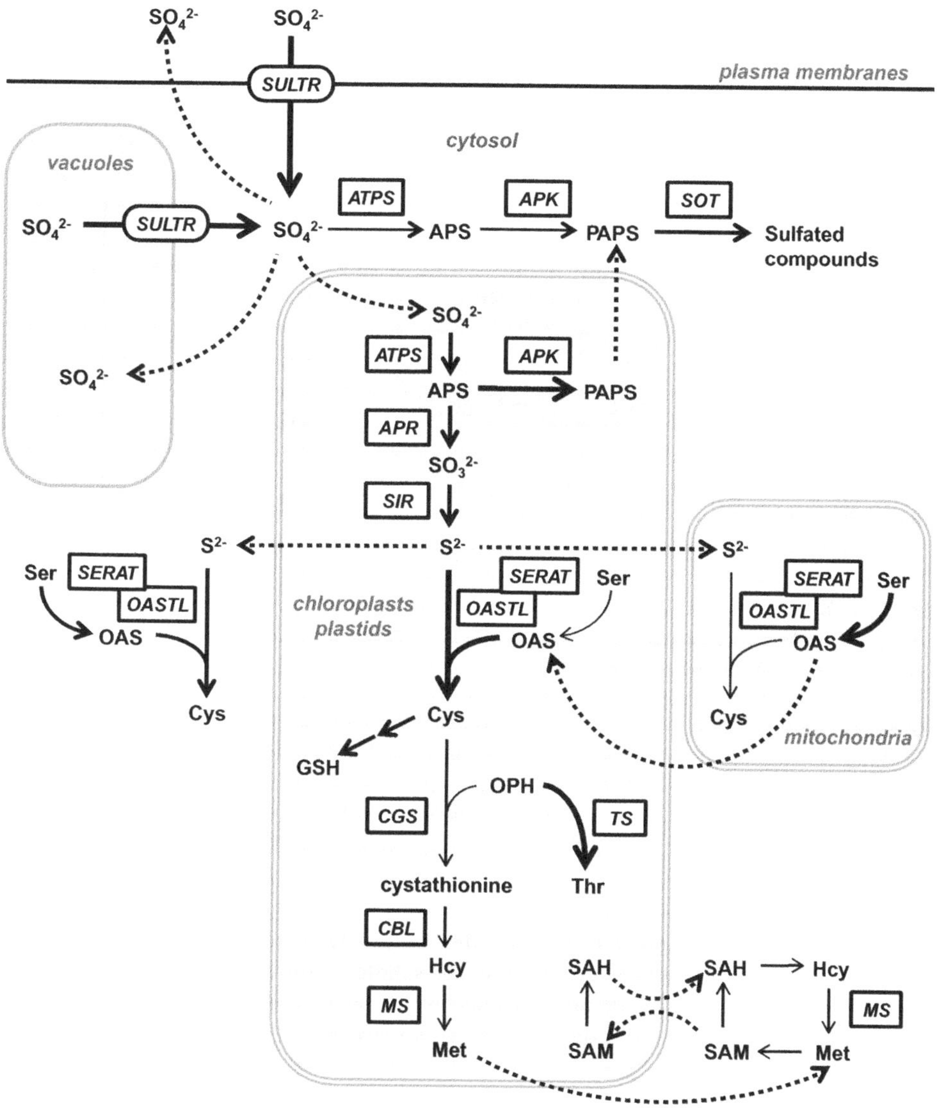

Figure 4.1 Pathways for primary sulfur metabolism from sulfate uptake to cysteine and methionine biosynthesis in higher plants. The figure illustrates pathways for sulfate transport across plasma membranes and tonoplast, and sulfur metabolism in cytosol, chloroplasts/plastids and mitochondria in higher plants. Fluxes of metabolites are indicated by thicknesses of the lines. Putative metabolite transport pathways are indicated by dashed lines. Abbreviations: APR, APS reductase; APS, adenosine 5′-phosphosulfate; ATPS, ATP sulfurylase; CBL, cystathionine β-lyase; CGS, cystathionine γ-synthase; Cys, cysteine; GSH, glutathione; Hcy, homocysteine; Met, methionine; MS, methionine synthase; OAS, O-acetylserine; OASTL, OAS(thiol)lyase; OPH, O-phosphohomoserine; SAH, S-adenosylhomocysteine; SAM, S-adenosylmethionine; Ser, serine; SERAT, serine acetyltransferase; SIR, sulfite reductase; SULTR, sulfate transporter; TS, threonine synthase

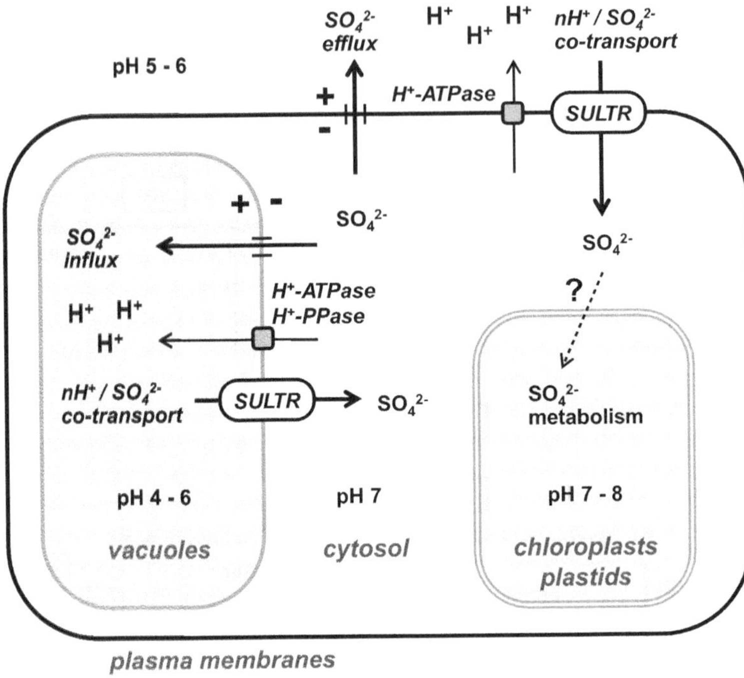

Figure 4.2 *Mechanisms of sulfate transport across plasma membrane and tonoplast. Sulfate transporters (SULTR) mediate import of sulfate to the cell across plasma membrane and efflux of sulfate from vacuoles across tonoplast. H^+-ATPase and H^+-pyrophosphatase (PPase) are indicated by grey squares. Proteins facilitating efflux of sulfate from the cell, and influx of sulfate to vacuoles and chloroplasts, are not known*

vacuoles is equally as significant as the influx of sulfate across plasma membranes, in the sense that similar mechanisms can provide sulfate to sulfur metabolic pathways. It is considered that proton gradients generated by tonoplast-bound H^+-ATPase and H^+-pyrophosphatase drive a proton/sulfate co-transport system to release vacuolar storage of sulfate to cytosol (Figure 4.2).

4.2.2 Sulfate Uptake System

In plant roots, the kinetics of sulfate uptake activity follows the Michealis-Menten equation (Leggett and Epstein, 1956). Analysis of sulfate uptake kinetics has indicated that a high-affinity sulfate transport protein mediates the influx of sulfate particularly under sulfur-limited conditions (Clarkson *et al.*, 1983; Deane-Drummond, 1987). This is equivalent to phase I transport systems for the uptake of other nutrients.

Molecular cloning of sulfate transporters was first demonstrated by complementation of a chromate/selenate resistant yeast mutant strain lacking sulfate transport activities (Smith *et al.*, 1995a; Cherest *et al.*, 1997). Subsequently, plant cDNAs encoding high-affinity sulfate transporters were identified from a leguminous plant, *Stylosanthes hamata*, by functional complementation of the yeast mutant (Smith *et al.*, 1995b). Up to now, numbers of orthologous sulfate transporters have been identified from various plant species (Smith *et al.*, 1997; Bolchi *et al.*, 1999; Vidmar *et al.*, 1999, 2000; Takahashi *et al.*, 2000; Shibagaki *et al.*, 2002; Yoshimoto *et al.*, 2002; Howarth *et al.*, 2003; Buchner *et al.*, 2004a; Hopkins *et al.*, 2005). Their major functions are suggested to be relevant to sulfate uptake as they are expressed in the roots of sulfur-starved plants. The *SULTR* gene family of a model plant, *Arabidopsis thaliana*, is the most well characterised group of sulfate transporters. The gene family consists of 12 distinct members classified into four subgroups (Buchner *et al.*, 2004b; Takahashi *et al.*, 2006; Takahashi, 2010). The components of the high-affinity sulfate uptake system of *Arabidopsis* roots are encoded by *SULTR1;1* and *SULTR1;2*, and their biochemical and physiological functions have been precisely demonstrated (Takahashi *et al.*, 2000; Vidmar *et al.*, 2000; Shibagaki *et al.*, 2002; Yoshimoto *et al.*, 2002).

In *Arabidopsis* roots, both *SULTR1;1* and *SULTR1;2* are expressed in epidermal and cortical cell layers where nutrients are absorbed to roots through the functions of transport proteins (Takahashi *et al.*, 2000; Shibagaki *et al.*, 2002; Yoshimoto *et al.*, 2002). *SULTR1;1* and *SULTR1;2* are regulated by sulfur at transcript and protein levels, and deletion of both genes results in loss of viability under low-sulfur conditions (Yoshimoto *et al.*, 2007; Barberon *et al.*, 2008). These results indicate the significance of two sulfate transport systems in sulfate uptake, although slight differences are present. SULTR1;2 is suggested to be the major form, while SULTR1;1 seems to represent a subsidiary or compensatory function (Yoshimoto *et al.*, 2007; Barberon *et al.*, 2008). The major contribution of SULTR1;2 was suggested from abundance of transcripts, characteristics in sulfate uptake, and tolerance of mutants to selenate (Shibagaki *et al.*, 2002; Yoshimoto *et al.*, 2002, 2007; Maruyama-Nakashita *et al.*, 2003; El Kassis *et al.*, 2007; Barberon *et al.*, 2008). In contrast, SULTR1;1 has been featured for its potential contribution to high-affinity sulfate uptake under extreme conditions. Its specialised functions are the low K_m value for sulfate, and strong and rapid induction of transcript accumulation in response to sulfur starvation (Takahashi *et al.*, 2000; Yoshimoto *et al.*, 2002).

4.2.3 Transport of Sulfate from Roots to Shoots

After entering epidermal and cortical cell layers, sulfate moves horizontally through the plasmodesmata between the cells and reaches xylem parenchyma cells (Figure 4.3). During this horizontal transfer through the symplastic pathway, sulfate can be leaked from the cells to apoplast (i.e. cell wall space) but quickly retrieved back to the cells as SULTR1;1 and SULTR1;2 are present (Figure 4.3). The sulfate efflux system remains unverified. The Casparian strip of the endodermal cell layer makes a barrier for apoplastic transfer of nutrients and water to the central cylinder. Once sulfate reaches pericycle layers inside the central cylinder, it can be released to apoplast. Transporters having retrieval functions would be necessary to bring sulfate back to symplastic pathways in parenchyma cells connected to xylem vessels. According to this model, the presence of barley HVST1 sulfate transporter in pericycle and xylem parenchyma cells suggests its function for this retrieval mechanism (Rae and Smith, 2002). In *Arabidopsis*, low-affinity sulfate transporters probably mediate this process. SULTR2;1 and SULTR3;5 are the components suggested to co-facilitate import of sulfate and prevent the leakage of sulfate to apoplastic space in the central cylinder (Kataoka *et al.*, 2004a) (Figure 4.3). Interestingly, the activity of SULTR2;1

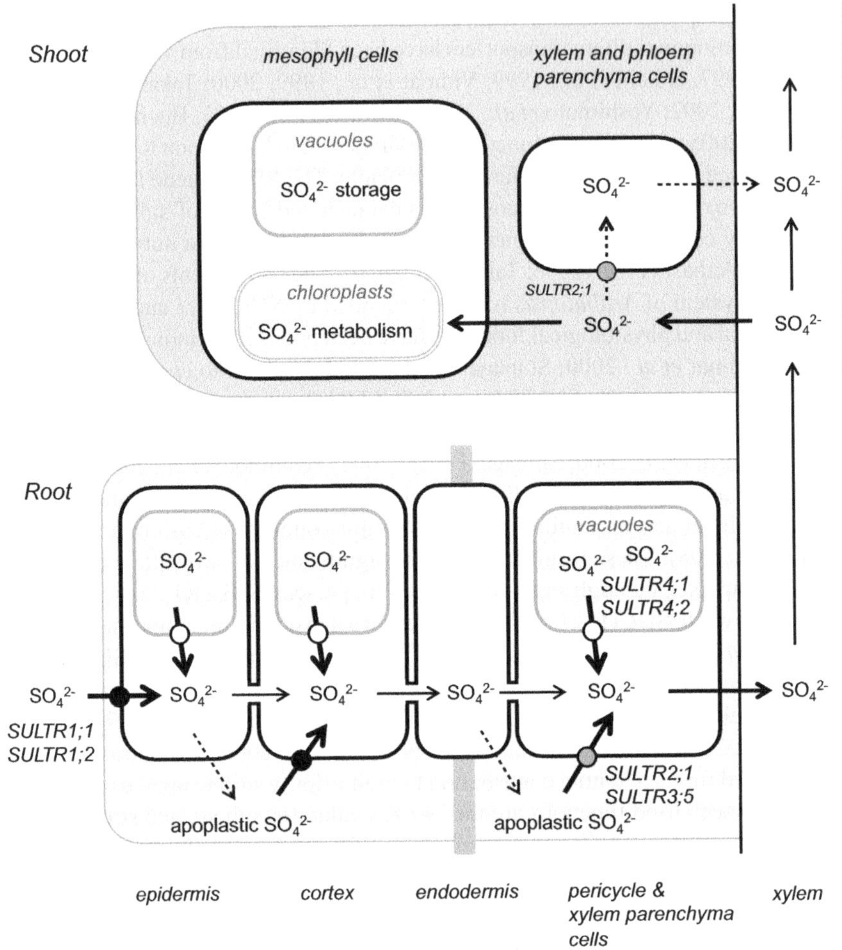

Figure 4.3 *Uptake and internal transport of sulfate in plants. Pathways operated for the uptake and root-to-shoot transport of sulfate in* Arabidopsis *are illustrated. Sulfate transport pathways upregulated are indicated by thick solid lines with arrowheads. Circles on the lines indicate sulfate transporters, SULTR1;1 and SULTR1;2 (black circles), SULTR4;1 and SULTR4;2 (white circles), and SULTR2;1 and SULTR3;5 (grey circles), respectively. Pathways downregulated under low-sulfur conditions or those unfavourable for sulfate uptake processes are indicated by dashed lines*

low-affinity sulfate transporter (Takahashi *et al.*, 1997, 2000) was shown to be enhanced in the presence of SULRT3;5, which by itself was non-functional (Kataoka *et al.*, 2004a). In addition to the proposed retrieval mechanism, release of sulfate from vacuoles has been shown to increase the flux of sulfate transported to shoots (Kataoka *et al.*, 2004b) (Figure 4.3). SULTR4;1 and SULTR4;2 are the tonoplast-localising transporters mediating sulfate remobilisation. Transporters involved in these two mechanisms are generally upregulated by sulfur limitation, suggesting that the processes are dependent upon the requirement of sulfur sources in shoots (Takahashi *et al.*, 1997, 2000; Kataoka *et al.*, 2004a,b).

4.2.4 Subcellular Transport of Sulfate

Vacuoles are the storage compartments for sulfate. The positive electrical charge gradient across the tonoplast is the most likely driving force to facilitate entry of sulfate to the vacuoles (Martinoia *et al.*, 2000, 2007). However, channel or carrier proteins mediating the influx of sulfate to vacuoles are not known. By contrast, export of sulfate from vacuoles requires SULTR4;1 and SULTR4;2 (Kataoka *et al.*, 2004b), suggesting that these components mediate proton/sulfate co-transport mechanisms at the tonoplast. Storage and release of the vacuolar sulfate pool depend upon demands for sulfur. Regulation of this process would be important for sulfur storage and efficient sulfur utilisation (Figures 4.2 and 4.3).

Sulfate in cytosol is metabolised by a cytosolic isoform of ATP sulfurylase or transported to chloroplasts/plastids to enter the reduction pathway. Transport systems for entry of sulfate to chloroplasts still remain elusive. It is known that sulfate inhibits the activity of phosphate/triose-phosphate translocator in chloroplasts, although the interference seems to occur under very high concentrations of sulfate (Gross *et al.*, 1990). The contribution of this system to sulfate influx is probably limited. In *Chlamydomonas reinhardii*, a bacterial-type sulfate transporter complex mediates the influx of sulfate to the chloroplast (Lindberg and Melis, 2008). This complex consists of a sulfate-binding protein, membrane-anchored proteins, and an ATP-binding cassette protein homologous to their bacterial origins (Sirko *et al.*, 1990; Laudenbach and Grossman, 1991). Orthologous proteins for this system have not been identified from higher plant species, suggesting that the mechanism is probably different from those in algae (Figure 4.2).

4.2.5 Redistribution of Sulfur

Redistribution of sulfur from source to sink organs occurs through phloem. Besides sulfate, glutathione and *S*-methylmethionine are present in phloem sap (Bourgis *et al.*, 1999; Herschbach *et al.*, 2000; Kuzuhara *et al.*, 2000). In *Arabidopsis* leaves, *SULTR2;1* low-affinity sulfate transporter gene is expressed in parenchyma cells around the phloem and xylem (Takahashi *et al.*, 2000). *SULTR2;1* was downregulated by sulfur limitation in leaf vasculature (Takahashi *et al.*, 2000). This would prevent retrieval of sulfate back to xylem and to phloem, which allows sulfate to be transferred to mesophyll cells (Figure 4.3). Eventually, distribution of sulfur via phloem can be restricted under low-sulfur conditions. The model suggests that the mobility of sulfate and/or other sulfur-compounds from old to young leaves declines during sulfur starvation. SULTR2;1 also functions for translocation of sulfur source to developing seeds where its expression is found in funiculus and vasculature of seed pods (Awazuhara *et al.*, 2005). It has been shown that developing seeds respond to sulfur deficiency and accumulate sulfurless seed storage proteins (Hirai *et al.*, 1995). However, it has not been defined whether deficits of sulfate or other sulfur-metabolites trigger this response in seeds. The metabolic pathways for synthesis of sulfur-compounds are present in developing cotyledons of lupin (Tabe and Droux, 2001). On the other hand, glutathione synthesis is active in embyos and funiculus of *Arabidopsis*, allowing the possibility that organic sulfur-compounds are synthesized from sulfate and transported to seeds (Cairns *et al.*, 2006).

Relevant to phloem transport, a high-affinity sulfate transporter, SULTR1;3, is shown to localise in companion cells of phloem (Yoshimoto *et al.*, 2003). The knockout of *SULTR1;3* restricted transport of sulfur from cotyledons to shoot meristem and roots in *Arabidopsis*, suggesting significance for this transporter in source-to-sink transport of sulfur (Yoshimoto *et al.*, 2003). Accumulation of *SULTR1;3* transcripts in sulfur-starved plants further suggests its importance under sulfur-limited conditions. The induction of *SULTR1;3* transcripts on low sulfur would make a considerable contribution to increasing the amount of sulfate loaded to phloem. It is also possible that transportable forms of sulfur-metabolites were decreased by repression of *SULTR2;1* in phloem parenchyma cells (Takahashi *et al.*, 2000), and the induction of *SULTR1;3* could have occurred in response to local sulfur deficiency. In addition to *SULTR1;3*, the low-affinity isoform, *SULTR2;2*, has been reported to localise in phloem companion cells (Takahashi *et al.*, 2000).

Vacuolar sulfate transporters are also suggested to participate in redistribution of sulfur. Accumulation of *SULTR4;1* and *SULTR4;2* transcripts in senescing leaves of *Brassica napus* plants suggests that sulfate pools are remobilised during senescence and most likely transferred to sink organs (Dubousset *et al.*, 2009). Sulfate in vacuole is considered to be temporal storage of sulfur, and will be released by induction of *SULTR4;1* and *SULTR4;2* during sulfur limitation or senescence to allow it to transfer to sink organs.

4.2.6 Regulation of Sulfate Uptake

Sulfate transporters in roots are primarily regulated by sulfate availability (Smith *et al.*, 1995b, 1997; Takahashi *et al.*, 2000; Vidmar *et al.*, 2000; Shibagaki *et al.*, 2002; Yoshimoto *et al.*, 2002; Maruyama-Nakashita *et al.*, 2004a). The amount of sulfate supply affects synthesis of downstream sulfur-metabolites such as cysteine and glutathione. These metabolites can feedback-regulate the expression of sulfate transporters (Herschbach and Rennenberg, 1991; Smith *et al.*, 1997; Bolchi *et al.*, 1999; Lappartient *et al.*, 1999; Vidmar *et al.*, 1999, 2000; Maruyama-Nakashita *et al.*, 2004b). It is further suggested that sulfate to glutathione ratios of phloem sap correlate with rates of sulfate acquisition in roots (Herschbach and Rennenberg, 1991; Herschbach *et al.*, 2000). In addition, interorgan repressive signals are possibly related to glutathione in phloem, regulating the amount of low-affinity sulfate transporter gene, *SULTR2;1*, expressed in roots (Lappartient and Touraine, 1996; Lappartient *et al.*, 1999).

As for the positive signals, *O*-acetylserine (OAS) can induce the expression of sulfate transporters by external application (Smith *et al.*, 1997; Maruyama-Nakashita *et al.*, 2004b, 2005). OAS is the precursor of cysteine. It is suggested that OAS can override negative feedback effects of cysteine and glutathione accumulated in the cells (Smith *et al.*, 1997). It is not known how these metabolites may stimulate signalling cascades, however, microarray studies suggest significant overlaps between transcriptomes of OAS treatment and sulfur starvation in *Arabidopsis* (Hirai *et al.*, 2003, 2004; Nikiforova *et al.*, 2003). The sulfur responsive element (SURE) present in the 5′-region of *SULTR1;1* has been shown to respond not only to sulfate but also cysteine, glutathione and OAS (Maruyama-Nakashita *et al.*, 2005). This provides additional evidence that metabolic effectors actually influence the sulfur-responsive regulatory pathway. However, as suggested by the absence of SURE in the 5′-region of *SULTR1;2*, the regulatory pathways are not always uniform among the two sulfur-responsive high-affinity sulfate transporters. Nevertheless, both of them show similar responses to sulfur limitation and effector metabolites.

In addition to sulfur specific regulations, sulfate transporters are known to be regulated by other general factors. As an intrinsic signal, a plant hormone cytokinin was shown to repress the expression of high-affinity sulfate transporters in *Arabidopsis* roots (Maruyama-Nakashita *et al.*, 2004b). Signals downstream of cytokinin receptor CRE1 (Inoue *et al.*, 2001) diminished the transcript levels of *SULTR1;1* and *SULRT1;2*; however, their sulfur-deficiency responsiveness was not substantially affected (Maruyama-Nakashita *et al.*, 2004b). It is suggested that cytokinin functions as a general repressive signal to modulate the levels of sulfate transporters and sulfate uptake rates in roots. Generality of the cytokinin signalling pathway in the regulation of nutrient acquisition has been suggested from its involvement in controlling phosphorus response (Martin *et al.*, 2000; Franco-Zorrilla *et al.*, 2002). Besides hormone signals, carbon and nitrogen status are general factors that may enhance the levels of high-affinity sulfate transporters in roots (Vidmar *et al.*, 1999; Lejay *et al.*, 2003; Wang *et al.*, 2003; Maruyama-Nakashita *et al.*, 2004c). In general, *SULTR1;1* is regulated more specifically by sulfur conditions, whereas *SULTR1;2* accepts broader environmental signals for regulation (Rouached *et al.*, 2008).

The promoter analysis of sulfur responsiveness of *SULTR1;1* gene expression indicates the presence of an auxin response factor binding sequence (Ulmasov *et al.*, 1999; Hagen and Guilfoyle, 2002) within the *cis*-element (Maruyama-Nakashita *et al.*, 2005). Although a potential binding sequence in the *cis*-element was relevant to a general growth regulator, the response mediated under this element was rather specific to sulfur (Maruyama-Nakashita *et al.*, 2005). Identification of SLIM1/EIL3 from *Arabidopsis* indicates another unique upstream pathway significant for regulation of both *SULTR1;1* and *SULTR1;2* during sulfur limitation (Maruyama-Nakashita *et al.*, 2006). The role of this transcription factor will be described in the last section of this chapter.

As for post-transcriptional regulation, *SULTR1;1* and *SULRT1;2* are shown to be regulated at protein levels (Yoshimoto *et al.*, 2007). Over-accumulation of SULTR proteins occurred during sulfur limitation even though the genes were expressed under a constitutive promoter. This provides evidence for the significance of SULTR1;2 whose transcript accumulates only moderately in response to sulfur limitation under the native promoter. Another mode of post-transcriptional regulation involves the function of its *C*-terminus region. Sulfate transporter has a hydrophilic STAS (sulfate transporter and anti-sigma factor antagonist) domain in its *C*-terminus (Aravind and Koonin, 2000). Plasma membrane localisation and function of sulfate transporter are suggested to be largely influenced when mutations are introduced around a putative phosphorylated residue in the STAS domain (Shibagaki and Grossman, 2004, 2006; Rouached *et al.*, 2005). Other than the STAS domain, polar residues present in the first and second predicted membrane spanning regions are shown to be necessary for the activity of sulfate transporter (Leves *et al.*, 2008).

4.3 Sulfate Reduction

4.3.1 ATP Sulfurylase

ATP sulfurylase is the first committing enzyme for reduction of sulfate (Figure 4.1). The enzyme catalyzes the reaction, generating adenosine 5′-phosphosulfate (APS) and pyrophosphate from sulfate and ATP. Removal of pyrophosphate drives the reaction, otherwise the enzyme catalyzes the reverse reaction. ATP sulfurylase activity is present in both cytosol and chloroplasts (Lunn *et al.*, 1990; Renosto *et al.*, 1993; Rotte and Leustek, 2000). cDNAs encoding these subcellular specific isoforms are identified from potato (Klonus *et al.*, 1994). *Arabidopsis* has four ATP sulfurylase genes, *ATPS1* to *ATPS4*. Proteins encoded by these four *ATPS* genes in *Arabidopsis* contain chloroplast-targeting transit peptides in their *N*-terminal regions (Leustek *et al.*, 1994; Murillo and Leustek, 1995; Logan *et al.*, 1996; Hatzfeld *et al.*, 2000a), although *ATPS2* is predicted to be translated also as a cytosolic isoform (Hatzfeld *et al.*, 2000a). It still remains unresolved how these four *ATPS* isoforms share their functions *in planta*.

4.3.2 APS Reductase

In higher plants, APS is directly reduced to sulfite by APS reductase (APR) (Gutierrez-Marcos *et al.*, 1996; Setya *et al.*, 1996; Prior *et al.*, 1999; Suter *et al.*, 2000) (Figure 4.1). The pathway is different from those operating in fungi where APS is first phosphorylated by APS kinase (APK) to form 3′-phosphoadenosine 5′-phosphosulfate (PAPS), then PAPS reductase subsequently serves for enzymatic conversion of PAPS to sulfite. Some bacteria and lower plants have both pathways (Koprivova *et al.*, 2002). Thus, plant APR is suggested to derive evolutionally from PAPS reductase (Kopriva and Korpivova, 2004). Three *APR* genes, *APR1*, *APR2* and *APR3*, are identified from *Arabidopsis* (Gutierrez-Marcos *et al.*, 1996; Setya *et al.*, 1996). APR is a nuclear encoded chloroplast/plastid-localising enzyme. Consistent with predicted subcellular localisations of APR isoforms, the enzyme activity was found only in chloroplasts/plastids (Rotte and Leustek, 2000). The mature APR enzyme consists of a catalytic domain similar to PAPS reductase, and a C-terminal thioredoxin-like domain which functions as a glutaredoxin and utilises reduced glutathione as a reductant (Bick *et al.*, 1998; Prior *et al.*, 1999; Koprivova *et al.*, 2002).

In sulfate reduction pathways of plants, APR is the only metabolic enzyme strongly upregulated by sulfur starvation at transcript levels (Gutierrez-Marcos *et al.*, 1996; Takahashi *et al.*, 1997). APR is suggested to be the flux-controlling enzyme (Vauclare *et al.*, 2002). The significance of this pathway has been suggested from studies with ectopic overexpression of a *Pseudomonas APR* gene as a chloroplast-targeted form in *Arabidopsis* (Tsakraklides *et al.*, 2002). Sulfate reduction and cysteine synthesis were enhanced in overexpressors, suggesting that the step catalyzed by APR is rate-limiting under normal conditions (Tsakraklides *et al.*, 2002). Furthermore, *APR2* is identified as a locus determining sulfate accumulation in *Arabidopsis* natural variation, suggesting significance of this enzyme in driving sulfate reduction pathways (Loudet *et al.*, 2007).

4.3.3 APS Kinase

In higher plants, APK and APR share a common substrate, APS, at the juncture of the sulfate reduction pathway (Leustek *et al.*, 2000; Saito, 2000; Kopriva and Korpivova, 2004; Kopriva, 2006) (Figure 4.1). PAPS is not a substrate of sulfate reduction, but instead it is used exclusively for sulfation reactions in plants (Klein and Papenbrock, 2004; Piotrowski *et al.*, 2004; Hirai *et al.*, 2005; Klein *et al.*, 2006). In fact, the amounts of sulfated metabolites including glucosinolates were reduced in the knockout of APK isoforms in *Arabidopsis* (Mugford *et al.*, 2009). As mentioned in the following subsections, APK and APR were oppositely regulated to direct the flux of sulfur for use in cysteine biosynthesis or secondary sulfur metabolism, depending on sulfur availabilities (Takahashi *et al.*, 1997; Maruyama-Nakashita *et al.*, 2003; Hirai *et al.*, 2003, 2004).

Arabidopsis has four APK isoforms (Lee and Leustek, 1998; Lillig *et al.*, 2001). APK1, APK2 and APK4 are localised in chloroplasts/plastids, while APK3 encodes a cytosolic isoform (Mugford *et al.*, 2009). Synthesis of its substrate, APS, occurs in both compartments. *Arabidopsis ATPS2* is predicted to encode both chloroplastic and cytosolic isoforms (Hatzfeld *et al.*, 2000a), and a cytosolic ATPS exists in potato (Klonus *et al.*, 1994). Considering subcellular localisation of APK and ATPS isoforms, conversion of sulfate to APS and further phosphorylation to PAPS may occur partly in cytosol; however, two chloroplast/plastid-localising APK isoforms, APK1 and APK2, are the major components that make significant contributions to the generation of PAPS for sulfation reactions in *Arabidopsis* (Mugford *et al.*, 2009). On the other hand, as for the use of PAPS in sulfation reactions, sulfotransferases are shown to localise in cytosol or are predicted to be in compartments other than chloroplasts (Klein and Papenbrock, 2004; Klein *et al.*, 2006). These observations still leave us to consider cytosolic pathways for ATPS and APK, and the relevance of transport of PAPS across chloroplast membranes (Figure 4.1).

4.3.4 Sulfite Reductase

Sulfite reductase (SIR) is a chloroplasts/plastids-localising enzyme that generates sulfide from sulfite using reduced ferredoxin as a reductant (Krüger and Siegel, 1982) (Figure 4.1). Photosynthetic electron transfer provides reduced ferredoxin in leaves, while NADPH generated from the oxidative pentose phosphate pathway reduces ferredoxin in roots (Yonekura-Sakakibara *et al.*, 1998, 2000). The *Arabidopsis SIR* is a single copy gene (Brühl *et al.*, 1996; Bork *et al.*, 1998), suggesting that the reaction is a rate-limiting step of sulfate reduction. However, *SIR* transcripts are not regulated significantly by sulfur conditions (Takahashi *et al.*, 1997). As mentioned previously, APR and APK are the enzymes regulated at transcript levels (Takahashi *et al.*, 1997; Maruyama-Nakashita *et al.*, 2003; Hirai *et al.*, 2003). SIR appears to be present as an abundant enzyme that helps rapid conversion of toxic sulfite to sulfide. SIR is also known as a chloroplast DNA-binding protein. Nucleoid attachment of this enzyme is relevant to compaction of DNA and is suggested to regulate transcription in chloroplasts (Sekine *et al.*, 2002, 2007; Chi-Ham *et al.*, 2002).

4.3.5 Regulation of Sulfate Reduction

Sulfate reduction pathways are regulated under complex mechanisms in plants. As for the entry step of sulfate reduction, *ATPS* transcript levels are not always regulated positively for reduction of sulfate in sulfur-starved plants. The regulatory mechanism for *ATPS* appears to be complicated. In *Arabidopsis*, *ATPS1* and *ATPS3* are known as isoforms whose mRNAs are moderately accumulated by sulfur limitation and repressed by feedback regulation responding to demands for sulfur (Takahashi *et al.*, 1997; Lappartient *et al.*, 1999). However, recent studies indicate an additional mechanism of post-transcriptional regulation. Under sulfur deficiency, microRNA-395 (miR395) accumulates to destabilise *ATPS1* and *ATPS3* transcripts (Jones-Rhoades and Bartel, 2004; Allen *et al.*, 2005; Kawashima *et al.*, 2009). *ATPS2* lacks the miR395 target sequence and is insensitive to this mechanism. *ATPS4* is an isoform repressed under sulfur deficiency and is subject to miR395-mediated regulation (Kawashima *et al.*, 2009). More importantly, SLIM1, a transcription factor that controls the sulfur limitation response (Maruyama-Nakashita *et al.*, 2006), is an upstream regulator of miR395 which in turn negatively controls *ATPS* transcript accumulation (Kawashima *et al.*, 2009).

Regulation of sulfate reduction essentially occurs at the juncture of primary and secondary sulfur metabolism in plants. As described in the previous subsection, APR and APK share a common substrate, APS. When supply of sulfate is limited, *APR* transcripts will increase (Gutierrez-Marcos *et al.*, 1996; Takahashi *et al.*, 1997), while those for *APK* decrease significantly (Maruyama-Nakashita *et al.*, 2003; Hirai *et al.*, 2003). Assuming that transcript levels correlate with abundance of enzymes being translated, the mechanism favours an increased sulfur flux in the reduction pathway. This would ensure the plants can primarily synthesize essential sulfur-metabolites such as cysteine and methionine under sulfur deficiency. By contrast, APK is likely to be an unfavourable enzyme for primary metabolism, as the metabolic pathway is branched for secondary metabolism. It appears that synthesis of PAPS is allowed mostly under sulfur-rich conditions.

APR is a key metabolic enzyme that drives the reduction of sulfate in plants. It is primarily regulated at transcript levels by the availability of sulfate (Gutierrez-Marcos *et al.*, 1996; Takahashi *et al.*, 1997). In addition, the isoforms of *APR* in *Arabidopsis* are responsive to OAS, cysteine and GSH as expected (Koprivova *et al.*, 2000; Vauclare *et al.*, 2002; Hesse *et al.*, 2003). Besides regulation by sulfur, *APR* transcript accumulations are dependent on supplies of nitrogen and carbon sources that provide the backbone of cysteine (Kopriva *et al.*, 1999, 2002; Koprivova *et al.*, 2000; Hesse *et al.*, 2003; Wang *et al.*, 2003). Supply of carbon may increase the amount of reductant used for the APR reaction, as NADPH can eventually reduce glutathione. In addition to general metabolic balancing, plant hormones participate in the regulation of *APR*. Upregulation of *APR* by cytokinin is suggested to be caused by sucrose accumulation (Ohkama *et al.*, 2002) or by reduced uptake of sulfate (Maruyama-Nakashita *et al.*, 2004b). Jasmonate upregulates pathways of sulfate reduction and metabolism in addition to *APR*, suggesting a mechanism to alleviate oxidative stress and/or to induce defence response (Jost *et al.*, 2005). At the level of enzyme activities, *Arabidopsis* APR1 was activated under oxidative conditions, suggesting an additional mode of regulation for this key enzyme (Bick *et al.*, 2001).

4.4 Cysteine Biosynthesis

4.4.1 *O*-Acetylserine(thiol)lyase

Following reduction of sulfate, sulfide is transferred to *O*-acetylserine to form cysteine and acetate. This reaction is catalyzed by *O*-acetylserine(thiol)lyase (OASTL) which localises in cytosol, chloroplasts/plastids and mitochondria (Lunn *et al.*, 1990) (Figure 4.1). Sulfide, the substrate of OASTL, needs to be transported between subcellular compartments as SIR is present exclusively in chloroplasts/plastids. Numbers of genes encoding OASTL have been identified and characterised (Saito *et al.*, 1992, 1994a; Youssefian *et al.*, 1993; Ruffet *et al.*, 1994; Hell *et al.*, 1994; Barroso *et al.*, 1995; Jost *et al.*, 2000). The chloroplast/plastid-localising isoform existed in excess amounts and was suggested to play major roles in cysteine biosynthesis, directly utilising sulfide from SIR (Ruffet *et al.*, 1994). Cytosolic isoforms of OASTL would also accept atmospheric sulfide (Gotor *et al.*, 1997). Loss-of-function mutants of OASTL isoforms further indicated the significance of cysteine biosynthesis in cytosol and chloroplasts (Heeg *et al.*, 2008; Watanabe *et al.*, 2008a). Importance was also suggested from the overexpression of OASTL that makes plants tolerant to sulfite and sulfide (Youssefian *et al.*, 1993; Saito *et al.*, 1994b; Noji *et al.*, 2001). The functions of OASTL families are diverse. In mitochondria, cysteine generated by OASTL serves as a substrate for β-cyanoalanine synthase, which is a close homologue of OASTL (Hatzfeld *et al.*, 2000b; Yamaguchi *et al.*, 2000; Maruyama *et al.*, 2001). Recent findings indicate that sulfocysteine synthase, which utilises thiosulfate as a specific substrate, belongs to this family (Bermúdes *et al.*, 2010). Knockout of this isoform results in accumulating reactive oxygen species and impaired growth.

4.4.2 Serine Acetyltransferase

O-acetylserine (OAS), the precursor of cysteine biosynthesis, is synthesized by serine acetyltransferase (SERAT). SERAT activity was found in cytosol, chloroplasts/plastids and mitochondria (Ruffet *et al.*, 1995) (Figure 4.1). *Arabidopsis* has five SERAT isoforms showing distinctive features for OAS biosynthesis and localisation (Ruffet *et al.*, 1995; Noji *et al.*, 1998; Hell *et al.*, 2002; Kawashima *et al.*, 2005; Haas *et al.*, 2008; Watanabe *et al.*, 2008b). As OASTL accumulates in excess in chloroplasts (Ruffet *et al.*, 1994), cysteine can be readily formed following synthesis of OAS. However, OASTL does not exist abundantly in mitochondria. This allows OAS, the product of SERAT, to be exported from mitochondria to cytosol and converted to cysteine by cytosolic OASTL. Genetic reconstitution of each *SERAT* isoform in quintuple mutant of *Arabidopsis* indicated that mitochondrial SERAT2;2 and cytosolic SERAT1;1 make major contributions to OAS synthesis (Watanabe *et al.*, 2008b). Current understandings indicate that mitochondrial SERAT2;2 (SAT3) most likely controls the rate of OAS synthesis and makes the largest contribution among the isoforms (Haas *et al.*, 2008; Watanabe *et al.*, 2008b). By contrast, contribution of chloroplastic SERAT2;1 was suggested to be limited (Watanabe *et al.*, 2008b). Consistent with these findings, it has been suggested that OAS can be limiting in chloroplasts. Some transgenic studies provided supporting evidence. Biosynthesis of cysteine and glutathione was enhanced by overexpression of bacterial SERAT as a chloroplast-targeted form in plants (Blaszczyk *et al.*, 1999; Harms *et al.*, 2000). Overexpression of *Arabidopsis* chloroplastic SERAT2;1 to lupin led to an enhanced production of OAS, cysteine and glutathione in developing seeds (Tabe *et al.*, 2010). Furthermore, external feeding of OAS significantly enhanced the rates of cysteine and glutathione production in transgenic plants overexpressing *Pseudomonas*-derived APR as a chloroplast-targeted form (Tsakraklides *et al.*, 2002). Accordingly, transport of OAS to chloroplasts is likely to be a necessary process for cysteine biosynthesis by OASTL (Figure 4.1).

Cytosolic isoforms of SERAT were either sensitive to feedback inhibition by cysteine, or less active unless sufficient amount of substrates were supplied (Noji *et al.*, 1998; Kawashima *et al.*, 2005). Specific functions are suggested for cytosolic SERAT3;1 and SERAT3;2 under sulfur deficiency and in reproductive organs (Kawashima *et al.*, 2005; Watanabe *et al.*, 2008b). On the other hand, it was evident that another cytosolic isoform, SERAT1;1, makes a substantial contribution to OAS synthesis (Watanabe *et al.*, 2008b), although the activity of this enzyme was feedback-inhibited by cysteine (Noji *et al.*, 1998). These observations may indicate that cysteine concentration is strictly maintained below the inhibitory levels in the cytosol. Alternatively, the enzyme could be functional as a cysteine synthase complex (described in the following subsection). For this model, tight control of OAS concentration would become necessary, as the complex will dissociate when OAS accumulates in excess (Droux *et al.*, 1998). OAS needs to be metabolised or sequestered to maintain the activities of SERAT isoforms in cytosol and mitochondria. OAS can be metabolised by OASTL in cytosol or transported to chloroplasts where a huge demand for cysteine biosynthesis should exist (Heeg *et al.*, 2008; Watanabe *et al.*, 2008a) (Figure 4.1).

4.4.3 Cysteine Synthase Complex

SERAT and OASTL form cysteine synthase complex, which tightly controls cysteine biosynthesis through changes of conformation (Droux *et al.*, 1998; Berkowitz *et al.*, 2002; Wirtz and Hell, 2006). Dissociation and reversible re-association of the complex occurs in the presence of OAS and sulfide, respectively (Droux *et al.*, 1998) (Figure 4.4). When the complex dissociates in the presence of excess OAS, SERAT aggregates and becomes inactive, while unbound free OASTL converts excess OAS to cysteine. Under this mechanism, OAS will soon be depleted and sulfide accumulates. The complex re-associates in the presence of sulfide and serves as SERAT to produce OAS. The entire cycle controls cysteine biosynthesis depending upon OAS and sulfide supplied to the system. In addition, SERAT itself was feedback-inhibited by cysteine unless bound to OASTL (Figure 4.4). The feedback inhibition occurred particularly for cytosolic SERAT, playing a major role in cysteine synthesis (Noji *et al.*, 1998; Watanabe *et al.*, 2008b). Besides these control mechanisms, the amounts of SERAT and OASTL accumulated in each subcellular compartment are likely to vary, suggesting the significance of compartmentation and enzymatic properties of each

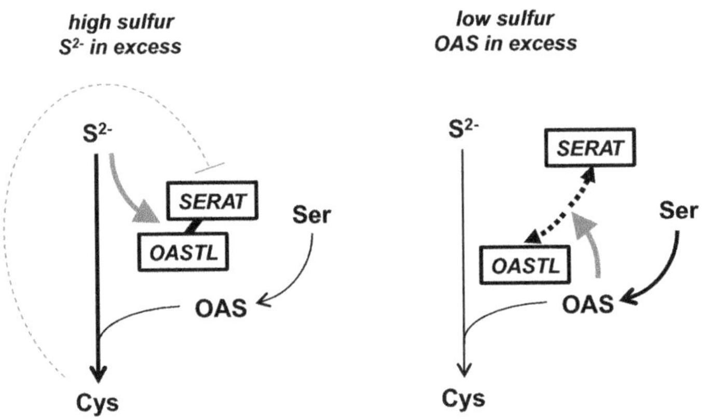

Figure 4.4 *Regulatory mechanisms of cysteine biosynthesis. Under high-sulfur conditions where sulfide is in excess, cysteine synthase complex assembles to function as SERAT. The activity of free SERAT can be feedback-inhibited by cysteine unless it is bound to OASTL. Under low-sulfur conditions, OAS will accumulate and dissociates the cysteine synthase complex to stop OAS production. Abbreviations: Cys, cysteine; OAS, O-acetylserine; OASTL, OAS(thiol)lyase; Ser, serine; SERAT, serine acetyltransferase*

isoform to fulfil the consecutive reactions for OAS and cysteine biosynthesis (Haas *et al.*, 2008; Heeg *et al.*, 2008; Watanabe *et al.*, 2008a,b) (Figure 4.1).

4.5 Methionine Biosynthesis

4.5.1 Biosynthetic Pathways

Cystathionine γ-synthase (CGS) catalyzes the first step of methionine biosynthesis (Figures 4.1 and 4.5). Cysteine and *O*-phosphohomoserine (OPH) are the substrates for synthesis of cystathionine in this step. Cystathionine β-lyase (CBL) subsequently catalytes the cleavage of cystathionine to form homocysteine. CGS and CBL are present only in chloroplasts (Figure 4.1). The significance of these chloroplast-localised metabolic pathways is suggested by the need for CGS and CBL for plant growth (Ravanel *et al.*, 1998a; Maimann *et al.*, 2000). CGS and CBL were encoded by single or few copies of genes, indicating that the pathways are indispensable (Kim and Leustek, 1996; Ravanel *et al.*, 1995). Following consecutive reactions catalyzed by CGS and CBL, homocysteine is subsequently methylated to form methionine by methionine synthase (MS) using methyltetrahydrofolate as a methyl donor (Eichel *et al.*, 1995; Ravanel *et al.*, 1998a, 2004; Zeh *et al.*, 2002). This final step is present in both chloroplasts and cytosol (Figure 4.1). *Arabidopsis* has three isoforms of MS showing distinct subcellular localisations. MS1 and MS2 are the cytosolic forms, while MS3 is located in chloroplasts (Ravanel *et al.*, 2004). Methionine in cytosol is further metabolised to *S*-adenosylmethionine (SAM) by cytosolic SAM synthetase, and

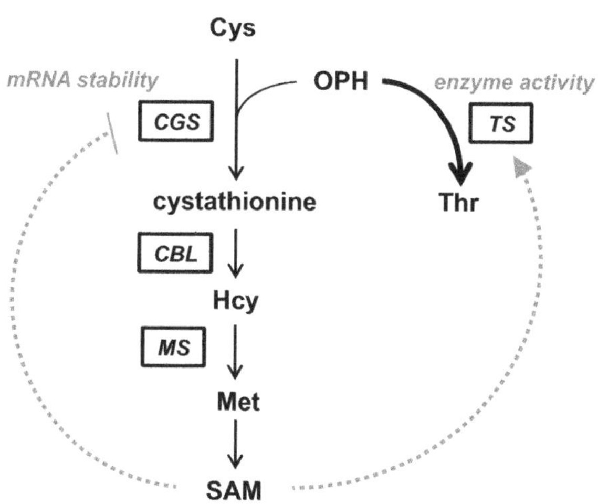

Figure 4.5 Regulatory mechanisms of methionine biosynthesis. SAM destabilises CGS mRNA and limits the flux of OPH for methionine biosynthesis. TS can utilise OPH more efficiently than CGS. In addition, SAM activates TS to increase the flux of OPH for threonine biosynthesis. Abbreviations: CBL, cystathionine β-lyase; CGS, cystathionine γ-synthase; Cys, cysteine; Hcy, homocysteine; Met, methionine; MS, methionine synthase; OPH, O-phosphohomoserine; SAM, S-adenosylmethionine; TS, threonine synthase

then recycled to homocysteine to serve as a substrate for cytosolic MS1 and MS2. The cytosolic isoforms of MS are suggested to be responsible for this recycling pathway in addition to their involvement in *de novo* synthesis of methionine (Ravanel *et al.*, 2004).

4.5.2 Regulation of Methionine Biosynthesis

Regulation occurs at the juncture of pathways for methionine and threonine biosynthesis (Figures 4.1 and 4.5). Threonine synthase (TS) and CGS share OPH as a common substrate, and the amount of SAM, the product downstream of methionine, controls the activity of TS. SAM enhances the activity of TS by increasing both the conversion rate and affinity for OPH (Curien *et al.*, 1996, 1998). In addition, TS showed extremely high affinity for OPH compared with CGS (Curien *et al.*, 1998; Ravanel *et al.*, 1998b). These enzymatic properties and regulation are significant for balancing the two branching pathways (Figure 4.5). When SAM accumulates, TS is activated and CGS will receive lesser amounts of OPH, which limits methionine biosynthesis. When the level of SAM declines, CGS may utilise OPH as TS will become less active. Accordingly, the amounts of methionine and SAM are tightly controlled though this mechanism, and the major proportion of OPH appears to be used for threonine biosynthesis (Amir *et al.*, 2002; Galili and Höfgen, 2002; Hesse and Hoefgen, 2003). Methionine over-accumulation occurs when the flux towards threonine biosynthesis is interrupted by suppression or deletion of TS (Bartlem *et al.*, 2000; Zeh *et al.*, 2001).

Another important aspect of regulation relates to mRNA stability of CGS (Figure 4.5). *Arabidopsis mto1* mutants having mutations in the first exon of *CGS1* gene were found to over-accumulate methionine (Chiba *et al.*, 1999). Mutations of amino acid sequences of this particular MTO1 region stabilised *CGS1* mRNA, suggesting that regulation may necessarily involve translation of its polypeptide (Chiba *et al.*, 1999; Suzuki *et al.*, 2001; Lambein *et al.*, 2003). It is shown that SAM is the effector for degradation of *CGS1* mRNA (Chiba *et al.*, 2003), and that the mechanism of RNA degradation involves a temporal SAM-dependent translation arrest of nascent CGS1 polypeptide at the downstream near-vicinity of the MTO1 region (Onouchi *et al.*, 2005). In spite of conservation of sequences, potato *CGS1* mRNA was insensitive to MTO1 regulation, suggesting that the regulatory mechanism is not universal, and that elements for regulation might be missing in some plant species to allow sufficient production of methionine (Kreft *et al.*, 2003; Hesse and Höfgen, 2003). Besides MTO1 regulation, alternative *CGS1* transcripts with 90 or 87 nucleotide deletions downstream of the site of translation arrest are involved in controlling methionine synthesis in *Arabidopsis* (Hacham *et al.*, 2006). The deleted form of *CGS1* transcript was stable in the presence of methionine, and over-produced methionine when expressed in tobacco plants (Hacham *et al.*, 2006). This adds a new feature of a mRNA stability control mechanism for CGS.

4.6 Regulators for Coordination of Sulfur Metabolism

4.6.1 Transcriptional Regulators

The sulfur-responsive *cis*-element in the 5′-region of *SULTR1;1* contains an auxin response factor (ARF) binding site; however, the element is specific for sulfur but not for auxin response (Maruyama-Nakashita *et al.*, 2005). These findings suggest that ARFs involved in this mechanism are controlled specifically by sulfur. However, potential ARF candidates were not identified from transcriptome analysis, suggesting that an upstream control mechanism other than transcriptional regulation is critical. Involvement of auxin in sulfur response is suggested from transcriptome analysis of sulfur limitation treatment (Nikiforova *et al.*, 2003). In addition, overexpression of sulfur-responsive auxin signalling components has been shown to influence a wide spectrum of metabolic pathways including sulfur metabolism (Falkenberg *et al.*, 2008). ARFs and auxin-related regulatory components that specifically control plant sulfur response or induction of *SULTR1;1* gene expression remain unspecified.

A key essential regulator of *SULTR1;2* has been identified from genetic screening of sulfur limitation response-less mutants showing aberrant responses of *SULTR1;2* promoter-GFP expression under low-sulfur conditions (Maruyama-Nakashita *et al.*, 2006). The *sulfur limitation1* (*slim1*) mutant identified from this study was impaired in enhancing *SULTR1;2* transcript levels under sulfur-limited conditions. The *slim1* mutant showed reduced sulfate uptake activity and its growth was affected when sulfur was not supplied (Maruyama-Nakashita *et al.*, 2006). The causal gene of *slim1* encoded a transcription factor, ETHYLENE-INSENSITIVE3-LIKE3 (EIL3). EIL3's function in ethylene response has not been clearly identified (Chao *et al.*, 1997; Guo and Ecker, 2004), although the finding of SLIM1 confirmed its function in sulfur response (Maruyama-Nakashita *et al.*, 2006). Transcriptome analysis of *slim1* mutant indicates that SLIM1 may balance the entire sulfur metabolic pathways. Genes for sulfate uptake and internal remobilisation of sulfur source were generally induced by SLIM1 under sulfur deficiency, while those for secondary metabolism were regulated oppositely (Maruyama-Nakashita *et al.*, 2006). Sulfate transporters and enzymes for glucosinolate biosynthesis were regulated oppositely by sulfur, and SLIM1 functioned as an upstream coordinator to take the balance of both sides. *APR* seems to be an exception as its transcript accumulation was induced by sulfur limitation even in *slim1* mutants. Significance of an independent mechanism is suggested for regulation of *APR*.

When focused upon regulation of glucosinolate biosynthesis, DOF and MYB transcription factors are known to control the expression of biosynthetic enzymes (Celenza *et al.*, 2005; Skirycz *et al.*, 2006; Hirai *et al.*, 2007; Gigolashvili *et al.*, 2007a,b). These factors can potentially stimulate biosynthesis of glucosinolates. Their application for disease protection and production of beneficial compounds would be expected to improve qualities of Brassica crop plant species. DOF and MYBs are inducers responsible for glucosinolate biosynthesis and are functional under sulfur sufficient conditions. They are probably independent of the regulatory signals downstream of SLIM1 that repressively control glucosinolate biosynthesis under sulfur deficiency.

4.6.2 MicroRNA-395

A microRNA species is found to regulate transcript levels of ATPS and sulfate transporter in plants. MicroRNAs are non-coding short RNAs that hybridise with specific target mRNAs having complementary sequences, and promote the degradation of target mRNAs in RNA-induced silencing complex (Jones-Rhoades *et al.*, 2006). It is known that microRNA-395 (miR395) accumulates under sulfur-starved conditions and targets *ATPS1, ATPS3, ATPS4* and *SULTR2;1* (Jones-Rhoades and Bartel, 2004; Allen *et al.*, 2005; Kawashima *et al.*, 2009). Accumulation of miR395 is dependent upon SLIM1 (Kawashima *et al.*, 2009), indicating another mode of post-transcriptional regulation in sulfate transport and metabolism.

In shoots where *SULTR2;1* mRNA levels decline under sulfur limitation, miR395 would limit the function of SULTR2;1 to absorb apoplastic sulfate to parenchyma cells and allow transfer of sulfate to mesophyll cells. In roots, absorption of sulfate to parenchyma cells is a necessary process to increase transfer of sulfate from roots to shoots, and in fact *SULTR2;1* mRNA is accumulated under low-sulfur conditions (Takahashi *et al*, 1997, 2000; Kataoka *et al.*, 2004a). This would suggest additional regulatory mechanisms that may override miR395-meidated regulation. As for regulation of *ATPS*, suppression of transcripts by sulfur deficiency is significant for *ATPS4*, suggesting that miR395 regulates this gene. However, *ATPS1* and *ATPS3* are not downregulated by sulfur deficiency (Takahashi *et al.*, 1997; Lappartient *et al.*, 1999); nevertheless miR395-targeted specific cleavage products are found (Kawashima *et al.*, 2009). Additional mechanisms that may override or interfere with miR395 are left to be considered.

4.6.3 OAS-Mediated Regulation

Dissociation and association of cysteine synthase complex controls cysteine biosynthesis by sensing the amounts of OAS and sulfide (Droux *et al.*, 1998; Berkowitz *et al.*, 2002; Wirtz and Hell, 2006). OAS, the product of SERAT activity of the complex, can modulate the transcript levels of *APR* and *SULTR* (Smith *et al.*, 1997; Koprivova *et al.*, 2002; Hesse *et al.*, 2003; Maruyama-Nakashita *et al.*, 2004b). The significance of this mechanism is suggested particularly for regulation of sulfate reduction, as knockdown of SERAT2;2 (SAT3), the major pathway for OAS synthesis in mitochondria, resulted in downregulating *APR2* but not modulating the levels of any of the *SULTR* isoforms in shoots (Haas *et al.*, 2008). Over-accumulation of OAS by DNA demethylation of a putative thiol reductase gene in *Arabidopsis osh1* mutant stimulated the transcript accumulations of both *APR1* and *SULTR2;2* in parallel with the induction of sulfur limitation-responsive β-conglycinin gene promoter activity (Ohkama-Ohtsu *et al.*, 2004). Involvement of OAS in sulfur signalling was further suggested by significant overlaps between transcriptomes of OAS treatment and sulfur starvation in *Arabidopsis* (Hirai *et al.*, 2003, 2004; Nikiforova *et al.*, 2003; Maruyama-Nakashita *et al.*, 2003, 2005). These lines of evidence indicate that OAS regulates key pathways of sulfate assimilation, although the modes of action are likely to be different depending on sulfur conditions, OAS concentrations, and downstream genes to be regulated. OAS can be interpreted as an effector metabolite that modulates or potentiates sulfur signalling pathways. However, linkages between cysteine synthase complex and signalling pathways remain unverified.

4.6.4 Outlook for Application

Improvement of sulfur use efficiency is an important biotechnological issue to obtain maximum crop yield and quality from a minimum input of sulfur. Biosynthesis of beneficial sulfur-metabolites is another important focus. Potential targets for engineering are sulfate transport systems and enzymes in key metabolic pathways. In terms of regulation, recent studies indicate that the uptake and metabolism of sulfate are tightly controlled by sulfur availability in the environment. In addition, some intrinsic signals including hormones and metabolites are suggested to be involved in regulation. The underlying mechanisms for regulation of sulfur metabolism are not fully understood. Researchers have just started to uncover some pieces of information relevant to regulation of critical steps in the systems. Findings of regulatory elements provide additional points for consideration. The entire metabolic system is tightly controlled under several different modes of regulatory mechanisms. In addition, primary and secondary metabolisms are balanced, responding to sulfur availabilities. Hierarchical engineering of key regulatory components would become necessary to totally break through these control mechanisms.

References

Allen, E., Xie, Z., Gustafson, A.M. and Carrington, J.C. (2005) MicroRNA-directed phasing during trans-acting siRNA biogenesis in plants. *Cell*, **121**, 207–221.

Amir, R., Hacham, Y. and Galili, G. (2002) Cystathionine γ-synthase and threonine synthase operate in concert to regulate carbon flow towards methionine in plants. *Trends Plant Sci.*, **7**, 153–156.

Aravind, L. and Koonin, E.V. (2000) The STAS domain – a link between anion transporters and antisigma-factor antagonists. *Curr. Biol.*, **10**, R53–55.

Awazuhara, M., Fujiwara, T., Hayashi, H., Watanabe-Takahashi, A., Takahashi, H. and Saito, K. (2005) The function of SULTR2;1 sulfate transporter during seed development in *Arabidopsis thaliana*. *Physiol. Plant.*, **125**, 95–105.

Barberon, M., Berthomieu, P., Clairotte, M., Shibagaki, N., Davidian, J-C. and Gosti, F. (2008) Unequal functional redundancy between the two *Arabidopsis thaliana* high-affinity sulphate transporters *SULTR1;1* and *SULTR1;2*. *New Phytol.*, **180**, 608–619.

Barroso, C., Vega, J.M. and Gotor, C. (1995) A new member of the cytosolic *O*-acetylserine(thiol)lyase gene family in *Arabidopsis thaliana*. *FEBS Lett.*, **363**, 1–5.

Bartlem, D., Lambein, I., Okamoto, T., *et al.* (2000) Mutation in the threonine synthase gene results in an over-accumulation of soluble methionine in Arabidopsis. *Plant Physiol.*, **123**, 101–110.

Benning, C. (1998) Biosynthesis and function of the sulfolipid sulfoquinovosyl diacylglycerol. *Ann. Rev. Plant Physiol. Plant Mol. Biol.*, **49**, 53–75.

Berkowitz, O., Wirtz, M., Wolf, A., Kuhlmann, J. and Hell, R. (2002) Use of biomolecular interaction analysis to elucidate the regulatory mechanism of the cysteine synthase complex from *Arabidopsis thaliana*. *J. Biol. Chem.*, **277**, 30629–30634.

Bermúdes, M.A., Páez-Ochoa, M.A., Gotor, C. and Romero, L.C. (2010) *Arabidopsis* *S*-sulfocysteine synthase activity is essential for chloroplast function and long-day light-dependent redox control. *Plant Cell*, **22**, 403–416.

Bianchini, F. and Vainio, H. (2001) Allium vegetables and organosulfur compounds: Do they help prevent cancer? *Environ. Health Perspect.*, **109**, 893–902.

Bick, J.A., Åslund, F., Chen, Y. and Leustek, T. (1998) Glutaredoxin function for the carboxyl-terminal domain of the plant-type 5′-adenylylsulfate reductase. *Proc. Natl. Acad. Sci. USA*, **95**, 8404–8409.

Bick, J.A., Setterdahl, A.T, Knaff, D.B., Chen, Y., Pitcher, L.H., Zilinskas, B.A. and Leustek, T. (2001) Regulation of the plant-type 5′-adenylyl sulfate reductase by oxidative stress. *Biochemistry*, **40**, 9040–9048.

Blaszczyk, A., Brodzik, R. and Sirko, A. (1999) Increased resistance to oxidative stress in transgenic tobacco plants overexpressing bacterial serine acetyltransferase. *Plant J.*, **20**, 237–243.

Bolchi, A., Petrucco, S., Tenca, P.L., Foroni, C. and Ottonello, S. (1999) Coordinate modulation of maize sulphate permease and ATP sulfurylase mRNAs in response to variations in sulfur nutritional status: stereospecific down-regulation by L-cysteine. *Plant Mol. Biol.*, **39**, 527–537.

Bork, C., Schwenn, J.D. and Hell, R. (1998) Isolation and characterization of a gene for assimilatory sulfite reductase from *Arabidopsis thaliana*. *Gene*, **212**, 147–153.

Bourgis, F., Roje, S., Nuccio, M.L., Fisher, D.B., Tarczynski, M.C., Li, C., Herschbach, C., Rennenberg, H., Pimenta, M.J., Shen, T.L., Gage, D.A. and Hanson, A.D. (1999) *S*-methylmethionine plays a major role in phloem sulfur transport and is synthesized by a novel type of methyltransferase. *Plant Cell*, **11**, 1485–1498.

Brühl, A., Haverkamp, T., Gisselmann, G. and Schwenn J.D. (1996) A cDNA clone from *Arabidopsis thaliana* encoding plastidic ferredoxin:sulfite reductase. *Biochim Biophys Acta.*, **1295**, 119–124.

Buchanan, B.B. and Balmer,Y. (2005) Redox regulation: a broadening horizon. *Ann. Rev. Plant Biol.*, **56**, 187–220.

Buchner, P., Stuiver, C.E., Westerman, S., Wirtz, M., Hell, R., Hawkesford, M.J. and De Kok, L.J. (2004a) Regulation of sulfate uptake and expression of sulphate transporter genes in *Brassica oleracea* as affected by atmospheric H_2S and pedospheric sulphate nutrition. *Plant Physiol.*, **136**, 3396–3408.

Buchner, P., Takahashi, H. and Hawkesford, M.J. (2004b) Plant sulphate transporters: co-ordination of uptake, intracellular and long-distance transport. J. Exp. Bot., **55**, 1765–1773.

Cairns, N.G., Pasternak, M., Wachter, A., Cobbett, C.S. and Meyer, A.J. (2006) Maturation of Arabidopsis seeds is dependent on glutathione biosynthesis within the embryo. *Plant Physiol.*, **141**, 446–455.

Celenza, J.L., Quiel, J.A., Smolen, G.A., Merrikh, H., Silvestro, A.R., Normanly, J. and Bender, J. (2005) The Arabidopsis ATR1 Myb transcription factor controls indolic glucosinolate homeostasis. *Plant Physiol.*, **137**, 253–262.

Chao, Q., Rothenberg, M., Solano, R., Roman, G., Terzaghi, W. and Ecker, J.R. (1997) Activation of the ethylene gas response pathway in *Arabidopsis* by the nuclear protein ETHYLENE-INSENSITIVE3 and related proteins. *Cell*, **89**, 1133–1144.

Cherest, H., Davidian, J-C., Thomas, D., Benes, V., Ansorge, W. and Surdin-Kerjan Y. (1997) Molecular characterization of two high affinity sulfate transporters in *Saccharomyces cerevisiae*. *Genetics*, **145**, 627–635.

Chiba, Y., Ishikawa, M., Kijima, F., Tyson, R.H., Kim, J., Yamamoto, A., Nambara, E., Leustek, T., Wallsgrove, R.M. and Naito, S. (1999) Evidence for autoregulation of cystathionine γ-synthase mRNA stability in *Arabidopsis*. *Science*, **286**, 1371–1374.

Chiba, Y., Sakurai, R., Yoshino, M., Ominato, K., Ishikawa, M., Onouchi, H. and Naito, S. (2003) *S*-adenosyl-L-methionine is an effector in the posttranscriptional autoregulation of the cystathionine γ-synthase gene in *Arabidopsis*. *Proc. Natl. Acad. Sci. USA*, **100**, 10225–10230.

Chi-Ham, C.L., Keaton, M.A., Cannon, G.C. and Heinhorst, S. (2002) The DNA-compacting protein DCP68 from soybean chloroplasts is ferredoxin:sulfite reductase and co-localizes with the organellar nucleoid. *Plant Mol Biol.*, **49**, 621–631.

Clarkson, D.T., Smith, F.W. and Vanden Berg, P.J. (1983) Regulation of sulphate transport in a tropical legume, *Macroptilium atropurpureum* cv. Siratro. *J. Exp. Bot.*, **34**, 1463–1483.

Crawford, N.M., Kahn, M.L., Leustek, T. and Long, S.R. (2000) Nitrogen and sulfur, in *Biochemistry and Molecular Biology of Plants* (eds B.B. Buchanan, W. Gruissem and R.L. Jones), American Society of Plant Biologists, Rockville, MD, pp. 824–849.

Curien, G., Dumas, R., Ravanel, S. and Douce, R. (1996) Characterization of an *Arabidopsis thaliana* cDNA encoding an *S*-adenosylmethionine-sensitive threonine synthase. Threonine synthase from higher plants. *FEBS Lett.*, **390**, 85–90.

Curien, G., Job, D., Douce, R. and Dumas, R. (1998) Allosteric activation of *Arabidopsis* threonine synthase by *S*-adenosylmethionine. *Biochemistry*, **37**, 13212–13221.

Deane-Drummond, C.E. (1987) The regulation of sulphate uptake following growth of *Pisum sativum* L. seedlings in S nutrient limiting conditions. Interaction between nitrate and sulphate transport. *Plant Sci.*, **50**, 27–35.

Droux, M., Ruffet, M.L., Douce, R. and Job, D. (1998) Interactions between serine acetyltransferase and *O*-acetylserine (thiol) lyase in higher plants – structural and kinetic properties of the free and bound enzymes. *Eur. J. Biochem.*, **255**, 235-245.

Dubousset, L., Abdallah, M., Desfeux, A.S., Etienne, P., Meuriot, F., Hawkesford, M.J., Gombert, J., Ségura, R., Bataillé, M.P., Rezé, S., Bonnefoy, J., Ameline, A.F., Ourry, A., Le Dily, F. and Avice, J.C. (2009) Remobilization of leaf S compounds and senescence in response to restricted sulphate supply during the vegetative stage of oilseed rape are affected by mineral N availability. *J. Exp. Bot.*, **60**, 3239–3253.

Eichel, J., González, J.C., Hotze, M., Matthews, R.G. and Schröder, J. (1995) Vitamin-B_{12}-independent methionine synthase from a higher plant (*Catharanthus roseus*). Molecular characterization, regulation, heterologous expression, and enzyme properties. *Eur. J. Biochem.*, **230**, 1053–1058.

El Kassis, E., Cathala, N., Rouached, H., Fourcroy, P., Berthomieu, P., Terry, N. and Davidian, J-C. (2007) Characterization of a selenate-resistant *Arabidopsis* mutant. Root growth as a potential target for selenate toxicity. *Plant Physiol.*, **143**, 1231–1241.

Falkenberg, B., Witt, I., Zanor, M.I., Steinhauser, D., Mueller-Roeber, B., Hesse, H. and Hoefgen, R. (2008) Transcription factors relevant to auxin signalling coordinate broad-spectrum metabolic shifts including sulphur metabolism. *J. Exp. Bot.*, **59**, 2831–2846.

Fisher, R.F. and Long, S.R. (1992) *Rhizobium*-plant signal exchange. *Nature*, **357**, 655–660.

Franco-Zorrilla, J.M., Martin, A.C., Solano, R., Rubio, V., Leyva, A. and Paz-Ares, J. (2002) Mutations at *CRE1* impair cytokinin-induced repression of phosphate starvation responses in *Arabidopsis*. *Plant J.*, **32**, 353–360.

Galili, G. and Höfgen, R. (2002) Metabolic engineering of amino acids and storage proteins in plants. *Metab. Eng.*, **4**, 3–11.

Gigolashvili, T., Berger, B., Mock, H.P., Müller, C., Weisshaar, B. and Flügge, U.I. (2007a) The transcription factor HIG1/MYB51 regulates indolic glucosinolate biosynthesis in *Arabidopsis thaliana*. *Plant J.*, **50**, 886–901.

Gigolashvili, T., Yatusevich, R., Berger, B., Müller, C. and Flügge, U.I. (2007b) The R2R3-MYB transcription factor HAG1/MYB28 is a regulator of methionine-derived glucosinolate biosynthesis in *Arabidopsis thaliana*. *Plant J.*, **51**, 247–261.

Gotor, C., Cejudo, F.J., Barroso, C. and Vega, J.M. (1997) Tissue-specific expression of *ATCYS-3A*, a gene encoding the cytosolic isoform of *O*-acetylserine(thiol)lyase in *Arabidopsis*. *Plant J.*, **11**, 347–352.

Gross, A., Brückner, G., Heldt, H.W. and Flügge, U.-I. (1990) Comparison of the kinetic properties, inhibition and labelling of the phosphate translocators from maize and spinach mesophyll chloroplasts. *Planta*, **180**, 262–271.

Grubb, C. and Abel, S. (2006) Glucosinolate metabolism and its control. *Trends Plant Sci.*, **11**, 89–100.

Guo, H. and Ecker, J.R. (2004) The ethylene signaling pathway: new insights. *Curr. Opin. Plant Biol.*, **7**, 40–49.

Gutierrez-Marcos, J.F., Roberts, M.A., Campbell, E.J. and Wray, J.L. (1996) Three members of a novel small gene-familiy from *Arabidopsis thaliana* able to complement functionally an *Escherichia coli* mutant defective in PAPS reductase activity encode proteins with a thioredoxin-like domain and "APS reductase" activity. *Proc. Natl Acad. Sci. USA*, **93**, 13377–13382.

Haas, F.H., Heeg, C., Queiroz, R., Bauer, A., Wirtz, M. and Hell, R. (2008) Mitochondrial serine acetyltransferase functions as a pacemaker of cysteine synthesis in plant cells. *Plant Physiol.*, **148**, 1055–1067.

Hacham, Y., Schuster, G. and Amir, R. (2006) An in vivo internal deletion in the N-terminus region of Arabidopsis cystathionine γ-synthase results in CGS expression that is insensitive to methionine. *Plant J.*, **45**, 955–967.

Hagen, G. and Guilfoyle, T.J. (2002) Auxin-responsive gene expression, genes, promoters and regulatory factors. *Plant Mol. Biol.*, **49**, 373–385.

Halkier, B.A. and Gerschenzon, J. (2006) Biology and biochemistry of glucosinolates. *Ann. Rev. Plant Biol.*, **57**, 303–333.

Harms, K., von Ballmoos, P., Brunold, C., Höfgen, R. and Hesse, H. (2000) Expression of a bacterial serine acetyltransferase in transgenic potato plants leads to increased levels of cysteine and glutathione. *Plant J.*, **22**, 335–343.

Hatzfeld, Y., Lee, S., Lee, M., Leustek, T. and Saito, K. (2000a) Functional characterization of a gene encoding a fourth ATP sulfurylase isoform from *Arabidopsis thaliana*. *Gene*, **248**, 51–58.

Hatzfeld, Y., Maruyama, A., Schmidt, A., Noji, M., Ishizawa, K. and Saito, K. (2000b) β-cyanoalanine synthase is a mitochondrial cysteine synthase-like protein in spinach and *Arabidopsis*. *Plant Physiol.*, **123**, 1163–1171.

Hawkesford, M.J., Davidian, J-C. and Grignon, C. (1993) Sulphate/proton cotransport in plasma-membrane vesicles isolated from roots of *Brassica napus* L.: increased transport in membranes isolated from sulphur-starved plants. *Planta*, **190**, 297–304.

Heeg, C., Kruse, C., Jost, R., Gutensohn, M., Ruppert, T., Wirtz, M. and Hell, R. (2008) Analysis of the *Arabidopsis O*-acetylserine(thiol)lyase gene family demonstrates compartment-specific differences in the regulation of cysteine synthesis. *Plant Cell*, **20**, 168–185.

Hell, R., Bork, C., Bogdanova, N., Frolov, I. and Hauschild, R. (1994) Isolation and characterization of two cDNAs encoding for compartment specific isoforms of *O*-acetylserine(thiol)lyase from *Arabidopsis thaliana*. *FEBS Lett.*, **351**, 257–262.

Hell, R., Jost, R., Berkowitz, O. and Wirtz, M. (2002) Molecular and biochemical analysis of the enzymes of cysteine biosynthesis in the plant *Arabidopsis thaliana*. *Amino Acids*, **22**, 245–257.

Herschbach, C. and Rennenberg, H. (1991) Influence of glutathione (GSH) on sulphate influx, xylem loading and exudation in excised tobacco roots. *J. Exp. Bot.*, **42**, 1021–1029.

Herschbach, C., van Der Zalm, E., Schneider, A., Jouanin, L., De Kok, L.J. and Rennenberg, H. (2000) Regulation of sulfur nutrition in wild-type and transgenic poplar over-expressing γ-glutamylcysteine synthetase in the cytosol as affected by atmospheric H_2S. *Plant Physiol.*, **124**, 461–473.

Hesse, H. and Hoefgen, R. (2003) Molecular aspects of methionine biosynthesis. *Trends Plant Sci.*, **8**, 259-262.

Hesse, H., Trachsel, N., Suter, M., Kopriva, S., von Ballmoos, P., Rennenberg, H. and Brunold, C. (2003) Effect of glucose on assimilatory sulphate reduction in *Arabidopsis thaliana* roots. *J. Exp. Bot.*, **54**, 1701–1709.

Hirai, M.Y., Fujiwara, T., Chino, M. and Naito, S. (1995) Effects of sulfate concentrations on the expression of a soybean seed storage protein gene and its reversibility in transgenic *Arabidopsis thaliana*. *Plant Cell Physiol.*, **36**, 1331–1339.

Hirai, M.Y., Fujiwara, T., Awazuhara, M., Kimura, T., Noji, M. and Saito, K. (2003) Global expression profiling of sulfur-starved *Arabidopsis* by DNA macroarray reveals the role of *O*-acetyl-L-serine as a general regulator of gene expression in response to sulphur nutrition. *Plant J.*, **33**, 651–663.

Hirai, M.Y., Yano, M., Goodenowe D.B., Kanaya, S., Kimura, T., Awazuhara, M., Arita, M., Fujiwara, T. and Saito, K. (2004) Integration of transcriptomics and metabolomics for understanding of global responses to nutritional stresses in *Arabidopsis thaliana*. *Proc. Natl. Acad. Sci. USA*, **101**, 10205–10210.

Hirai, M.Y., Klein, M., Fujikawa, Y., Yano, M., Goodenowe, D.B., Yamazaki, Y., Kanaya, S., Nakamura, Y., Kitayama, M., Suzuki, H., Sakurai, N., Shibata, D., Tokuhisa, J., Reichelt, M., Gershenzon, J., Papenbrock, J. and Saito, K. (2005) Elucidation of gene-to-gene and metabolite-to-gene networks in arabidopsis by integration of metabolomics and transcriptomics. *J. Biol. Chem.*, **280**, 25590–25595.

Hirai, M.Y., Sugiyama, K., Sawada, Y., Tohge, T., Obayashi, T., Suzuki, A., Araki, R., Sakurai, N., Suzuki, H., Aoki, K., Goda, H., Nishizawa, O.I., Shibata, D. and Saito, K. (2007) Omics-based identification of *Arabidopsis* Myb transcription factors regulating aliphatic glucosinolate biosynthesis. *Proc. Natl. Acad. Sci. USA*, **104**, 6478–6483.

Hopkins, L., Parmar, S., Blaszczyk, A., Hesse, H., Hoefgen, R. and Hawkesford, M.J. (2005) *O*-acetylserine and the regulation of expression of genes encoding components for sulfate uptake and assimilation in potato. *Plant Physiol.*, **138**, 433–440.

Howarth, J.R., Fourcroy, P., Davidian, J-C., Smith, F.W. and Hawkesford, M.J. (2003) Cloning of two contrasting high-affinity sulphate transporters from tomato induced by low sulphate and infection by the vascular pathogen *Verticillium dahlia*. *Planta*, **218**, 58–64.

Inoue, T., Higuchi, M., Hashimoto, Y., Seki, M., Kobayashi, M., Kato, T., Tabata, S., Shinozaki, K. and Kakimoto, T. (2001) Identification of CRE1 as a cytokinin receptor from *Arabidopsis*. *Nature*, **409**, 1060–1063.

Jones, M.G., Hughes, J., Tregova, A., Milne, J., Tomsett, A.B. and Collin, H.A. (2004) Biosynthesis of the flavour precursors of onion and garlic. *J. Exp. Bot.*, **55**, 1903–1918.

Jones-Rhoades, M.W. and Bartel, D.P. (2004) Computational identification of plant microRNAs and their targets, including a stress-induced miRNA. *Mol. Cell*, **14**, 787–799.

Jones-Rhoades, M.W., Bartel, D.P. and Bartel, B. (2006) MicroRNAs and their regulatory roles in plants. *Ann. Rev. Plant Biol.*, **57**, 19–53.

Jost, R., Berkowitz, O., Wirtz, M., Hopkins, L., Hawkesford, M.J. and Hell, R. (2000) Genomic and functional characterization of the *oas* gene family encoding *O*-acetylserine (thiol) lyases, enzymes catalyzing the final step in cysteine biosynthesis in *Arabidopsis thaliana*. *Gene*, **253**, 237–247.

Jost, R., Altschmied, L., Bloem, E., Bogs, J., Gershenzon, J., Hähnel, U., Hänsch, R., Hartmann, T., Kopriva, S., Kruse, C., Mendel, R.R., Papenbrock, J., Reichelt, M., Rennenberg, H., Schnu,g E., Schmidt, A., Textor, S., Tokuhisa, J., Wachter, A., Wirtz, M., Rausch, T. and Hell, R. Bogs, J., Gershenzon, J., Hähnel, U., Hänsch, R., Hartmann, T., Kopriva, S., Kruse, C., Mendel, R.R., Papenbrock, J., Reichelt, M., Rennenberg, H., Schnu,g E., Schmidt, A., Textor, S., Tokuhisa, J., Wachter, A., Wirtz, M., Rausch, T. and Hell, R. (2005) Expression profiling of metabolic genes in response to methyl jasmonate reveals regulation of genes of primary and secondary sulfur-related pathways in *Arabidopsis thaliana*. *Photosynth Res.*, **86**, 491–508.

Kaiser, G., Martinoia, E., Schroppelmeier, G. and Heber, U. (1989) Active-transport of sulfate into the vacuole of plant-cells provides halotolerance and can detoxify SO_2. *J. Plant Physiol.*, **133**, 756–763.

Kataoka, T., Watanabe-Takahashi, A., Hayashi, N., Ohnishi, M., Mimura, T., Buchner, P., Hawkesford, M.J., Yamaya, T. and Takahashi, H. (2004a) Vacuolar sulfate transporters are essential determinants controlling internal distribution of sulfate in Arabidopsis. *Plant Cell*, **16**, 2693–2704.

Kataoka, T., Hayashi, N., Yamaya, T. and Takahashi, H. (2004b) Root-to-shoot transport of sulfate in Arabidopsis: evidence for the role of SULTR3;5 as a component of low-affinity sulphate transport system in the root vasculature. *Plant Physiol.*, **136**, 4198–4204.

Kawashima, C.G., Berkowitz, O., Hell, R., Noji, M. and Saito, K. (2005) Characterization and expression analysis of a serine acetyltransferase gene family involved in a key step of the sulfur assimilation pathway in Arabidopsis. *Plant Physiol.*, **137**, 220–230.

Kawashima, C.G., Yoshimoto, N., Maruyama-Nakashita, A., Tsuchiya, Y.N., Saito, K., Takahashi, H. and Dalmay, T. (2009) Sulphur starvation induces the expression of microRNA-395 and one of its target genes but in different cell types. *Plant J.*, **57**, 313–321.

Kim, J. and Leustek, T. (1996) Cloning and analysis of the gene for cystathionine γ-synthase from *Arabidopsis thaliana*. *Plant Mol. Biol.*, **32**, 1117–1124.

Klein, M. and Papenbrock, J. (2004) The multi-protein family of *Arabidopsis* sulphotransferases and their relatives in other plant species. *J. Exp. Bot.*, **55**, 1809–1820.

Klein, M., Reichelt, M., Gershenzon, J. and Papenbrock, J. (2006) The three desulfoglucosinolate sulfotransferase proteins in *Arabidopsis* have different substrate specificities and are differentially expressed. *FEBS J.*, **273**, 122–136.

Klonus, D., Höfgen, R., Willmitzer, L. and Riesmeier, J.W. (1994) Isolation and characterization of two cDNA clones encoding ATP-sulfurylases from potato by complementation of a yeast mutant. *Plant J.*, **6**, 105–112.

Kopriva, S. (2006) Regulation of sulfate assimilation in *Arabidopsis* and beyond. *Ann. Bot.*, **97**, 479–495.

Kopriva, S., Muheim, R., Koprivova, A., Trachsel, N., Catalano, C., Suter, M. and Brunold, C. (1999) Light regulation of assimilatory sulphate reduction in *Arabidopsis thaliana*. *Plant J.*, **20**, 37–44.

Kopriva, S., Suter, M., von Ballmoos, P., Hesse, H., Krähenbühl, U., Rennenberg, H. and Brunold, C. (2002) Interaction of sulfate assimilation with carbon and nitrogen metabolism in *Lemna minor*. *Plant Physiol.*, **130**, 1406–1413.

Kopriva, S. and Koprivova, A. (2004) Plant adenosine 5′-phosphosulphate reductase: the past, the present, and the future. *J. Exp. Bot.*, **55**, 1775–1783.

Koprivova, A., Suter, M., Op den Camp, R., Brunold, C. and Kopriva, S. (2000) Regulation of sulfur assimilation by nitrogen in *Arabidopsis*. *Plant Physiol.*, **122**, 737–746.

Koprivova, A., Meyer, A.J., Schween, G., Herschbach, C., Reski, R. and Kopriva, S. (2002) Functional knockout of the adenosine 5′-phosphosulfate reductase gene in *Physcomitrella patens* revives an old route of sulphate assimilation. *J. Biol. Chem.*, **277**, 32195–321201.

Kreft, O., Hoefgen, R. and Hesse, H. (2003) Functional analysis of cystathionine γ-synthase in genetically engineered potato plants. *Plant Physiol.*, **131**, 1843–1854.

Krüger, R.J. and Siegel, L.M. (1982) Evidence for siroheme-Fe_4S_4 interaction in spinach ferredoxin-sulfite reductase. *Biochemistry*, **21**, 2905–2909.

Kuzuhara, Y., Isobe, A., Awazuhara, M., Fujiwara, T., Hayashi, H. (2000) Glutathione levels in phloem sap of rice plants under sulfur deficient conditions. *Soil Sci. Plant Nutr.*, **46**, 265–270.

Lambein, I., Chiba, Y., Onouchi, H. and Naito, S. (2003) Decay kinetics of autogenously regulated *CGS1* mRNA that codes for cystathionine γ-synthase in *Arabidopsis thaliana*. *Plant Cell Physiol.*, **44**, 893–900.

Lappartient A.G. and Touraine, B. (1996) Demand-driven control of root ATP sulfurylase activity and SO_4^{2-} uptake in intact canola. The role of phloem-translocated glutathione. *Plant Physiol.*, **111**, 147–157.

Lappartient, A.G., Vidmar, J.J., Leustek, T., Glass, A.D.M. and Touraine, B. (1999) Inter-organ signaling in plants: regulation of ATP sulfurylase and sulfate transporter genes expression in roots mediated by phloem-translocated compound. *Plant J.*, **18**, 89–95.

Lass, B. and Ullrich-Eberius, C.L. (1984) Evidence for proton/sulfate cotransport and its kinetics in *Lemna gibba* G1. *Planta*, **161**, 53–60.

Laudenbach, D.E. and Grossman, A.R. (1991) Characterization and mutagenesis of sulfur-regulated genes in a cyanobacterium: evidence for function in sulfate transport. *J. Bacteriol.*, **173**, 2739–2750.

Lee, S. and Leustek, T. (1998) APS kinase from *Arabidopsis thaliana*: genomic organization, expression, and kinetic analysis of the recombinant enzyme. *Biochem. Biophys. Res. Commun.*, **247**, 171–175.

Leggett, J.E. and Epstein, E. (1956) Kinetics of sulfate absorption by barley roots. *Plant Physiol.*, **31**, 222–226.

Lejay, L., Gansel, X., Cerezo, M., Tillard, P., Müller, C., Krapp, A., von Wirén, N., Daniel-Vedele, F. and Gojon, A. (2003) Regulation of root ion transporters by photosynthesis: functional importance and relation with hexokinase. *Plant Cell*, **15**, 2218–2232.

Leustek, T., Murillo, M. and Cervantes, M. (1994) Cloning of a cDNA encoding ATP sulfurylase from *Arabidopsis thaliana* by functional expression in *Saccharomyces cerevisiae*. *Plant Physiol.*, **105**, 897–902.

Leustek, T., Martin, M.N., Bick, J.A. and Davies, J.P. (2000) Pathways and regulation of sulfur metabolism revealed through molecular and genetic studies. *Ann. Rev. Plant Physiol. Plant Mol. Biol.*, **51**, 141–165.

Leves, F.P., Tierney, M.L. and Howitt, S.M. (2008) Polar residues in a conserved motif spanning helices 1 and 2 are functionally important in the SulP transporter family. *Int. J. Biochem. Cell Biol.*, **40**, 2596–2605.

Lillig, C.H., Schiffmann, S., Berndt, C., Berken, A., Tischka, R. and Schwenn, J.D. (2001) Molecular and catalytic properties of *Arabidopsis thaliana* adenylyl sulfate (APS)-kinase. *Arch. Biochem. Biophys.*, **392**, 303–310.

Lindberg, P. and Melis, A. (2008) The chloroplast sulfate transport system in the green alga *Chlamydomonas reinhardtii*. *Planta*, **228**, 951–961.

Logan, H.M., Cathala, N., Grignon, C. and Davidian, J.C. (1996) Cloning of a cDNA encoded by a member of the *Arabidopsis thaliana* ATP sulfurylase multigene family. Expression studies in yeast and in relation to plant sulfur nutrition. *J. Biol. Chem.*, **271**, 12227–12233.

Loudet, O., Saliba-Colombani, V., Camilleri, C., Calenge, F., Gaudon, V., Koprivova, A., North, K.A., Kopriva, S. and Daniel-Vedele, F. (2007) Natural variation for sulfate content in *Arabidopsis thaliana* is highly controlled by APR2. *Nat. Genet.*, **39**, 896–900.

Lunn, J.E., Droux, M., Martin, J. and Douce, R. (1990) Localization of ATP sulfurylase and *O*-acetylserine(thiol)lyase in spinach leaves. *Plant Physiol.*, **94**, 1345–1352.

Maimann, S., Wagner, C., Kreft, O., Zeh, M., Willmitzer, L., Höfgen, R. and Hesse, H. (2000) Transgenic potato plants reveal the indispensable role of cystathionine β-lyase in plant growth and development. *Plant J.*, **23**, 747–758.

Marschner, H. (1995) Introduction, definition and classification of mineral nutrients, in *Mineral Nutrition of Higher Plants* (2nd edn), Academic Press, London, pp. 3–5.

Martin, A.C., del Pozo, J.C., Iglesias, J., Rubio, V., Solano, R., de la Pena, A., Leyva, A. and Paz-Ares, J. (2000) Influence of cytokinins on the expression of phosphate starvation responsive genes in *Arabidopsis*. *Plant J.*, **24**, 559–567.

Martinoia, E., Massonneau, A. and Frangne, N. (2000) Transport processes of solutes across the vacuolar membrane of higher plants. *Plant Cell Physiol.*, **41**, 1175–1186.

Martinoia, E., Maeshima, M. and Neuhaus, H.E. (2007) Vacuolar transporters and their essential role in plant metabolism. *J. Exp. Bot.*, **58**, 83–102.

Maruyama, A., Saito, K. and Ishizawa, K. (2001) β-cyanoalanine synthase and cysteine synthase from potato: molecular cloning, biochemical characterization, and spatial and hormonal regulation. *Plant Mol. Biol.*, **46**, 749–760.

Maruyama-Nakashita, A., Inoue, E., Watanabe-Takahashi, A., Yamaya, T. and Takahashi, H. (2003) Transcriptome profiling of sulfur-responsive genes in Arabidopsis reveals global effect on sulfur nutrition on multiple metabolic pathways. *Plant Physiol.*, **132**, 597–605.

Maruyama-Nakashita, A., Nakamura, Y., Watanabe-Takahashi, A., Yamaya, T. and Takahashi, H. (2004a) Induction of SULTR1;1 sulfate transporter in *Arabidopsis* roots involves protein phosphorylation/dephosphorylation circuit for transcriptional regulation. *Plant Cell Physiol.*, **45**, 340–345.

Maruyama-Nakashita, A., Nakamura, Y., Yamaya, T. and Takahashi, H. (2004b) A novel regulatory pathway of sulfate uptake in *Arabidopsis* roots: Implication of CRE1/WOL/AHK4-mediated cytokinin-dependent regulation. *Plant J.*, **38**, 779–789.

Maruyama-Nakashita, A., Nakamura, Y., Yamaya, T. and Takahashi, H. (2004c) Regulation of high-affinity sulphate transporters in plants: towards systematic analysis of sulphur signalling and regulation. *J. Exp. Bot.*, **55**, 1843–1849.

Maruyama-Nakashita, A., Nakamura, Y., Watanabe-Takahashi, A., Inoue, E., Yamaya, T. and Takahashi, H. (2005) Identification of a novel *cis*-acting element conferring sulfur deficiency response in *Arabidopsis* roots. *Plant J.*, **42**, 305–314.

Maruyama-Nakashita, A., Nakamura, Y., Tohge, T., Saito, K. and Takahashi, H. (2006) *Arabidopsis* SLIM1 is a central transcriptional regulator of plant sulfur response and metabolism. *Plant Cell*, **18**, 3235–3251.

Mugford, S.G., Yoshimoto, N., Reichelt, M., Wirtz, M., Hill, L., Mugford, S.T., Nakazato, Y., Noji, M., Takahashi, H., Kramell, R., Gigolashvili, T., Flügge, U.I., Wasternack, C., Gershenzon, J., Hell, R., Saito, K. and Kopriva, S. (2009) Disruption of adenosine-5′-phosphosulfate kinase in *Arabidopsis* reduces levels of sulfated secondary metabolites. *Plant Cell*, **21**, 910–927.

Murillo, M. and Leustek, T. (1995) Adenosine-5′-triphosphate-sulfurylase from *Arabidopsis thaliana* and *Escherichia coli* are functionally equivalent but structurally and kinetically divergent: nucleotide sequence of two adenosine-5′-triphosphate-sulfurylase cDNAs from *Arabidopsis thaliana* and analysis of a recombinant enzyme. *Arch. Biochem. Biophys.*, **323**, 195–204.

Nikiforova, V., Freitag, J., Kempa, S., Adamik, M., Hesse, H. and Hoefgen, R. (2003) Transcriptome analysis of sulfur depletion in *Arabidopsis thaliana*: interlacing of biosynthetic pathways provides response specificity. *Plant J.*, **33**, 633–650.

Noji, M., Inoue, K., Kimura, N., Gouda, A. and Saito, K. (1998) Isoform-dependent differences in feedback regulation and subcellular localization of serine acetyltransferase involved in cysteine biosynthesis from *Arabidopsis thaliana*. *J. Biol. Chem.*, **273**, 32739–32745.

Noji, M., Saito, M., Nakamura, M., Aono, M., Saji, H. and Saito, K. (2001) Cysteine synthase overexpression in tobacco confers tolerance to sulfur-containing environmental pollutants. *Plant Physiol.*, **126**, 973–980.

Ohkama, N., Takei, K., Sakakibara, H., Hayashi, H., Yoneyama, T. and Fujiwara, T. (2002) Regulation of sulfur-responsive gene expression by exogenously applied cytokinins in *Arabidopsis thaliana*. *Plant Cell Physiol.*, **43**, 1493–1501.

Ohkama-Ohtsu, N., Kasajima, I., Fujiwara, T. and Naito, S. (2004) Isolation and characterization of an *Arabidopsis* mutant that overaccumulates *O*-acetyl-L-serine. *Plant Physiol.*, **136**, 3209–3222.

Onouchi, H., Nagami, Y., Haraguchi, Y., Nakamoto, M., Nishimura, Y., Sakurai, R., Nagao, N., Kawasaki, D., Kadokura, Y. and Naito, S. (2005) Nascent peptide-mediated translation elongation arrest coupled with mRNA degradation in the *CGS1* gene of *Arabidopsis*. *Genes Dev.*, **19**, 1799–1810.

Piotrowski, M., Schemenewitz, A., Lopukhina, A., Müller, A., Janowitz, T., Weiler, E.W. and Oecking, C. (2004) Desulfoglucosinolate sulfotransferases from *Arabidopsis thaliana* catalyze the final step in the biosynthesis of the glucosinolate core structure. *J. Biol. Chem.*, **279**, 50717–50725.

Prior, A., Uhrig, J.F., Heins, L., Wiesmann, A., Lillig, C.H., Stoltze, C., Soll, J. and Schwenn, J.D. (1999) Structural and kinetic properties of adenylyl sulfate reductase from *Catharanthus roseus* cell cultures. *Biochim. Biophys. Acta*, **1430**, 25–38.

Rae, A.L. and Smith, F.W. (2002) Localisation of expression of a high-affinity sulfate transporter in barley roots. *Planta*, **215**, 565–568.

Ravanel, S., Ruffet, M.L. and Douce, R. (1995) Cloning of an *Arabidopsis thaliana* cDNA encoding cystathionine β-lyase by functional complementation in *Escherichia coli*. *Plant. Mol. Biol.*, **29**, 875–882.

Ravanel, S., Gakière, B., Job, D. and Douce, R. (1998a) The specific features of methionine biosynthesis and metabolism in plants. *Proc. Natl. Acad. Sci. USA*, **95**, 7805–7812.

Ravanel, S., Gakière, B., Job, D. and Douce, R. (1998b) Cystathionine γ-synthase from *Arabidopsis thaliana*: purification and biochemical characterization of the recombinant enzyme overexpressed in *Escherichia coli*. *Biochem J.*, **331**, 639–648.

Ravanel, S., Block, M.A., Rippert, P., Jabrin, S., Curien, G., Rébeillé, F. and Douce, R. (2004) Methionine metabolism in plants: chloroplasts are autonomous for *de novo* methionine synthesis and can import *S*-adenosylmethionine from the cytosol. *J. Biol. Chem.*, **279**, 22548–22557.

Renosto, F., Patel, H.C., Martin, R.L., Thomassian, C., Zimmerman, G. and Segel, I.H. (1993) ATP sulfurylase from higher plants: kinetic and structural characterization of the chloroplast and cytosol enzymes from spinach leaf. *Arch. Biochem. Biophys.*, **307**, 272–285.

Rotte, C. and Leustek, T. (2000) Differential subcellular localization and expression of ATP sulfurylase and 5′-adenylylsulfate reductase during ontogenesis of *Arabidopsis* leaves indicates that cytosolic and plastid forms of ATP sulfurylase may have specialized functions. *Plant Physiol.*, **124**, 715–724.

Rouached, H., Berthomieu, P., El Kassis, E., Cathala, N., Catherinot, V., Labesse, G., Davidian, J-C. and Fourcroy, P. (2005) Structural and functional analysis of the C-terminal STAS (sulfate transporter and anti-sigma antagonist) domain of the *Arabidopsis thaliana* sulfate transporter SULTR1.2. *J. Biol. Chem.*, **280**, 15976–15983.

Rouached, H., Wirtz, M., Alary, R., Hell, R., Arpat, A.B., Davidian, J-C., Fourcroy, P. and Berthomieu, P. (2008) Differential regulation of the expression of two high-affinity sulfate transporters, SULTR1.1 and SULTR1.2, in Arabidopsis. *Plant Physiol.*, **147**, 897–911.

Ruffet, M.L., Droux, M. and Douce, R. (1994) Purification and kinetic properties of serine acetyltransferase free of *O*-acetylserine(thiol)lyase from spinach chloroplasts. *Plant Physiol.*, **104**, 597–604.

Ruffet, M.L., Lebrun, M., Droux, M. and Douce, R. (1995) Subcellular distribution of serine acetyltransferase from *Pisum sativum* and characterization of an *Arabidopsis thaliana* putative cytosolic isoform. *Eur. J. Biochem.*, **227**, 500–509.

Saito, K. (2000) Regulation of sulfate transport and synthesis of sulfur-containing amino acids. *Curr. Opin. Plant Biol.*, **3**, 188–195.

Saito, K. (2004) Sulfur assimilatory metabolism. The long and smelling road. *Plant Physiol.*, **136**, 2443–2450.

Saito, K., Miura, N., Yamazaki, M., Hirano, H. and Murakoshi, I. (1992) Molecular cloning and bacterial expression of cDNA encoding a plant cysteine synthase. *Proc. Natl. Acad. Sci. USA*, **89**, 8078–8082.

Saito, K., Tatsuguchi, K., Takagi, Y. and Murakoshi, I. (1994a) Isolation and characterization of cDNA that encodes a putative mitochondrion-localizing isoform of cysteine synthase (*O*-acetylserine(thiol)-lyase) from *Spinacia oleracea*. *J. Biol. Chem.*, **269**, 28187–28192.

Saito, K., Kurosawa, M., Tatsuguchi, K., Takagi, Y. and Murakoshi, I. (1994b) Modulation of cysteine biosynthesis in chloroplasts of transgenic tobacco overexpressing cysteine synthase [*O*-acetylserine(thiol)-lyase]. *Plant Physiol.*, **106**, 887–895.

Sekine, K., Hase, T. and Sato, N. (2002) Reversible DNA compaction by sulfite reductase regulates transcriptional activity of chloroplast nucleoids. *J. Biol. Chem.*, **277**, 24399–24404.

Sekine, K., Fujiwara, M., Nakayama, M., Takao, T., Hase, T. and Sato, N. (2007) DNA binding and partial nucleoid localization of the chloroplast stromal enzyme ferredoxin:sulfite reductase. *FEBS J.*, **274**, 2054–2069.

Setya, A., Murillo, M. and Leustek, T. (1996) Sulfate reduction in higher plants: molecular evidence for a novel 5′-adenylylsulfate reductase. *Proc. Natl Acad. Sci. USA*, **93**, 13383–13388.

Shibagaki, N., Rose, A., Mcdermott, J.P., Fujiwara, T., Hayashi, H., Yoneyama, T. and Davies, J.P. (2002) Selenate-resistant mutants of *Arabidopsis thaliana* identify SULTR1;2, a sulfate transporter required for efficient transport of sulfate into roots. *Plant J.*, **29**, 475–486.

Shibagaki, N. and Grossman, A.R. (2004) Probing the function of STAS domains of the *Arabidopsis* sulfate transporters. *J. Biol. Chem.*, **279**, 30791–30799.

Shibagaki, N. and Grossman, A.R. (2006) The role of the STAS domain in the function and biogenesis of a sulfate transporter as probed by random mutagenesis. *J. Biol. Chem.*, **281**, 22964–22973.

Sirko, A., Hryniewicz, M., Hulanicka, D. and Böck, A. (1990) Sulfate and thiosulfate transport in *Escherichia coli* K-12: nucleotide sequence and expression of the *cysTWAM* gene cluster. *J. Bacteriol.*, **172**, 3351–3357.

Skirycz, A., Reichelt, M., Burow, M., Birkemeyer, C., Rolcik, J., Kopka, J., Zanor, M.I., Gershenzon, J., Strnad, M., Szopa, J., Mueller-Roeber, B. and Witt, I. (2006) DOF transcription factor AtDof1.1 (OBP2) is part of a regulatory network controlling glucosinolate biosynthesis in *Arabidopsis*. *Plant J.*, **47**, 10–24.

Smith, F.W., Hawkesford, M.J., Prosser, I.M. and Clarkson, D.T. (1995a) Isolation of a cDNA from *Saccharomyces cerevisiae* that encodes a high affinity sulphate transporter at the plasma membrane. *Mol. Gen. Genet.*, **247**, 709–715.

Smith, F.W., Ealing, P.M., Hawkesford, M.J. and Clarkson, D.T. (1995b) Plant members of a family of sulfate transporters reveal functional subtypes. *Proc. Natl. Acad. Sci. USA*, **92**, 9373–9377.

Smith, F.W., Hawkesford, M.J., Ealing, P.M., Clarkson, D.T., Vanden Berg, P.J., Belcher, A.R. and Warrilow, A.G.S. (1997) Regulation of expression of a cDNA from barley roots encoding a high affinity sulfate transporter. *Plant J.*, **12**, 875–884.

Suter, M., von Ballmoos, P., Kopriva, S., den Camp, R.O., Schaller, J., Kuhlemeier, C., Schürmann, P. and Brunold. C. (2000) Adenosine 5′-phosphosulfate sulfotransferase and adenosine 5′-phosphosulfate reductase are identical enzymes. *J. Biol. Chem.*, **275**, 930–936.

Suzuki, A., Shirata, Y., Ishida, H., Chiba, Y., Onouchi, H. and Naito, S. (2001) The first exon coding region of cystathionine γ-synthase gene is necessary and sufficient for downregulation of its own mRNA accumulation in transgenic *Arabidopsis thaliana*. *Plant Cell Physiol.*, **42**, 1174–1180.

Tabe, L.M. and Droux, M. (2001) Sulfur assimilation in developing lupin cotyledons could contribute significantly to the accumulation of organic sulfur reserves in the seed. *Plant Physiol.*, **126**, 176–187.

Tabe, L., Wirtz, M., Molvig, L., Droux, M. and Hell, R. (2010) Overexpression of serine acetlytransferase produced large increases in *O*-acetylserine and free cysteine in developing seeds of a grain legume. *J. Exp. Bot.*, **61**, 721–733.

Takahashi, H. (2010) Sulfate transport and assimilation in plants. *Int. Rev. Cell Mol. Biol.*, **281**, 129–159.

Takahashi, H., Yamazaki, M., Sasakura, N., Watanabe, A., Leustek, T., de Almeida Engler, J., Engler, G., Van Montagu, M. and Saito, K. (1997) Regulation of sulfur assimilation in higher plants: a sulfate transporter induced in sulfate starved roots plays a central role in *Arabidopsis thaliana*. *Proc. Natl. Acad. Sci. USA*, **94**, 11102–11107.

Takahashi, H., Watanabe-Takahashi, A., Smith, F.W., Blake-Kalff, M., Hawkesford, M.J. and Saito, K. (2000) The role of three functional sulfate transporters involved in uptake and translocation of sulfate in *Arabidopsis thaliana*. *Plant J.*, **23**, 171–182.

Takahashi, H., Yoshimoto, N. and Saito, K. (2006) Anionic nutrient transport in plants: the molecular bases of sulfate transporter gene family, in *Genetic Engineering* Vol. **27** (ed. J.K. Setlow), Springer, New York, pp. 67–80.

Talalay, P. and Fahey, J. W. (2001) Phytochemicals from cruciferous plants protect against cancer by modulating carcinogen metabolism. *J. Nutr.*, **131**, 3027S–3033S.

Tsakraklides, G., Martin, M., Chalam, R., Tarczynski, M.C., Schmidt, A. and Leustek, T. (2002) Sulfate reduction is increased in transgenic *Arabidopsis thaliana* expressing 5′-adenylylsulfate reductase from *Pseudomonas aeruginosa*. *Plant J.*, **32**, 879–889.

Ulmasov, T., Hagen, G. and Guilfoyle, T.J. (1999) Dimerization and DNA binding of auxin response factors. *Plant J.*, **19**, 309–319.

Vauclare, P., Kopriva, S., Fell, D., Suter, M., Sticher, L., von Ballmoos, P., Krähenbühl, U., den Camp, R.O. and Brunold, C. (2002) Flux control of sulphate assimilation in *Arabidopsis thaliana*: adenosine 5′-phosphosulphate reductase is more susceptible than ATP sulphurylase to negative control by thiols. *Plant J.*, **31**, 729–740.

Vidmar, J.J., Schjoerring, J.K., Touraine, B. and Glass, A.D.M. (1999) Regulation of the *hvst1* gene encoding a high-affinity sulfate transporter from *Hordeum vulgare*. *Plant Mol. Biol.*, **40**, 883–892.

Vidmar, J.J., Tagmount, A., Cathala, N., Touraine, B. and Davidian, J-C.E. (2000) Cloning and characterization of a root specific high-affinity sulfate transporter from *Arabidopsis thaliana*. *FEBS Lett.*, **475**, 65–69.

Wang, R., Okamoto, M., Xing, X. and Crawford, N.M. (2003) Microarray analysis of the nitrate response in *Arabidopsis* roots and shoots reveals over 1,000 rapidly responding genes and new linkages to glucose, trehalose-6-phosphate, iron, and sulfate metabolism. *Plant Physiol.*, **132**, 556–567.

Watanabe, M., Kusano, M., Oikawa, A., Fukushima, A., Noji, M. and Saito, K. (2008a) Physiological roles of the β-substituted alanine synthase gene family in *Arabidopsis*. *Plant Physiol.*, **146**, 310–320.

Watanabe, M., Mochida, K., Kato, T., Tabata, S., Yoshimoto, N., Noji, M. and Saito, K. (2008b) Comparative genomics and reverse genetics analysis reveal indispensable functions of the serine acetyltransferase gene family in *Arabidopsis*. *Plant Cell*, **20**, 2484–2496.

Wirtz, M. and Hell, R. (2006) Functional analysis of the cysteine synthase protein complex from plants: Structural, biochemical and regulatory properties. *J. Plant Physiol.*, **163**, 273–286.

Yamaguchi, Y., Nakamura, T., Kusano, T. and Sano, H. (2000) Three *Arabidopsis* genes encoding proteins with differential activities for cysteine synthase and β-cyanoalanine synthase. *Plant Cell Physiol.*, **41**, 465–476.

Yonekura-Sakakibara, K., Ashikari, T., Tanaka, Y., Kusumi, T. and Hase, T. (1998) Molecular characterization of tobacco sulfite reductase: enzyme purification, gene cloning, and gene expression analysis. *J. Biochem.*, **124**, 615–621.

Yonekura-Sakakibara, K., Onda, Y., Ashikari, T., Tanaka, Y., Kusumi, T. and Hase, T. (2000) Analysis of reductant supply systems for ferredoxin-dependent sulfite reductase in photosynthetic and nonphotosynthetic organs of maize. *Plant Physiol.*, **122**, 887–894.

Yoshimoto, N., Takahashi, H., Smith, F.W., Yamaya, T. and Saito, K. (2002) Two distinct high-affinity sulfate transporters with different inducibilities mediate uptake of sulfate in *Arabidopsis* roots. *Plant J.*, **29**, 465–473.

Yoshimoto, N., Inoue, E., Saito, K., Yamaya, T. and Takahashi, H. (2003) Phloem-localizing sulfate transporter, Sultr1;3, mediates re-distribution of sulfur from source to sink organs in *Arabidopsis*. *Plant Physiol.*, **131**, 1511–1517.

Yoshimoto, N., Inoue, E., Watanabe-Takahashi, A., Saito, K. and Takahashi, H. (2007) Post-transcriptional regulation of high-affinity sulfate transporters in *Arabidopsis* by sulfur nutrition. *Plant Physiol.*, **145**, 378–388.

Youssefian, S., Nakamura, M. and Sano, H. (1993) Tobacco plants transformed with the *O*-acetylserine (thiol) lyase gene of wheat are resistant to toxic levels of hydrogen sulfide gas. *Plant J.*, **4**, 759–769.

Zeh, M., Casazza, A.P., Kreft, O., Roessner, U., Bieberich, K., Willmitzer, L., Hoefgen, R. and Hesse, H. (2001) Antisense inhibition of threonine synthase leads to high methionine content in transgenic potato plants. Plant Physiol., **127**, 792–802.

Zeh, M., Leggewie, G., Hoefgen, R. and Hesse, H. (2002) Cloning and characterization of a cDNA encoding a cobalamin-independent methionine synthase from potato (*Solanum tuberosum* L.). *Plant Mol. Biol.*, **48**, 255–265.

5

Nucleotide Metabolism

Rita Zrenner and Hiroshi Ashihara

5.1 Introduction

Nucleotides are important nitrogen compounds in all living organisms. Purine and pyrimidine are essential precursors for nucleic acids, as well as metabolites participating in bio-energetic processes and in the synthesis of macromolecules, including polysaccharides, phospholipids and glycolipids (Ross, 1981). Pyridine nucleotides, NAD and NADP, are coenzymes that participate in oxido-reduction reactions. Nucleotides also serve as direct precursors for the synthesis of vitamins, such as riboflavin, thiamine and folates (Crozier *et al.*, 2000; Herz *et al.*, 2000; Hanson and Gregory, 2002), flavin adenine dinucleotide (FAD), and *S*-adenosylmethionine.

Some reviews on plant nucleotide metabolism have been published (Ross, 1981; Wagner and Backer, 1992; Moffatt and Ashihara, 2002; Zrenner *et al.*, 2006). In the present review, we briefly outline metabolic pathways of nucleotides, and recent molecular studies on nucleotide metabolism are summarised. Further, we review biotechnological approaches concerning plant nucleotide-related metabolism. There we focus on herbicide targets, growth-related aspects, and the influence of modulated nucleotide pools on starch accumulation and yield in potato tubers. Physiological and biochemical aspects of nucleotide metabolism during growth and organised development in plants have been summarised and published elsewhere (Stasolla *et al.*, 2003).

Plant Metabolism and Biotechnology, First Edition. Edited by Hiroshi Ashihara, Alan Crozier, and Atsushi Komamine.

5.2 Pyrimidine Metabolism

5.2.1 *De Novo* Biosynthetic Pathway

The *de novo* pyrimidine biosynthetic pathway (orotate pathway) is defined as the formation of UMP from carbamoyl phosphate. The orotate pathway consists of six reactions as shown in Figure 5.1. In mammals and many other eukaryotes, the first three enzymes, carbamoylphosphate synthetase, aspartate transcarbamoylase and dihydroorotase (steps 1–3, Figure 5.1) are present as a multifunctional protein called the CAD protein, after the initial letters of the three constituent enzymes (Christopherson and Szabados, 1997). However, no such complex has thus far been detected in plants (see Moffatt and Ashihara, 2002). Although there are two different types of carbamoyl phosphate synthetases that provide substrates for pyrimidine and arginine biosynthesis in most eukaryotes, including mammals, the first complete plant genome sequence (the *Arabidopsis* Genome Initiative 2000) together with biochemical studies (Wasternack, 1982) reveal that higher plants possess only a single form of CPSase like *E. coli* that supplies carbamoylphosphate to both the pyrimidine and the arginine pathway. Plant CPSase consists of two subunits. The small subunit exhibits a glutamine-amidotransferase activity and provides ammonia to the large subunit. The large subunit catalyzes the CPSase reaction and possesses regulatory properties. Cloning and characterisation of the respective subunit cDNAs from various plants revealed that the subunits are encoded by individual genes (Giermann *et al.*, 2002). The second enzyme, aspartate transcarbamoylase (ATCase) is, therefore, the first committed step in pyrimidine biosynthesis. ATCase catalyzes the condensation of aspartate and carbamyl phosphate. This reaction is reversible, but the equilibrium is much in favour of carbamoyl aspartate. Sequences encoding proteins with ATCase activity have been characterised from various plants (Williamson and Slocum, 1994; Nasr *et al.*, 1994; Giermann *et al.*, 2002). The next enzyme, dihydroorotase (DHOase), catalyzes the conversion of carbamoyl aspartate into 4,5-dihydroorotate. Sequences encoding proteins with DHOase activity have been characterised from *Arabidopsis* and potato (Giermann *et al.*, 2002).

The fourth enzyme, dihydroorotate dehydrogenase (DHODH), catalyzes the conversion of dihydroorotate to orotate (step 4, Figure 5.1). Sequences encoding proteins with DHODH activity have been characterised in *Arabidopsis* and tobacco (Minet *et al.*, 1992; Giermann *et al.*, 2002). Plant DHODHs sequenced to date belong to the membrane-bound family 2 of dihydroorotate dehydrogenases, which are flavoproteins. The activity of the *Arabidopsis* DHODH shows significant differences in substrate specificity and inhibition from the animal enzyme (Ullrich *et al.*, 2002). The fifth and sixth enzymes, orotate phosphoribosyltransferase (step 5, Figure 5.1) and orotidine-5′-monophosphate decarboxylase (step 6, Figure 5.1), are present as distinct domains of a bifunctional protein named UMP synthase (UMPSase), which is observed in both plants and animals (Santoso and Thornburg, 1998). Sequences encoding proteins with UMPSase activity have been characterised from various plants (Santoso and Thornburg, 1992; Nasr *et al.*, 1994).

From amino acid sequences of the *N*-terminal transit peptides together with mining of subcellular proteome databases (http://suba.plantenergy.uwa.edu.au), it has been assumed that the reactions in the *de novo* pyrimidine biosynthetic pathway take place in a number of different subcellular locations. Whereas the first three steps of the pathway take place in the plastids, the DHODH reaction is carried out at the outer surface of the inner mitochondrial

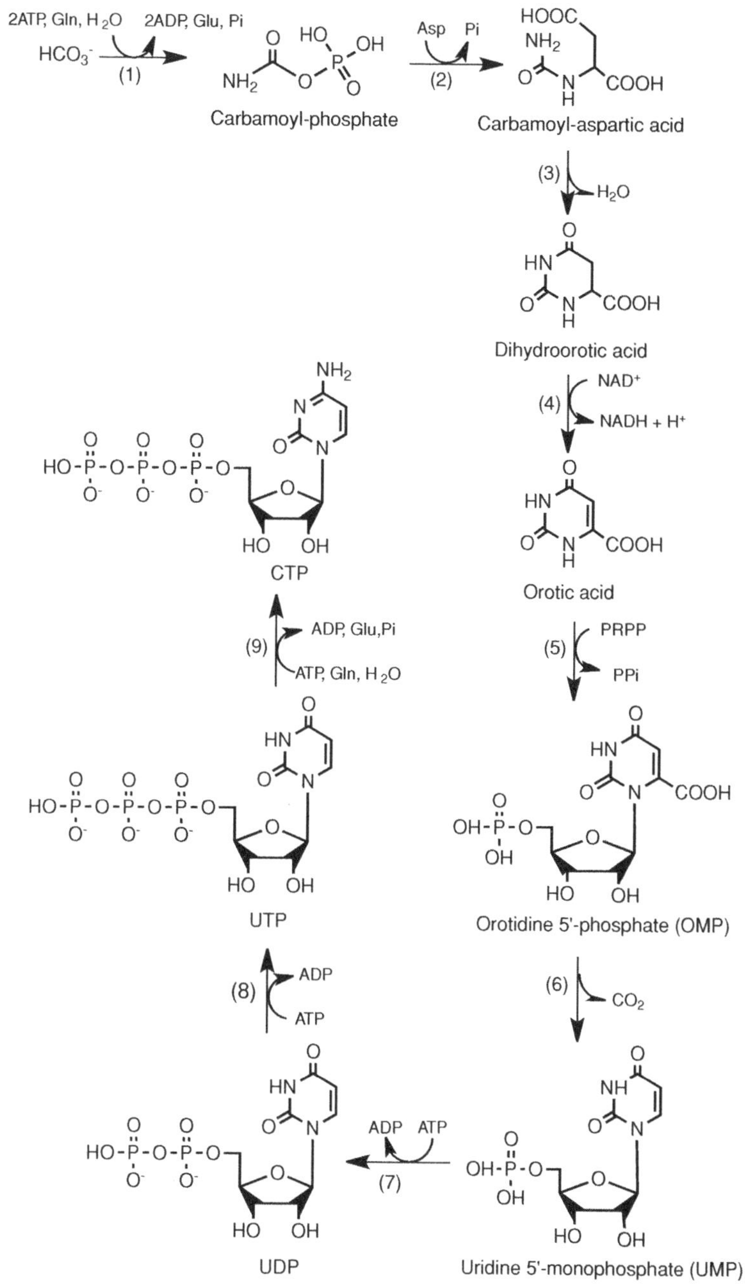

Figure 5.1 De novo *biosynthetic pathway of pyrimidine nucleotides in plants. Enzymes (EC numbers) shown are: (1) carbamoyl phosphate synthetase (6.3.5.5); (2) aspartate transcarbamoylase (2.1.3.2); (3) dihydroorotase (3.5.2.3); (4) dihydroorotate dehydrogenase (1.3.99.11); (5)–(6) UMP synthase [orotate phosphoribosyltransferase (2.4.2.10) plus orotidine-5'-phosphate decarboxylase (4.1.1.23)]; (7) UMP kinase (2.7.4.4); (8) nucleoside diphosphate kinase (2.7.4.6); (9) CTP synthetase (6.3.4.2). Asp, aspartic acid; Gln, glutamine; Glu, glutamic acid; PRPP, 5-phosphoribosyl-1-pyrophosphate*

membrane and is coupled to the respiratory chain. The localisation of the last two steps, catalyzed by UMPSase, has not been determined yet. They may take place in the plastids or may be located in the cytosol. Therefore, to produce UMP, dihydroorotate must leave the plastid and access the outer surface of the inner mitochondrial membrane, where it is converted to orotate. Subsequently the orotate must move back across the outer membrane of the mitochondria and probably re-enter the plastids, where it is converted to UMP.

There is no strong control of gene expression of the *de novo* synthesis pathway, with the exception of a developmental gradient (Giermann *et al.*, 2002). Further transcriptional regulation is also found during phosphate limitation (Hewitt *et al.*, 2005) or after feeding the metabolic inhibitor 5-fluoroorotic acid (Santoso and Thornburg, 1998). Regulation on enzyme activity level is exerted via feedback and feed-forward loops on CPSase and ATCase. Plant CPSase is allosterically inhibited by UMP and feed-forward activated by 5-phosphoribosyl-1-pyrophosphate (PRPP) (Kanamori *et al.*, 1980; Ong and Jackson, 1972), while UMP and CP influence plant ATCase activity *in vivo* (Khan *et al.*, 1999).

UMP, the end product of the orotate pathway, is further phosphorylated by UMP kinase (step 7, Figure 5.1) and nucleoside diphosphate kinase (step 8, Figure 5.1) to UTP via UDP. CTP is formed from UTP by a one-step reaction by CTP synthetase, which catalyzes amination of UMP (step 9, Figure 5.1).

5.2.2 Biosynthesis of Thymidine Nucleotides

The reduction of the ribose moiety of the ribonucleotide diphosphates (NDP) is catalyzed by a single plant ribonucleotide reductase (RNR, step 1, Figure 5.2), and deoxyribonucleoside diphosphates (dNDPs) are produced (see Wagner and Backer, 1992). With the exception of dUDP, dNDPs are further phosphorylated and the resultants, dCTP, dATP and dGTP, are utilised as direct precursors for DNA synthesis.

RNR consists of two large subunits (R1) and two small subunits (R2). The R2 subunit houses the diiron tyrosyl radical cofactor essential for the reduction of NDP to dNDP. The R1 subunit binds the nucleoside diphosphate substrates and allosteric effectors to ensure the production of a balanced deoxyribonucleoside triphosphate (dNTP) pool. R1 is the target of feedback regulation, which ensures that dNTPs are not overproduced and that enough NDPs are left for RNA synthesis (Wang and Liu, 2006). In plants, dUDP formed by the ribonucleotide reductase is first converted to dUTP, and then hydrolyzed to dUMP (step 3, Figure 5.2) (Pardo and Gutiérrez, 1990).

Thymidine 5′-monophospate (dTMP) is synthesized from dUMP by thymidylate synthase (step 4, Figure 5.2). In this reaction, *N5*, *N10*-methylenetetrahydrofolate produced by dihydrofolate reductase acts both as donor of the methyl group and reducing agent. A bifunctional protein that consists of thymidylate synthase and dihydrofolate reductase has been detected in plants (Lazar *et al.*, 1993). cDNAs encoding the bifunctional dihydrofolate reductase-thymidylate synthase from *Glycine max* were isolated and sequenced. The encoded protein has two domains: a 226 residue dihydrofolate reductase domain in the *N*-terminus and a 304 residue thymidylate synthase domain (Wang *et al.*, 1995). Conversion of dTMP to dTTP occurs by sequential reactions catalyzed by nucleoside monophosphate kinase (step 5, Figure 5.2) and nucleoside diphosphate kinase (step 2, Figure 5.2).

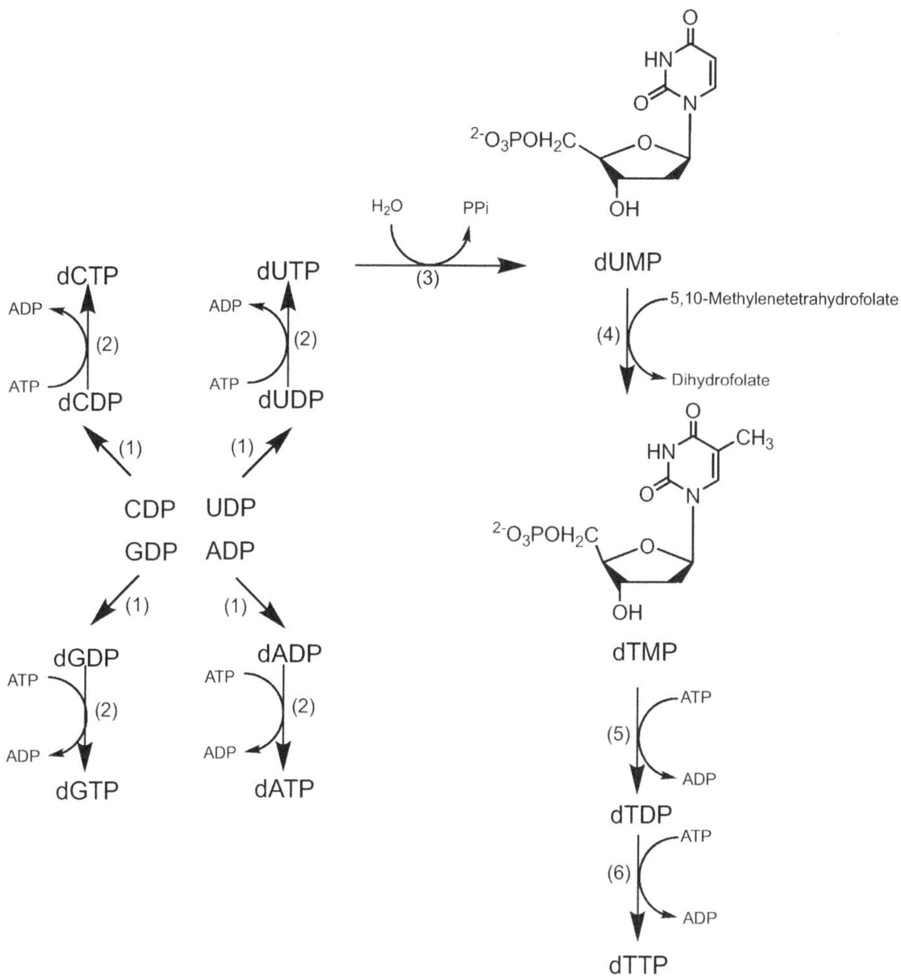

Figure 5.2 De novo *biosynthetic pathway of deoxynucleotides in plants. Enzymes (EC numbers) shown are: (1) Ribonucleotide reductase (1.17.4.10); (2) nucleoside diphosphate kinase (2.7.4.6); (3) dUTP pyrophosphatase (3.6.1.23); (4) thymidylate synthase (2.1.1.45); (5) nucleoside monophosphate kinase (2.7.4.14) and/or dTMP kinase (2.7.4.9); (6) nucleoside diphosphate kinase (2.7.4.6)*

5.2.3 Salvage Pathways

Some of the pyrimidine bases and nucleosides produced as degradation products of nucleotide and nucleic acids are re-utilised for generation of pyrimidine nucleotides (see Ross, 1981; Wasternack, 1982; Wagner and Backer, 1992). Uracil, one of the pyrimidine bases, is salvaged by the PRPP-dependent uracil phosphoribosyltransferase (UPRT). The completion of the *Arabidopsis* genome sequence revealed that UPRT is encoded by a small gene family. Functional analysis of the six genes annotated as UPRTs in the *Arabidopsis* genome showed that a single nuclear gene encoding a protein targeted to plastids is responsible for almost all UPRT activity (Mainguet *et al.*, 2009). Sequence analysis of the other five related genes revealed that the N-terminal region of the encoded proteins contains a uridine kinase (UK) domain and the C-terminal region consists of a UPRT domain.

No cytosine salvage activity is found in plants as well as animals (Ross, 1981; Katahira and Ashihara, 2002), although cytosine is converted to uracil by cytosine deaminase and salvaged in some microorganisms (Ipata *et al.*, 1971; Sakai *et al.*, 1975, Katsuragi *et al.*, 1986).

Pyrimidine nucleosides, uridine, cytidine, deoxycytidine and thymidine are salvaged to their respective nucleotides, UMP, CMP, dCMP and dTMP. A single enzyme, uridine/cytidine kinase which is present in all plants investigated to date, phosphorylates both uridine and cytidine. A recent report proposed *AtUK/UPRT1* of *Arabidopsis*, one of the genes encoding a dual domain protein with uridine kinase and uracil phosphoribosyltransferase signatures. Functional analysis of the separate domains in *E. coli* suggests that AtUK/UPRT1 can use uracil and uridine as substrates for the production of UMP (Islam *et al.*, 2007).

Deoxycytidine kinase and thymidine kinase may be present in plants, but details of their properties have not yet been elucidated. Non-specific nucleoside phosphotransferase activity also participates in the salvage of pyrimidine nucleosides, as well as purine nucleosides. To date, no functional analysis of candidate genes coding for these activities has been reported.

5.2.4 Catabolism

In plants the pyrimidine bases uracil and thymine are catabolized by a reductive pathway (Wasternack, 1978), whereas no catabolic pathway of cytosine has been observed (Katahira and Ashihara, 2002). Thus, catabolism of CMP must take place after conversion of cytidine to uridine (step 2, Figure 5.3). Plants have cytidine deaminase, but not cytosine deaminase, so conversion at the nucleoside level is essential. Uracil and thymine are degraded by the same three sequential reactions (steps 4–6, Figure 5.3) catalyzed by dihydrouracil dehydrogenase (PYD1), dihydropyrimidinase (PYD2) and β-ureidopropionase (PYD3). The end-products of this catabolic pathway are either β-alanine or β-aminoisobutyrate. In both cases, CO_2 and NH_3 are produced as byproducts. The expression of *PYD* genes is coordinated with the expression of genes encoding *de novo* and salvage synthesis pathway enzymes, presumably to maintain pyrimidine homeostasis in response to changing metabolic demands (Zrenner et al., 2009). It has been proposed that uracil degradation might be an important source of β-alanine as a precursor for pantothenate of coenzyme A (Walsh *et al.*, 2001; Katahira and Ashihara, 2006). But functional analyses of *pyd* mutants unable to degrade uracil exhibit no obvious phenotype, thus demonstrating that the catabolic pathway does not appear to be essential under normal growing conditions (Zrenner *et al.*, 2009). The recent identification and partial characterisation of an Arabidopsis uridine nucleosidase (URH1) showed its participation and function as regulative in pyrimidine degradation. Further studies with plants altered in *URH1* expression indicated that this enzyme activity must be well balanced especially in the early phase of plant development (Jung *et al.*, 2009).

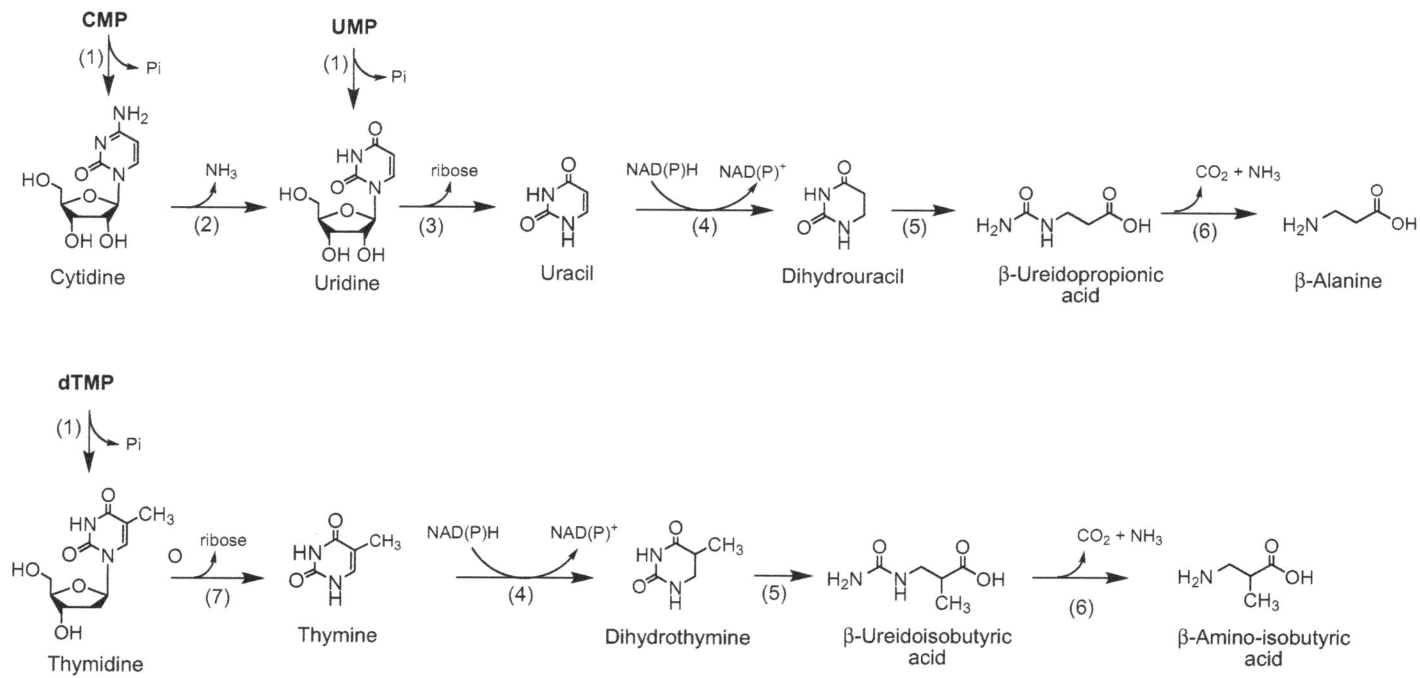

Figure 5.3 *Catabolism of pyrimidine nucleotides in plants. Enzymes (EC numbers) shown are: (1) 5′-nucleotidase (3.1.3.5) and/or various phosphatases (3.1.3.1, etc.); (2) cytidine deaminase (3.5.4.5); (3) uridine nucleosidase (3.2.2.3); (4) dihydrouracil dehydrogenase (1.3.1.2); (5) dihydropyriminase (3.5.2.2); (6) β-ureidopropionase (3.5.1.6); (7) thymidine phosphorylase (2.4.2.4) and/or thymidine (pyrimidine) nucleosidase (3.2.2.3)*

5.2.5 Secondary Metabolites

A variety of secondary metabolites derived from pyrimidine have been reported (Kafer *et al.*, 2004). 5-aminouracil was discovered in *Mimosoideae* species. This compound is presumably synthesized from uracil by way of isobarbituric acid (Brown and Turan, 1995). As this compound blocks the mitotic cycle, it may function as a defensive compound. 2,3-diaminopropionic acid is also found in these plants as a product of 5-aminouracil catabolism (Brown and Turan, 1996a). Lathyrine, unusual pyrimidinyl amino acid, initially isolated from Tangier peas (*Lathyrus tingitanus*), is produced from 2-amino-4-carboxypyrimidine (Brown and Turan, 1996b). This compound has cytokinin-like function in plant cells (Purse *et al.*, 1985).

There are non-protein amino acids that contain uracil in the side chain. Willardiine and isowillardiine appear to be synthesized from *O*-acetyl serine coupled with either the N1 or N3 of the pyrimidine ring (Ikegami and Murakoshi, 1994). Vicine and convicine are pyrimidine glucosides and accumulate in the stems and roots of some leguminous plants, such as *Vicia faba*. In the guts of herbivores, vicine and convicine are hydrolysed by β-glucanases into divicine and isouramil (Ahmmad *et al.*, 1984), which become toxic for animals.

5.3 Purine Metabolism

5.3.1 De Novo Biosynthetic Pathway

Purine nucleotides are synthesized *de novo* from small molecules, such as glycine, glutamine, and aspartate, 5-phosphoribosyl-1-pyrophosphate (PRPP), 10-formyl tetrahydrofolate, and carbon dioxide. The *de novo* biosynthetic pathway of purine nucleotides, AMP and GMP, from PRPP is shown in Figure 5.4. Detailed molecular and biochemical studies (Moffatt and Ashihara, 2002; Smith and Atkins, 2002; Hung *et al.*, 2004; van der Graaff *et al.*, 2004), the complete *Arabidopsis* genome (Boldt and Zrenner, 2003; *Arabidopsis* Genome Initiative, 2000), and the available sequence information from the rice genome (International Rice Genome Sequencing Project, 2005) reveal that plants synthesize IMP, AMP and GMP using similar reactions to those found in microorganisms and animals. The 14 enzymatic reactions of the pathway of the purine biosynthesis are summarised in Figure 5.4.

Figure 5.4 De novo *biosynthetic pathway of purine nucleotides in plants. Enzymes (EC numbers) shown are: (1) PRPP amidotransferase (2.4.2.14); (2) GAR synthetase (6.3.4.13); (3) GAR formyl transferase (2.1.2.2); (4) FGAM synthetase (6.3.5.3); (5) AIR synthetase (6.3.3.1); (6) AIR carboxylase (4.1.1.21); (7) SAICAR synthetase (6.3.2.6); (8) adenylosuccinate lyase (4.3.2.2); (9) AICAR formyl transferase (2.1.2.3); (10) IMP cyclohydrolase (3.5.4.10); (11) SAMP synthetase (6.3.4.4); (12) IMP dehydrogenase (1.1.1.205); (13) GMP synthetase (6.3.5.2). Metabolites: PRA, 5-phosphoribosyl amine; GAR, glycineamide ribonucleotide; FGAR, formylglycineamide ribonucleotide; FGRAM, formylglycine amidine ribonucleotide; AIR, 5-aminoimidazole ribonucleotide; CAIR, 5-aminoimidazole 4-carboxylate ribonucleotide; SCAIR, 5-aminoimidazole-4-N-succinocarboxyamide ribonucleotide; AICAR, 5-aminoimidazole-4-carboxyamide ribonucleotide; FAICAR, 5-formamidoimidazole-4-carboxyamide ribonucleotide; SAMP, adenylosuccinate; XMP, xanthosine-5′-monophosphate*

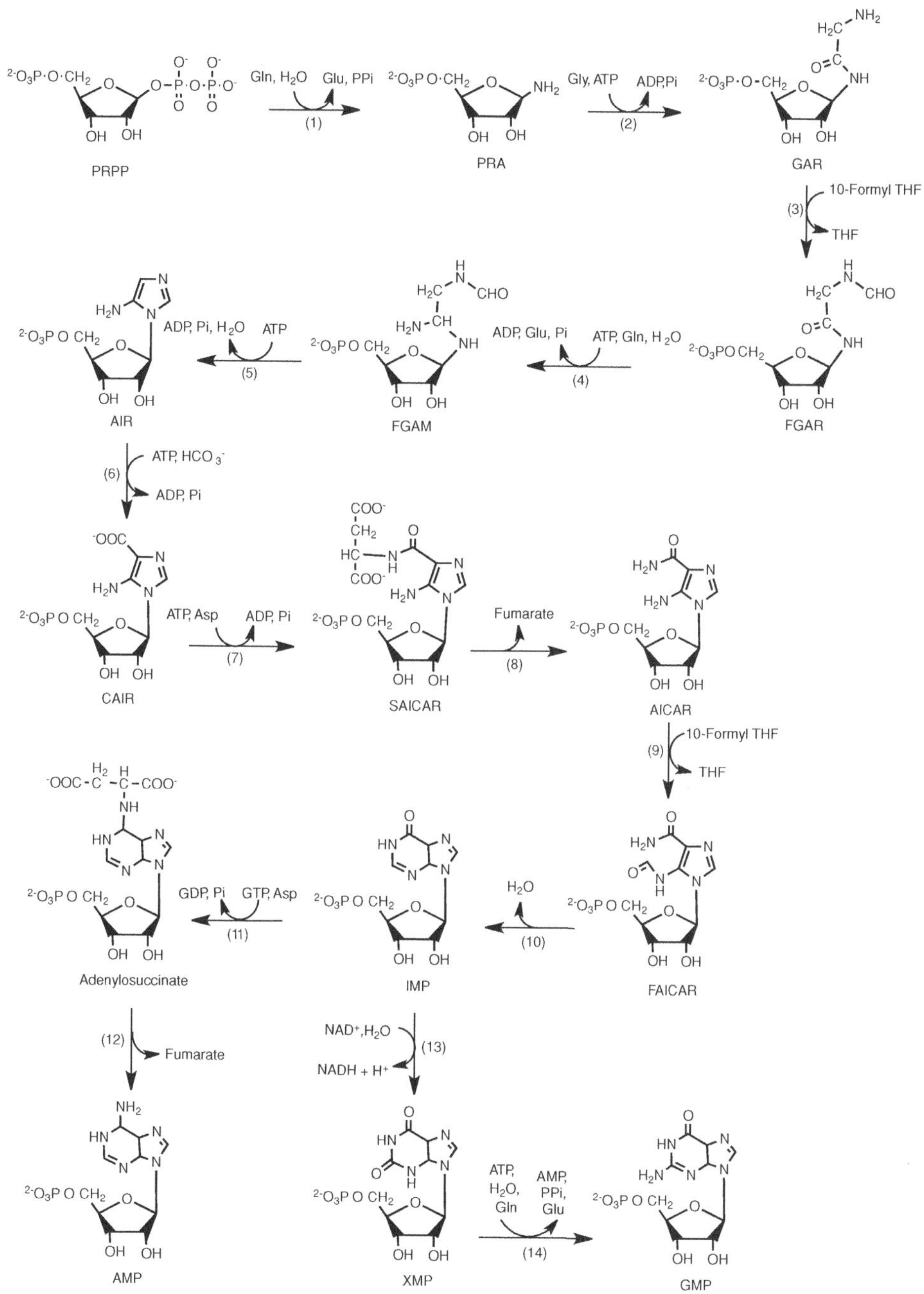

Figure 5.4 (Continued)

The purine biosynthesis starts with the formation of phosphoribosylamine (PRA) from PRPP and glutamine (step 1, Figure 5.4). This reaction is catalyzed by PRPP amidotransferase (PRPP-ATase). GAR synthetase catalyzes the ATP-dependent formation of glycine amide ribonucleotide (GAR) by attaching glycine via an amide bond to PRA (step 2, Figure 5.4). GAR is subsequently transformylated by the enzyme GAR transformylase (GART) using 10-formyltetrahydrofolate (10-Formyl-THF) to generate formylglycinamide ribonucleotide (FGAR) (step 3, Figure 5.4). The next step is catalyzed by formylglycinamidine ribonucleotide synthetase (FGAMS), consumes ATP and glutamine, and leads to the formation of formylglycinamidine ribonucleotide (FGAM) (step 4, Figure 5.4). FGAM then undergoes ring closure to form 5-aminoimidazole ribonucleotide (AIR) by consuming an additional molecule of ATP (step 5, Figure 5.4). This reaction is catalyzed by AIR-synthase (AIRS). To build the second ring of the purine skeleton, CO_2, aspartate and a second molecule of 10-Formyl-THF are inserted. AIR is carboxylated in a two-step reaction by AIR carboxylase (AIRC) to 4-carboxy aminoimidazole ribonucleotide (CAIR) (step 6, Figure 5.4). Adding aspartate and using a further molecule of ATP, *N*-succinyl-5-aminoimidazole-4-carboxamide ribonucleotide (SAICAR) is formed (step 7, Figure 5.4). This step is catalyzed by SAICAR synthase. Fumarate is released to build 5-aminoimidazole-4-carboxamide ribonucleotide (AICAR) (step 8, Figure 5.4), catalyzed by adenylosuccinatelyase (ASL). The last two steps to form the first complete purine molecule IMP are catalyzed by the bifunctional enzyme 5-aminoimidazole-4-carboxamide ribonucleotide formyltransferase/inosine monophosphate cyclohydrolase (ATIC). In the first part of this reaction the final carbon of the purine ring is provided by 10-Formyl-THF to form 5-formaminoimidazole-4-carboxamide ribonucleotide (FAICAR) (step 9, Figure 5.4). FAICAR undergoes dehydration and ring closure to generate IMP (step 10, Figure 5.4).

AMP and GMP are synthesized from IMP. AMP is created by replacing the carboxyl group at C6 by an amino group. The amino group is provided by aspartate, and GTP is the donor for the energy-rich phosphate bond to form adenylosuccinate (SAMP) (step 11, Figure 5.4). This reaction is catalyzed by the enzyme SAMP synthase (ASS). The removal of fumarate to form AMP is catalyzed by ASL (step 12, Figure 5.4). The synthesis of GMP is initiated by the oxidation of IMP followed by the insertion of an amino group that is provided by glutamine. Xanthosine 5′-phosphate (XMP) is formed by IMP dehydrogenase (IMPDH) using NAD^+ as the hydrogen acceptor (step 13, Figure 5.4). The final step to form GMP is catalyzed by GMP synthase (GMPS) (step 14, Figure 5.4).

In animals, enzymes of the *de novo* biosynthetic pathway and those of hydrofolate metabolism are present as a multi-enzyme complex (Christopherson and Szabados, 1997). In contrast, such a complex has not been observed in higher plants (Zrenner *et al.*, 2006). The organisation of the *de novo* purine biosynthesis pathway in plants is more similar to prokaryotes, with monofunctional proteins, except for the bifunctional enzymes AIRC and ATIC. AMP and GMP are phosphorylated to nucleoside diphosphates and finally to the nucleoside triphosphates.

All of the genes encoding enzymes required for the ten-step synthesis of IMP contain sequences that are predicted to encode *N*-terminal plastid-transit peptides. This observation indicates a chloroplast localisation of this portion of the purine biosynthesis. The cDNAs encoding the enzymes for AMP synthesis from IMP (ASL and ASS) also contain putative plastid transit sequences. This observation suggests that the AMP synthesis from IMP is located in chloroplasts (Fonne-Pfister *et al.*, 1996; van der Graaff *et al.*, 2004). In contrast, the genes encoding IMPDH and GMPS, which are involved in the pathway leading to GMP (Cao and Schubert, 2001; van der Graaff *et al.*, 2004), do not contain *N*-terminal transit peptides. Therefore, the synthesis of XMP and GMP from IMP may occur in the cytosol.

The expression analyses (Senecoff *et al.*, 1996), particularly of the Arabidopsis *PURC* gene encoding SAICARS, revealed that genes of the *de novo* purine biosynthesis are basically expressed in mitotically active tissues and are primarily involved in the process of cell division. The expression of the complete pathway has been analysed in tobacco (*Nicotiana tabacum*) plants by van der Graaff and co-workers (2004). All genes of the purine biosynthesis show their strongest expression in ovaries; this observation implicates their function in cell division. Most of those genes are expressed at lower levels in a constitutive manner; this observation indicates housekeeping functions for these genes. In contrast, the genes encoding GMPS, SAICAS and IMPDH were specifically expressed in sink leaves and floral organs. Different expression patterns during flower development could also be observed for the *NtPURF*, *NtPURL* and *NtPURH* genes encoding PRPP-ATase (step 1, Figure 5.4), FGAMS (step 4, Figure 5.4) and ATIC (steps 9–10, Figure 5.4), respectively. This expression analysis clearly showed transcriptional regulation of purine biosynthesis genes during plant development and, furthermore, indicates that the expression of purine biosynthesis genes is not confined solely to meristematic tissues.

Feedback control of the *de novo* purine biosynthetic pathway involves at least three steps: 5-phosphoribosylamine synthase activity (step 1, Figure 5.4) is inhibited by IMP, AMP and GMP; activities of ASS (step 11, Figure 5.4) and IMPDH (step 12, Figure 5.4) are inhibited by AMP and GMP, respectively (Stasolla *et al.*, 2003).

5.3.2 Salvage Pathways

Utilisation of preformed purine bases and nucleosides for nucleotide synthesis occurs through the salvage pathway. Purine nucleosides and nucleobases arise from the intercellular breakdown of unstable RNA and nucleotides, and in some cases, originate from exogenous sources as catabolic products of nucleic acids and nucleotides in decaying cells (Wasternack, 1982; Bray, 1983). Adenosine released from hydrolysis of *S*-adenosyl-L-homocysteine in the *S*-adenosyl-L-methionine cycle, and adenine released in the methionine cycle of ethylene synthesis, are also utilised as substrates for purine salvage (Wasternack, 1982; Ashihara and Crozier, 1999; Crozier et al., 2000; Moffatt and Ashihara, 2002). Several enzymes involved in purine salvage have been found in higher plants. The three purine bases adenine, guanine and hypoxanthine are salvaged to AMP, GMP and IMP, respectively. Two distinct enzymes, adenine phosphoribosyltransferase (APRT) and hypoxanthine/guanine phosphoribosyltransferase (HGPRT) participate in these salvage reactions. Purine bases may also be salvaged to nucleotides via nucleosides. Adenosine phosphorylase and inosine-guanosine phosphorylase convert adenine, hypoxanthine and guanine to their respective ribonucleosides using ribose-1-phosphate. However, activities of these enzymes are very low in higher plants (Wagner and Backer, 1992; Ashihara and Crozier, 1999; Moffatt and Ashihara, 2002; Stasolla *et al.,* 2003). Thus, the production of purine ribonucleotides by these routes may only play a minor role in purine salvage. Purine nucleosides, adenosine, guanosine and inosine, are salvaged by kinases and/or nucleoside phosphotransferase. Adenosine kinase (AK) is distributed ubiquitously and its activity is usually high, whereas inosine-guanosine kinase (IGK) has been found in only a limited number of plant species. In some plants, inosine and guanosine are mainly salvaged by the non-specific nucleoside phosphotransferase (Wagner and Backer, 1992, Stasolla *et al.*, 2003).

A comprehensive biochemical and molecular analysis of APRT and AK has been carried out in *Arabidopsis* (Moffatt and Ashihara, 2002). APRT is represented by a multi-gene family (*AtAPT*) encoding five isoenzymes in *Arabidopsis*. None of the *AtAPTs* contain *N*-terminal signal sequences, and by immunolocalisation it could be demonstrated that APT1, APT2 and APT3 are localised in the cytosol (Allen *et al.*, 2002). The *Arabidopsis* AK is encoded by two genes (*ADK*). *ADK1* and *ADK2* are constitutively expressed, which classifies the *ADKs* as housekeeping enzymes (Allen *et al.*, 2002; Moffatt and Ashihara, 2002). The APRT and AK also play a crucial role in the inter-conversion of cytokinins (Moffatt and Ashihara, 2002). It is generally assumed that the free base and the riboside derivatives are the biologically active forms of the cytokinins (Astot *et al.*, 2000).

5.3.3 Catabolism

Plants possess the complete oxidative purine catabolic pathway to CO_2 and NH_3 via uric acid and allantoin (Schubert and Boland, 1990; Ashihara and Crozier, 1999; Stasolla *et al.*, 2003). A key starting compound of purine catabolism is xanthine; thus, all purine nucleotides must be converted to xanthine before their purine ring cleavage is initiated (Figure 5.5). Deamination, dephosphorylation and glycosidic bond cleavage are involved in this process. In contrast to animals, adenosine deaminase has generally not been found in most plants, although very low levels of adenosine deaminase was detected in roots and foliage of alfalfa (*Medicago sativa*) plants (Edwards, 1996). Thus, AMP deaminase (AMPD; step 1, Figure 5.5) and guanosine deaminase (step 5, Figure 5.5) are the predominant deamination enzymes for adenine and guanine nucleotides respectively. Guanine deaminase (step 6, Figure 5.5) is also detected in plants, but its activity is generally low. GMP reductase, which catalyzes the conversion of GMP to IMP, is not present in plants. There are many enzymes for the dephosphorylation reaction. Various phosphatases, 3′-nucleotidase and 5′-nucleotidase appear to produce nucleosides. Adenosine nucleosidase and inosine-guanosine nucleosidase seem to participate in glycosidic bond cleavage.

Xanthine is converted to uric acid by xanthine dehydrogenase (XDH; step 7, Figure 5.5). Uricase catalyzes the formation of allantoin, and allantoic acid is produced by an allantoinase-catalyzed reaction (steps 8 and 9, Figure 5.5). Some plant organs, such as roots of the tropical legumes and maple trees, accumulate allantoin and/or allantoic acid, which play an important role in the storage and translocation of nitrogen (Schubert and Boland, 1990). Different metabolic fates of allantoic acid are proposed in plants. In the allantoicase pathway (Piedras *et al.*, 2000; Munoz *et al.*, 2006), allantoic acid is degraded to CO_2, NH_3 and glyoxylate via urea and ureidoglycolate (steps 10–12, Figure 5.5). An alternative route, the allantoic acid amidohydrolase pathway, has been proposed by Winkler *et al.* (1988) in which allantoic acid is initially converted to ureidoglycine, CO_2 and NH_3 (steps 13–15, Figure 5.5). The NH_3 is released directly and urea formation is not involved in the subsequent catabolism to glyoxylate.

Recently molecular analyses on enzymes and genes of some purine catabolism have been carried out. The amino acid sequence of the *AtAMPD* does not contain an *N*-terminal

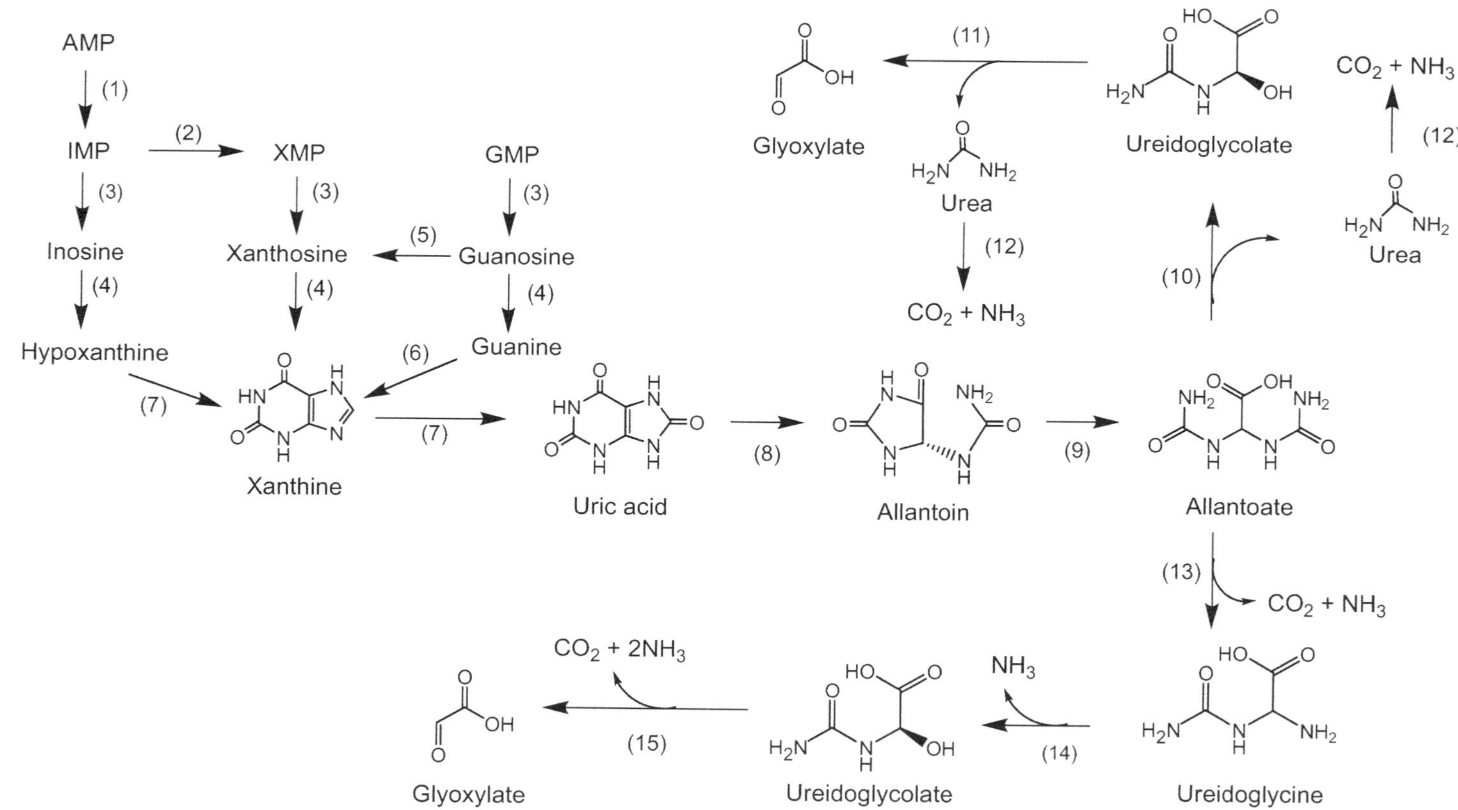

Figure 5.5 *Catabolic pathways of purine nucleotides in plants. Enzymes (EC numbers) shown are: (1) AMP deaminase (3.5.4.6); (2) IMP dehydrogenase (1.1.1.205); (3) 5′-nucleotidase (3.1.3.5) and/or various phosphatises (3.1.3.1, etc.); (4) inosine-guanosine nucleosidase (3.2.2.2); (5) guanosine deaminase (3.5.4.15); (6) guanine deaminase (3.5.4.3); (7) xanthine dehydrogenase (1.1.1.204); (8) uricase (1.7.3.3); (9) allantoinase (3.5.2.4); (10) allantoicase (3.5.3.4); (11) ureidoglycolate lyase (4.3.2.3); (12) urease (3.5.1.5); (13) allantoin deaminase (3.5.3.9); (14) ureidoglycine amidohydrolase (not yet listed); (15) ureidoglycolate amidohydrolase (3.5.3.9)*

transit peptide, thus the protein is expected to accumulate in the cytosol. XDH is a metalloflavo enzyme requiring FAD and the molybdenum cofactor sulfurase. In *Arabidopsis*, two isoforms were discovered and characterised at the molecular and biochemical level (Hesberg *et al*., 2004). XDH is encoded by two genes that are orientated in tandem and separated by 704 bp on chromosome 4 in the *Arabidopsis* genome. Both *XDHs* are expressed differently. Whereas *AtXDH2* is expressed constitutively, *AtXDH1* transcription is induced by drought, salt, and abscisic acid treatment. XDH belongs to the class of xanthine oxidoreductases, which in mammals exist as xanthine dehydrogenases or as O_2-dependant xanthine oxidases. The *Arabidopsis* XDHs are strict dehydrogenases, but produce superoxide radicals; this observation indicates that plant XDHs might also be involved in stress responses that require reactive oxygen species (Hesberg *et al*., 2004). Uricase catalyzes the formation of allantoin and has been intensively investigated in nitrogen-fixing legumes. In these studies it was discovered as a nodule-specific gene from soybean and designated as Nod-35 (Tajima *et al*., 2004). In soybean, which is a specialised symbiotic species, uricase plays an important role to produce ureides as nitrogen storage and transport intermediates, and may have a function in the process of plant microbe interaction (Tajima *et al*., 2004). The next reactions in purine degradation are performed by allantoinase, which converts allantoin into allantoate (Stasolla *et al*., 2003). Allantoinases from tropical legumes (Schubert and Boland, 1990) and more recently from *Arabidopsis* (Yang and Han, 2004) have been characterised at the biochemical and molecular level. In *Arabidopsis* the single copy gene and the corresponding cDNA have been characterised. The encoded polypeptide contains an *N*-terminal signal peptide for the secretory pathway, and this observation corresponds to biochemical results that locate allantoinase activity to peroxisomes or the endoplasmatic reticulum (Hanks *et al*., 1981). Recently the identification and cloning of an allantoate amidohydrolase (*AtAAH*) (Todd and Polacco, 2006) demonstrates the presence of the allantoic acid amidohydrolase pathway with direct release of NH_3 in *Arabidopsis thaliana*. In a comparative genomic approach coupled with biochemical analyses, this AAH pathway was completed by identifying the final reactions and cloning of cDNAs coding for ureidoglycine amidohydrolase (UGlyAH) and ureidoglycolate amidohydrolase (UAH) of *Arabidopsis thaliana* (Werner *et al*., 2010). Several studies on subcellular localisation of purine catabolic enzymes finally demonstrate that after the generation of allantoin in the peroxisome, plant purine degradation continues in the endoplasmic reticulum (Werner *et al*., 2008; Serventi *et al*., 2010).

5.3.4 Secondary Metabolites

Methylxanthines and methyluric acids, known as purine alkaloids, are synthesized from purine nucleotides. Caffeine (1,3,7-trimethylxanthine) and theobromine (3,7-dimethylxanthine) are the well-known purine alkaloids found in some species including tea (*Camellia sinensis*), coffee (*Coffea arabica, Coffea canephora*), maté (*Ilex paraguariensis*) and cacao (*Theobroma cacao*) plants. In addition to purine alkaloids, higher plants possess the metabolic pathways for the production of some specialised purine compounds. Among the most important are the pathways that lead to production of cytokinins, such as zeatin, which are of widespread occurrence in higher plants. The biosynthesis and metabolism of purine alkaloids are shown in Chapter 6. Cytokinins are reviewed by Crozier *et al.* (2000) and Sakakibara (2006).

5.4 Pyridine Metabolism

5.4.1 *De Novo* Biosynthetic Pathway

Pyridine nucleotides (NAD and NADP) are synthesized *de novo* from amino acid precursors. Two distinct pathways called the aspartate pathway and the kynurenine pathway are present in different organisms (Figure 5.6; Rongvaux *et al.*, 2003; Katoh and Hashimoto, 2004). In plants, the aspartate pathway is operative. In the first step of this pathway, aspartate is oxidised to α-iminosuccinic acid by aspartate oxidase (step 1, Figure 5.6). In the next step, α-iminosuccinate is condensed with glyceraldehyde-3-P and cyclised to produce quinolinic acid by quinolinate synthase (step 2, Figure 5.6). The third step is catalyzed by quinolinic acid phosphoribosyltransferase, which forms nicotinic acid mononucleotide (NaMN) from quinolinic acid and PRPP (step 3, Figure 5.6). The subcellular distribution of functional GFP-fused enzymes and import of precursor proteins into isolated chloroplasts indicates that these three enzymes are located in the plastids of *Arabidopsis* (Katoh *et al.*, 2006). The two subsequent enzymatic steps then convert NaMN to NAD (steps 4 and 5, Figure 5.6). NaMN is synthesized either from quinolinic acid via the *de novo* pathway or from the salvage reaction (see below). NaMN reacts with ATP by NaMN adenylyltransferase (NaMN-AT) to form a condensed compound, nicotinic acid adenine dinucleotide (NaAD). This is then converted to NAD by NAD synthetase. The steps between NaMN and NAD are referred to as the 'Preiss–Handler pathway' in older literature (Moat and Foster, 1987). NADP is formed from NAD by NAD kinase.

Enzyme activities of NaMN-AT and NAD synthase have been detected in crude protein extracts from several tissues of tobacco (Wagner and Wagner, 1985; Wagner *et al.*, 1986a,b). *Arabidopsis* contains only a single NaMN-AT gene (Hunt *et al.*, 2004), although mammals and fungi possess isoforms. The tobacco NAD synthase activity was detected with glutamine or asparagine but not with ammonia. The aspartate pathway of NAD synthesis is also found in most bacteria, including *Escherichia coli*. In contrast, mammals, fungi and some bacteria produce NAD by the kynurenine pathway. The steps leading from quinolinic acid to NAD are conserved among prokaryotes and eukaryotes.

5.4.2 Pyridine Nucleotide Cycle

In many organisms, nicotinamide and nicotinic acid, degradation products of NAD and NADP, are re-utilised for the synthesis of pyridine nucleotides by salvage pathways. The route, which consists of several reactions involved in the degradation and re-synthesis of pyridine nucleotides, has been referred to as the pyridine nucleotide cycle (Waller *et al.*, 1966; Gholson, 1966). Pyridine nucleotide cycles show wide variations between species and even organs (Moat and Foster, 1987; Wagner and Backer, 1992; Ashihara *et al.*, 2005; Noctor *et al.*, 2006). The simplest cycle (PNC II) appears to be a NAD → nicotinamide mononucleotide (NMN) → NAD pathway. Wagner and Wagner (1985) found activity of NAD pyrophosphatase and NMN adenylyltransferase in tobacco plants, so this cycle might be operative in potato tubers. However, a more complicated cycle seems to be operative in most plants, because tracer experiments using radioactive precursors, such as [^{3}H]quinolinic acid, show the incorporation of radioactivity into some catabolites of NAD (Ashihara *et al.*, 2005). It has been considered that NAD is converted to nicotinamide via at least two routes

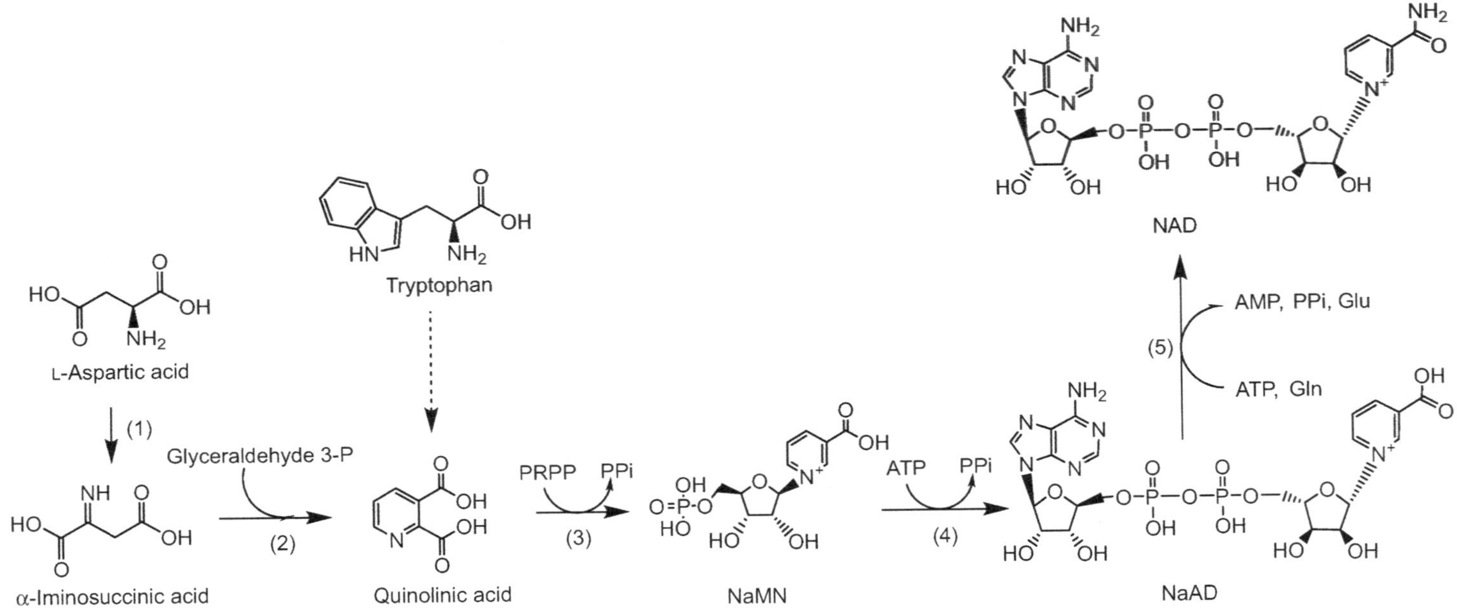

Figure 5.6 De novo *biosynthetic pathway of pyridine nucleotides in plants. Enzymes (EC numbers) shown are: (1)* L*-aspartate oxidase (1.4.3.16); (2) quinolinate synthase (2.5.1.72); (3) quinolinate phosphoribosyltransferase (decarboxylating) (2.4.2.19); (4) nicotinic acid mononucleotide adenylyltransferase (2.7.7.18); (5) NAD synthase (6.3.5.1)*

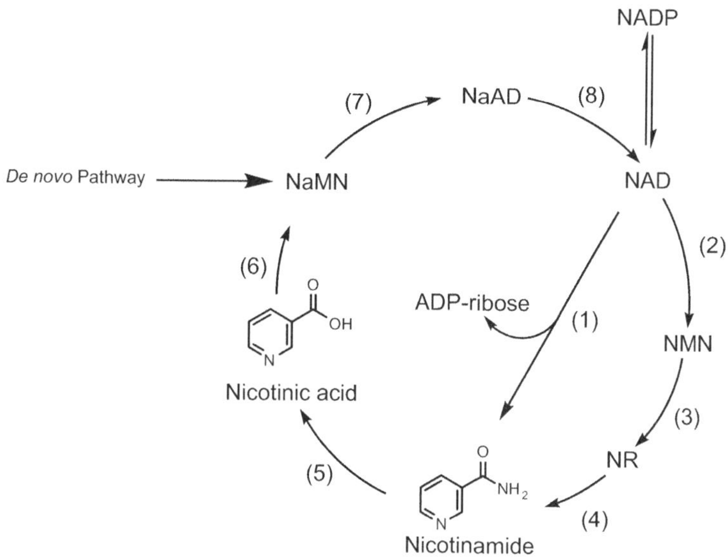

Figure 5.7 *Pyridine nucleotide cycle in plants. Enzymes (EC numbers) shown are: (1) poly (ADP-ribose) polymerases (2.4.2.30), mono (ADP-ribosyl) transferase (2.4.2.31) or NAD-dependent deacetylase (Sir 2) (3.5.1.-); (2) NAD pyrophosphatase (3.6.1.22); (3) 5'-nucleotidase (3.1.3.5); (4) nicotinamide riboside nucleosidase (3.2.2.-); (5) nicotinamidase (3.5.1.19); (6) nicotinate phosphoribosyltransferase (2.4.2.11); (7) nicotinic acid mononucleotide adenylyltransferase (2.7.7.18); (8) NAD synthase (6.3.5.1). Metabolites: NaAD, nicotinic acid adenine dinucleotide; NaMN, nicotinic acid mononucleotide; NMN, nicotinamide mononucleotide; NR, nicotinamide riboside*

in plants. The first is direct formation from NAD, releasing ADP-ribose by poly (ADP-ribose) polymerases, mono (ADP-ribosyl) transferase or NAD-dependent deacetylase (Sir 2) (Noctor *et al.*, 2006) (step 1, Figure 5.7). The second is three-step formation via NMN and nicotinamide riboside (NR) (steps 2–4, Figure 5.7) (Ashihara *et al.*, 2005; Noctor *et al.*, 2006). Unlike in animals (Revollo *et al.*, 2004; van der Veer *et al.*, 2007), nicotinamide is not directly salvaged to NMN, because plants have no nicotinamide phosphoribosyltransferase (Ashihara *et al.*, 2005; Zheng *et al.*, 2005; Wang and Pichersky, 2007). Instead, plants readily convert nicotinamide to nicotinic acid with plant-specific nicotinamidase (step 5, Figure 5.7) (Ashihara *et al.*, 2005; Wang and Pichersky, 2007), and salvage nicotinic acid for NaMN synthesis using nicotinic acid phosphoribosyltransferase (NaPRT) (step 6, Figure 5.7). Distinct pyridine nucleotide cycles have been proposed in different plants (Wagner *et al.*, 1986a,b,c; Zheng *et al.*, 2004; Ashihara *et al.*, 2005; Matsui and Ashihara, 2008). A seven-component pyridine nucleotide cycle (PNC VII), NAD → NMN → NR → nicotinamide → nicotinic acid → NaMN → NaAD → NAD route (steps 2–8, Figure 5.7), has been proposed in several plants (Matsui *et al.*, 2007). In addition, a further route that includes newly discovered NR deaminase and/or NaR kinase is also operative in some plant species including potato and mungbean (Katahira and Ashihara, 2009; Matsui and Ashihara, 2008). These findings imply that a new six-component pyridine nucleotide cycle

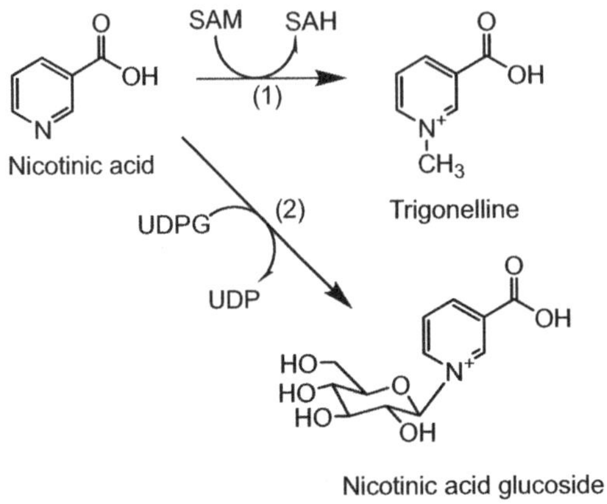

Figure 5.8 *Synthesis of nicotinic acid conjugates in plants. Enzymes (EC numbers) shown are: (1) nicotinate* N*-methyltransferase (trigonelline synthase, 2.1.1.7); (2) nicotinate glucosyltransferase (2.4.1.196)*

(PNC VI), namely a NMN → NR → NaR → NaMN → NaAD → NAD route, and the cycle PNC VIII, NMN → NR → nicotinamide → nicotinic acid → NaR → NaMN → NaAD → NAD route, can be at work in some organs of plant species. The most likely rate-limiting step of the pyridine cycle (PNC VII) appears to be the one catalyzed by nicotinate phosphoribosyltransferase (NaPRT). This enzyme requires ATP for its activity (Zheng *et al.*, 2005). Thus, availability of ATP as well as PRPP regulates the formation of NaMN. Purification and characterisation of plant NaPRT has not yet been performed, but feedback inhibition by NAD of NaPRT purified from *Brevibacterium* has been reported (Dulyaninova *et al.*, 2000). In contrast, other steps including nicotinamidase and NaMN adenylyltransferase appear not to be the rate-limiting steps on the PNC, since there was little accumulation of the substrates nicotinamide and NaMN found in most plants (Matsui *et al.*, 2007; Katahira and Ashihara, 2009). These facts also suggest an adequate supply of ATP for NaMN adenylyltransferase to form an adenine nucleotide moiety of NAD.

5.4.3 Secondary Metabolites

Nicotine is the most well-known secondary metabolite derived from NAD in tobacco plants (see Chapter 7). In contrast to nicotine, many plants produce the nicotinic acid conjugates, trigonelline (*N*-methyl nicotinic acid) and/or nicotinic acid-*N*-glucoside (Figure 5.8). Trigonelline accumulates in seeds of leguminous plants and *Coffea* sp. The detailed distribution and biosynthesis of these nicotinic acid conjugates is reviewed elsewhere (Ashihara, 2008).

5.5 Biotechnological Approaches

Nucleotides are crucial cellular components for plant growth, development and metabolism. Because they are required for primary and secondary metabolism, and also for information storage and gene expression, they are important for the majority of fundamental cellular and biochemical processes that are essential for cell growth and storage compound accumulation. While in plant sciences we are still in the infancy of using biotechnological applications with relation to nucleotide metabolism, in medical sciences nucleotide biosynthesis is a well-known target for many antineoplastic and antiviral agents designed for potential pharmaceutical use (Christopherson *et al.*, 2002).

5.5.1 New Herbicide Targets

The natural product phytotoxins hydantocidin and ribofuranosyl triazolonone (Figure 5.9) have been shown to exert their toxic effect by bioconversion via phosphorylation into inhibitors of adenylosuccinate synthetase (step 11, Figure 5.4; Siehl *et al.*, 1996; Schmitzer *et al.*, 2000), the enzyme catalyzing the penultimate step in AMP biosynthesis. These results identify the site of action of hydantocidin and establish adenylosuccinate synthetase as a herbicide target of commercial potential. Furthermore it is demonstrated that nucleotide *de novo* synthesis is a fundamental biochemical pathway that is essential for cellular growth.

In another case, AMP deaminase (step 1, Figure 5.5; AMPD) has been recognised as a potent herbicide target (Dancer *et al.*, 1997). The microbial compound carbocyclic coformycin (Figure 5.9) in its phosphorylated form is a strong inhibitor of AMPD. When applied to plants, it causes a dramatic increase in ATP levels, indicating a perturbation of purine metabolism accompanied by decreased extractable activity of AMPD by tight binding of the inhibitor. It is proposed that inhibition of AMPD leads to the death of the plant through perturbation of the intracellular ATP pool. Thus, inhibition of AMPD will, through its effects on adenylate metabolism, have consequences for almost every aspect of metabolism, which explains why this enzyme should represent such a potent herbicide target. This crucial role for AMPD in plant development was recently re-discovered by the characterisation of the embryogenic factor1 (FAC1) in *Arabidopsis* (Xu *et al.*, 2005). FAC1 was identified by screening an EMS-mutagenised *Arabidopsis* population and by positional cloning of the respective locus. The gene encodes AMPD and, when mutagenised, causes a zygote-lethal embryonic phenotype. Another *Arabidopsis* T-DNA insertion mutant also showed this phenotype (Xu *et al.*, 2005).

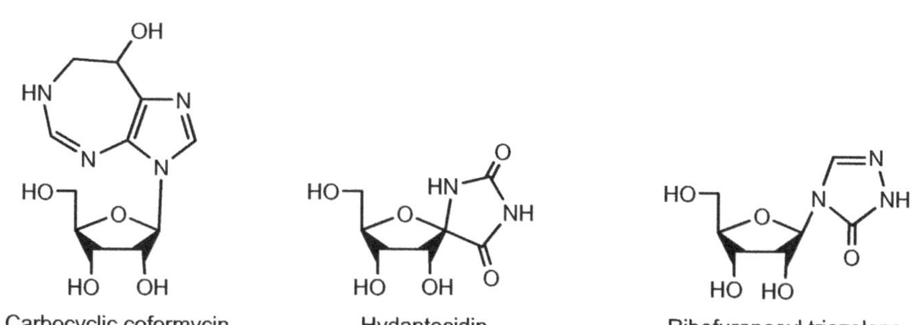

Figure 5.9 *Chemical structures of carbocyclic coformycin, hydantocidin and ribofuranosyl triazolonone*

New chemical genomics approaches using plant phenotype-based chemical screens with cell cultures and seedlings have identified small molecules targeting a variety of processes (Robert *et al.*, 2009). In recent years, strategies of herbicide discovery have switched from the testing of chemicals for efficacy on whole plants towards the use of *in vitro* assays against molecular targets (Lein *et al.*, 2004). Many different approaches have been developed to identify *bona fide* targets for *in vitro* screening. Thus, targeted antisense-inhibition of each single step in an entire pathway (Schröder *et al.*, 2005), or high-throughput unbiased gene silencing and visual phenotyping, were used to identify new genes in which partial inhibition of expression leads to marked phenotypic changes (Lein *et al.*, 2008). Using both approaches, genes of purine and pyrimidine nucleotide *de novo* synthesis have been identified as targets for *in vitro* screening for new herbicides, because reduced expression of genes of *de novo* nucleotide synthesis is compensated for by strong reduction in growth rates with or without further phenotypic alterations.

Proof of concept of this strategy has finally shown in the work on PRPP-ATase, the first committed enzyme of *de novo* purine biosynthesis (step 1, Figure 5.4). Independent experimental approaches with transgenic tobacco plants with reduced PRPP-ATase activity and the Arabidopsis *atd2* knock-out mutant revealed a strong impact on whole plant growth and development. Both findings indicate the central function of PRPP-ATase in plant *de novo* purine biosynthesis as the probable rate-limiting step (van der Graaff *et al.*, 2004). In another approach using a combination of genetic, chemical and biochemical techniques, the molecular target of the phenyltriazole acetate compound was elucidated as PRPP-ATase (Walsh *et al.*, 2007). Functional expression of Arabidopsis PRPP-ATase in *Escherichia coli* at high levels showed tight-binding inhibitor with the enzyme. Thus, a novel specific chemical probe of the first step in plant purine biosynthesis has been identified and its mode of interaction with its cognate target has been characterised.

5.5.2 Increased Growth by Increased Nucleotide Precursor Availability

As outlined above, reduced expression of genes of purine and pyrimidine nucleotide *de novo* synthesis is compensated by a strong reduction in plant growth. Therefore it is assumed that increasing levels of nucleotides might lead to higher metabolic fluxes thus fuelling plant growth. The strategy of increasing the availability of precursors of basic metabolic processes, such as nucleotide biosynthesis, has obvious implications. It can be regarded as part of the biotechnological approach to increase plant biomass as an alternative renewable energy source. This was achieved by expression of *Ashbya gossypii* genes coding for PRPP synthetase (PRS), the enzyme catalyzing the synthesis of PRPP needed for *de novo* and salvage synthesis of nucleotides in *A. thaliana* and *N. tabacum*. Increased PRS activity leads to a substantial increase in plant biomass accumulation under different growth conditions in both plant species (Koslowsky *et al.*, 2008). This analysis provides evidence that the supply of PRPP co-limits growth rates and it is shown that increased PRS activity in the cytosol promotes nucleotide availability by enhancing nucleotide salvage processes.

Another interesting and unexpected result on growth and yield was achieved with transgenic potato plants with tuber-specific silencing of UMPSase (steps 5–6, Figure 5.1; Geigenberger *et al.*, 2005). These tubers showed decreased pyrimidine *de novo* synthesis which was compensated by a stimulation of the salvage pathway. This shift in the pathways of UMP synthesis was accompanied by increased levels of tuber uridine nucleotides, and increased amounts of starch and cell wall components in the tubers. Although there is no obvious explanation for the overall increased nucleotide content in plants with partial reduction of *de novo* synthesis, this result also indicates that increased uridine nucleotide pool levels in tubers improve biosynthetic performance.

5.5.3 Increased Potato Tuber Yield by Modulating Adenylate Pools

The enzyme adenylate kinase is a nucleoside monophosphate kinase catalyzing the interconversion of ATP and AMP into ADP. Because adenylate kinase is described as a crucial enzyme in maintaining the pool sizes of various adenylates at equilibrium (Pradet and Raymond, 1983), it represents an interesting target for modulating adenylate pools. Down-regulation of the activity of a plastidial isoform of adenylate kinase has substantial effects on the pool size of the various adenylates and, most importantly, leads to an increase in total tuber yield above that found in wild-type plants (of $\leq$39%) and an increased starch content per gram fresh weight (by $\leq$60%) above that found in wild-type plants (Regierer *et al.*, 2002). These data unequivocally show that modulation of nucleotide pools represents a useful strategy for increasing formation of one of the most important resources for both human and animal diet. Further analysis of the underlying mechanism elucidated increased rates of respiratory oxygen consumption and increased carbon fluxes into starch (Oliver *et al.*, 2008). Increased rates of starch synthesis were accompanied by post-translational redox-activation of ADP-glucose pyrophosphorylase (AGPase), catalyzing the key regulatory step of starch synthesis in the plastid.

However, the general strategy to increase starch synthesis by decreasing adenylate kinase activity in plastids may not be transferred to other plants or even tissues. For example, the *Arabidopsis* genome contains seven genes identified as putative adenylate kinases, at least two of them coding for enzymes located in plastids. When the individual genes are disrupted, mutant phenotypes are strikingly different. While absence of one isoform has no effect on plant performance, absence of the other isoform causes only 30% reduction of total adenylate kinase activity in leaves. However, there is loss of chloroplast integrity leading to small, pale-looking plantlets from embryo to seedling development (Lange *et al.*, 2008), indicating at least the different roles of plastid adenylate kinase in leaves and tuber tissues.

Since nucleotide pools are different in the various subcellular compartments of the cell transport processes, the presence and activity of diverse transport proteins may become important for adjustment of nucleotide pools. The wealth of functionally characterised intracellular plant membrane transporters has recently been summarised (Linka and Weber, 2010). Molecular studies have further demonstrated that the adenylate supply to the plastid is of fundamental importance to starch biosynthesis in storage organs such as potato tubers (Tjaden *et al.*, 1998). Over-expression of the amyloplastidial ATP/ADP translocator resulted in an increased starch accumulation, whereas antisense inhibition of the same protein resulted in a reduced starch yield. These studies indicate that the increased starch content in tubers from sense plants is due to an increased rate of ATP import from the cytosol into the ATP-consuming stroma. Such an increased flux depends upon a sufficient cytosolic ATP supply. The subcellular adenylate levels in potato tubers indicate (Farre *et al.*, 2001) that a high cytosolic ATP/ADP ratio would permit an increased ATP import into the stroma after increasing the activity of the plastidic ATP/ADP transporter. Comparable results were found with transgenic potato plants simultaneously over-expressing a *Pisum sativum* glucose-6-phosphate/phosphate translocator and an *Arabidopsis thaliana* adenylate translocator in tubers (Zhang *et al.*, 2008).

References

Ahmmad, M.A.S., Maskall, C.S. and Brown, E.G. (1984) Partial-purification and properties of willardiine and isowillardiine synthase activity from *Pisum sativum*. *Phytochemistry*, **23**, 265–270.

Allen, M., Qin, W.S., Moreau, F. and Moffatt, B. (2002) Adenine phosphoribosyltransferase isoforms of Arabidopsis and their potential contribution to adenine and cytokinin metabolism. *Physiol. Plant.* **115**, 56–68.

Arabidopsis Genome Initiative (2000) Analysis of the genome sequence of the flowering plant *Arabidopsis thaliana*. *Nature*, **408**, 796–815.

Ashihara, H. (2008) Trigonelline (*N*-methylnicotinic acid) biosynthesis and its biological role in plants. *Nat. Prod. Commun.* **3**, 1423–1428.

Ashihara, H. and Crozier, A. (1999) Biosynthesis and metabolism of caffeine and related purine alkaloids in plants. *Adv. Bot. Res.* **30**, 117–205.

Ashihara, H., Stasolla, C., Yin, Y., Loukanina, N. and Thorpe, T.A. (2005) De novo and salvage biosynthetic pathways of pyridine nucleotides and nicotinic acid conjugates in cultured plant cells. *Plant Sci.*, **169**, 107–114.

Astot, C., Dolezal, K., Nordstrom, A., Wang, Q., Kunkel, T., Moritz, T., Chua, N.H. and Sandberg, G. (2000) An alternative cytokinin biosynthesis pathway. *Proc. Nat. Acad. Sci. USA*, **97**, 14778–14783.

Boldt, R. and Zrenner, R. (2003) Purine and pyrimidine biosynthesis in higher plants. *Physiol. Plant.*, **117**, 297–304.

Bray, C.M. (1983) *Nitrogen Metabolism in Plants*, Longman, London.

Brown, E.G. and Turan, Y. (1995) Pyrimidine metabolism and secondary product formation – biogenesis of albizziine, 4-hydroxyhomoarginine and 2,3-diaminopropanoic acid. *Phytochemistry*, **40**, 763–771.

Brown, E.G. and Turan, Y. (1996a) Lathyrine biosynthesis and the origin of the precursor 2-amino-4-carboxypyrimidine in *Lathyrus tingitanus*. *Phytochemistry* **43**, 1029–1031.

Brown, E.G. and Turan, Y. (1996b) Formation of albizziine and 2,3-diaminopropanoic acid from uracil in *Albizia* seedlings. *Phytochemistry* **41**, 1491–1495.

Cao, Y. and Schubert, R.K. (2001) Molecular cloning and characterization of a cDNA encoding soybean nodule IMP dehydrogenase. *Biochim. Biophys. Acta*, **1520**, 242–246.

Christopherson, R.I. and Szabados, E. (1997) Nucleotide biosynthesis in mammals, in *Channelling in Intermediary Metabolism* (eds L. Agius and H.S.A. Sherratt), Portland Press, London, **23**, 315–335.

Christopherson, R.I., Lyons, S.D. and Wilson, P.K. (2002) Inhibitors of *de novo* nucleotide biosynthesis as drugs. *Acc. Chem. Res.*, **35**, 961–971.

Crozier, A., Kamiya, J., Bishop, G. and Yokota, T. (2000) Biosynthesis of hormones and elicitor molecules, in *Biochemistry and Molecular Biology of Plants* (eds. B. Buchanan, W. Gruissem and R.L. Jones), American Society of Plant Physiology, Rockville, pp. 850–929.

Dancer, J.E., Hughes, R.G. and Lindell, S.D. (1997) Adenosine-5′-phosphate deaminase. A novel herbicide target. *Plant Physiol.*, **114**, 119–129.

Dulyaninova, N.G., Podlepa, E.M., Toulokhonova, L.V. and Bykhovsky, V.Y. (2000) Salvage pathway for NAD biosynthesis in *Brevibacterium ammoniagenes*: regulatory properties of triphosphate-dependent nicotinate phosphoribosyltransferase. *Biochim. Biophys. Acta – Protein Struc. Molecular Enzymol.*, **1478**, 211–220.

Edwards, R. (1996) *S*-adenosyl-L-methionine metabolism in alfalfa cell cultures following treatment with fungal elicitors. *Phytochemistry*, **43**, 1163–1169.

Farre, E.M., Tiessen, A., Roessner, U., Geigenberger, P., Trethewey, R.N. and Willmitzer, L. (2001) Analysis of the compartmentation of glycolytic intermediates, nucleotides, sugars, organic acids, amino acids, and sugar alcohols in potato tubers using a nonaqueous fractionation method. *Plant Physiol.*, **127**, 685–700.

Fonne-Pfister, R., Chemla, P., Ward, E., Girardet, M. and Kreuz, K.E. (1996) The mode of action and the structure of a herbicide in complex with its target: binding of activated hydantocidin to the feedback regulation site of adenylosuccinate synthase. *Proc. Nat. Acad. Sci. USA*, **93**, 9431–9436.

Geigenberger, P., Regierer, B., Nunes-Nesi, A., Leisse, A., Urbanczyk-Wochniak, E., Springer, F., van Dongen, J.T., Kossmann, J. and Fernie, A.R. (2005) Inhibition of *de novo* pyrimidine synthesis in growing potato tubers leads to a compensatory stimulation of the pyrimidine salvage pathway and a subsequent increase in biosynthetic performance. *Plant Cell*, **17**, 2077–2088.

Gholson, R.K. (1966) Pyridine nucleotide cycle. *Nature*, **212**, 933–935.

Giermann, N., Schroder, M., Ritter, T. and Zrenner, R. (2002) Molecular analysis of de novo pyrimidine synthesis in solanaceous species. *Plant Mol. Biol.*, **50**, 393–403.

Hanks, J.F., Tolbert, N.E. and Schubert, K.R. (1981) Localization of enzymes of ureide biosynthesis in peroxisomes and microsomes of nodules. *Plant Physiol.*, **68**, 65–69.

Hanson, A.D. and Gregory, J.F. (2002) Synthesis and turnover of folates in plants. *Curr. Opin. Plant Biol.*, **5**, 244–249.

Herz, S., Eberhard, S. and Bacher, A. (2000) Biosynthesis of riboflavin in plants. The ribA gene of *Arabidopsis thaliana* specifies a bifunctional GTP cyclohydrolase II/3,4-dihydroxy-2-butanone 4-phosphat synthase. *Phytochemistry*, **53**, 723–731.

Hesberg, C., Hansch, R., Mendel, R.R. and Bittner, F. (2004) Tandem orientation of duplicated xanthine dehydrogenase genes from *Arabidopsis thaliana* – Differential gene expression and enzyme activities. *J. Biol. Chem.*, **279**, 13547–13554.

Hewitt, M.M., Carr, J.M., Williamson, C.L. and Slocum, R.D. (2005) Effects of phosphate limitation on expression of genes involved in pyrimidine synthesis and salvaging in Arabidopsis. *Plant Physiol. Biochem.*, **43**, 91–99.

Hung, W.F., Chen, L.J., Boldt, R., Sun, C.W. and Li, H.M. (2004) Characterization of Arabidopsis glutamine phosphoribosyl pyrophosphate amidotransferase-deficient mutants. *Plant Physiol.*, **135**, 1314–1323.

Hunt, L., Lerner, F. and Ziegler, M. (2004) NAD – new roles in signalling and gene regulation in plants. *New Phytol.*, **163**, 31–44.

Ikegami, F. and Murakoshi, I. (1994) Enzymatic-synthesis of nonprotein β-substituted alanines and some higher homologs in plants. *Phytochemistry*, **35**, 1089–1104.

International Rice Genome Sequencing Project (2005) The map-based sequence of the rice genome. *Nature*, **436**, 793–800.

Ipata, P.L., Marmocchi, F., Magni, G., Felicioli, R. and Polidoro, G. (1971) Bakers yeast cytosine deaminase – some enzymic properties and allosteric inhibition by nucleosides and nucleotides. *Biochemistry*, **10**, 4270–4276.

Islam, M.R., Kim, H., Kang, S.W., Kim, J.S., Jeong, Y.M., Hwang, H.J., Lee, S.Y., Woo, J.C. and Kim, S.G. (2007) Functional characterization of a gene encoding a dual domain for uridine kinase and uracil phosphoribosyltransferase in *Arabidopsis thaliana*. *Plant Mol. Biol.*, **63**, 465–477.

Jung, B., Florchinger, M., Kunz, H.H., Traub, M., Wartenberg, R., Jeblick, W., Neuhaus, H.E. and Mohlmann, T. (2009) Uridine-ribohydrolase is a key regulator in the uridine degradation pathway of Arabidopsis. *Plant Cell*, **21**, 876–891.

Kafer, C., Zhou, L., Santose, D., Guirgis, A. and Weers, B. (2004) Regulation of pyrimidine metabolism in plants. *Front. Biosci.*, **9**, 1611–1625.

Kanamori, I., Ashihara, H. and Komamine, A. (1980) Subcellular-distribution and activity of enzymes involved in uridine-5′-monophosphate synthesis in *Vinca rosea* cells. *Z. Pflanzenphysiol.*, **96**, 7–16.

Katahira, R. and Ashihara, H. (2002) Profiles of pyrimidine biosynthesis, salvage and degradation of disks of potato (*Solanum tuberosum* L.) tubers. *Planta*, **215**, 821–828.

Katahira, R. and Ashihara, H. (2006) Dual function of pyrimidine metabolism in potato (*Solanum tuberosum*) plants: pyrimidine salvage and supply of β-alanine to pantothenic acid synthesis. *Physiol. Plant.*, **127**, 38–43.

Katahira, R. and Ashihara, H. (2009) Profiles of the biosynthesis and metabolism of pyridine nucleotides in potatoes (*Solanum tuberosum* L.). *Planta*, **231**, 35–45.

Katoh, A. and Hashimoto, T. (2004) Molecular biology of pyridine nucleotide and nicotine biosynthesis. *Front. Biosci.*, **9**, 1577–1586.

Katoh, A., Uenohara, K., Akita, M. and Hashimoto, T. (2006) Early steps in the biosynthesis of NAD in Arabidopsis start with aspartate and occur in the plastid. *Plant Physiol.*, **141**, 851–857.

Katsuragi, T., Sakai, T., Matsumoto, K. and Tonomura, K. (1986) Cytosine deaminase from *Escherichia coli* – production, purification, and some characteristics. *Agric. Biol. Chem.*, **50**, 1721–1730.

Khan, A.I., Chowdhry, B.Z. and Yon, R.J. (1999) Wheat-germ aspartate transcarbamoylase: revised purification, stability and re-evaluation of regulatory kinetics in terms of the Monod-Wyman-Changeux model. *Eur. J. Biochem.*, **259**, 71–78.

Koslowsky, S., Riegler, H., Bergmuller, E. and Zrenner, R. (2008) Higher biomass accumulation by increasing phosphoribosylpyrophosphate synthetase activity in *Arabidopsis thaliana* and *Nicotiana tabacum*. *Plant Biotechnol. J.*, **6**, 281–294.

Lange, P.R., Geserick, C., Tischendorf, G. and Zrenner, R. (2008) Functions of chloroplastic adenylate kinases in Arabidopsis. *Plant Physiol.*, **146**, 492–504.

Lazar, G., Hong, Z. and Goodman, H.M. (1993) The origin of the bifunctional dihydrofolate-reductase thymidylate synthase isogenes of *Arabidopsis thaliana*. *Plant J.*, **3**, 657–668.

Lein, W., Bornke, F., Reindl, A., Ehrhardt, T., Stitt, M. and Sonnewald, U. (2004) Target-based discovery of novel herbicides. *Curr. Opin. Plant Biol.*, **7**, 219–225.

Lein, W., Usadel, B., Stitt, M., Reindl, A., Ehrhardt, T., Sonnewald, U. and Bornke, F. (2008) Large-scale phenotyping of transgenic tobacco plants (*Nicotiana tabacum*) to identify essential leaf functions. *Plant Biotechnol. J.*, **6**, 246–263.

Linka, N. and Weber, A.P.M. (2010) Intracellular metabolite transporters in plants. *Mol. Plant*, **3**, 21–53.

Mainguet, S.E., Gakiere, B., Majira, A., Pelletier, S., Bringel, F., Guerard, F., Caboche, M., Berthome, R. and Renou, J.P. (2009) Uracil salvage is necessary for early Arabidopsis development. *Plant J.*, **60**, 280–291.

Matsui, A., Yin, Y.L., Yamanaka, K., Iwasaki, M. and Ashihara, H. (2007) Metabolic fate of nicotinamide in higher plants. *Physiol. Plant.*, **131**, 191–200.

Matsui, A. and Ashihara, H. (2008) Nicotinate riboside salvage in plants: presence of nicotinate riboside kinase in mungbean seedlings. *Plant Physiol. Biochem.*, **46**, 104–108.

Minet, M., Dufour, M.E. and Lacroute, F. (1992) Complementation of *Saccharomyces cerevisiae* auxotrophic mutants by *Arabidopsis thaliana* cDNAs. *Plant J.*, **2**, 417–422.

Moat, G.A. and Foster, J.W. (1987) *Pyridine Nucleotide Coenzymes*, John Wiley & Sons Ltd, New York.

Moffatt, B.A. and Ashihara, H. (2002) Purine and pyrimidine nucleotide synthesis and metabolism, in *The Arabidopsis Book* (eds C.R. Somerville and E.M. Meyerowitz), American Society of Plant Biology, Rockville. DOI: 10.1199/tab.0018. http://www.aspb.org/publications/arabidopsis/

Munoz, A., Raso, M.J., Pineda, M. and Piedras, P. (2006) Degradation of ureidoglycolate in French bean (*Phaseolus vulgaris*) is catalysed by a ubiquitous ureidoglycolate urea-lyase. *Planta*, **224**, 175–184.

Nasr, F., Bertauche, N., Dufour, M.E., Minet, M. and Lacroute, F. (1994) Heterospecific cloning of *Arabidopsis thaliana* cDNAs by direct complementation of pyrimidine auxotrophic mutants of *Saccharomyces cerevisiae*. I. Cloning and sequence analysis of two cDNAs catalysing the second, fifth and sixth steps of the de novo pyrimidine biosynthesis pathway. *Mol. Gen. Genet.*, **244**, 23–32.

Noctor, G., Queval, G. and Gakiere, B. (2006) NAD(P) synthesis and pyridine nucleotide cycling in plants and their potential importance in stress conditions. *J. Ex. Bot.*, **57**, 1603–1620.

Oliver, S.N., Tiessen, A., Fernie, A.R. and Geigenberger, P. (2008) Decreased expression of plastidial adenylate kinase in potato tubers results in an enhanced rate of respiration and a stimulation of starch synthesis that is attributable to post-translational redox-activation of ADP-glucose pyrophosphorylase. *J. Exp. Bot.*, **59**, 315–325.

Ong, B.L. and Jackson, J.F. (1972) Pyrimidine nucleotide biosynthesis in *Phaseolus aureus*. Enzymatic aspects of the control of carbamoyl phosphate synthesis and utilization. *Biochem. J.*, **129**, 583–593.

Pardo, E.G. and Gutierrez, C. (1990) Cell cycle-dependent and differentiation stage-dependent variation of dutpase activity in higher-plant cells. *Exp. Cell Res.*, **186**, 90–98.

Piedras, P., Munoz, A., Aguilar, M. and Pineda, M. (2000) Allantoate amidinohydrolase (allantoicase) from *Chlamydomonas reinhardtii*: its purification and catalytic and molecular characterization. *Arch. Biochem. Biophys.*, **378**, 340–348.

Pradet, A. and Raymond, P. (1983) Adenine-nucleotide ratios and adenylate energy-charge in energy-metabolism. *Ann. Rev. Plant Physiol. Plant Mol. Biol.* **34**, 199–224.

Purse, J.G., Lee, T.S. and Pryce, R.J. (1985) Stimulation of soybean callus growth by *S*-lathyrine. *Phytochemistry*, **24**, 897–900.

Regierer, B., Fernie, A.R., Springer, F., Perez-Melis, A., Leisse, A., Koehl, K., Willmitzer, L., Geigenberger, P. and Kossmann, J. (2002) Starch content and yield increase as a result of altering adenylate pools in transgenic plants. *Nature Biotechnol.*, **20**, 1256–1260.

Revollo, J.R., Grimm, A.A. and Imai, S. (2004) The NAD biosynthesis pathway mediated by nicotinamide phosphoribosyltransferase regulates Sir2 activity in mammalian cells. *J. Biol. Chem.*, **279**, 50754–50763.

Robert, S., Hicks, G.R. and Raikhel, N.V. (2009) Powerful partners: Arabidopsis and chemical genomics, in *The Arabidopsis Book* (eds C.R. Somerville and E.M. Meyerowitz), American Society of Plant Biology, Rockville. http://www.aspb.org/publications/arabidopsis/

Rongvaux, A., Andris, F., Van Gool, F. and Leo, O. (2003) Reconstructing eukaryotic NAD metabolism. *Bioessays*, **25**, 683–690.

Ross, C.W. (1981) Biosynthesis of nucleotides, in *The Biochemistry of Plants: A Comprehensive Treatise, Vol. 6. Proteins and Nucleic Acids* (ed. A. Marcus), Academic Press, New York, pp. 169–206.

Sakai, T., Yu, T., Tabe, H. and Omata, S. (1975) Metabolism of pyrimidine nucleotides in bacteria 4. Purification of cytosine deaminase from *Serratia marcescens*. *Agric. Biol. Chem.* **39**, 1623–1629.

Sakakibara, H. (2006) Cytokinins: activity, biosynthesis, and translocation. *Ann. Rev. Plant Biol.*, **57**, 431–449.

Santoso, D. and Thornburg, R.W. (1992) Isolation and characterization of UMP synthase mutants from haploid cell suspensions of *Nicotiana tabacum*. *Plant Physiol.*, **99**, 1216–1225.

Santoso, D. and Thornburg, R.W. (1998) Uridine 5′-monophosphate synthase is transcriptionally regulated by pyrimidine levels in *Nicotiana plumbaginifolia*. *Plant Physiol.*, **116**, 815–821.

Schmitzer, P.R., Graupner, P.R., Chapin, E.L., Fields, S.C., Gilbert, J.R., Gray, J.A., Peacock, C.L. and Gerwick, B.C. (2000) Ribofuranosyl triazolone: A natural product herbicide with activity on adenylosuccinate synthetase following phosphorylation. *J. Nat. Prod.* **63**, 777–781.

Schroder, M., Giermann, N. and Zrenner, R. (2005) Functional analysis of the pyrimidine de novo synthesis pathway in solanaceous species. *Plant Physiol.*, **138**, 1926–1938.

Schubert, K.R. and Boland, M.J. (1990) The urides, in *The Biochemistry of Plants, A Comprehensive Treatise*, Vol. **16** (eds B.J. Miflin and P.J. Lea), Academic Press, San Diego, pp. 197–282.

Senecoff, J.F., McKinney, E.C. and Meagher, R.B. (1996) *De novo* purine synthesis in *Arabidopsis thaliana*. II. The PUR7 gene encoding 5′-phosphoribosyl-4-(*N*-succinocarboxamide)-5-aminoimidazole synthetase is expressed in rapidly dividing tissues. *Plant Physiol.*, **112**, 905–917.

Serventi, F., Ramazzina, I., Lamberto, I., Puggioni, V., Gatti, R. and Percudani, R. (2010) Chemical basis of nitrogen recovery through the ureide pathway: formation and hydrolysis of *S*-ureidoglycine in plants and bacteria. *ACS Chem. Biol.*, **5**, 203–214.

Siehl, D.L., Subramanian, M.V., Walters, E.W., Lee, S.F., Anderson, R.J. and Toschi, A.G. (1996) Adenylosuccinate synthetase: site of action of hydantocidin, a microbial phytotoxin. *Plant Physiol.*, **110**, 753–758.

Smith, P.M.C. and Atkins, C.A. (2002) Purine biosynthesis. Big in cell division, even bigger in nitrogen assimilation. *Plant Physiol.*, **128**, 793–802.

Stasolla, C., Katahira, R., Thorpe, T.A. and Ashihara, H. (2003) Purine and pyrimidine nucleotide metabolism in higher plants. *J. Plant Physiol.*, **160**, 1271–1295.

Tajima, S., Nomura, M. and Kouchi, H. (2004) Ureide biosynthesis in legume nodules. *Front. Biosci.*, **9**, 1374–1381.

Tjaden, J., Mohlmann, T., Kampfenkel, K., Henrichs, G. and Neuhaus, H.E. (1998) Altered plastidic ATP/ADP-transporter activity influences potato (*Solanum tuberosum* L.) tuber morphology, yield and composition of tuber starch. *Plant J.*, **16**, 531–540.

Todd, C.D. and Polacco, J.C. (2006) AtAAH encodes a protein with allantoate amidohydrolase activity from *Arabidopsis thaliana*. *Planta*, **223**, 1108–1113.

Ullrich, A., Knecht, W., Piskur, J. and Löffler, M. (2002) Plant dihydroorotate dehydrogenase differs significantly in substrate specificity and inhibition from the animal enzymes. *FEBS Lett.*, **529**, 346–350.

van der Graaff, E., Hooykaas, P., Lein, W., Lerchl, J. and Kunze, G. (2004) Molecular analysis of the *de novo* purine biosynthesis in solanaceous species and in *Arabidopsis thaliana*. *Front. Biosci.*, **9**, 1803–1816.

van der Veer, E., Ho, C., O'Neil, C., Barbosa, N., Scott, R., Cregan, S.P. and Pickering, J.G. (2007) Extension of human cell lifespan by nicotinamide phosphoribosyltransferase. *J. Biol. Chem.*, **282**, 10841–10845.

Wagner, K.G. and Backer, A.I. (1992) Dynamics of nucleotides in plants studied on a cellular basis. *Int. Rev. Cytol.*, **134**, 1–84.

Wagner, R. and Wagner, K.G. (1985) The pyridine-nucleotide cycle in tobacco. Enzyme activities for the *de novo* synthesis of Nad. *Planta*, **165**, 532–537.

Wagner, R., Feth, F. and Wagner, K.G. (1986a) The pyridine-nucleotide cycle in tobacco – enzyme activities for the recycling of Nad. *Planta*, **167**, 226–232.

Wagner, R., Feth, F. and Wagner, K.G. (1986b) The regulation of enzyme-activities of the nicotine pathway in tobacco. *Physiol. Plant.*, **68**, 667–672.

Wagner, R., Feth, F. and Wagner, K.G. (1986c) Regulation in tobacco callus of enzyme-activities of the nicotine pathway 2. The pyridine-nucleotide cycle. *Planta*, **168**, 408–413.

Waller, G.R., Yang, K.S., Gholson, R.K., Hadwiger, L.A. and Chaykin, S. (1966) Pyridine nucleotide cycle and its role in biosynthesis of ricinine by *Ricinus communis* l. *J. Biol. Chem.*, **241**, 4411–4418.

Walsh, T.A., Green, S.B., Larrinua, I.M. and Schmitzer, P.R. (2001) Characterisation of plant β-ureidopropionase and functional over expression in *Escherichia coli*. *Plant Physiol.*, **125**, 1001-1011.

Walsh, T.A., Bauer, T., Neal, R., Merlo, A.O., Schmitzer, P.R., Hicks, G.R., Honma, M., Matsumura, W., Wolff, K. and Davies, J.P. (2007) Chemical genetic identification of glutamine phosphoribosylpyrophosphate amidotransferase as the target for a novel bleaching herbicide in arabidopsis. *Plant Physiol.*, **144**, 1292–1304.

Wang, C.X. and Liu, Z.C. (2006) Arabidopsis ribonucleotide reductases are critical for cell cycle progression, DNA damage repair, and plant development. *Plant Cell*, **18**, 350–365.

Wang, G.D. and Pichersky, E. (2007) Nicotinamidase participates in the salvage pathway of NAD biosynthesis in Arabidopsis. *Plant J.*, **49**, 1020–1029.

Wang, M.H., Ratnam, S. and Freisheim, J.H. (1995) Cloning, nucleotide-sequence and expression of the bifunctional dihydrofolate reductase-thymidylate synthase from *Glycine max*. *Biochim. Biophys. Acta-Gene Struct. Express.*, **1261**, 325–336.

Wasternack, C. (1978) Degradation of pyrimidines – enzymes, localization and role in metabolism. *Biochem. Physiol. Pflanzen*, **173**, 467–499.

Wasternack, C. (1982) Metabolism of pyrimidines and purines, in *Encyclopedia of Plant Physiology* (eds A. Pirson and M.H. Zimmermann), Springer-Verlag, Berlin **14B**, 263–301.

Werner, A.K., Sparkes, I.A., Romeis, T., and Witte, C.P. (2008) Identification, biochemical characterization, and subcellular localization of allantoate amidohydrolases from Arabidopsis and soybean. *Plant Physiol.*, **146**, 418–430.

Werner, A.K., Romeis, T., and Witte, C.P. (2010) Ureide catabolism in *Arabidopsis thaliana* and *Escherichia coli*. *Nature Chem. Biol.*, **6**, 19–21.

Williamson, C.L. and Slocum, R.D. (1994) Molecular cloning and characterization of the pyrB1 and pyrB2 genes encoding aspartate transcarbamoylase in pea (*Pisum sativum* L.). *Plant Physiol.*, **105**, 377–384.

Winkler, R.G., Blevins, D.G., Polacco, J.C. and Randall, D.D. (1988) Ureide catabolism in nitrogen-fixing legumes. *Trends Biochem. Sci.*, **13**, 97–100.

Xu, J., Zhang, H.Y., Xie, C.H., Xue, H.W., Dijkhuis, P. and Liu, C.M. (2005) Embryonic factor 1 encodes an AMP deaminase and is essential for zygote to embryo transition in Arabidopsis. *Plant J.*, **42**, 743–756.

Yang, J. and Han, K.H. (2004) Functional characterization of allontoinase genes from Arabidopsis and a nonureide type legume black locust. *Plant Physiol.*, **134**, 1039–1049.

Zhang, L., Hausler, R.E., Greiten, C., Hajirezaei, M.R., Haferkamp, I., Neuhaus, H.E., Flugge, U.I. and Ludewig, F. (2008) Overriding the co-limiting import of carbon and energy into tuber amyloplasts increases the starch content and yield of transgenic potato plants. *Plant Biotechnol. J.*, **6**, 453–464.

Zheng, X.Q., Nagai, C. and Ashihara, H. (2004) Pyridine nucleotide cycle and trigonelline (*N*-methylnicotinic acid) synthesis in developing leaves and fruits of *Coffea arabica*. *Physiol. Plant.*, **122**, 404–411.

Zheng, X.Q., Hayashibe, E. and Ashihara, H. (2005) Changes in trigonelline (*N*-methylnicotinic acid) content and nicotinic acid metabolism during germination of mungbean (*Phaseolus aureus*) seeds. *J. Exp. Bot.*, **56**, 1615–1623.

Zrenner, R., Stitt, M., Sonnewald, U. and Boldt, R. (2006) Pyrimidine and purine biosynthesis and degradation in plants. *Ann. Rev. Plant Biol.*, **57**, 805–836.

Zrenner, R., Riegler, H., Marquard, C.R., Lange, P.R., Geserick, C., Bartosz, C.E., Chen, C.T. and Slocum, R.D. (2009) A functional analysis of the pyrimidine catabolic pathway in Arabidopsis. *New Phytol.*, **183**, 117–132.

6

Purine Alkaloid Metabolism

Hiroshi Ashihara, Shinjiro Ogita and Alan Crozier

6.1 Introduction

Purine alkaloids, such as methylxanthines and methyluric acids, are derived from purine nucleotides (Zulak *et al.*, 2006). Caffeine, theobromine, theophylline and theacrine are found in sizeable amounts in plants, and structures of these and other naturally occurring purine alkaloids are illustrated in Figure 6.1. Interestingly, almost all traditional non-alcoholic beverages, tea (*Camellia sinensis*), coffee (*Coffea arabica*), yerba-maté (*Ilex paraguarensis*), cocoa (*Theobroma cacao*) and guaraná (*Paullinia cupana*), contain purine alkaloids (Table 6.1). Ancient civilisations probably selected these plants because of the pleasurable effects of caffeine acting as a stimulant of the central nervous system.

Possible pathways of caffeine biosynthesis were first suggested after ^{14}C-tracer experiments and studies with crude enzyme extracts were published (see Suzuki *et al.*, 1992). Later, the caffeine biosynthetic pathway was investigated in more detail using various ^{14}C- and ^{15}N-precursors with samples being analysed by HPLC-radio-counting and GC-mass spectrometry (Ashihara *et al.*, 1996, 1997; Ito and Ashihara, 1999). The results obtained confirmed that the traditional route to caffeine from xanthosine via theobromine is the main purine alkaloid biosynthetic pathway in tea and coffee plants (Figure 6.2). These studies also provided information on the operation of a number of minor pathways leading to caffeine. Highly purified caffeine synthase, an *N*-methyltransferase which catalyzes the last two steps of the four-step caffeine biosynthesis pathway, was obtained from tea leaves (Kato *et al.*, 1999). Using information obtained on the amino acid sequence of the *N*-terminal of this enzyme, a full-length cDNA of caffeine synthase was cloned and a recombinant enzyme produced which had caffeine synthase activity (Kato *et al.*, 2000). Subsequently, cDNAs of

---

Plant Metabolism and Biotechnology, First Edition. Edited by Hiroshi Ashihara, Alan Crozier, and Atsushi Komamine.

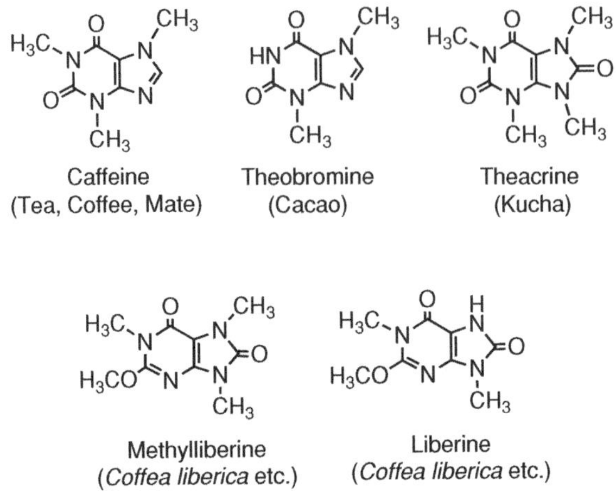

Figure 6.1 *Structures of methylxanthines and methyluric acids occurring in plants*

Table 6.1 *Distribution of purine alkaloids in selected plants*

Species	Common name	Organ	Major alkaloids (% dry weight)	References
Camellia sinensis	Tea	Leaves	Caffeine (2–3%)	Nagata and Sakai (1984)
Camellia sinensis var. Kucha	Kucha	Leaves	Theacrine (2.8%)	Zheng *et al.* (2002)
Camellia ptilophylla	Cocoa tea	Leaves	Theobromine (5.9%)	Ye *et al.* (1997)
Camellia irrawadiensis		Leaves	Theobromine (0.9%)	Nagata and Sakai (1984)
Coffea arabica	Arabica coffee	Seeds	Caffeine (1%)	Koshiro *et al.* (2006)
Coffea canephora	Robusta coffee	Seeds	Caffeine (2%)	Koshiro *et al.* (2006)
Coffea eugenioides		Seeds	Caffeine (0.4%)	Mazzafera and Carvalho (1992)
Theobroma cacao	Cacao	Seeds	Theobromine (2.5%), caffeine (0.6%)	Senanayake and Wijesekera (1971)
Theobroma grandiflorum	Cupu	Seeds	Liberine (0.25%)	Baumann and Wanner (1980)
Ilex paraguariensis	Maté	Leaves	Caffeine (0.9%)	Mazzafera (1994)
Paullinia cupana	Guaraná	Seeds	Caffeine (4%)	Baumann *et al.* (1995)

Figure 6.2 *The caffeine biosynthesis pathway. Purine alkaloids are synthesized via xanthosine which is produced via at least four routes: from IMP originating from* de novo *purine synthesis (*de novo *route); from adenosine released from the SAM cycle (SAM route); from the cellular adenine nucleotide pool (AMP route); and from the guanine nucleotide pool (GMP route). Abbreviations of enzymes are as follows: AMPDA, AMP deaminase; APRT, adenine phosphoribosyltransferase; ARK, adenosine kinase; ARN, adenosine nucleosidase; CS, caffeine synthase; GRDA, guanosine deaminase; mXRN, methylxanthine nucleosidase; 7mXRS, 7-methylxanthine synthase; 5′NT, 5′-nucleotidase; TS, theobromine synthase*

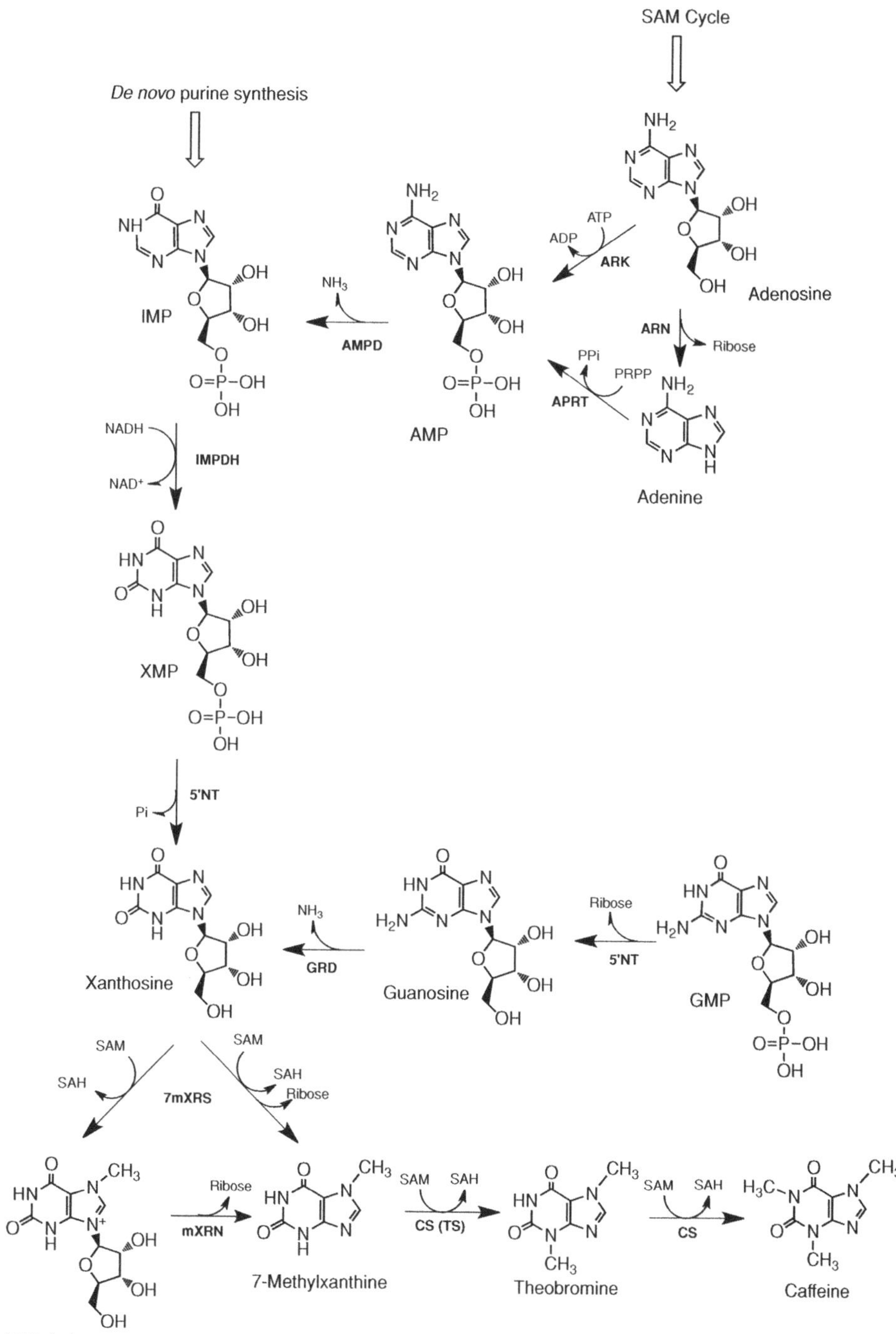

Figure 6.2 *(Continued)*

N-methyltransferases involved in caffeine biosynthesis have been cloned from leaves and seeds of coffee, and the molecular properties of the genes independently characterised by two Japanese groups (see Ashihara *et al.*, 2008).

In this chapter, we describe the distribution of purine alkaloids in plants and the caffeine biosynthetic pathways recently confirmed using the *N*-methyltransferase-encoding genes. Possible pathways of caffeine catabolism are discussed, along with physiological studies related to caffeine biosynthesis, including the emerging role of purine alkaloids *in planta*. In the final section we describe examples of metabolic engineering of caffeine which have produced decaffeinated coffee plants and transgenic caffeine-producing tobacco plants with pest-resistant properties.

6.2 Classification of Purine Alkaloids

Methylxanthines and methyluric acids are classified as purine alkaloids, since they are nitrogen-containing secondary metabolites with a purine skeleton (Figures 6.1 and 6.2). Mono- and di-methylxanthines are biosynthetic intermediates (e.g. 7-methylxanthine) or catabolites (e.g. 3-methylxanthine) of caffeine (1,3,7-trimethylxanthine). Methyluric acids such as theacrine (1,3,7,9-tetramethyluric acid) appear to be derived from caffeine. The xanthine structure is oxidised to uric acid by xanthine oxidase/dehydrogenase.

6.3 Occurrence of Purine Alkaloids

Compared with other alkaloids, such as nicotine, morphine and strychnine (see Zulak *et al.*, 2006), purine alkaloids are distributed more widely throughout the plant kingdom (Ashihara and Crozier, 1999). According to Kihlman (1977), caffeine is present in more than 63 species belong to 28 genera over 17 families. Subsequently, caffeine and other purine alkaloids have been detected in new species of *Herreania*, *Camellia* and *Citrus*, and have now been shown to occur in more than 100 species of the plant kingdom, although they accumulate in quantity in relatively few species (Ashihara and Crozier, 1999; Ashihara *et al.*, 2008). The distributions and concentrations of purine alkaloids in selected species are shown in Table 6.1.

The major purine alkaloid is caffeine in tea (*Camellia sinensis*), but other *Camellia* species, *Camellia ptilophylla* and *Camellia irrawadiensis*, accumulate theobromine to a greater extent than caffeine. A Chinese tea known as kucha (*Camellia assamica* var. kucha) contains theacrine (Zheng *et al.*, 2002). The stamens and petals of flowers of several *Camellia* species contain caffeine and/or theobromine (Suzuki, 1985; Fujimori and Ashihara 1990, 1993), and the species-to-species pattern of purine alkaloid distribution in flowers is the same as observed in leaves (Ashihara and Kubota, 1987). In the mature dry tea fruit, the pericarp contains most caffeine, but there are also considerable amounts in the seed coat and, to a lesser extent, the fruit stalk and the seed. Germinating seeds also contain substantial quantities of the alkaloids (Suzuki and Waller, 1985).

The caffeine content of seeds of different *Coffea* species varies from 0.4 to 2.4% of dry weight (Mazzafera and Carvalho, 1992). Most cultivars of *Coffea arabica* contain *ca.* 1%

dry weight of caffeine, while *Coffea canephora* contains *ca.* 2%. Caffeine occurs mainly in leaves and seeds and is all but absent in roots and the older brown parts of shoots (Zheng and Ashihara, 2004). Mature leaves of *Coffea liberica, Coffea dewevrei* and *Coffea abeokutae* contain the methyluric acids theacrine, liberine (*O*(2),1,9-trimethyluric acid) and methylliberine (*O*(2),1,7,9-tetramethyluric acid) (Baumann *et al.*, 1976; Petermann and Baumann, 1983; Figure 1). The major purine alkaloid in the cotyledons of mature cacao seeds is theobromine, with lower amounts of caffeine. Theobromine also occurs in the seed coat and placenta of cacao fruits and young leaves (Zheng *et al.*, 2004a). Cupu (*Theobroma grandiflorum*), which originated from the Amazon region, contains 0.25% liberine in cotyledons and 0.08% in the nut shells (Baumann and Wanner, 1980).

6.4 Biosynthesis of Caffeine

6.4.1 Biosynthetic Pathway from Purine Nucleotides

Xanthosine, the initial substrate of purine alkaloid synthesis, is supplied by at least four different pathways: *de novo* purine biosynthesis (*de novo* route), the degradation pathways of adenine nucleotides (AMP route), and guanine nucleotides (GMP route), and adenosine released from the *S*-adenosyl-L-methionine (SAM) cycle (Figure 6.2; see also Chapter 7).

The production of purine nucleotides is the result of *de novo* synthesis from several precursors, namely, CO_2, 10-formyltetrahydrofolate, 5-phosphoribosyl-1-pyrophosphate (PRPP) and the amino acids, glycine, glutamine and aspartate (see Chapter 7). Utilisation of IMP, which is formed as the first nucleotide by the *de novo* purine biosynthetic pathway, for caffeine biosynthesis was demonstrated in experiments with young tea leaves using ^{15}N-glycine and some ^{14}C-labelled precursors and inhibitors of *de novo* purine biosynthesis (Ito and Ashihara, 1999). Xanthosine is produced by an IMP → XMP → xanthosine pathway. IMP dehydrogenase (EC 1.1.1.205) and 5′-nucleotidase (EC 3.1.3.5) may catalyze these reactions, IMP dehydrogenase being implicated since ribavirin, an inhibitor of IMP dehydrogenase, reduces the rate of caffeine biosynthesis in tea and coffee leaves (Keya *et al.*, 2003).

A portion of the xanthosine used for caffeine biosynthesis is derived from the adenine and guanine nucleotide pools which are produced by the *de novo* and salvage pathways. There are several potential pathways for xanthosine synthesis from AMP, but the AMP → IMP → XMP → xanthosine route is likely to predominate (Fujimori and Ashihara, 1993). All three enzymes involved in the conversion, AMP deaminase (EC 3.5.4.6), IMP dehydrogenase and 5′-nucleotidase, have been detected in tea leaves (Koshiishi *et al.*, 2001). Xanthosine for caffeine biosynthesis is also derived from guanine nucleotides by a GMP → guanosine → xanthosine pathway. Guanosine deaminase (EC 3.5.4.15) activity catalyzing the last of the two conversions has been demonstrated in cell-free extracts from young tea leaves (Negishi *et al.*, 1994). SAM is the methyl donor for methylation steps in many cellular reactions including those catalyzed by *N*-methyltransferases in the caffeine biosynthetic pathway. SAM is converted to *S*-adenosyl-L-homocysteine (SAH), which is in turn hydrolysed to homocysteine and adenosine. Adenosine released from the SAM cycle (also known as the activated-methyl cycle) is metabolised to AMP which is converted to xanthosine as mentioned above (Koshiishi *et al.*, 2001). Because three moles of SAH are

Table 6.2 N-methyltransferases and their genes of caffeine biosynthesis in coffee, tea and cacao plants

Common name	Systematic name	EC number	Gene name	Accession number
7-Methylxanthosine synthase	*S*-adenosyl-L-methionine:xanthosine N^7-methyltransferase	2.1.1.158	*CmXRS1*	AB034699
			CaXMT	AB048793
Theobromine synthase	*S*-adenosyl-L-methionine:7-methylxanthine N^3-methyltransferase	2.1.1.159	*CTS1*	AB034700
			CaMXMT1	AB048794
			CTS2	AB054841
			CaMXMT2	AB084126
			BTS1	AB096699
Caffeine synthase	*S*-adenosyl-L-methionine:3,7-dimethylxanthine N^1-methyltransferase	2.1.1.160	*CCS1*	AB086414
			CaDXMT1	AB984125

produced via the SAM cycle for each mole of caffeine that is synthesized, this pathway has the capacity to be the sole source of both the purine skeleton and the methyl groups required for caffeine biosynthesis in young tea leaves.

6.4.2 Caffeine Biosynthesis from Xanthosine

In the narrow sense, caffeine biosynthesis means the conversion of xanthosine to caffeine using specific *N*-methyltransferases. A list of participating enzymes and genes are summarised in Table 6.2. The first step is the conversion of xanthosine to 7-methylxanthosine (Figure 6.2). This reaction is catalyzed by 7-methylxanthosine synthase (EC 2.1.1.158). Full length cDNAs encoding 7-methylxanthosine synthase, *CmXRS1* and *CaXMT1*, were isolated from *Coffea arabica* (Mizuno *et al.*, 2003a; Uefuji *et al.*, 2003). These encode a polypeptide consisting of 372 amino acids with an apparent molecular mass of 41.8 kDa. It is expressed almost uniformly in aerial green tissues of *Coffea arabica*, including leaves, floral buds and immature but not mature beans. The recombinant proteins obtained from these cDNAs exhibit 7-methylxanthosine synthase activity *in vitro*.

The second step of caffeine biosynthesis involves the hydrolysis of 7-methylxanthosine to 7-methylxanthine. Although *N*-methylnucleosidase (EC 3.2.2.25) has been detected in cell-free preparations from tea leaves (Negishi *et al.*, 1988), detailed structural studies on 7-methylxanthosine synthase from coffee suggest that the methyl transfer and nucleoside cleavage may be coupled and catalyzed by a single enzyme (McCarthy and McCarthy, 2007).

The last two steps of the caffeine biosynthetic pathway are also catalyzed by SAM-dependent *N*-methyltransferases, but this enzyme is distinct from 7-methylxanthosine synthase. A highly purified native *N*-methyltransferase preparation has been obtained from young *Camellia sinensis* leaves (Kato *et al.*, 1999). The enzyme, assigned the name caffeine synthase (EC 2.1.1.160), catalyses the last two steps of caffeine biosynthesis, the conversion of 7-methylxanthine to caffeine via theobromine (Figure 6.2). The gene encoding caffeine synthase has been cloned from young tea leaves (Kato *et al.*, 2000). Since then, plural genes encoding coffee *N*-methyltransferases with different substrate specificities have been reported (Table 6.2).

Plural genes encoding theobromine synthase have been also reported (Table 6.2). These genes encode a 42.7 kDa or 43.4 kDa polypeptide consisting of 378 or 384 amino acids. They differ by insertion or deletion of blocks of several amino acid residues in the *C*-terminal region. The recombinant theobromine synthase (EC 2.1.1.159) preferentially catalyzes the conversion of 7-methylxanthine to theobromine, although the detailed catalytic properties of recombinant enzymes are apparently distinct from each other. Nevertheless, these genes are expressed in young leaves, floral buds and immature but not mature seeds of *Coffea arabica*.

The genes, *CCS1* and *CaDXMT1*, encoding caffeine synthase have been reported (Table 6.2) (Ogawa *et al.*, 2001; Mizuno *et al.*, 2003b; Uefuji *et al.*, 2003). They encode a 43 kDa polypeptide consisted of 384 amino acids. The recombinant caffeine synthases (CCS1 and CaDMXT1, EC 2.1.1.160) can utilise paraxanthine (1,7-dimethylxanthine), theobromine and 7-methylxanthine as substrates in a similar manner to tea caffeine synthase (TCS1). Although paraxanthine is the most active substrate of this recombinant enzyme, only limited amounts of paraxanthine are synthesized in coffee cells, indicating that *in planta* caffeine synthase is involved principally in the conversion of 7-methylxanthine to caffeine via theobromine. Profiles of expression of these genes in different organs are distinct, *DXMT* being expressed exclusively in immature beans, while *CCS1* expression is ubiquitous, occurring in all tissues. The presence of isoforms of theobromine synthase and caffeine synthase with different substrate specificities and kinetic properties suggests that caffeine is synthesized using different isoforms of these enzymes in different organs and/or in different growth stages.

Recently, genes encoding a second *N*-methyltransferase associated with purine alkaloid synthesis were characterised in some *Camellia* plants, although the gene encoding the first enzyme, 7-methylxanthosine synthase, has not, as yet, been cloned in any *Camellia* species including tea. Yoneyama *et al.* (2006) reported the isolation and characterisation of *N*-methyltransferase genes from theobromine-accumulating *Camellia irrawadiensis* and *Camellia ptilophylla*. The total lengths of *ICS1* and *ICS2* from *Camellia irrawadiensis* (AB056108 and AB207816), *PCS1* and *PCS2* from *Camellia ptilophylla* (AB207817 and AB207818), are respectively 1432, 1413, 1394 and 1357 bp, and encode 363–365 amino acids. The predicted amino acid sequences of *N*-methyltransferases in these theobromine-accumulating *Camellia* species share 80% homology with caffeine synthase (*TCS1*) of *Camellia sinensis*. The recombinant enzymes of ICS1 and PCS1 exhibit only 3-*N*-methyltransferase activity; 7-methylxanthine is the best methyl acceptor, followed by paraxanthine, which is ten-fold less active. The accumulation of theobromine seems to be dependent upon the substrate specificity of *N*-methyltransferase. From the sequence analysis, it has been postulated that the substrate specificity is determined by a single amino acid residue in the central part of the protein.

The full-length cDNA of a theobromine synthase gene (*BTS1*) was isolated from cacao leaves by Yoneyama *et al.* (2006). It consists of 1264 bp and contains a single open reading frame of 1092 bp with a first ATG initiation codon at position 48 coding for a protein of 364 amino acids. *BTS1* has an overall 55% homology to *N*-methyltransferase genes of *Camellia* plants and 40% homology to the genes of *Coffea arabica*. The recombinant BTS1 is specific for 7-methylxanthine, and, in contrast to ICS1 and PCS1, no activity was detected with paraxanthine. In other words, the recombinant protein is a theobromine synthase. Therefore, in cacao plants, a specific theobromine synthase is present. Along with theobromine, lower but still considerable quantities of caffeine, 17% and 22% of theobromine content, were observed in young leaves and mature seeds, respectively (Koyama *et al.*, 2003; Zheng *et al.*, 2004a). Therefore, in addition to the theobromine synthase gene, caffeine synthase gene(s) may also be present in cacao plants.

In this section, the main pathway of caffeine biosynthesis from xanthosine was outlined. Some additional minor routes, such as 7-methyxanthine → paraxanthine → caffeine, and xanthine → 3-methylxanthine → theobromine → caffeine, may also operate simultaneously. These minor pathways may be a consequence of the broad methyl acceptor specificity of the *N*-methyltransferases (Kato *et al.*, 1999).

6.4.3 Cellular Localisation of Caffeine Biosynthesis

Theobromine and caffeine are synthesized exclusively in the green chlorophyll-containing buds, leaves and shoots of coffee and tea seedlings. There is an absence of biosynthetic activity in roots and cotyledons (Ashihara and Kubota, 1986; Zheng and Ashihara, 2004; Li *et al.*, 2007a). Li *et al.* (2007b) reported that RNA *in situ* hybridisation indicated that the *CS* gene was expressed mainly in the palisade parenchyma and the epicuticle of tea leaves and less so in the spongy parenchyma and hypoderm. The activity of *N*-methyltransferases involved in caffeine biosynthesis, and part of the activities of SAH hydrolase, adenosine nucleosidase, adenine phosphoribosyltransferase and adenosine kinase, were associated with a purified chloroplast preparation from young tea leaves (Kato *et al.*, 1998; Koshiishi *et al.*, 2001). This suggests that caffeine biosynthesis is localised in chloroplasts where *de novo* and salvage purine nucleotide-synthesis occur, and most members of the SAM cycle enzymes are present (see Chapter 7).

There are currently two reports on the localisation of caffeine synthase in coffee plants. Ogawa *et al.* (2001) investigated localisation using a cDNA fragment covering the entire coding region of the caffeine synthase gene (*CaMXMT*) fused to pGFP2. When the resulting plasmid was introduced into the epidermal layer of onions by particle bombardment, green fluorescence was detected in the cytoplasm. As the onion epidermal cells had not developed chloroplasts, this study does not necessarily give an accurate picture of the cellular localisation of CaMXMT.

Satyanarayana *et al.* (2005) cloned the promoter for one of the *N*-methyltransferase gene families involved in caffeine biosynthesis. Using the promoter, Kumar *et al.* (2007) investigated the localisation of caffeine synthase in coffee endosperm cells. These constructs and pCAMBIA 1301 bearing the intron *uidA* gene driven by the cauliflower mozaic virus (CaMV) *35S* promoter were electroporated into coffee endosperm, and the activity of β-glucuronidase (GUS) localised. In tissues transformed with the construct-containing promoter and first exon, enzymatic activity was localised on the outer surface of the vacuole. Antibodies to the coffee CS were also specifically localised in the same region. In tissues bearing either the CS-GUS construct without the first exon, or pCAMBIA 1301 with intron GUS, GUS activity was spread throughout the cytoplasm. The results suggest that NMT is targeted to the external surface of the vacuole.

Caffeine and other purine alkaloids may be stored in vacuoles along with many other secondary metabolites, although direct evidence has yet to be demonstrated. Likewise, mechanisms for uptake and sequestration of caffeine in vacuoles remain to be investigated. Active/passive transport by channels and/or transporters, and membrane/vesicle fusion may be involved in these processes.

6.5 Catabolism of Caffeine

Caffeine is produced in young leaves and immature fruits, and continues to accumulate gradually during the maturation of these organs. However, it is very slowly degraded with cleavage of the three methyl groups, resulting in the formation of xanthine. Catabolism of caffeine in coffee leaves was first reported by Kalberer (1965). Since then a number of tracer experiments using ^{14}C-labelled purine alkaloids have demonstrated that the major catabolic pathway is caffeine → theophylline → 3-methyxanthine → xanthine (Suzuki and Waller, 1984; Ashihara *et al.*, 1996, 1997; Vitoria and Mazzafera, 1998). Xanthine is further degraded by the conventional purine catabolism pathway to CO_2 and NH_3 via uric acid, allantoin and allantoate (Figure 6.3) (Ashihara and Crozier, 1999; Stasolla *et al.*, 2003; Zrenner *et al.*, 2006).

[8-^{14}C]Theophylline is degraded to CO_2 far more rapidly than [8-^{14}C]caffeine, so the conversion of caffeine to theophylline is the major rate-limiting step of caffeine catabolism and the reason why caffeine accumulates in high concentrations in tissues of *Camellia sinensis* and *Coffea arabica* (Ashihara *et al.*, 1996; Ito *et al.*, 1997).

In leaves of *Coffea eugenioides*, a low caffeine-containing species, [8-^{14}C]caffeine is degraded rapidly, and much of the radioactivity recovered as $^{14}CO_2$ (Ashihara and Crozier, 1999). *Coffea eugenioides* therefore possesses far higher levels of caffeine demethylase activity than *Coffea arabica*, and is able to efficiently convert caffeine to theophylline, which is rapidly further degraded. The N^7-demethylase involved in the conversion of caffeine to theophylline is thought to be a P450-dependent mono-oxygenase (Huber and Baumann, 1998; Mazzafera, 2004). However, the activity of this enzyme has not yet been demonstrated even in cell-free extracts.

Catabolism of [8-^{14}C]theophylline is prevalent in purine alkaloid-forming species. Its activity in *Camellia sinensis, Camellia irrawadiensis* and *Ilex paraguariensis* is greater than in *Avena sativa, Vigna mungo* and *Catharanthus roseus*, species that do not contain purine alkaloids (Ito *et al.*, 1997).

In microorganisms, catabolism of caffeine commences via two possible mechanisms: demethylation and oxidation. With the demethylation route, the major metabolite formed in fungi is theophylline, whereas theobromine is the predominant metabolite in bacteria. In certain bacterial species, caffeine is oxidised directly to trimethyluric acid in a single step. The conversion of caffeine to its metabolites is brought about primarily by *N*-demethylases, caffeine oxidase and xanthine oxidase which are produced by several caffeine-degrading bacterial species including *Pseudomonas putida* and species within the genera *Alcaligenes, Rhodococcus* and *Klebsiella.* Catabolism of caffeine in microorganisms and their potential use as bio-decaffeination techniques has been reviewed by Mazzafera (2004), Gokulakrishnan *et al.* (2005) and Dash and Gummadi (2006).

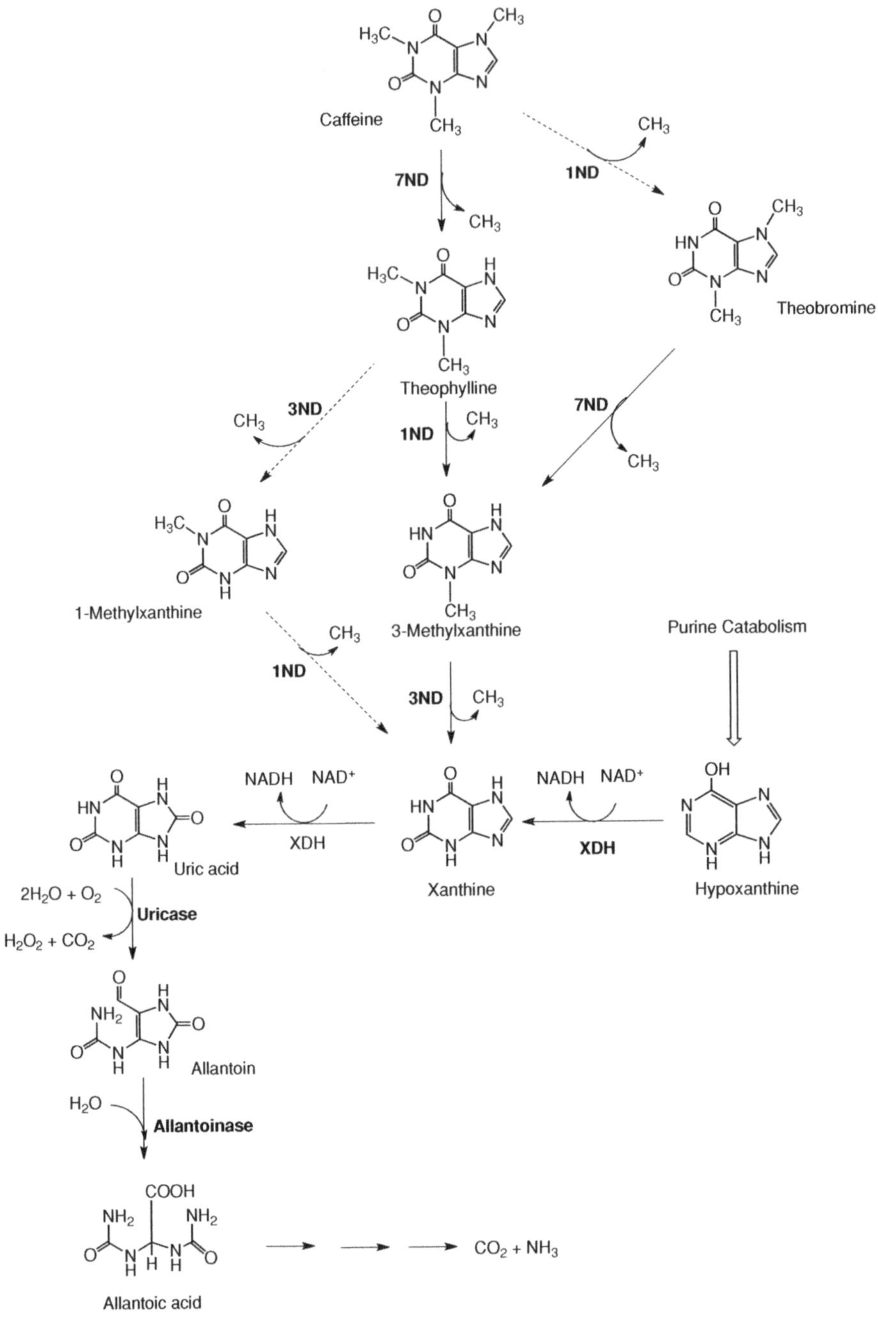

Figure 6.3 Catabolism of caffeine. Caffeine is catabolised mainly to xanthine via theophylline and 3-methylxanthine. Xanthine is further degraded to CO_2 and NH_3 by the conventional oxidative purine catabolic pathway. Dotted arrow shows minor routes. Typically, the conversion of caffeine to theophylline is the major rate-limiting step, and as a consequence caffeine accumulates in species such as Coffea arabica and Camellia sinensis. Abbreviations of enzymes are as follows: 1ND, 1N-demethylase; 3ND, 3N-demethylase; 7ND, 7N-demethylase; XDH, xanthine dehydrogenase.

6.6 Physiological and Ecological Aspects of Purine Alkaloid Metabolism in Plants

6.6.1 Tissue Age and Caffeine Metabolism

There are many publications related to tissue age and caffeine biosynthesis capacity. Some examples are listed in Table 6.3. There are four different types of investigations involving (i) determinations of caffeine content, (ii) *in situ* tracer experiments using segments of tissues, (iii) measurement of enzyme activity, and (iv) determination of gene expression (transcript level). In general, caffeine biosynthesis occurs in young tissues, in particular leaves of flush shoots, young fruits and flower buds. Although exceptions have been noted, caffeine synthesis activity is dependent upon the expression of *CS* genes and high activity of the resultant enzyme. Theobromine synthesis from purine nucleotides appears to be restricted to young tissues, although conversion of theobromine to caffeine occurs even in the later stages of tissue development. Transient accumulation of theobromine is, therefore, often observed in young tissue in caffeine-accumulating species such as tea and coffee plants. Caffeine appears to be produced from purine precursors in the same tissues as it accumulates. In general, the highest concentrations of caffeine are in young leaves and fruits, but total content per organ increases gradually during development and maturation. After that the amounts in leaves and pericarp decrease with age, especially in leaves. Zheng *et al.* (2004b) reported an absence of caffeine in senescent leaves following abscission from the parent plant. Catabolism of purine alkaloids appears to be more rapid in mature and aged leaves than in young leaves. For example, in tea, [8-^{14}C]theophylline is extensively degraded in mature and aged leaves. However, in young leaves, sizable amounts of [8-^{14}C]theophylline are salvaged for the synthesis of caffeine via a 3-methylxanthine → theobromine → caffeine pathway (Ashihara *et al.*, 1997).

6.6.2 Stress Response of Caffeine Biosynthesis

Formation and accumulation of caffeine in young, actively growing leaves of *Theobroma cacao* following wounding or infection was reported by Aneja and Gianfagna (2001). *Crinipellis perniciosa*, a fungus causing Witches' broom disease, attacks young actively growing tissues. Infected stem tissue contains up to eight times more caffeine than healthy stems. Wounding young, actively growing leaves induces the production of considerable amounts of caffeine. Caffeine production is stimulated by treatment of young leaves with salicylic acid and its synthetic analogue, benzothiadiazole, compounds that induce pathogen defence responses in plants (Crozier *et al.*, 2000). In culture, growth of *Crinipellis perniciosa* is inhibited significantly by caffeine. Overall these findings suggest that caffeine biosynthesis in *Theobroma cacao* is inducible by pathogen attack, wounding and salicylic acid, and may be part of the plant's defence response to herbivory and infection. Unfortunately, Aneja and Gianfagna (2001) did not analyse theobromine, which is the major purine alkaloid in cacao leaves (Koyama *et al.*, 2003), and as a consequence they failed to ascertain the impact of infection on theobromine content and whether or not conversion of theobromine to caffeine is induced by stress.

The effects of mechanical wounding, ethylene, and methyl jasmonate on gene expression have been investigated in cacao leaves (Bailey *et al.*, 2005). The gene putatively encoding

Table 6.3 *Physiological studies of purine alkaloid metabolism*

Plant materials	Phenomena	Experiments	Remarks	Reference
Camellia sinensis				
Leaves	Development	C, M	Caffeine synthesis is high in young tissues	Ashihara and Kubota (1986)
		M, E	Caffeine synthesis is high in freshly emerged shoots in April	Fujimori *et al.* (1991)
		GE	*TCS* gene expression is high in young tissues	Li *et al.* (2007a), Kato *et al.* (2010)
		GE	*TCS* gene expression is high in young tissues	Mohanpuria *et al.* (2009)
		M	Salvage of theophylline and xanthine is found in young leaves	Ashihara *et al.* (1997)
	Light	C	Shading increases caffeine content	Anan and Nakagawa (1974)
		M	No direct effect of light on caffeine synthesis is found	Koshiishi *et al.* (2000)
		GE	*TCS* expression is not changed by light	Kato *et al.* (2010)
	Environmental stress	GE	*TCS* expression decreased by Cd^{2+} ions and drought	Kato *et al.* (2010)
Flowers	Development	M	Caffeine biosynthesis is high prior to the opening of petals	Fijimori and Ashihara (1990)
		GE	*TCS* gene expression is constant during flowering	Kato *et al.* (2010)
Fruits	Development	C, M	Caffeine synthesis is high in young pericarp and cotyledon tissues	Terrasaki *et al.* (1994)

Camellia sinensis var. Kucha				
Leaves	Development	C, M	Theacrine synthesis is high in expanding buds	Zheng *et al.* (2002)
Camellia ptilophylla				
Leaves	Development	M	Theobromine synthesis is high in young leaves	Ashihara *et al.* (1998)
Coffea arabica				
Leaves	Development	C	Caffeine concentration is very high in young leaves	Frischknecht *et al.* (1986)
		C, M	Biosynthesis of caffeine is restricted in young leaves	Fujimori and Ashihara (1994); Zheng and Ashihara (2004)
Fruits	Development	C, M, GE	High biosynthesis in pericarp expansion to endosperm formation	Koshiro *et al.* (2006)
Coffea dewevrei etc.				
Leaves	Development	C	Methyluric acids are formed in old leaves	Petermann and Baumann (1983)
Theobroma cacao				
Fruits	Development	C, M	Theobromine synthesis is found in young pericarps and cotyledons	Zheng *et al.* (2004a)
Leaves	Development	C, M	Theobromine synthesis is found only in young leaves	Koyama *et al.* (2003)
	Development	GE	*CS* gene is only expressed in young leaves	Bailey *et al.* (2005)

a caffeine synthase (*TcCaf-1*) was induced by methyl jasmonate and repressed by ethylene in young leaves, but was not affected by mechanical wounding.

Kato *et al.* (2010) reported the effect of environmental stress on the caffeine synthase (*TCS*) gene in tea leaves. Expression was repressed significantly by drought but recovered upon rehydration. The physiological significance of the drought repression of the *TCS* gene has not yet been elucidated. A heavy metal (Cd^{2+}) also suppressed expression of the gene, but this appears not to be specific for *TCS* as a similar effect was found with almost all other genes examined. The level of *TCS* transcripts increased slightly when the plants were supplied with NH_4^+, but little or no differences were observed with other treatments including light, low temperature, salt, wounding, salicylic acid and methyl jasmonate.

6.6.3 Ecological Role of Purine Alkaloids

Until recently the role of purine alkaloids *in planta* remained largely undetermined, and it was thought that they could be functionless end-products produced during the course of evolution in a limited number of plant species. There are, however, two hypotheses concerning the role of caffeine in plants, the 'chemical defence' and 'allelopathic function' theories. The 'chemical defence' theory proposes that the high concentrations of caffeine in young leaves, fruits and flower buds of species such as *Coffea arabica* and *Camellia sinensis* act as a defence to protect young soft tissues from pathogens and herbivores. It has been shown that spraying tomato leaves with a 1% solution of caffeine deters feeding by tobacco hornworms, while treatment of cabbage leaves and orchids with 0.01–0.1% solutions of caffeine acts as a neurotoxin and kills or repels slugs and snails (Hollingsworth *et al.*, 2002). This work has now been extended, and convincing evidence for the chemical defence theory has been obtained with transgenic caffeine-producing tobacco plants (Uefuji *et al.*, 2005; Kim *et al.*, 2006). Likewise, caffeine synthesized in response to infection of cacao leaves might inhibit the growth of fungal and bacterial pathogens and restrict them spreading beyond the original site of infection (Aneja and Gianfagna, 2001)

The 'allelopathic or autotoxic function' theory proposes that caffeine in seed coats and falling leaves is released into the soil and inhibits germination of seeds around the parent plants (see Anaya *et al.*, 2006). Although there is evidence from laboratory studies to support this suggestion, it is unclear to what extent caffeine is involved in allelopathy in natural ecosystems, especially as soil bacteria such as *Pseudomonas putida* can degrade purine alkaloids (Hohnloser *et al.*, 1980; Gluck and Lingens, 1988).

6.7 Metabolic Engineering of Caffeine *In Planta*

Since cDNAs of *N*-methyltransferases involved in caffeine biosynthesis have been isolated (Mizuno *et al.*, 2001, 2003a,b; Ogawa *et al.*, 2001; Uefuji *et al.*, 2003) (see Section 6.4 and Table 6.2), it is possible to engineer caffeine biosynthesis not only in coffee plants but also in other agriculturally important plant species. Two approaches are feasible: one is construction of genetically modified (GM) coffee plants, in which caffeine production is suppressed; and the other is introduction of caffeine biosynthesis into what are normally non-caffeine producing plants. The first approach involves introducing antisense or the double-stranded RNA interference (RNAi) constructs for *CaMXMT1* into *Coffea canephora* and *Coffea arabica* (Ogita *et al.*, 2003, 2004, 2005). The second has been achieved using a multi-gene transfer system, involving cDNAs for all three *N*-methyltransferases, and the resulting transgenic tobacco plants successfully produced caffeine in leaves that proved unpalatable to tobacco cutworms (Uefuji *et al.*, 2005; Kim *et al.*, 2006). In this section, we briefly summarise the current status of caffeine metabolic engineering *in planta*.

6.7.1 Tissue Culture Technologies of Coffee Plants

Conventional breeding of coffee species has been performed by seeding or grafting after selecting superior varieties. However, to introduce new traits can take several decades due to the long life-cycle of this woody plant species (Spiral *et al.*, 1999). In order to shorten the breeding period, which is one of the objectives for metabolic engineering, tissue culture technologies have been developed. There are several approaches for coffee tissue culture, with protoplast culture, adventitious bud formation, and somatic embryogenesis procedures having been established (see Berthouly and Etienne, 1999). Currently, somatic embryogenesis is the best method. Indirect somatic embryogenesis, that is, embryogenic callus-induction, facilitates the production of large quantities of somatic embryos, and is generally recommended for micropropagation and/or transformation of higher plants. However, the intricacies of plant regeneration from somatic embryos derived from embryogenic calli sometimes prevents efficient genetic transformation. This can be overcome by using direct somatic embryogenesis methodology as follows:

Collect newly flushed young leaves *ca.* 5–9 cm long from coffee plants growing under a natural photoperiod. Surface-sterilise the leaves with 70% ethanol for 1 min followed by a 2% NaClO solution for 10 min. Then wash the leaves three times with sterile water, dry them on a sterile paper, and use as explants. The growth of yellow-whitish somatic embryos is induced on the cut edges of young leaves of both *Coffea arabica* and *Coffea canephora* cultured on a modified half concentration of Murashige-Skoog medium (M-1/2MS) containing 20 μM 2-isopentenyladenine. Without any visible callus formation, secondary somatic embryos proliferate in the same M1/2MS medium after a period of around three weeks. In order to induce further development, somatic embryos are transferred to M-1/2MS medium, hormone-free or containing 5 μM of N^6-benzyladenine. In these media the somatic embryos grow vigorously developing into small plantlets which can be transplanted and grown successfully in soil (Figure 6.4).

6.7.2 Suppression of Caffeine Biosynthesis in Coffee Plants

Although dependent upon volume and strength, a typical cup of coffee can contain up to 200 mg of caffeine. A minority of coffee drinkers suffer unpleasant short-term side-effects following caffeine consumption including palpitations, gastrointestinal disturbances, anxiety, tremor and increased blood pressure (International Food Information Foundation, 2008). Caffeine-induced insomnia is more widespread, and many people who need a cup or two of coffee in the morning to 'kick-start' their day will not partake of the beverage in the evening because of the prospects of a sleepless night. Because of the side-effects of caffeine there has been a growing demand for decaffeinated coffee, with 'decaf' now accounting for around 10–15% of coffee consumption worldwide. These days 'decaf' is manufactured mainly by supercritical fluid extraction with liquid CO_2, but as well as reducing the caffeine

Figure 6.4 *Coffee somatic embryogenesis for GM coffee production. (A) Direct somatic embryogenesis from leaf segments. (B) The surface structures of secondary somatic embryos of globular stages. (C) Histological characteristics of the embryogenic tissues. (D) Mature somatic embryos of torpedo-shaped stages with juvenile cotyledons. (E) Transgenic coffee seedling directly regenerated on a selection medium. (F) 4-year-old GM decaffeinated coffee trees, RNAi (right side) and wild-type (left side). (G) Flowering of RNAi transgenic coffee plant. Reprinted with permission from Ogita* et al. *(2005), Copyright 2005 Japanese Society for Plant Cell and Molecular Biology*

content, other ingredients relating to fragrance and taste can also be removed, resulting in a taste that does not suit the palate of many coffee connoisseurs. A molecular breeding approach for genetic engineering of caffeine biosynthesis can overcome this problem by producing an 'aromatic, full-bodied' coffee, in which only the caffeine content is reduced. The potential use of molecular breeding for producing GM 'decaf' was first mentioned by Ashihara and Crozier (2001), although the relevant genes to facilitate such an approach were not available at that time. Subsequently, the cloning of genes encoding key enzymes in the caffeine biosynthetic pathway (see Section 6.4.2) has made it possible to suppress caffeine production using either RNAi (Ogita *et al.*, 2003, 2004, 2005) or antisense methodology (Ashihara *et al.*, 2006).

In order to establish a method to stably transform and regenerate coffee plants, Ogita *et al.* (2003) first performed a pilot experiment with *sGFP* as the reporter gene, driven by the CaMV 35S promoter. Somatic embryos prepared as described above were transformed with a vector containing *sGFP* and *HPT* genes through *Agrobacterium tumefaciens* infection. Somatic embryos of *Coffea arabica* and *Coffea canephora* were co-cultured with a bacterial suspension for 10 min in liquid M-1/2MS medium containing plant growth regulators and 50 mg/l of acetosyringon. After aspiration of the bacterial suspension, the infected tissues were transferred to an induction-maintenance medium containing 300 mg/l of cefotaxim and 50 mg/l of hygromycin. They were subcultured at three-weekly intervals onto fresh medium containing the same components for a period of 2–4 months. During the course of subculturing, most infected tissues gradually turned brown and necrotic, after which transformed tissues regenerated from the surface of the tissues. The resulting transformed somatic embryos were transferred to M1/2MS medium containing 100 mg/l of hygromycin and grown to maturity. Expression of *sGFP* could be detected in all parts of both the regenerated somatic embryos and the regenerated seedlings, indicating that the introduced gene was stably expressed.

The 3′-untranslated region (UTR) and the coding region of *CaMXMT1* cDNA were first selected to design the *RNAi* constructs. Two particular *RNAi* constructs, *RNAi-S* having a short insert with 150 bp (positions 1117–1277 of the cDNA), and *RNAi-L* with a long insert of 360 bp (positions 946–1277) separated by a 517 bp DNA fragment, derived from the *GUS* gene, at positions 3436–3952 as the spacer, were inserted into the pIG121Hm vector (Hiei *et al.*, 1994; Ohta *et al.*, 1990), which was introduced into the EHA101 strain of *Agrobacterium tumefaciens* to transform *Coffea arabica* and *Coffea canephora*. The resulting transformed lines were assayed for expression of *N*-methyltransferase genes by RT-PCR, and it was found that transcripts for not only *CaMXMT1* but also *CaXMT1* and *CaDXMT1* were suppressed with the *CaMXMT1*-RNAi construct. These results are consistent with the finding that RNAi effects spread from the initiator region into the adjacent regions of the target gene, as well as other genes whose sequences are closely related (Vaistij *et al.*, 2002). The homology of *CaMXMT1*, *CaXMT1* and *CaDXMT1* is over 90% in the coding region (Uefuji *et al.*, 2003), suggesting that the primary small double-stranded *CaMXMT1*-RNA progressively produces many secondary small double-stranded RNAs spanning its coding region to the adjacent sequence of the initiator region, which in turn destroys mRNAs from *N*-methyltransferase genes of caffeine biosynthesis. The reduced level of transcripts suggested decreased activities of the corresponding enzymes, and this was confirmed by analysing their products, theobromine and caffeine. The caffeine content in the controls was approximately 8.4 mg/g of leaf tissue, while that in both RNAi-S and RNAi-L was 4.0 mg, an average reduction of 50%. However, the amount was variable depending on the line, with one notable example of *Coffea canephora* showing up to a 70% reduction (Ogita *et al.*, 2003; Figure 6.5), and almost a 100% reduction in the case of *Coffea arabica* by introducing the *RNAi-L* construct (Ogita, Unpublished results).

Since both the RNAi and antisense GM 'decaf' plants, have the same profile of endogenous metabolites as wild-type tissue (Ashihara *et al.*, 2006), it is anticipated that they will produce essentially normal coffee beans except for a low caffeine content. Before the decaffeinated plants can be cultivated on a commercial scale, some technical improvements are required. For example, the RNAi should preferably be expressed only in beans so that other parts of the plant retain wild-type caffeine production. Targeting of multiple genes is also necessary to achieve 100% decaffeinated beans. In this context, construction of a limited expression system might be useful, employing, for example, the promoter of a seed storage protein gene.

An alternative route to decaffeinated coffee may be the naturally caffeine-deficient *Coffea arabica* detected in a screening programme in Brazil by Silvarolla *et al.* (2004). Recently, a new low-caffeine hybrid coffee was established in Madagascar from local tetraploid inter-specific hybrids from *Coffea eugenioides*, *C. canephora* and *C. arabica* (Nagai *et al.*, 2008).

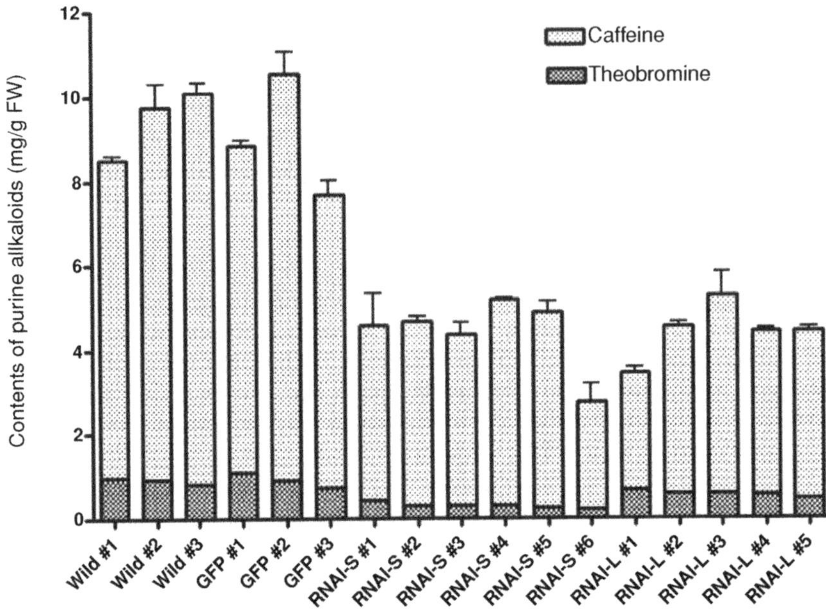

Figure 6.5 *Purine alkaloid contents in GM coffee plants. Endogenous theobromine and caffeine levels (mg/g fresh weight) in young leaves of wild type, transgenic sGFP (control), and indicated transgenic RNAi plants of* Coffea canephora

6.7.3 Production of Caffeine in Tobacco Plants

The physiological function of caffeine has been proposed to constitute part of a chemical defence system against pathogens and herbivores (see Section 6.4.3). Exogenously applied caffeine has been found to markedly increase the resistance of plants against several pests, and thereby enhanced their growth and survival (Nathanson, 1984). Based on these observations, caffeine-producing transgenic plants were constructed using tobacco (*Nicotiana tabacum*), a species that naturally does not synthesize the purine alkaloid (Uefuji *et al.*, 2005; Kim *et al.*, 2006).

6.7.4 Construction of Transgenic Caffeine-Producing Tobacco Plants

Firstly, a system for multiple transfer of the three coffee *N*-methyltransferase genes, *CaXMT1*, *CaMXMT1/2* and *CaDXMT1*, was established. Each gene was initially introduced independently into pBI221, and then individual expression cassettes containing the CaMV *35S* promoter, cDNA clone and *NOS* terminator were removed and successively inserted into the multiple cloning site of pBluescript II SK(-). The three connected cassettes were finally replaced with the GUS coding sequence and *NOS* terminator of pIG121Hm (Hiei *et al.*, 1994; Ohta *et al.*, 1990) and designated as pBIN-NMT777. Tobacco (cv. Xanthi) leaf discs were transformed with pBIN-NMT777 using the *Agrobacterium*-transformation method (LBA4404 strain). After appropriate culture and selection, 23 kanamycin-resistant transgenic plantlets were obtained, of which 15 were shown by RT-PCR to express all three

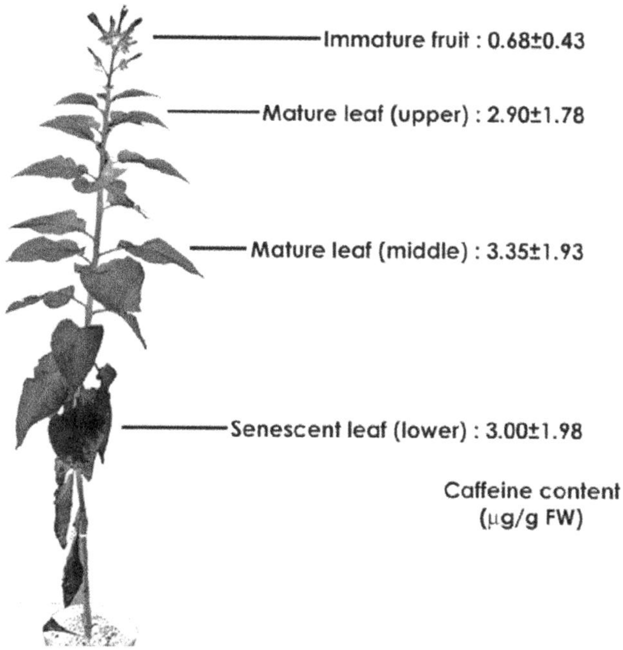

Figure 6.6 *Caffeine contents of mature tobacco plants. Transgenic line #5 was grown to maturity, and caffeine was extracted from indicated organs and quantified by HPLC. Caffeine content is expressed in μg/g fresh weight (FW) with standard deviation. Reprinted with permission from Ogita* et al. *(2005), Copyright 2005 Japanese Society for Plant Cell and Molecular Biology*

N-methyltransferase genes. The selected lines were grown to maturity and the accumulation of purine alkaloids in leaves was monitored. Initial analysis using mature leaves showed the transgenic plants accumulated caffeine and theobromine, in contrast to control plants containing an empty vector. Subsequently, the caffeine content of individual leaves at different developmental stages was determined (Figure 6.6). Plants in the vegetative phase of growth contained only limited amounts of caffeine, and in immature leaves the alkaloid was often undetectable. In mature leaves, however, the average caffeine content was 0.2 μg/g fresh weight, and when plants aged and entered the reproductive phase forming flower buds, the caffeine content increased to >5 μg/g. Immature fruits contained caffeine at a lower level of <1.3 μg/g. No caffeine was detected in control plants. The results thus indicated that caffeine was indeed synthesized in the transgenic tobacco leaves, and that accumulation was highest in the older leaves during the reproductive phase of growth.

6.7.5 Repelling Effects on Tobacco Cutworms

Since tobacco cutworms (*Spodoptera litura*) damage a wide range of crops, including tobacco, the feeding behaviour of caterpillars using transgenic tobacco lines which produced different levels of caffeine was investigated. Third-instar larvae were starved for several hours and then allowed to select and feed on leaf discs prepared from transgenic or control plants. The larvae selectively fed on the control tissues and positively avoided the transgenic leaves. Quantitative estimations indicated that consumed areas were up to 1.1 cm^2 for control leaves, with values of <0.02 cm^2 for the transgenic discs containing caffeine at 5 μg/g. Transgenic discs accumulating as little as 0.4 μg caffeine/g were also effective in repelling caterpillars (Figure 6.7).

These observations indicate that caffeine has a clear repelling effect on insect pests. However, if there was no alternative food supply the caterpillars did eat transgenic leaves

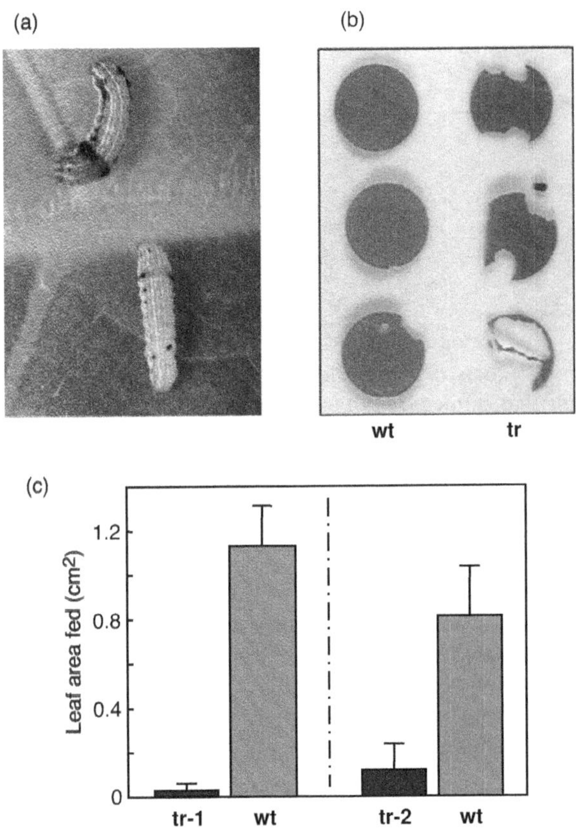

Figure 6.7 *Repellent effects of transgenic caffeine-producing tobacco plants against tobacco cutworm (*Spodoptera litura*). (A) Tobacco cutworm larvae at the third instar stage of development. (B) Larvae were starved for 3 h, then allowed to feed for 3 h in darkness on three wild-type (wt) and three transgenic leaf discs producing 5 μg/g caffeine/g fresh weight (tr), after which the leaf discs were photographed. (C) Quantification of feeding behaviour. Twenty replicate tests were performed, and the leaf areas consumed by the larvae (y axis) calculated with the aid of an image analyser and expressed in cm^2 ± standard error. The x axis indicates discs from transgenic plants containing 5 μg caffeine/g fresh weight (tr-1) and 0.4 μg/g caffeine/g (tr-2), and from wild-type plants (wt). Reprinted from Ashihara* et al. *(2008) Copyright 2008, with permission from Elsevier*

containing caffeine at 5 μg/g, and this was not lethal as the caterpillars grew normally up to the pupating stage. Further experiments in which caffeine was added to artificial food established that >10 mg of caffeine/g was toxic to the caterpillars. This high concentration implies that caffeine will not be a powerful pest controller and is of limited use in agriculture. However, earlier research has shown that the reproductive potential of caterpillars is greatly reduced when they are fed a diet containing a non-lethal level of caffeine, achieved by soaking leaves for 1 h in 0.2–0.4% caffeine (Mathavan *et al.*, 1985). Mature moths laid fewer eggs with a lower protein content compared with controls fed a caffeine-free diet. Thus, the production of caffeine by transgenic agricultural crops may be a practical approach for overall pest suppression, acting as a repellent at a low dose, and being toxic at higher concentrations such as those that prevail in young leaves of *Coffea arabica* and *Camellia sinensis*.

To be of practical value, caffeine levels *in planta* will have to be increased 2–3 fold over what has been achieved to date with the transgenic tobacco plants. Several approaches are conceivable: (i) increasing the pool of the starting material of the caffeine biosynthesis pathway, xanthosine; (ii) application of other genes encoding enzymes with more efficient catalytic properties; and (iii) manipulation of caffeine transportation and assimilation in vacuoles. The medium-to-long term possibility of producing relatively large amounts of caffeine in appropriate crop species has profound implications as, in an environmentally friendly manner, it will provide a means of conferring herbivore resistance without the need for routine, expensive pesticide treatments.

6.7.6 Perspectives

In order to synthesize a target compound *in planta*, multiple steps are required, in which different enzymes successively catalyze the individual reactions. For example, scopolamine is synthesized through more than ten steps, while berberine is produced via a 13-step pathway (Sato *et al.*, 2001). In contrast, caffeine biosynthesis which comprises three methylations and one ribose removal is a mere four-step pathway from xanthosine (Section 6.4.2). It is just over ten years since the first cloning of caffeine synthase from tea, the gene encoding the enzyme regulating the last two methylation steps in the caffeine biosynthesis pathway (Kato *et al.*, 2000). Since this pioneering study, extensive research has led to the successful cloning of a number of *N*-methyltransferase-encoding genes from coffee, tea and cacao (Table 6.2). Much of the widespread interest in this topic has been fuelled by the possibilities of using genetic engineering to obtain transgenic, low-caffeine coffee and tea plants that could be used to produce full-bodied, tasty decaffeinated beverages. To date, transgenic *Coffea canephora* plants with leaves containing a 50–70% reduced caffeine content have been obtained (Ogita *et al.*, 2003) but as yet there are no reports on the caffeine content of beans produced by such plants. When this does occur, a more substantial suppression of caffeine production will be required, as at least a 90% reduction is necessary for commercial coffees to qualify for sale under the 'decaffeinated' label. It will also be more appropriate to produce transgenic decaffeinated *Coffea arabica* which yields a more superior beverage than Robusta coffee from *Coffea canephora*.

Because, worldwide, coffee is such a popular beverage, and decaffeinated coffee has a market share of around 10% in Europe and around 12–15% in the US, research on decaffeinated GM coffee has attracted enormous media attention (http://news.bbc.co.uk/1/hi/sci/

tech/903308.stm; http://news.bbc.co.uk/1/hi/sci/tech/3002112.stm; http://www.abc.net.au/science/news/stories/s882676.htm; http://www.abc.net.au/science/news/stories/s882676.htm) with much interest from potential consumers, and occasional adverse comment about the practical benefits from manufacturers of coffee who have made recent multimillion dollar investments in the construction of supercritical fluid decaffeination plants.

The recent reports that tobacco plants engineered to produce caffeine both repel insect pests and exhibit resistance to viral and bacterial infection (Uefuji *et al.*, 2005; Kim *et al.*, 2006; Kim and Sano, 2008) give an example of the potential benefits of GM crops. Worldwide, attack by herbivores and diseases results in about 40% loss of agricultural production (Baker *et al.*, 1997; Jouanin *et al.*, 1998). Practical and economic means of pest control is, therefore, one of the most urgent measures required to obtain a reliable and increasing supply of foodstuffs for the world's expanding population. The production of a low level of caffeine by agriculturally important crops is one potential route to protection against both insect pests and pathogens and, from an environmental prospective, it would certainly be safer and cheaper than treatment with pesticides and fungicides.

Much biochemical and molecular information has now been obtained on the biosynthesis of caffeine and, although it has not progressed to the same extent, an increasing amount of data are becoming available on the catabolism of caffeine in a variety of purine alkaloid-containing plants. Information on the cellular metabolic machinery of caffeine biosynthesis and catabolism, links to purine nucleotide metabolism, intercellular translocation and accumulation mechanisms at specific cellular sites, such as chloroplasts and vacuoles, have yet to be fully revealed. Cell-, tissue-, and organ-specific synthesis and catabolism of purine alkaloids may be regulated by unique and as yet unknown developmental and environmental control mechanisms. A great deal of fascinating purine alkaloid biology remains to be discovered.

References

Anan, T. and Nakagawa, M. (1974) Effect of light on chemical constituents in the tea leaves. *Nippon Nogeikagaku Kaishi* **48**, 91–98.

Anaya, A.L., Cruz-Ortega, R. and Waller, G.R. (2006) Metabolism and ecology of purine alkaloids. *Front. Biosci.*, **11**, 2354–2370.

Aneja, M. and Gianfagna,T. (2001) Induction and accumulation of caffeine in young, actively growing leaves of cocoa (*Theobroma cacao* L.) by wounding or infection with *Crinipellis perniciosa. Physiol. Mol. Plant Pathol.*, **59**, 13–16.

Ashihara, H. and Kubota, H. (1986) Patterns of adenine metabolism and caffeine biosynthesis in different parts of tea seedlings. *Physiol. Plant.*, **68**, 275–281.

Ashihara, H. and Kubota, H. (1987) Biosynthesis of purine alkaloids in *Camellia* plants. *Plant Cell Physiol.*, **28**, 535–539.

Ashihara, H., Monteiro, A.M., Gillies, F.M. and Crozier, A. (1996) Biosynthesis of caffeine in leaves of coffee. *Plant Physiol.*, **111**, 747–753.

Ashihara, H., Gillies, F.M. and Crozier, A. (1997) Metabolism of caffeine and related purine alkaloids in leaves of tea (*Camellia sinensis* L.). *Plant Cell Physiol.*, **38**, 413–419.

Ashihara, H., Kato, M. and Ye, C.-X. (1998) Biosynthesis and metabolism of purine alkaloids in leaves of cocoa tea (*Camellia ptilophylla*). *J. Plant Res.*, **111**, 599–604.

Ashihara, H. and Crozier, A. (1999) Biosynthesis and metabolism of caffeine and related purine alkaloids in plants. *Adv. Bot. Res.*, **30**, 117–205.

Ashihara, H. and Crozier, A. (2001) Caffeine: a well known but little mentioned compound in plant science. *Trends Plant Sci.*, **6**, 407–413.

Ashihara, H., Sano, H. and Crozier, A. (2008) Caffeine and related purine alkaloids: biosynthesis, catabolism, function and genetic engineering. *Phytochemistry*, **69**, 841–856.

Bailey, B.A., Bae, H., Strem, M.D., Antnez De Mayolo, G., Guiltinan, M.J., Verica, J.A., Maximova, S.N. and Bowers, J.H. (2005) Developmental expression of stress response genes in *Theobroma cacao* leaves and their response to Nep1 treatment and a compatible infection by *Phytophthora megakarya. Plant Physiol. Biochem.*, **43**, 611–622.

Baker, B., Zambryski, P., Staskawicz, B. and Dinesh-Kumar, S.P. (1997) Signaling in plant–microbe interactions. *Science*, **267**, 726–733.

Baumann, T.W., Oechslin, M. and Wanner, H. (1976) Caffeine and methylated uric acids: chemical patterns during vegetative development of *Coffea liberica. Biochem. Physiol. Pflanzen.*, **170**, 217–225.

Baumann, T.W. and Wanner, H. (1980) The 1,3,7,9-tetramethyluric acid content of cupu *(Theobroma grandiflorum). Acta Amaz,.* **10**, 425.

Baumann, T.W., Schulthess, B.H. and Hanni, K. (1995) Guarana (*Paullinia cupana*) rewards seed dispersers without intoxicating them by caffeine. *Phytochemistry*, **39**, 1063–1070.

Berthouly, M. and Etienne, H. (1999) Somatic embryogenesis of coffee, in *Somatic Embryogenesis in Woody Plants*, Vol. **5** (eds S.M. Jain, P.K. Gupta and R.J. Newton), Kluwer Academic Publishers, Dordrecht, pp. 259–288.

Crozier, A., Kamiya, Y., Bishop, G.J. and Yokota, T. (2000) Biosynthesis of hormones and elicitor molecules, in *Biochemistry and Molecular Biology of Plants* (eds B.B. Buchanan, W. Gruissem and R.L. Jones), American Society of Plant Physiologists, Rockville, pp. 850–929.

Dash, S. and Gummadi, S. (2006) Catabolic pathways and biotechnological applications of microbial caffeine degradation. *Biotechnol. Lett.*, **28**, 1993–2002.

Frischknecht, P.M., Ulmer-Dufek, J. and Baumann, T.W. (1986) Purine alkaloid formation in buds and developing leaflets of *Coffea arabica*: Expression of an optimal defence strategy? *Phytochemistry*, **25**, 613–616.

Fujimori, N. and Ashihara, H. (1990) Adenine metabolism and the synthesis of purine alkaloids in flowers of *Camellia. Phytochemistry*, **29**, 3513–3516.

Fujimori, N., Suzuki, T. and Ashihara, H. (1991) Seasonal variations in biosynthetic capacity for the synthesis of caffeine in tea leaves. *Phytochemistry*, **30**, 2245–2248.

Fujimori, N. and Ashihara, H. (1993) Biosynthesis of caffeine in flower buds of *Camellia sinensis. Ann. Bot.*, **71**, 279–284.

Fujimori, N. and Ashihara, H. (1994) Biosynthesis of theobromine and caffeine in developing leaves of *Coffea arabica. Phytochemistry*, **36**, 1359–1361.

Gluck, M. and Lingens, F. (1988) Heteroxanthine demethylase, a new enzyme in the degradation of caffeine by *Pseudomonas putida. Appl. Microbiol. Biotechnol.*, **28**, 59–62.

Gokulakrishnan, S., Chandraraj, K. and Gummadi, S.N. (2005) Microbial and enzymatic methods for the removal of caffeine. *Enzym. Microbial. Technol.*, **37**, 225–232.

Hiei, Y., Ohta, S., Komari, T. and Kumashiro, T. (1994) Efficient transformation of rice (*Oryza sativa* L.) mediated by *Agrobacterium* and sequence analysis of the boundaries of the T-DNA. *Plant J.*, **6**, 271–282.

Hohnloser, W., Osswald, B. and Lingens, F (1980) Enzymological aspects of caffeine demethylation and formaldehyde oxidation by *Pseudomonas putida* C1. *Hoppe Seyler's Z. Physiol. Chem.*, **361**, 1763–1766.

Hollingsworth, R.G., Armstrong, J.W. and Campbell, E. (2002) Caffeine as a repellent for slugs and snails. *Nature*, **417**, 915–916.

Huber, M. and Baumann, T.W. (1998) The first step of caffeine degradation in coffee – still a mystery. Symposium on Future Trends in Phytochemistry, Phytochemical Society of Europe, Rolduc.

International Food Information Foundation (2008) *Caffeine and Health: Clarifying the Controversies*, International Food Information Foundation, Washington, DC.

Ito, E., Crozier, A. and Ashihara, H. (1997) Theophylline metabolism in higher plants. *Biochim. Biophys. Acta*, **1336**, 323–330.

Ito, E. and Ashihara, H. (1999) Contribution of purine nucleotide biosynthesis *de novo* to the formation of caffeine in young tea *(Camellia sinencis)* leaves. *J. Plant Physiol.*, **254**, 145–151.

Jouanin, L., Bonade-Bottino, M., Girard, C., Morrot, G. and Giband, M. (1998) Transgenic plants for insect resistance. *Plant Sci.*, **131**, 1–11.

Kalberer, P. (1965) Breakdown of caffeine in the leaves of *Coffea arabica* L. *Nature*, **205**, 597–598.

Kato, A., Crozier, A. and Ashihara, H. (1998) Subcellular localisation of the *N*-3 methyltransferase involved in caffeine biosynthesis in tea. *Phytochemistry*, **48**, 777–779.

Kato, M., Mizuno, K., Fujimura, T., Iwama, M., Irie, M., Crozier, A. and Ashihara, H. (1999) Purification and characterization of caffeine synthase from tea leaves. *Plant Physiol.*, **120**, 579–586.

Kato, M., Mizuno, K., Crozier, A., Fujimura, T. and Ashihara, H. (2000) A gene encoding caffeine synthase from tea leaves. *Nature*, **406**, 956–957.

Kato, M., Kitao, N., Ishida, M., Morimoto, H., Irino, F. and Mizuno, K. (2010) Expression analysis of the genes for caffeine biosynthesis and its related enzymes in *Camellia sinensis*. *Z. Naturforsch.*, **64c**, 245–256.

Keya, C.A., Crozier, A. and Ashihara, H. (2003) Inhibition of caffeine biosynthesis in tea (*Camellia sinensis*) and coffee (*Coffea arabica*) plants by ribavirin. *FEBS Lett.*, **554**, 473–477.

Kihlman, B.A. (1977) *Caffeine and Chromosomes*, Elsevier, Amsterdam.

Kim, Y.S., Uefuji, H., Ogita, S. and Sano, H. (2006) Transgenic tobacco plants producing caffeine: a potential new strategy for insect pest control. *Transgen. Res.*, **15**, 667–672.

Kim, Y.-S. and Sano, H. (2008) Pathogen resistance in transgenic tobacco plants producing caffeine. *Phytochemistry*, **69**, 882–888.

Koshiishi, C., Ito, E., Kato, A., Yoshida, Y., Crozier, A. and Ashihara, H. (2000) Purine alkaloid biosynthesis in young leaves of *Camellia sinensis* in light and darkness. *J. Plant Res.*, **113**, 217–221.

Koshiishi, C., Kato, A., Yama, S., Crozier, A. and Ashihara, H. (2001) A new caffeine biosynthetic pathway in tea leaves: utilisation of adenosine released from the *S*-adenosyl-L-methionine cycle. *FEBS Lett.*, **499**, 50–54.

Koshiro, Y., Zheng, X.Q., Wang, M., Nagai, C. and Ashihara, H. (2006) Changes in content and biosynthetic activity of caffeine and trigonelline during growth and ripening of *Coffea arabica* and *Coffea canephora* fruits. *Plant Sci.*, **171**, 242–250.

Koyama, Y., Tomoda, Y., Kato, M. and Ashihara, H. (2003) Metabolism of purine bases, nucleosides and alkaloids in theobromine-forming *Theobroma cacao* leaves. *Plant Physiol. Biochem.*, **41**, 977–984.

Kumar, V., Satyanarayana, K.V., Ramakrishna, A., *et al.* (2007) Evidence for localization of *N*-methyltransferase (MMT) of caffeine biosynthetic pathway in vacuolar surface of *Coffea canephora* endosperm elucidated through localization of GUS reporter gene driven by NMT promoter. *Curr. Sci.*, **93**, 383–386.

Li, Y., Ogita, S., Keya, C.A. and Ashihara, H. (2007a) Expression of caffeine biosynthesis genes in tea (*Camellia sinensis*). *Z. Naturforsch.*, **63C**, 267–270.

Li, Y.H., Gu, W. and Ye, S. (2007b) Expression and location of caffeine synthase in tea plants. *Russ. J. Plant Physiol.*, **54**, 698–701.

Mathavan S., Premalatha Y. and Christopher, M.S.M. (1985) Effects of caffeine and theophylline on the fecundity of four lepidopteran species. *J. Exp. Biol.*, **44**, 133–138.

Mazzafera, P. (1994) Caffeine, theobromine and theophylline distribution in *Ilex paraguariensis*. *Rev. Brasil. Fisiol. Veg.*, **6**, 149–151.

Mazzafera, P. (2004) Catabolism of caffeine in plants and microorganisms. *Front. Biosci.*, **9**, 1348–1359.

Mazzafera, P. and Carvalho, A. (1992) Breeding for low seed caffeine content of coffee (*Coffea* L.) by interspecific hybridization. *Euphytica*, **59**, 55–60.

McCarthy, A.A. and McCarthy, J.G. (2007) The structure of two *N*-methyltransferases from the caffeine biosynthetic pathway. *Plant Physiol.*, **144**, 879–889.

Mizuno, K., Tanaka, H., Kato, M., Ashihara, H. and Fujimura, T. (2001) cDNA cloning of caffeine (theobromine) synthase from coffee (*Coffea arabica* L.), in Proceedings of the 19th International Conference on Coffee Science, Trieste. ASIC, Paris, pp. 815–818.

Mizuno, K., Kato, M., Irino, F., Yoneyama, N., Fujimura, T., and Ashihara, H. (2003a) The first committed step reaction of caffeine biosynthesis: 7-methylxanthosine synthase is closely homologous to caffeine synthases in coffee (*Coffea arabica* L.). *FEBS Lett.*, **547**, 56–60.

Mizuno, K., Okuda, A., Kato, M., Yoneyama, N., Tanaka, H., Ashihara, H., and Fujimura, T. (2003b) Isolation of a new dual-functional caffeine synthase gene encoding an enzyme for the conversion of 7-methylxanthine to caffeine from coffee (*Coffea arabica* L.). *FEBS Lett.*, **534**, 75–81.

Mohanpuria, P., Kumar, V., Joshi, R., Gulati, A., Ahuja, P. and Yadav, S. (2009) Caffeine biosynthesis and degradation in tea [*Camellia sinensis* (L.) O. Kuntze] is under developmental and seasonal regulation. *Mol. Biotechnol.*, **43**, 104–111.

Nagai, C., Rakotomalala, J.J., Katahira, R., Li, Y., Yamagata, K. and Ashihara, H. (2008) Production of a new low-caffeine hybrid coffee and the biochemical mechanism of low caffeine accumulation. *Euphytica* **164**, 133–142.

Nagata, T. and Sakai, S. (1984) Differences in caffeine, flavonols and amino acids contents in leaves of cultivated species of *Camellia*. *Jpn. J. Breed.*, **34**, 459–467.

Nathanson, J.A. (1984) Caffeine and related methylxanthines: possible naturally occurring pesticides. *Science*, **226**, 184–187.

Negishi, O., Ozawa, T. and Imagawa, H. (1988) *N*-methylnucleosidase from tea leaves. *Agric. Biol. Chem.*, **52**, 169–175.

Negishi, O., Ozawa, T. and Imagawa, H. (1994) Guanosine deaminase and guanine deaminase from tea leaves. *Biosci. Biotechnol. Biochem.*, **58**, 1277–1281.

Ogawa, M., Herai, Y., Koizumi, N., Kusano, T. and Sano, H. (2001) 7-Methylxanthine methyltransferase of coffee plants. Gene isolation and enzymatic properties. *J. Biol. Chem.*, **276**, 8213–8218.

Ogita, S., Uefuji, H., Yamaguchi, Y., Koizumi, N., and Sano, H. (2003) Production of decaffeinated coffee plants by genetic engineering. *Nature,* **423**, 823.

Ogita, S., Uefuji, H., Morimoto, M. and Sano, H. (2004) Application of RNAi to confirm theobromine as the major intermediate for caffeine biosynthesis in coffee plants with potential for construction of decaffeinated varieties. *Plant Mol. Biol.*, **54**, 931–941.

Ogita, S., Uefuji, H., Morimoto, M. and Sano, H. (2005) Metabolic engineering of caffeine production. *Plant Biotech.*, **22**, 461–468.

Ohta, S., Mita, S., Hattori, T., and Nakamura, K. (1990) Construction and expression in tobacco of a β-glucuronidase (GUS) reporter gene containing an intron within the coding sequence. Plant Cell Physiol., **31**, 805–813.

Petermann, J. and Baumann, T.W. (1983) Metabolic relations between methylxanthines and methyluric acids in *Coffea. Plant Physiol.*, **73**, 961–964.

Sato, F., Hashimoto, T., Hachiya, A., Tamura, K., Choi, K., Morishige, T., Fujimoto, H., and Yamada, Y. (2001) Metabolic engineering of plant alkaloid biosynthesis. *Proc. Natl. Acad. Sci. USA*, **98**, 367–372.

Satyanarayana, K.V., Kumar, V., Chandrashekar, A. and Ravishankar, G.A. (2005) Isolation of promoter for *N*-methyltransferase gene associated with caffeine biosynthesis in *Coffea canephora. J. Biotechnol.*, **119**, 20–25.

Senanayake, U.M. and Wijesekera, R.O.B. (1971) Theobromine and caffeine content of the cocoa bean during its growth. *J. Sci. Food Agric.*, **22**, 262–263.

Silvarolla, M.B., Mazzafera, P. and Fazuoli, L.C. (2004) A naturally decaffeinated arabica coffee. *Nature*, **429**, 826.

Spiral, J., Leroy, T., Paillard, M. and Petiard, V. (1999) Transgenic coffee, in *Transgenic Trees* (ed. Y.P.S. Bajaj), Springer, Berlin, pp. 55–76.

Stasolla, C., Katahira, R., Thorpe, T.A. and Ashihara, H. (2003) Purine and pyrimidine nucleotide metabolism in higher plants. *J. Plant Physiol.*, **160**, 1271–1295.

Suzuki, T. (1985) Purine alkaloids in *Camellia sinensis* flowers. *Agric. Biol. Chem.*, **49**, 2803–2805.

Suzuki, T. and Waller, G.R. (1984) Biodegradation of caffeine: formation of theophylline and caffeine in mature *Coffea arabica* fruits. *J. Sci. Food Agric.*, **35** 66–70.

Suzuki, T. and Waller, G.R. (1985) Effects of light on the production and degradation of caffeine in *Camellia sinensis* L. seedlings. *Plant Cell Physiol.*, **26**, 765–768.

Suzuki, T., Ashihara, H. and Waller, G.R. (1992) Purine and purine alkaloid metabolism in *Camellia* and *Coffea* plants. *Phytochemistry*, **31**, 2575–2584.

Terrasaki, Y., Suzuki, T. and Ashihara, H. (1994) Purine metabolism and the biosynthesis of purine alkaloids in tea fruits during development. *Plant Physiol., (Life Sci. Adv.)* **13**, 135–142.

Uefuji, H., Ogita, S., Yamaguchi, Y., Koizumi, N. and Sano, H. (2003) Molecular cloning and functional characterization of three distinct *N*-methyltransferases involved in the caffeine biosynthetic pathway in coffee plants. *Plant Physiol.*, **132**, 372–380.

Uefuji, H., Tatsumi, Y., Morimoto, M., Kaothien-Nakayama, P., Ogita, S. and Sano, H. (2005) Caffeine production in tobacco plants by simultaneous expression of three coffee *N*-methyltrasferases and its potential as a pest repellent. *Plant Mol. Biol.*, **59**, 221–227.

Vaistij, F.E., Jones, L. and Baulcombe, D.C. (2002) Spreading of RNA targeting and DNA methylation in RNA silencing requires transcription of the target gene and a putative RNA-dependent RNA polymerase. *Plant Cell*, **14**, 857–867.

Vitoria, A.P. and Mazzafera, P. (1998) Caffeine degradation in leaves and fruits of *Coffea arabica* and *Coffea dewevrei*. *Pesquisa Agropecuaria Brasileira*, **33**, 1957–1961.

Ye, C., Lin, Y., Zhou, H., Cheng, F. and Li, X. (1997) Isolation and analysis of purine alkaloids from *Camellia ptilophylla* Chang. *Acta Sci. Natur. Univ. Sunyatseni*, **36**, 30–33.

Yoneyama, N., Morimoto, H., Ye, C.X., Ashihara, H., Mizuno, K. and Kato, M. (2006) Substrate specificity of N-methyltransferase involved in purine alkaloids synthesis is dependent upon one amino acid residue of the enzyme. *Mol. Genet. Genom.*, **275**, 125–135.

Zheng, X.Q., Ye, C.X., Kato, M., Crozier, A. and Ashihara, H. (2002) Theacrine (1,3,7,9-tetramethyluric acid) synthesis in leaves of a Chinese tea, kucha (*Camellia assamica* var. kucha). *Phytochemistry*, **60**, 129–134.

Zheng, X.Q. and Ashihara, H. (2004) Distribution, biosynthesis and function of purine and pyridine alkaloids in *Coffea arabica* seedlings. *Plant Sci.*, **166**, 807–813.

Zheng, X.Q., Koyama, Y., Nagai, C. and Ashihara, H. (2004a) Biosynthesis, accumulation and degradation of theobromine in developing *Theobroma cacao* fruits. *J. Plant Physiol.*, **161**, 363–369.

Zheng, X.Q., Nagai, C. and Ashihara, H. (2004b) Pyridine nucleotide cycle and trigonelline (*N*-methylnicotinic acid) synthesis in developing leaves and fruits of *Coffea arabica*. *Physiol. Plant.*, **122**, 404–411.

Zrenner, R., Stitt, M., Sonnewald, U. and Boldt, R. (2006) Pyrimidine and purine biosynthesis and degradation in plants. *Ann. Rev. Plant Biol.*, **57**, 805–836.

Zulak, K.G., Liscome, D.K., Ashihara, H. and Facchini, P.J. (2006) Alkaloids, in *Plant Secondary Metabolites: Occurrence, Structure and Role in the Human Diet* (eds A. Crozier, M.N. Clifford and H. Ashihara), Blackwell, Oxford, pp. 102–136.

7

Nicotine Biosynthesis

Tsubasa Shoji and Takashi Hashimoto

7.1 Introduction

Alkaloids are nitrogen-containing, mostly alkaline organic compounds and constitute the second most diverse group of secondary metabolites in plants after terpenoids (Roberts and Wink, 1998). Alkaloids are found in approximately 20% of flowering plants and have more than 12,000 chemical structures. Alkaloids mainly play defensive roles against pathogens and herbivores, conferring adaptive advantages to the plants producing them. Due to their pharmacological activities, alkaloids have been exploited as poisons, narcotics and medicines for thousands of years. Alkaloids, typically derived from amino acids or their derivatives, can be classified into biogenically related groups, and some of their biosynthetic pathways have been characterised (Hashimoto and Yamada, 1994, 2003; Kutchan, 1995; DeLuca and Laflamme, 2001; Facchini, 2001; Ziegler and Facchini, 2008). Recent applications of expression sequence tags (ESTs), cDNA microarrays and metabolite profiling, have led to discoveries of novel genes and mechanisms in the alkaloid pathways (Goossens *et al.*, 2003; Rischer *et al.*, 2006; Ziegler *et al.*, 2006).

Nicotine is the predominant alkaloid in tobacco plants (*Nicotiana tabacum*) as well as tobacco products, such as cigars, cigarettes, pipe tobacco and chewing tobacco. It is an important compound determining smoke quality and causing addiction to smoking (Davis and Nielsen, 1999; Gorrod and Jacob, 1999). When Columbus discovered the New World he found that the leaves of *Nicotiana*, indigenous to tropical America, were smoked by the native inhabitants. Tobacco use was implemented culturally and became popular in most countries of the world (Goodman, 1993). Nowadays, however, it is widely recognised that

Plant Metabolism and Biotechnology, First Edition. Edited by Hiroshi Ashihara, Alan Crozier, and Atsushi Komamine.

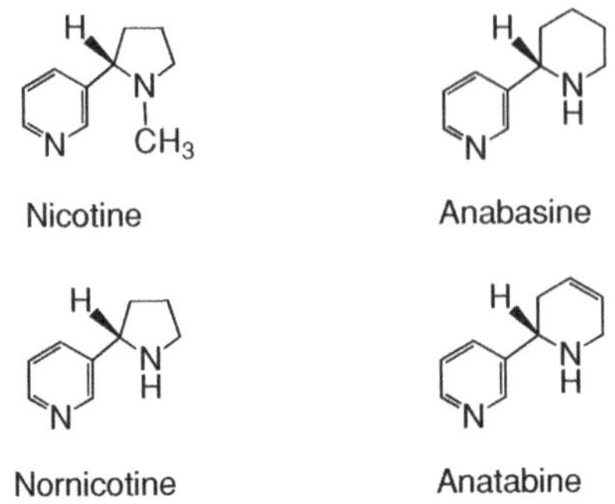

Figure 7.1 *Alkaloids in* Nicotiana *species*

smoking has a marked and deleterious effect on health (Hecht, 2003), and laws designed to reduce smoking are appearing with increasing frequency, with smoking being banned in cinemas, bars and public buildings in an increasing number of countries (Stead and Lancaster, 2007). Nicotine acts on nicotinic acetylcholine receptors widely distributed in the nervous system, stimulating ganglia through depolarisation of the postsynaptic membranes at low doses, or blocking them through permanent depolarisation at high doses (Dome *et al.*, 2009). As a potent toxin, nicotine had also been used as an effective insecticide until the advent of synthetic counterparts, such as DDT.

First isolated in 1828 by Posselt and Reimann, nicotine was named after Jean Nicot de Villemain, French ambassador to Portugal, who introduced tobacco to Europe in 1560 for medicinal use. Its chemical structure was later determined by Garry Pinner in 1893. Naturally-occurring nicotine is an optically pure (–)-form, in which the configuration at the C′-2 chiral centre is (*S*) (Figure 7.1). Nicotine is the most abundant alkaloid in tobacco, typically accounting for 90–95% of total alkaloid content, with the remainder of the alkaloid pool made up primarily of secondary alkaloids, such as nornicotine, anabasine and anatabine (Figure 7.1). The alkaloid profile within the genus *Nicotiana* is highly variable and in most species a single alkaloid is predominant (Saitoh *et al.*, 1985). For example, *N. tomentosiformis* largely accumulates nornicotine, while in *N. glauca* (tree tobacco), anabasine is a major alkaloid.

7.2 Pathways and Enzymes

In tobacco, nicotine and related alkaloids are synthesized exclusively in roots and transported throughout the plant via the xylem (Dawson, 1942). Roots of intact plants, root cultures or cell cultures have been used as experimental materials, and the nicotine pathway in tobacco has been studied through precursor feeding and enzymatic experiments followed by molecular identification of the structural genes (Leete, 1983; Bush, 1999; Bush *et al.*, 1999; Katoh *et al.*, 2005). Molecular approaches to the study of tobacco and related *Nicotiana* species, such as EST (http://www.estabacco.info/), Estarray (http://www.estarray.org/) (Katoh *et al.*, 2003), and genome (TGI: http://www.tobaccogenome.org/) (Rushton *et al.*,

2008) sequencing projects, are advancing our knowledge of the genes involved and their regulation. The relevant pathways are depicted in Figures 7.2 and 7.3.

7.2.1 Pyrrolidine Formation

Nicotine is composed of two heterocyclic rings, pyrrolidine and pyridine rings. The pyrrolidine ring is derived from ornithine and possibly from arginine via the symmetric diamine putrescine (Figure 7.2). This branch of the nicotine pathway is shared with tropane alkaloids, including well-known anticholinergic compounds, hyoscyamine and scopolamine, in medicinal plants of *Atropa*, *Hyoscyamus* and *Datura* species, and polyhydroxylated nortropane alkaloids calystegines, which act as selective glucosidase inhibitors and occur mainly in Solanaceae and Convolvulaceae species (Dräger, 2004). Feeding of [2-^{14}C]-ornithine or its precursor [2-^{14}C]-glutamate affords nicotine equally labelled at C-2′ and C-5′ in the *N*-methylpyrrolidine ring (Dewey *et al.*, 1955; Leete and Siegfried, 1957; Lamberts and Byerrum, 1958; Leete and Yu, 1980), and such symmetrical labelling of the ring can be readily accounted for by the involvement of free putrescine (not bound to an enzyme) in the pathway (Figure 7.2).

Putrescine is formed directly from ornithine by decarboxylation catalyzed by ornithine decarboxylase (ODC) (Mizusaki *et al.*, 1973). Additionally, arginine is converted to putrescine, via agmatine and *N*-carbamylputrescine (Yoshida and Mitake, 1966), with argine decarboxylase (ADC) involved in the first conversion. Transgenic research and gene expression analysis support the involvement of ODC in nicotine's formation (see below), while the contribution of the ADC pathway to nicotine synthesis has remained obscure (Burtin and Michael, 1997). The ADC pathway exists in bacteria and plants but not in animals (Kusano *et al.*, 2007), whereas ODC is present in nearly all plant lineages but is absent in *Arabidopsis* (Hanfrey *et al.*, 2001). Putrescine is converted to higher polyamines, such as spermidine and spermine, or conjugated with cinnamate or other derivatives in all higher plants, while it is also converted to *N*-methylputrescine for secondary metabolite formation (Kusano *et al.*, 2007) (Figure 7.2). Both ODC and ADC belong to the group IV of pyridoxal 5′-phosphate-dependent decarboxylases (Sandemeier *et al.*, 1994) and their genes in various plants have been cloned mainly based on homology (Bell and Malmberg, 1990; Michael *et al.*, 1996; Imanishi *et al.*, 1998). Unlike mammalian ODCs, plant ODCs do not possess the PEST sequence thought to be involved in the rapid degradation of the proteins (Murakami *et al.*, 2000). *N. glutinosa* ODC, like some of the other ODCs, was shown to decarboxylate both ornithine and lysine at the same catalytic site but with different optimal pH (Lee and Cho, 2001).

Putrescine is converted to *N*-methylputrescine through *S*-adenosylmethionine-dependent *N*-methylation catalyzed by putrescine *N*-methyltransferase (PMT) at the first committed step in the formation of pyrrolidine (Mizusaki *et al.*, 1971, 1973). Molecular cloning of tobacco *PMT* cDNA by subtraction between cDNA pools of the wild type and a low-nicotine mutant was one of the first examples of the integration of metabolite and gene expression profiles as a strategy to isolate genes involved in plant secondary metabolism (Hibi *et al.*, 1994). Homologous *PMT* cDNAs from *Atropa belladonna*, *Hyoscyamus niger* and other species have also been cloned (Suzuki *et al.*, 1999; Stenzel *et al.*, 2006; Teuber *et al.*, 2007). *PMT* cDNAs encode proteins with distinct sequence similarity to spermidine synthase (SPDS), which transfers the aminopropyl moiety of decarboxylated *S*-adenosylmethionine

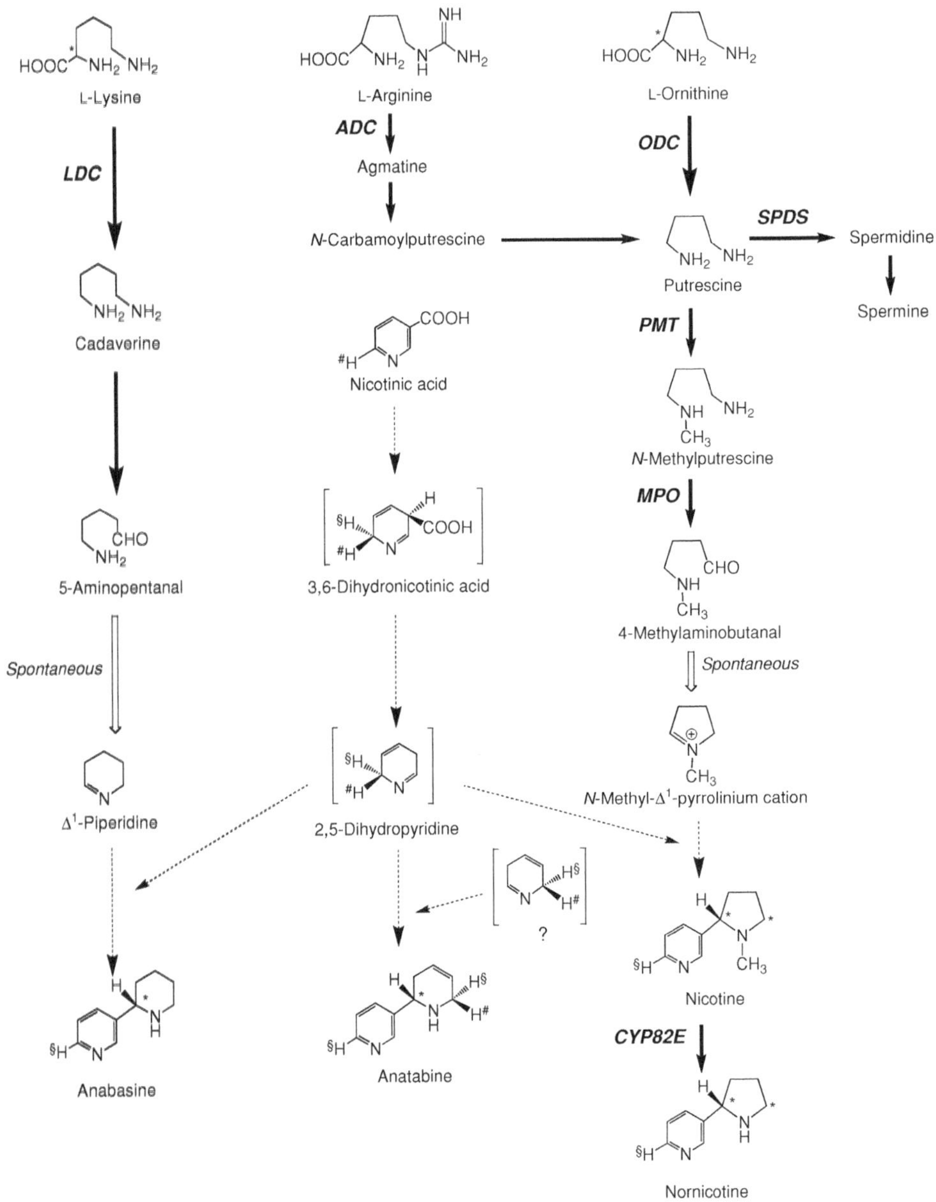

Figure 7.2 *Biosynthetic pathway of nicotine alkaloids. The C-2 carbons of ornithine and lysine, the hydrogen at C-6 of nicotinic acid, and another hydrogen introduced at this position are indicated using symbols to show their fates. Hypothetical intermediates derived from nicotinic acid (an intermediate of the NAD pathway; see Figure 7.3) are included in brackets, and undefined steps including the ring couplings are shown with broken arrows. Enzyme abbreviations are as follows: ODC, ornithine decarboxylase; ADC, arginine decarboxylase; SPDS, spermidine synthase; PMT, putrescine* N*-methyltransferase; MPO,* N*-methylputrescine oxidase; LDC, lysine decarboxylase; DAO, diamine oxidase; CYP82E, cytochrome P450 monooxygenase for nicotine* N*-demethylase (NND) activity*

to putrescine, producing spermidine (Hibi *et al.*, 1994; Hashimoto *et al.*, 1998b). Thus, *PMT* genes presumably evolved from *SPDS* genes during the diversification of alkaloid-forming plants. Protein modelling of PMT based on the crystal structure of SPDS suggests that overall, the protein folds of PMT and SPDS are comparable to each other, although their substrates and co-substrates seem to be positioned differently at active sites (Teuber *et al.*, 2007). PMTs from *Nicotiana* species differ from SPDSs by the addition of hydrophilic *N*-terminal extensions, which include various numbers of direct tandem repeats of 11 amino-acid residues (Hibi *et al.*, 1994; Hashimoto *et al.*, 1998a; Riechers and Timko, 1999). The repeated element is not required for the enzymatic activity (Hashimoto *et al.*, 1998a) and does not resemble any known functional motifs.

N-methylputrescine is oxidatively deaminated by *N*-methylputrescine oxidase (MPO) to 4-methylaminobutanal, which spontaneously cyclises to *N*-methyl-Δ^1-pyrrolinium cation (Mizusaki *et al.*, 1972, 1973). The reactive cation is incorporated into the pyrrolidine moiety of nicotine (Leete, 1967), possibly as an immediate precursor. MPO belongs to a group of diamine oxidases (DAOs) widely distributed in nature. However, in contrast to DAOs from pig and pea which bind *N*-methylputrescine with low affinity, MPOs partially purified from the roots of tobacco and other alkaloid-producing plants showed markedly specific binding to *N*-methylputrescine rather than symmetric diamines, such as putrescine and cadaverine (Hashimoto *et al.*, 1990; Walton and McLauchlan, 1990; Haslam and Young, 1992). Thus, MPOs may have evolved from a DAO through the optimisation of substrate specificity. *MPO* cDNA was eventually isolated through homology-based (Heim *et al.*, 2007) and expression profile-based (Katoh *et al.*, 2007) approaches. The MPO protein, like typical DAOs, contains three histidine residues that coordinate a copper ion and a tyrosine residue which is post-translationally modified to a redox cofactor, topaquione (Heim *et al.*, 2007; Katoh *et al.*, 2007). In tobacco, the association of MPO with *S*-adenosylhomocysteine hydrolase as part of a multi-enzyme complex was suggested based on experiments using MPO antiserum (Heim and Jelesko, 2004), yet direct biochemical evidence is lacking.

7.2.2 Pyridine Formation

The pyridine ring of nicotine is synthesized from nicotinic acid or its derivatives. Nicotinic acid is an intermediate of a pathway synthesizing nicotinamide adenine dinucleotide (NAD), a ubiquitous cofactor used in oxido-reduction and other reactions (Katoh and Hashimoto, 2004) (Figure 7.3). When administered to tobacco tissues, nicotinic acid and quinolinic acid, another intermediate of the NAD pathway, were readily incorporated into the pyridine ring of nicotine (Yang *et al.*, 1965; Frost *et al.*, 1967). The NAD pathway was reviewed in Chapter 5.

In dicotyledonous plants including *Arabidopsis* and tobacco, the NAD pathway starts from aspartate (Figure 7.3). First, aspartate is oxidised by aspartate oxidase (AO) to form α-iminosuccinic acid. Next, α-iminosuccinic acid is condensed by glyceraldehyde-3-phosphate and cyclised by quinolinic acid synthase (QS), yielding quinolinic acid, which has a pyridine ring. The third reaction is the formation of nicotinic acid mononucleotide from quinolinic acid and phosphoriboxyl pyrophosphate by quinolinic acid phosphoribosyl transferase (QPT). Nicotinic acid mononucleotide is converted to NAD and also to nicotinic acid in subsequent cyclic steps (see Chapter 5).

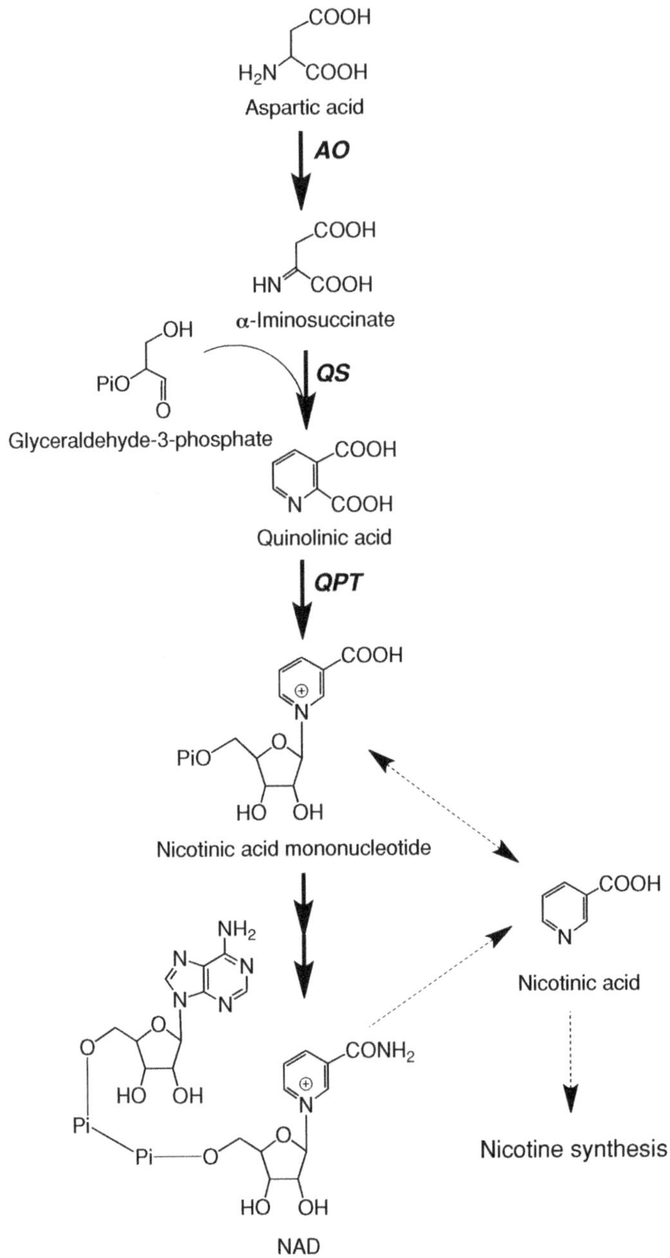

Figure 7.3 *Pathway of NAD synthesis supplying nicotinic acid for nicotine biosynthesis. Abbreviations: AO, aspartate oxidase; QS, quinolinate synthase; QPT, quinolinate phosphoribosyl transferase. Details of how NAD and nicotinic acid form from nicotinic acid mononucleotide through cyclic routes are omitted (see Chapter 5)*

Arabidopsis AO, QS and *QPT* complemented *E. coli* mutants defective in the respective genes, and the knock-out of any one of these genes is embryonic lethal, indicating their essential function in *Arabidopsis* growth (Katoh *et al.*, 2006). Based on the functional expression of GFP-fused proteins and *in vitro* import assays with isolated chloroplasts, the localisation of these three enzymes to plastids has been demonstrated (Katoh *et al.*, 2006), suggesting the compartmentation of the aspartate pathway in plastids.

Tobacco *AO, QS* and *QPT* genes are known to be regulated in accordance with nicotine biosynthesis. Their transcripts are most abundant in nicotine-producing roots but expressed at basal levels in leaves, and also subjected to regulation by *NIC* loci and activation by jasmonate treatment together with other genes involved in nicotine biosynthesis (Sinclair *et al.*, 2000; Goossens *et al.*, 2003; Shoji *et al.*, 2010). Such co-regulation indicates the aspartate pathway to be under the same transcriptional control as nicotine biosynthesis in tobacco.

7.2.3 Coupling of the Pyrrolidine and Pyridine Rings

Nicotinic acid or its derivative is presumably coupled with *N*-methyl-Δ^1-pyrrolinum at the final stage in the formation of nicotine (Figure 7.2). Studies of nicotinic acid labelled with ^{14}C and ^3H at various positions have revealed how the pyridine moiety is incorporated into nicotine. The conversion of nicotinic acid into nicotine does not involve a symmetrical intermediate (Scott and Glynn, 1967; Leete and Liu, 1973); the *N*-methylpyrrolidine moiety is attached to the C-3 position of the pyridine ring, from which the carboxyl group is lost (Scott and Glynn, 1967). Also, loss of a hydrogen at C-6 of nicotinic acid occurs during nicotine's formation (Leete and Liu, 1973), possibly preceded by reduction introducing a hydrogen atom at the C-6 position, which is retained in the final nicotine (Figure 7.2). Indeed, in the case of anatabine, the reduction of nicotinic acid by the introduction of hydrogen at C-6 in the *pro-(R)* position was demonstrated (Leete, 1978). In explanation of these observations, 3,6-dihydronicotinic acid and its decarboxylated form 2,5-dihydropyridine have been postulated as potential intermediates derived from nicotinic acid (Figure 7.2). As for enzymatic activity, 'nicotine synthase' catalyzing condensation in the presence of oxygen was reported (Friesen and Leete, 1990), but the results have not been replicated.

It is intriguing that a novel reductase *A622* gene isolated initially as a gene repressed in the low-nicotine mutant (Hibi *et al.*, 1994) might be required for the coupling step, although the enzymatic reaction has yet to be defined (Kajikawa *et al.*, 2009; DeBoer *et al.*, 2009). *A622* is an orphan member of the PIP family of NADPH-dependent reductases, whose members include pinoresinol-laricresinol reductase, isoflavone reductase, and phenylcoumaran benzylic ether reductase (Gang *et al.*, 1999). Expression profiles of the *A622* transcript and protein are closely associated with other genes involved in nicotine biosynthesis in terms of tissue specificity and jasmonate response, as well as genetic control by *NIC* loci (Shoji *et al.*, 2002). *A622* suppression by RNA interference, presumably inhibiting the coupling step, severely reduces production of all of the tobacco pyridine alkaloids while compensating for the over-accumulation of nicotinic acid *N*-glycoside, a detoxified form of free nicotinic acid and *N*-methyl pyrrolinium cation (Kajikawa *et al.*, 2009; DeBoer *et al.*, 2009). Therefore, *A622* may synthesize a coupling-competent derivative of nicotinic acid or may be directly involved in the coupling reaction itself.

7.2.4 Nornicotine Formation

Accumulation of nicotine depends upon a balance between synthesis and degradation. The half-life of nicotine was estimated to be about 22 hours in tobacco plants (Robinson, 1974). Nicotine, not being an inert end product, is oxidatively *N*-demethylated by nicotine *N*-demethylase (NND) (Hao and Yeoman, 1996a, b, 1998) (Figure 7.2), which belongs to the CYP82E subfamily of cytochrome P450 monooxygenases (Siminszky *et al.*, 2005; Xu *et al.*, 2007). However, the conversion of nornicotine back to nicotine by methylation does not occur (Alworth and Rapoport, 1965). Nornicotine is further converted to myosmine (Leete and Chedekel, 1974). The stepwise degradation of nicotine is also known to occur in bacteria and animals (Hukkanen *et al.*, 2005; Brandsch, 2006).

Nornicotine is typically a minor alkaloid in tobacco, representing about 3–5% of the total alkaloid content. In many tobacco populations, however, a percentage of individuals known as 'converters' can convert as much as 97% of the nicotine to nornicotine during leaf senescence and curing, in contrast to 'non-converters' which accumulate nicotine as the predominant alkaloid. The occasional transition of non-converters to converters, termed 'conversion', is unique to modern cultivated tobacco, particularly burley tobacco where it occurs in up to 20% of the population per generation. The genetic basis of the conversion has been studied for nearly half a century and revealed to be caused by reactivation of a converter locus silenced transcriptionally in non-converters (Griffith *et al.*, 1955; Mann *et al.*, 1964; Wernsman and Matzinger, 1970). The converter locus was eventually found to be one of the tobacco *NND* genes, *CYP82E4*, through a cDNA microarray-based analysis aimed to isolate genes regulated differently between converter and non-converter lines (Siminszky *et al.*, 2005) (Figure 7.4). The *CYP82E4* gene is highly expressed in the senescing leaves of converters and encodes a protein with NND activity when expressed in yeast and tobacco (Siminszky *et al.*, 2005; Xu *et al.*, 2007). The unstable nature of the locus implies the reactivation of *CYP82E4* to be mediated through epigenetic changes, the activation of a transposon, or small-RNA mediated interactions.

A natural allotetraploid species, *N. tabacum*, was relatively recently obtained from the hybridisation of *N. sylvestris* and *N. tomentosiformis* (Clarkson *et al.*, 2005). Interestingly, the alkaloid profile of *N. tabacum*, accumulating more nicotine than nornicotine (in non-converters), is different from that of either of its two progenitors; nornicotine accumulates in senescing leaves of *N. sylvestris* and in both green and senescing leaves of *N. tomentosiformis* (Figure 7.4). To explain such variation within a closely related lineage, *CYP82E* subfamily genes from tobacco and the two parental *Nicotiana* species were isolated and functionally characterised (Siminszky *et al.*, 2005; Chakrabarti *et al.*, 2007; Gavilano *et al.*, 2007) (Figure 7.4). In tobacco, functional NNDs are encoded by *CYP82E4* (Siminszky *et al.*, 2005; Chakrabarti *et al.*, 2008) and *E5* (Gavilano and Siminszky, 2007), which are differentially expressed, and *CYP82E4* is silenced at least in non-converters. Additionally, *CYP82E2* and *E3* are present in the tobacco genome, but they encode inactive enzymes and are transcriptionally silenced (Siminszky *et al.*, 2005; Gavilano *et al.*, 2007). The tobacco *CYP82E3* and *E4* genes, both derived from *N. tomentosiformis*, are genetically linked (C_T locus) (Gavilano *et al.*, 2007), while *CYP82E2* from *N. sylvestris* resides at a separate locus (C_S locus) (Chakrabarti *et al.*, 2007). The genetic origin and other details of *CYP82E5* have yet to be determined. In contrast to the inactive and/or silenced tobacco genes, all ancestral orthologues of *CYP82E2*, *E3* and *E4* encode active NNDs and are expressed respectively

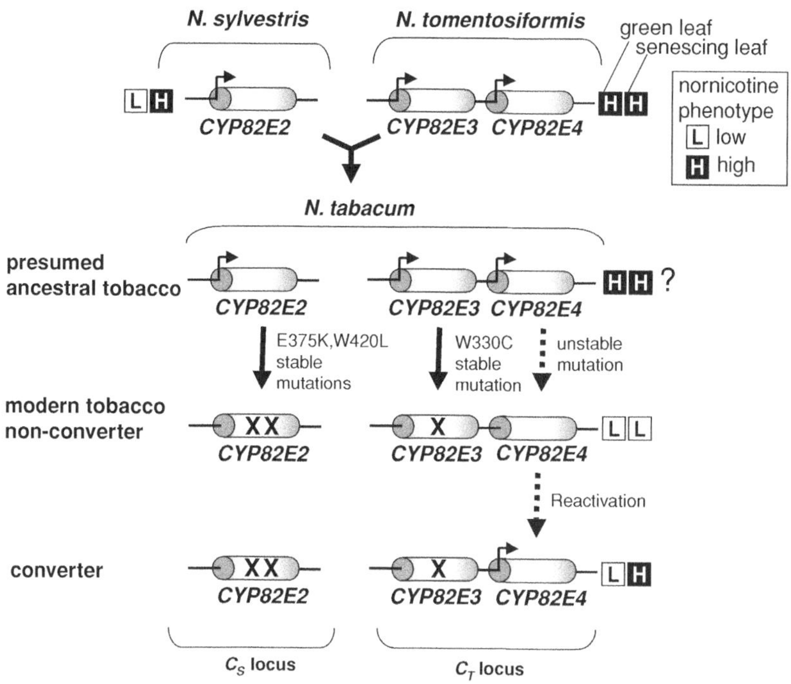

Figure 7.4 *Evolution of* CYP82E *encoding nicotine* N*-demethylase (NND) in* N. tabacum *and its progenitors* N. tomentosiformis *and* N. sylvestris. *Nornicotine phenotypes in green and senescing leaves are shown. Two converter loci,* C_T *(*CYP82E3 *and* CYP82E4*) and* C_S *(*CYP82E2*), are derived from* N. tomentosiformis *and* N. sylvestris*, respectively. Because its genetic origin and other details have not been defined, the tobacco* CYP82E5 *gene is not included in this figure*

in *N. sylvestris* senescing leaves, *N. tomentosiformis* green leaves, and *N. tomentosiformis* senescing leaves, consistent with nornicotine's accumulation in these tissues (Gavilano *et al.*, 2007; Chakrabarti *et al.*, 2007). Comparing sequences among these *CYP82E* genes, stable mutations causing substitutions of functionally important residues were found only in the tobacco *CYP82E2* and *E3* genes encoding inactive NNDs. Accordingly, we can assume that the ancestral *CYP82E* genes, which confer the ability to accumulate nornicotine, gain stable mutations in *CYP82E2* and *E3* and unstable mutation in *CYP82E4*, which collectively cause the inactivation of NNDs after the hybridisation of the progenitors to produce the nicotine-dominating profile in modern tobacco, as summarised in Figure 7.4.

7.2.5 Anabasine and Anatabine Formation

Anabasine, a major alkaloid in *N. glauca*, is composed of pyridine and piperidine rings. Like in nicotine and anatabine, nicotianic acid is incorporated into the pyridine ring in anabasine with elimination of a hydrogen at C-6 and loss of a carboxyl group (Figure 7.2). On the other hand, the piperidine ring of anabasine, corresponding to the pyrrolidine ring of nicotine, is synthesized from lysine through decarboxylation by lysine decarboxylase (LDC) to produce cadaverine, and then oxidation by DAO to produce 5-aminopentanal followed by spontaneous intramolecular cyclisation resulting in Δ^1-peperidine (Figure 7.2). A putative *LDC* gene of tobacco, whose protein has a domain conserved in other LDCs but has not been proved to exhibit LDC activity, was reported as one of the genes repressed in a low-nicotine mutant (Häkkinen *et al.*, 2007). *N. glauca* fed with [2-^{14}C] lysine gave anabasine labelled only at C′-2 (Leete, 1956), indicating an asymmetrical incorporation of lysine in contrast to the mode of incorporation of ornithine into nicotine (Figure 7.2).

The formation of anatabine is quite different from that of anabasine, which differs from anatabine only by two hydrogens (Figure 7.1). The piperidine ring of anatabine is not derived from lysine. The administration of [2-^{14}C]- and [6-^{14}C]-nicotinic acids to *N. tabacum and N. glauca* afforded anatabine labelled equally at C-2/C′-2 and C-6/C′-6, respectively (Leete and Slattery, 1976) (Figure 7.2). Such a pattern of incorporation indicates that both rings of anatabine originate from nicotinic acids. During the coupling of two molecules of nicotinic acid, one hydrogen atom at the C-6 position and both carboxyl groups are eliminated (Figure 7.2), analogous to the coupling for nicotine synthesis.

The ring coupling step of anabasine and anatabine, like that of nicotine, has remained elusive. Given the similar patterns of incorporation of nicotinic acid into all the pyridine alkaloids, the underlying reaction mechanisms seem to be nearly identical and also to be mediated by overlapping or shared enzymatic systems. Since the production of all the pyridine alkaloids was inhibited by A622 suppression (DeBoer *et al.*, 2009; Kajikawa *et al.*, 2009), A622 could be the enzyme or one of the enzymes acting in the coupling step(s).

7.3 Compartmentation and Trafficking

In multicellular plants, to ensure the proper functioning of secondary metabolites, their biosynthesis and accumulation are organised spatially and temporally and so are usually highly associated with differentiation. Indeed, undifferentiated cell cultures synthesize only limited amounts of secondary products without appropriate elicitation in many cases. In intact plants with a variety of differentiated cells, however, each step from synthesis to storage is executed in particular organelles, cells or tissues, and metabolites or intermediates are transported among multiple compartments usually by passing across membrane systems (Kutchan, 2005). To accomplish metabolic engineering, an understanding of such compartmentation and trafficking is required.

7.3.1 Long-Distance Transport from Roots to Leaves

One example of the long-distance transport of secondary metabolites is that of nicotine from roots to shoots, which was clearly shown by grafting between nicotine-producing tobacco and non-producing tomato; nicotine accumulated in tomato scions grafted on tobacco stocks (Dawson, 1942). Nicotine is produced exclusively in roots, loaded into the xylem, a vasculature connecting roots and shoots, and easily moved upwards along with the transpiration stream to leaves, where it plays defensive roles mainly against insect herbivores (Figure 7.5). Actually, nicotine can be detected in tobacco xylem sap at relatively high concentrations (Baldwin, 1989). To allow the xylem loading and unloading, nicotine

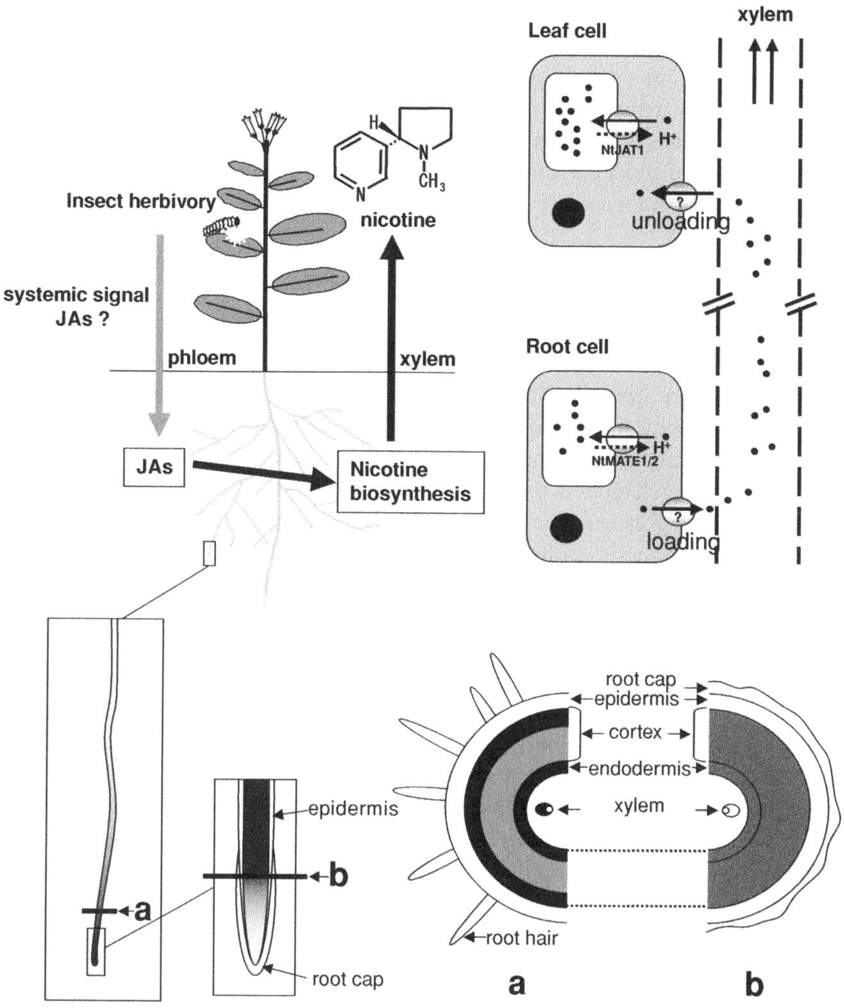

Figure 7.5 Sites of nicotine biosynthesis and trafficking in tobacco plants. Nicotine is produced in specific root cells, which are predicted based on expression patterns of PMT and A622. Nicotine transport via the xylem from root to leaf and putative membrane carriers, NtJAT1, NtMATE1/2 and unknown proteins, are presented schematically. NtJAT1 and NtMATE1/2 are MATE-type proton antiporters involved in the vacuolar sequestration of nicotine in leaves and roots, respectively. Insect herbivory in leaves induces nicotine production in roots by transmitting the JA signal systemically through phloem

efflux in root cells and influx in leaf cells should be carried out at plasma membranes (Figure 7.5). But until now, no nicotine transporters involved in these processes have been reported.

Nicotine is not the only leaf alkaloid derived from roots; tropane alkaloids in Solanaceae and pyrrolizine alkaloids in Asteraceae are also examples (Erb *et al.*, 2009). It remains unclear why plants synthesize such leaf-stored chemicals in roots. One possible explanation is the availability of nitrogen-containing precursors in roots, which are a source organ for nitrogen mainly absorbed from soil. Moreover, production underground protected from damage above ground may be of value when large portions of shoots are lost to herbivores. In contrast to leaves, which are not connected directly by vascular systems, roots can easily deliver defensive compounds through the xylem in a systemic manner, in response to signals delivered from aerial parts via the phloem, as exemplified in the jasmonate-induced formation of nicotine in tobacco (Figure 7.5).

7.3.2 Cell-Specific Nicotine Biosynthesis in Roots

Root-specific nicotine production has been proved by root cultures producing nicotine at high levels, as well as the localisation of the enzymatic activities. Moreover, the molecular localisation of relevant genes and proteins provides more precise information on the cell types involved in nicotine biosynthesis. PMT and A622 have been thoroughly characterised with immuno-localisation of the proteins and with promoter analyses, and shown to be expressed in the same cell types in the roots (Shoji *et al.*, 2000b, 2002) (Figure 7.5). Growing root tips often contain greater concentrations of nicotine than leaf lamina and are considered the main site of biosynthesis. Consistent with this, PMT and A622 accumulated at higher levels in the apical parts of the roots. In the root apex just behind the meristematic region, both proteins were abundant throughout the endodermis and cortex. In the differentiated region of the roots where root hairs were present, the cortex, endodermis and xylem were associated with the promoter activities of *PMT* and *A622*, and activities were strongest in the outermost cortex layer, the endodermis layer, and parenchyma cells around xylem vessels. The gene expression in the xylem tissues may facilitate nicotine loading into the xylem stream. If nicotine is produced in the cortex, the alkaloid needs to move symplastically to the stele since the Casparian strip of the endodermis will block apoplastic transport (Clarkson, 1996). It is also possible that part of the nicotine formed in the cortex moves out towards the epidermis and then out of the root. A significant proportion of the nicotine produced in *Nicotiana* root cultures is known to be secreted into the culture medium (Hamill *et al.*, 1986).

In the biosynthesis of other alkaloids, the enzyme subsets were shown to be present in different, non-overlapping cells and organelles, implying the intercellular and intracellular translocation of intermediates (Facchini and St-Pierre, 2005; Ziegler and Facchini, 2008). The subcellular distribution of all enzymes in the nicotine pathway remains to be defined, although *Arabipidosis* AO, QS and QPT are known to be located in the plastid (Katoh *et al.*, 2005). Understanding the spatial expression and distribution of the enzymes in the pathway will help to determine the biosynthetic sites.

7.3.3 Nicotine Transporters Involved in Vacuolar Sequestration

To prevent cytotoxic effects at relatively high concentrations, alkaloids, like other secondary metabolites, are usually sequestered into the vacuole, a plant cell organelle typically used for bulk storage of chemicals. Nicotine, a weakly basic alkaloid, is not charged and lipophilic under slightly alkaline conditions, and can partially pass through the tonoplast by simple diffusion. But, once in the acidic vacuole, nicotine can be protonated and trapped as a membrane-impermeable hydrophilic molecule. Such a scheme, termed the ion-trap mechanism, has been postulated for weak basics including nicotine. In addition to the ion-trap mechanism, active transport mediated by membrane-localised carriers also plays a major part in the vacuolar sequestration. Many transporters classified as ATP-binding cassette (ABC) transporters and pH-dependent proton antiporters are known to be located in the tonoplast (Yazaki, 2005).

Multidrug And Toxic compound Extrusion (MATE)-type transporters, Jasmonate-inducible Alkaloid Transporter 1 (NtJAT1), and a pair of homologous proteins NtMATE1 and NtMATE2, have been identified as tonoplast-localised nicotine transporters in tobacco (Morita *et al.*, 2009; Shoji *et al.*, 2009) (Figure 7.5). The MATE transporters are one of five families composing the multidrug transporter superfamily and have been shown to efflux low-molecular weight compounds ranging from organic cations to metal ions as drug/H^+ or drug/Na^+ antiport systems (Omote *et al.*, 2006). Plants in particular are abundant in MATE transporters, possibly reflecting their vast variety of secondary metabolites.

NtJAT1 was identified through cDNA-amplified fragment length polymorphism-based profiling as a gene co-induced by methyl jasmonate in tobacco BY2 cultures with other genes involved in nicotine biosynthesis (Goossens *et al.*, 2003). NtJAT1 shows relatively high similarity to *Arabidopsis* DTX1 mediating the efflux of xenobiotics at plasma membranes (Li *et al.*, 2002). NtJAT1 was expressed in leaves, stems and roots, and localised to the tonoplast in leaves (Morita *et al.*, 2009) (Figure 7.5). NtJAT1 showed nicotine transport activity when expressed in yeast, and further biochemical analysis using proteoliposomes reconstituted with NtJAT1 and pH-generating bacterial F_0F_1-ATPase demonstrated that NtJAT1 functioned as a H^+-antiporter, transporting nicotine and other alkaloids but not flavonoids (Morita *et al.*, 2009).

NtMATE1 and NtMATE2 (Shoji *et al.*, 2009) are phylogenically related to *Arabidopsis* TT12 (Marinova *et al.*, 2007) and tomato MTP77 (Mathews *et al.*, 2007), both of which have been implicated in the vacuolar sequestration of flavonoids. In contrast to *NtJAT1*, which is expressed in nearly every organ, *NtMATE* genes are specifically expressed in nicotine-producing root cells and regulated by *NIC* genes and jasmonate in concert with structural genes (Shoji *et al.*, 2009). The localisation of NtMATEs to tonoplasts was shown by subcellular fractionation, immunoelectron micrography, and GFP-tagged protein expression. Downregulation of *NtMATEs* rendered tobacco roots more sensitive to an exogenous application of nicotine, implying NtMATE's involvement in the movement of nicotine *in planta*. In contrast, *NtMATE1* overexpression in cultured tobacco cells induced cytoplasmic acidification after jasmonate elicitation, which might be followed by *de novo* nicotine production, or the addition of nicotine. When expressed in yeast cells, NtMATE1 reduced the amount of nicotine incorporated into the cells from the medium, where it was added, indicating its nicotine transport activity. NtMATEs might be involved in the vacuolar sequestration of nicotine in nicotine-synthesizing root cells (Shoji *et al.*, 2009) (Figure 7.5).

7.4 Gene Regulation

7.4.1 Jasmonate

Nicotine acts as a defensive toxin largely against insect herbivores, and its chemical ecology has been studied intensively using transgenic plants with altered nicotine contents (Baldwin, 1998; Steppuhn *et al.*, 2004). Nicotine functions as a nectar repellent (Kessler and Baldwin,

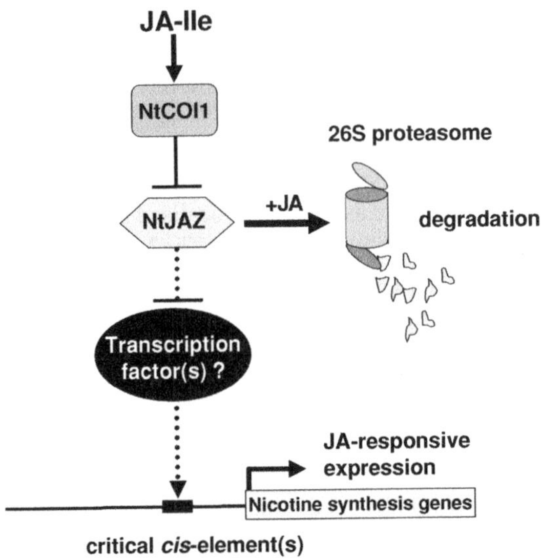

Figure 7.6 *Model of jasmonate (JA) signalling and transcriptional regulation in nicotine biosynthesis. The binding of JA-Ile to the NtCOI1 receptor induces the recruitment of NtJAZs, followed by ubiquitination of the NtJAZs by a SCF complex containing NtCOI1 and degradation by the 26S proteasome. The removal of NtJAZs in response to JAs allows downstream transcription factor(s) to activate the genes for nicotine biosynthesis directly by binding to the corresponding promoters or indirectly through other factors (not shown)*

2006; Kessler *et al.*, 2008). In addition to basal production under normal conditions, nicotine is readily produced in response to attacks from insects (Baldwin, 1989). Leaf damage caused by insect herbivory activates the wound response pathway mainly mediated by jasmonates (JAs) (Baldwin *et al.*, 1994). Jasmonic acid (JA) and its derivatives, collectively referred to as JAs, play signalling roles widely associated with developmental regulation and defensive responses to stress including herbivory and wounding (Wasternack, 2007; Browse, 2009). JAs can elicit production of a wide variety of secondary metabolites involved in plant defence (Gundlach *et al.*, 1992). The damage-induced formation of JAs in leaves and their translocation to roots via the phloem correlated with nicotine production in *Nicotiana* plants (Baldwin *et al.*, 1994; Zhang and Baldwin, 1997) (Figure 7.5), and thus JAs themselves may be the chemical signal transmitted systemically, although other possibilities cannot be excluded. In *Nicotiana* roots and cultured cells, JAs coordinate to activate most of the known genes for nicotine biosynthesis and transport: *ODC, PMT, MPO, AO, QS, QPT, A622, NtMATE1/2* and *NtJAT1* (Imanishi *et al.*, 1998; Shoji *et al.*, 2000b; Goossens *et al.*, 2003).

JAs are synthesized from linolenic acid via the octadecanoid pathway, in which 13-lipoxygenase (LOX) catalyzes the oxygenation of linolenic acid at the first step (Wasternack, 2007). The application of pharmacological inhibitors of the octadecanoid pathway (Baldwin *et al.*, 1996) as well as transgenic silencing of *LOX* expression (Halitschke and Baldwin, 2003), both suppressing JA production, can effectively reduce damage-induced production of nicotine. JA-isoleucine conjugate (JA-Ile) is the bioactive form of JA and its formation is catalyzed by a JA-Ile-conjugating enzyme encoded by *Jasmonate Resistant 1 (JAR1)* in *Arabidopsis* (Staswick and Tiryaki, 2004) and its orthologues *JAR4* and *JAR6* in *N. attenuate* (Wang *et al.*, 2008). Transgenic suppression of the conjugation step in *N. attenuate* by simultaneous silencing of *JAR4* and *JAR6* significantly reduced JA-induced nicotine accumulation, and the reduction was complemented by application of JA-Ile (Wang *et al.*, 2008), supporting an essential role for JA-Ile in nicotine's regulation.

In *Arabidopsis*, the molecular framework of JA signalling from perception to transcriptional activation has been elucidated mainly through molecular genetic approaches (Browse, 2009; Chung *et al.*, 2009). Reception of the bioactive JA-Ile by the JA receptor Coronatine Insenstive1 (COI1), a F-box component of a SCF-type E3-ubiqutin ligase complex (SCF^{COI1}), can induce the recruitment of jasmonate ZIM-domain (JAZ) proteins to the SCF^{COI1} complex through COI1-JAZ interaction, and the JAZ proteins ubiquitinated by the ligase activity are removed by 26S proteasome-mediated degradation. The JAZs with signature Jas and TIFY motifs are encoded by 12 distinct genes overlapping functionally in *Arabidopsis* and able to form homo- and heterodimers depending on interaction through the TIFY motifs. The JAZs act as transcriptional repressors by directly binding to transcriptional activators, such as the bHLH family's AtMYC2, and so the COI1-dependent removal of the JAZ repressors in response to the JA signal allows the release of transcription factors from inhibition by the JAZs and activation of their target genes. In *Nicotiana* plants, suppression of *NaCOI1* from *N. attenuate* (Paschold *et al.*, 2007) and *NtCOI1* from tobacco (Shoji *et al.*, 2008) effectively impaired the JA- and wound-induced formation of nicotine as well as other typical JA responses. Tobacco *JAZ* genes, *NtJAZ1, 2* and *3*, have been isolated based on the presence of the Jas and TIFY motifs (Shoji *et al.*, 2008). Like the *Arabidopsis* counterparts, *NtJAZs* were induced by JA, while their proteins were degraded in response to JA possibly by the ubiquitin-dependent proteasome system (Shoji *et al.*, 2008). The expression of NtJAZΔJas, truncated forms of NtJAZs without the C-terminal Jas motifs important functionally (involved in COI1-JAZ and JAZ-AtMYC2 interactions in *Arabidopsis*) but not for the dimerisation, which act in a dominant-negative manner probably by forming non-functional JAZ dimers, also clearly inhibited JA's induction of nicotine biosynthesis (Shoji *et al.*, 2008). Taken together, NtCOI1 and NtJAZs are involved in the regulation of nicotine production by JA (Figure 7.6). It remains to be revealed whether the tobacco orthologue of AtMYC2, NtMYC2, or other transcription factors act downstream of the NtJAZ repressors.

The genes for nicotine biosynthesis are transcriptionally activated by JA in *Nicotiana* plants. Promoter regions about 250 bp in length and highly conserved in multiple members of the *PMT* gene family in *N. tabacum* and *N. sylvestris* conferred JA-inducible gene expression to a *GUS* reporter gene (Shoji *et al.*, 2000b; Xu and Timko, 2004). These regions commonly contain G-box like and GCC-box like motifs, both of which are necessary for the responsiveness to JA (Xu and Timko, 2004; Oki and Hashimoto, 2004) (Figure 7.6). Furthermore, two AP2/ERF family transcription factors NtORC1 and NtJAP1 were found to positively regulate the JA-responsive *PMT* promoter region in transient assays in tobacco protoplasts (De Sutter *et al.*, 2005). Interestingly, these factors are closely related to ORCA3 from *Catharanthus roseus*, known to function in the JA-dependent activation of terpenoid indole alkaloid biosynthesis (van der Fits and Memelink, 2000). These transcription factors might function downstream of NtCOI1/NtJAZ to control the expression of genes for nicotine synthesis (Figure 7.6).

7.4.2 Ethylene

As anticipated from their overlapping roles in wound response and disease resistance, ethylene and JA interact synergistically or antagonistically in various signalling contexts. In nicotine biosynthesis, activation of a number of structural genes, such as *ODC*, *PMT* and *A622*, by JA is effectively suppressed by simultaneous treatment with ethylene or its natural precursor, and the suppressive effect is abrogated when ethylene perception is blocked with specific inhibitors (Shoji *et al.*, 2000a; Winz and Baldwin, 2001). A nicotine-tolerant herbivore *Manduca sexta* can induce production of ethylene preventing an increase in nicotine (Kahl *et al.*, 2000; Winz and Baldwin, 2001, von Dahl *et al.*, 2007). The altered response to herbivores by antagonistic interaction between ethylene and JA may ensure the re-allocation of resources from an ineffective nicotine-based defence.

7.4.3 Auxin

Auxin, usually used in plant cell cultures at relatively high concentrations, is known to have a negative effect on alkaloid production, although whether it downregulates the relevant gene expression directly, or indirectly by changing the state of differentiation of cells, is largely unclear. Low nicotine biosynthesis in cell and tissue cultures can be restored by depleting auxin in the culture medium, accompanied by the concomitant activation of *ODC*, *PMT* and *A622* (Hibi *et al.*, 1994; Imanishi *et al.*, 1998). To increase leaf mass as well as nicotine levels, removal of tobacco shoot apexes, termed topping or decapitation, is a common practice performed just before harvest in tobacco cultivation. Topping may reduce the supply of auxin from young apices and so remove apical dominance for leaf growth and enhance nicotine production in roots probably through gene activation by cancelling the auxin-mediated downregulation.

7.4.4 NIC Regulatory Genes

Genetic studies are valuable to elucidate the biosynthetic pathways and regulation of secondary metabolism. In the early 1930s, a Cuban cigar variety was found to have very low alkaloid levels. To meet demand for low-nicotine cigarettes, the alkaloid trait was introduced into cigarette varieties through repeated backcrosses, resulting in the development of stable lines such as LA Burley21 and LAFC53 with low alkaloid levels. These breeding lines show normal growth and development comparable to parental varieties, but are more susceptible to insect attack presumably due to their low alkaloid levels (Legg *et al.*, 1970). Genetic studies have revealed that semi-dominant mutant alleles at two unlinked genetic loci, designated *NIC1* and *NIC2* (originally called *A* and *B*), act synergistically to confer the low-alkaloid trait (Legg *et al.*, 1969; Legg and Collins, 1971). The effect of the *nic1* mutant allele on leaf nicotine levels is 2.4 times stronger than that of *nic2* (Legg and Collins, 1971) Reflecting the different dose effects of *nic1* and *nic2* and their synergistic effects, the alkaloid content decreases in tobacco lines with different *NIC* genotypes in the order wild-type > *NIC1nic2* > *nic1NIC2* > *nic1nic2*.

Feeding and enzymatic studies indicated the repression of multiple reactions in the nicotine pathway in the *nic* mutants (Saunders and Bush, 1979), implying a regulatory role of the *NIC* loci. Subsequently, a number of the structural genes encoding biosynthetic enzymes, *ODC*, *PMT*, *MPO*, *AO*, *QS*, *QPT* and *A622*, as well as transporter genes *NtMATE1* and *NtMATE2*, were found to be downregulated in proportion to the decrease in nicotine levels in the mutants (Hibi *et al.*, 1994; Kidd *et al.*, 2006; Katoh *et al.*, 2007; Shoji *et al.*, 2009, 2010). *NIC*-regulated genes have been identified using cDNA subtraction (Hibi *et al.*, 1994), differential display (Kidd *et al.*, 2006; Shoji *et al.*, 2009), and cDNA microarrays (Shoji *et al.*, 2010), by comparing steady-state mRNA levels between wild-type and *nic1nic2* mutant roots. Interestingly, *NIC*-regulated genes were responsive to jasmonate in wild-type plants but their responses were diminished or abolished in the *nic1nic2* mutant (Shoji *et al.*, 2002, 2010). Since the expression of other jasmonate-inducible genes not involved in nicotine synthesis was not affected in the *nic* genotypes, *NIC* genes may be one of the many downstream components of jasmonate signalling only regulating a set of genes specific to nicotine biosynthesis. The molecular identification of the *NIC* genes is awaited.

7.5 Metabolic Engineering

Several structural genes of the nicotine pathway are currently available for genetic engineering through overexpression with strong constitutive promoters, such as the CaMV 35S promoter, and through downregulation via antisense, cosuppression, or RNA interference (Sato *et al.*, 2007). Overexpression of either *ODC* (Hamill *et al.*, 1990), *ADC* (Burtin and Michael, 1997) or *PMT* (Sato *et al.*, 2001) has been shown to increase nicotine levels only moderately, while the direct products of these reactions increased considerably. The accumulation of intermediates rather than the final metabolite indicates the presence of multiple rate-limiting steps. To markedly increase nicotine accumulation, all major bottleneck reactions should be overcome concomitantly by the simultaneous expression of multiple structural genes. Downregulation of pyrrolidine-branch enzymes, such as *PMT* (Sato *et al.*, 2001; Chintapakorn and Hamill, 2003) and *MPO* (Shoji and Hashimoto, 2008), effectively decreases nicotine accumulation with concomitant increases in polyamines and other tobacco alkaloids that do not have a pyrrolidine ring, such as anatabine and anabasine. The apparent coordination of the pyrrolidine- and pyridine-branch pathways is required for efficient synthesis of nicotine. In contrast, downregulation of *A622* expression decreases levels of not only nicotine and nornicotine, but also anatabine and anabasine, probably by blocking a step shared by all tobacco alkaloids (Kajikawa *et al.*, 2009; DeBoer *et al.*, 2009).

Rate-limiting steps vary among tissues or conditions. Anatabine rather than nicotine was the predominant alkaloid in tobacco BY-2 cultured cells elicited by JAs (Goossens *et al.*, 2003; Shoji and Hashimoto, 2008). This profile is caused by a very low level of *MPO* expression. Accordingly, the overexpression of *MPO* restored the nicotine-accumulating profile (Shoji and Hashimoto, 2008).

Nornicotine production is inhibited by the downregulation of *CYP82E* encoding nornicotine *N*-demethylase (Siminszky *et al.*, 2005; Gavilano *et al.*, 2006; Lewis *et al.*, 2008). Since nornicotine may be converted to a more harmful carcinogen, tobacco-specific nitrosamine *N′*-nitrosonornicotine, during the curing and processing of harvested tobacco leaves (Bush *et al.*, 2001), nornicotine reduction may be a promising target of metabolic engineering in tobacco.

7.6 Recent Developments

After this manuscript was submitted, Shoji *et al.* (2010) reported that a group of highly homologous AP2/ERF family transcription factors including *NtORC1* clusters at the *NIC2* locus, and at least seven such *ERF* genes, are deleted in the *nic2* mutant genome. The *NIC2*-locus ERF proteins recognised a GCC-box element in the *PMT* promoter and specifically activated all known structural genes in the nicotine pathway.

7.7 Summary

In *Nicotiana* species, nicotine and related alkaloids are synthesized in the roots and then translocated via the xylem mainly to leaves, where they are sequestered into central vacuoles and play defensive roles against insect pests. A number of structural genes encoding enzymes of the nicotine pathway have been identified, mainly based on their homology or expression profiles, and shown to be expressed in certain types of root cells. Still missing are the elusive enzyme(s) involved in the final ring coupling reaction. Genes involved in the synthesis and transport of nicotine are coordinately activated by JA or a wounding signal mediated by NtCOI1 and NtJAZs. Cross-talk among JA, ethylene, and auxin signals may fine-tune nicotine's accumulation in tobacco plants attacked by herbivores. *NIC1* and *NIC2* may act at the end of the JA-signalling cascade, specifically regulating multiple structural genes for nicotine biosynthesis and transport.

References

Alworth, W.L. and Rapoport, H. (1965) Biosynthesis of the nicotine alkaloid in *Nicotiana glutinosa*: interrelationships among nicotine, nornicotine, anabasine, and anatabine. *Arch. Biochem. Biophys.*, **112**, 45–53.

Baldwin, I.T. (1989) Mechanism of damage-induced alkaloid production in wild tobacco. *J. Chem. Ecol.*, **15**, 1661–1680.

Baldwin, I.T. (1998) Jasmonate-induced responses are costly but benefit plants under attack in native populations. *Proc. Natl. Acad. Sci., USA*, **95**, 8113–8118.

Baldwin, I.T., Schmelz, E.A. and Ohnmeiss, T.E. (1994) Wound-induced changes in root and shoot jasmonic acid pools correlated with induced nicotine synthesis in *Nicotiana sylvestris* Spegazzini and Comes. *J. Chem. Ecol.*, **20**, 2139–2157.

Baldwin, I.T., Schmelz, E.A. and Zhang, Z.-P. (1996) Effects of octadecanoid metabolites and inhibitors on induced nicotine accumulation in *Nicotiana sylvestris*. *J. Chem. Ecol.*, **22**, 61–74.

Bell, E. and Malmberg, R.L. (1990) Analysis of a cDNA encoding arginine decarboxylase from oat reveals similarity to the *Escherichia coli* arginine decarboxylase and evidence of protein processing. *Mol. Gen. Genet.*, **224**, 431–436.

Brandsch, R. (2006) Microbiology and biochemistry of nicotine degradation. *Appl. Microbiol. Biotechnol.*, **69**, 493–498.

Browse, J. (2009) Jasmonate passes muster: a receptor and targets for the defense hormone. *Ann. Rev. Plant Biol.*, **60**, 183–205.

Burtin, D. and Michael, A.J. (1997) Over-expression of arginine decarboxylase in transgenic plants. *Biochem. J.*, **325**, 331–337.

Bush, L.P. (1999) Alkaloid biosynthesis, in *Tobacco: Production, Chemistry and Technology* (eds D.L. Davis and M.T. Nielsen), Blackwell Science, Oxford, pp. 285–291.

Bush, L., Hempfling, W.P. and Burton, H. (1999) Biosynthesis of nicotine and related compounds, in *Analytical Determination of Nicotine and Related Compounds and Their Metabolites* (eds J.W. Gorrod and P. Jacob), Elsevier Science, Amsterdam, pp. 13–44.

Bush, L.P., Cui, M., Shi, H. and Burton, H.R. (2001) Formation of tobacco specific nitrosamines in air-cured tobacco. *Rec. Adv. Tob. Sci.*, **27**, 23–46.

Chakrabarti, M., Meekins, K.M., Gavilano, L.B. and Siminszky, B. (2007) Inactivation of the cytochrome P450 gene *CYP82E2* by degenerative mutations was a key event in the evolution of the alkaloid profile of modern tobacco. *New Phytol.*, **175**, 565–574.

Chakrabarti, M., Bowen, S.W., Coleman, N.P., Meekins, K.M., Dewey, R.E. and Siminszky, B. (2008) *CYP82E4*-mediated nicotine to nornicotine conversion in tobacco is regulated by a senescence-specific signaling pathway. *Plant Mol. Biol.*, **66**, 415–427.

Chintapakorn, Y. and Hamill, J.D. (2003) Antisense-mediated down regulation of putrescine *N*-methyltransferase activity in transgenic *Nicotiana tabacum* L. can lead to elevated levels of anatabine at the expense of nicotine. *Plant Mol. Biol.*, **53**, 87–105.

Chung, H.S., Niu, Y., Browse, J. and Howe, G.A. (2009) Top hits in contemporary JAZ: an update on jasmonate signaling. *Phytochemistry*, **70**, 1547–1559.

Clarkson, D.T. (1996) Root structure and sites of ion uptake, in *Plant Roots: The Hidden Half* (eds Y. Waisei, A. Eshel and U. Kafkafi), Marcel Dekker Inc., New York, pp. 483–510.

Clarkson, J.J., Lim, K.Y., Kovarik, A., Chase, M.W., Knapp, S. and Leitch, A.R. (2005) Long-term genome diploidization in allopolyploid *Nicotiana* section Repandae (Solanaceae). *New Phytol.*, **168**, 241–252.

Davis, D.L. and Nielsen, M.T. (eds) (1999) *Tobacco: Production, Chemistry and Technology*, Blackwell Science, Oxford.

Dawson, R.F. (1942) Accumulation of nicotine in reciprocal grafts of tomato and tobacco. *Am. J. Bot.*, **29**, 66–71.

DeBoer, K.D., Lye, J.C., Aitken, C.D., Su, A.K. and Hamill, J.D. (2009) The *A622* gene in *Nicotiana glauca* (tree tobacco): evidence for a functional role in pyridine alkaloid synthesis. *Plant Mol. Bol.*, **69**, 299–312.

DeLuca, V. and Laflamme, P. (2001) The expanding universe of alkaloid biosynthesis. *Curr. Opin. Plant Biol.*, **4**, 225–233.

De Sutter, V., Vanderhaeghen, R., Tilleman, S., Lammertyn, F., Vanhoutte, I., Karimi, M., Inze, D., Goossens, A. and Hilson, P. (2005) Exploration of jasmonate signaling via automated and standardized transient expression assays in tobacco cells. *Plant J.*, **44**, 1065–1076.

Dewey, L.J., Byerrum, R.U. and Ball, C.D. (1955) Biosynthesis of the pyrrolidine ring of nicotine. *Biochim. Biophys. Acta*, **18**, 141–142.

Dome, P., Lazary, J., Kalapos, M.P. and Rihmer, Z. (2009) Smoking, nicotine, and neuropsychiatric disorders. *Neurosci. Biobehav. Rev.*, **34**, 295–342.

Dräger, B. (2004) Chemistry and biology of calystegines. *Nat. Prod. Rep.*, **21**, 211–223.

Erb, M., Lenk, C., Degenhardt, J. and Turlings, T.C.J. (2009) The underestimated role of roots in defense against leaf attackers. *Trends Plant Sci.*, **14**, 653–659.

Facchini, P.J. (2001) Alkaloid biosynthesis in plants: biochemistry, cell biology, molecular regulation, and metabolic engineering applications. *Ann. Rev. Plant Physiol. Plant Mol. Biol.*, **52**, 29–66.

Facchini, P.J. and St-Pierre, B. (2005) Synthesis and trafficking of alkaloid biosynthetic enzymes. *Curr. Opin. Plant Biol.*, **8**, 657–666.

Friesen, J.B. and Leete, E. (1990) Nicotine synthase- an enzyme from *Nicotiana* species which catalizes the formation of (*S*)-nicotine from nicotinic acid and 1-methyl-Δ^1-pyrrolinium chloride. *Tetrahedron Lett.*, **31**, 6295–6298.

Frost, G.M., Yang, K.S. and Waller, G.R. (1967) Nicotinamide adenine dinucleotide as a precursor of nicotine in *Nicotiana rustica* L. *J. Biol. Chem.*, **242**, 887–888.

Gang, D.R., Kasahara, H., Xia, Z.Q., Vander Mijnsbrugge, K., Bauw, G., Boerjan, W., Van Montagu, M., Davis, L.B. and Lewis, N.G. (1999) Evolution of plant defense mechanisms: relationships of phenylcoumaran benzylic ether reductases to pinoresinol-lariciresinol and isoflavone reductases. *J. Biol. Chem.*, **274**, 7516–7527.

Gavilano, L.B., Coleman, N.P., Burnley, L.E., Bowman, M.L., Kalengamaliro, N.E., Hayes, A., Bush, L. and Siminszky, B. (2006) Genetic engineering of *Nicotiana tabacum* for reduced nornicotine content. *J. Agric. Food Chem.*, **54**, 9071–9078.

Gavilano, L.B. and Siminszky, B. (2007) Isolation and characterization of the cytochrome P450 gene *CYP82E5v2* that mediates nicotine to nornicotine conversion in the green leaves of tobacco. *Plant Cell Physiol.*, **48**, 1567–1574.

Gavilano, L.B., Coleman, N.P., Bowman, M.L. and Siminszky, B. (2007) Functional analysis of nicotine demethylase genes reveals insights into the evolution of modern tobacco. *J. Biol. Chem.*, **282**, 249–256.

Goodman, J. (1993) *Tobacco in History: The Cultures of Dependence*, Routledge, Abingdon, UK.

Goossens, A., Heakkinen, S.T., Laakso, I., Seppaenen-Laakso, T., Biondi, S., Sutter, V.D., Lammertyn, F., Nuutila, A.M., Soederlund, H., Zabeau, M., Inze, D. and Oksman-Caldentey, K.M. (2003) A functional genomics approach toward the understanding of secondary metabolism in plant cells. *Proc. Natl. Acad. Sci. USA*, **100**, 8595–8600.

Gorrod, J.W. and Jacob, P. (eds) (1999) *Analytical Determination of Nicotine and Related Compounds and Their Metabolites*, Elsevier Science, Amsterdam.

Griffith, R.B, Valleau, W.D. and Stokes, G.W. (1955) Determination and inheritance of nicotine to nornicotine conversion in tobacco. *Science*, **121**, 343–344.

Gundlach, H., Müller, M.J., Kutchan, T.M. and Zenk, M.H. (1992) Jasmonic acid is a signal transducer in elicitor-induced plant cell cultutes. *Proc. Natl. Acad. USA*, **89**, 2389–2393.

Häkkinen, S.T., Tilleman, S., Swiatek, A., De Sutter, V., Rischer, H., Vanhoutte, I., Van Onckelen, H., Hilson, P., Inze, D., Oksman-Caldentey, K.M. and Goossens, A. (2007) Functional characterization of genes involved in pyridine alkaloid biosynthesis in tobacco. *Phytochemistry*, **68**, 2773–2785.

Halitschke, R. and Baldwin, I.T. (2003) Antisense LOX expression increases herbivore performance by decreasing defense responses and inhibiting growth-related transcriptional reorganization in *Nicotiana attenuate*. *Plant J.*, **36**, 794–807.

Hamill, J.D., Parr, A.J., Robin, R.J. and Rhodes, M.J.C. (1986) Secondary product formation by cultures of *Beta vulgaris* and *Nicotiana rustica* transformed with *Agrobacterium rhizogenes*. *Plant Cell Rep.*, **5**, 111–114.

Hamill, J.D., Robins, R.J., Parr, A.J., Evans, D.M., Furze, J.M. and Rhodes, M.J.C. (1990) Over-expressing a yeast ornithine decarboxylase gene in transgenic roots of *Nicotiana rustica* can lead to enhanced nicotine accumulation. *Plant Mol. Biol.*, **15**, 27–38.

Hanfrey, C., Sommer, S., Mayer, M.J., Burtin, D. and Michael, A.J. (2001) *Arabidopsis* polyamine biosynthesis: absence of ornithine decarboxylase and the mechanism of arginine decarboxylase activity. *Plant J.*, **27**, 551–560.

Hao, D.Y. and Yeoman, M.M. (1996a) Mechanism of nicotine *N*-demethylation in tobacco cell suspension cultures. *Phytochemistry*, **41**, 477–482.

Hao, D.Y. and Yeoman, M.M. (1996b) Nicotine *N*-demethylase in cell-free preparations from tobacco cell cultures. *Phytochemistry*, **42**, 325–329.

Hao, D.Y. and Yeoman, M.M. (1998) Evidence in favour of an oxidative *N*-demethylation of nicotine to nornicotine in tobacco cell cultures. *J. Plant Physiol.*, **152**, 420–426.

Hashimoto, T., Mitani, A. and Yamada, Y. (1990) Diamine oxidase from cultured roots of *Hyoscyamus niger:* its function in tropane alkaloid biosynthesis. *Plant Physiol.*, **93**, 216–221.

Hashimoto, T. and Yamada, Y. (1994) Alkaloid biogenesis: molecular aspects. *Ann. Rev. Plant Physiol. Plant Mol Biol.*, **45**, 257–285.

Hashimoto, T., Shoji, T., Mihara, H., Oguri, K., Tamaki, K., Suzuki, K. and Yamada, Y. (1998a) Intraspecific variability of the tandem repeats in *Nicotiana* putrescine *N*-methyltransferases. *Plant Mol. Biol.*, **37**, 25–37.

Hashimoto, T., Tamaki, K., Suzuki, K. and Yamada, Y. (1998b) Molecular cloning of plant spermidine synthase. *Plant Cell Physiol.*, **39**, 73–79.

Hashimoto, T. and Yamada, Y. (2003) New genes in alkaloid metabolism and transport. *Curr. Opin. Biotechnol.* **14**, 163–168.

Haslam, S.C. and Young, T.W. (1992) Purification of *N*-methylputrescine oxidase from *Nicotiana rustica*. *Phytochemistry*, **31**, 4075–4079.

Hecht, S.S. (2003) Tobacco carcinogens, their biomarkers and tobacco-induced cancers. *Nat. Rev. Cancer*, **3**, 733–744.

Heim, W.G. and Jelesko, J.G. (2004) Association of diamine oxidase and *S*-adenosylhomocysteine hydrolase in *Nicotiana tabacum* extracts. *Plant Mol. Biol.*, **56**, 299–308.

Heim, W.G., Sykes, K.A., Hildreth, S.B., Sun, J., Lu, R.H. and Jelesko, J.G. (2007) Cloning and characterization of a *Nicotiana tabacum* methylputrescine oxidase transcript. *Phytochemistry*, **68**, 454–463.

Hibi, N., Higashiguchi, S., Hashimoto, T. and Yamada, Y. (1994) Gene expression in tobacco low-nicotine mutants. *Plant Cell*, **6**, 723–735.

Hukkanen, J., Jacob, P. and Benowitz, N.L. (2005) Metabolism and disposition kinetics of nicotine. *Pharmacol. Rev.*, **57**, 79–115.

Imanishi, S., Hashizume, K., Nakakita, M., Kojima, H., Matsubayashi, Y., Hashimoto, T., Sakagami, Y., Yamada, Y. and Nakamura, K. (1998) Differential induction of methyl jasmonate of genes encoding ornithine decarboxylase and other enzymes involved in nicotine biosynthesis in tobacco cell cultures. *Plant Mol. Biol.*, **38**, 1101–1111.

Kahl, J., Siemens, D.H., Aerts, R.J., Gaebler, R., Kuehnemann, F., Preston, C.A. and Baldwin, I.T. (2000) Herbivore-induced ethylene suppresses a direct defense but not a putative indirect defense against an adapted herbivore. *Planta*, **210**, 336–342.

Kajikawa, M., Hirai, N. and Hashimoto, T. (2009) A PIP-family protein is required for biosynthesis of tobacco alkaloids. *Plant Mol Biol.*, **69**, 287–298.

Katoh, A., Yamaguchi, Y., Sano, H. and Hashimoto, T. (2003) Analysis of expression sequence tags from *Nicotiana sylvestris*. *Proc. Japan Acad. Ser. B.*, **79**, 151–154.

Katoh, A. and Hashimoto, T. (2004) Molecular biology of pyridine nucleotide and nicotine biosynthesis. *Front. Biosci.*, **9**, 1577–1586.

Katoh, A., Oki, H., Inai, K. and Hashimoto, T. (2005) Molecular regulation of nicotine biosynthesis. *Plant Biotechnol.*, **22**, 389–392.

Katoh, A., Uenohara, K., Akita, M. and Hashimoto, T. (2006) Early steps in the biosynthesis of NAD in Arabidopsis start with aspartate and occur in the plastid. *Plant Physiol.*, **141**, 851–857.

Katoh, A., Shoji, T. and Hashimoto, T. (2007) Molecular cloning of *N*-methylputrescine oxidase from tobacco. *Plant Cell Physiol.*, **48**, 550–554.

Kessler, D. and Baldwin, I.T. (2006) Making sense of nectar scents: the effects of nectar secondary metabolites on floral visitors of *Nicotiana attenuate*. *Plant J.*, **49**, 840–854.

Kessler, D., Gase, K. and Baldwin, I.T. (2008) Field experiments with transformed plants reveal the sense of floral scents. *Science*, **321**, 1200–1202.

Kidd, S.H., Melillo, A.A., Lu, R.-H., Reed, D.G., Kuno, N., Uchida, K., Furuya, M. and Jelesko, J.G. (2006) The *A* and *B* loci in tobacco regulate a network of stress response genes, few of which are associated with nicotine biosynthesis. *Plant Mol. Biol.*, **60**, 699–716.

Kusano, T., Berberich, T., Tateda, C. and Takahashi, Y. (2007) Polyamine: essential factors for growth and survival. *Planta*, **228**, 367–381.

Kutchan, T.M. (1995) Alkaloid biosynthesis: the basis for metabolic engineering of the medicinal plants. *Plant Cell*, **7**, 1059–1070.

Kutchan, T.M. (2005) A role for intra- and intercellular translocation in natural product biosynthesis. *Curr. Opin. Plant Biol.*, **8**, 292–300.

Lamberts, B.L. and Byerrum, R.U. (1958) Glutamate as a precursor for the pyrrolidine ring of nicotine. *J. Biol. Chem.*, **233**, 939–943.

Lee, Y.S. and Cho, Y.D. (2001) Identification of essensial active-site residues in ornithine decarboxylase of *Nicotiana glutinosa* decarboxylating both L-ornithine and L-lysine. *Biochem. J.*, **360**, 657–665.

Leete, E. (1956) The biogenesis of nicotine and anabasine. *J. Am. Chem. Soc.*, **78**, 3520–3523.

Leete, E. (1967) Biosynthesis of nicotine alkaloids XI: investigation of tautomerism in *N*-methyl-Δ^1-pyrrolinium chloride and its incorporation into nicotine. *J. Am. Chem. Soc.*, **89**, 7081–7084.

Leete, E. (1978) Stereochemistry of the reduction of nicotinic acid when it serves as a precursor of anatabine. *J. Chem. Soc., Chem. Commun.*, **1978**, 610–611.

Leete, E. (1983) Biosynthesis and metabolism of the tobacco alkaloids, in *Alkaloids: Chemical and Biological Perspectives* (ed. S.W. Pelletier), John Wiley & Sons Ltd, New York, pp. 85–152.

Leete, E. and Siegfried, K.J. (1957) Biogenesis of nicotine III, incorporation of ornithine into the pyrrolidine ring. *J. Am. Chem. Soc.*, **79**, 4529–4531.

Leete, E. and Liu, Y.Y. (1973) Metabolism of [2-^{3}H]- and [6-^{3}H]-nicotinic acid in intact *Nicotiana tabacum* plants. *Phytochemistry*, **12**, 593–596.

Leete, E. and Chedekel, M.R. (1974) Metabolism of nicotine in *Nicotiana gluca. Phytochemistry*, **13**, 1853–1859.

Leete, E. and Slattery, S.A. (1976) Incorporation of [2-^{14}C]- and [2-^{14}C]-nicotinic acid into tobacco alkaloids. Biosynthesis of anatabine and α, β-dipyridl. *J. Am. Chem. Soc.*, **98**, 6326–6330.

Leete, E. and Yu, M.L. (1980) The incorporation of ornithine-[2,3-$^{13}C_2$] into nicotine and nornicotine established by NMR. *Phytochemistry*, **19**, 1093–1097.

Legg, P.G., Chaplin, J.F. and Collins, G.B. (1969) Inheritance of percent total alkaloids in *Nicotiana tabacum* L. *J. Hered.*, **60**, 213–217.

Legg, P.G., Collins, G.B. and Litton, C.C. (1970) Registration of LA Burley 21 tobacco germplasm. *Crop Sci.*, **10**, 212.

Legg, P.G. and Collins, G.B. (1971) Inheritance of percent total alkaloids in *Nicotiana tabacum* L. II. genetic effects of two loci in Burley21 X LA Burley21 populations. *Can. J. Genet. Cytol.*, **13**, 287–291.

Lewis, R.S., Jack, A.M., Morris, J.W., Robert, V.J.M., Gavilano, L.B., Siminszky, B., Bush, L.P., Hayes, A.J. and Dewey, R.E. (2008) RNA interference (RNAi)-induced suppression of nicotine demethylase activity reduces levels of a key carcinogen in cured tobacco leaves. *Plant Bioth. J.*, **6**, 346–354.

Li, L., He, Z., Girdhar, K.P., Tsuchiya, T. and Luan, S. (2002) Functional cloning and characterization of multidrug and heavy metal detoxification. *J. Biol. Chem.*, **277**, 5360–5368.

Mann, T.J., Weybrew, J.A., Matzinger, D.F. and Hall, J.L. (1964) Inheritance of the conversion of nicotine to nornicotine in varieties of *Nicotiana tabacum* L. and related amphiploids. *Crop Sci.*, **4**, 349–353.

Marinova, K., Pourcel, L., Weder, B., Schwarz, M., Barron, D., Routaboul, J.M. and Klein, M. (2007) The Arabidopsis MATE transporter TT12 acts as a vacuolar flavonoid/H^+-antiporter active in proanthocyanidin-accumulating cells of the seed coat. *Plant Cell*, **19**, 2023–2038.

Mathews, H., Clendennen, S.K., Caldwell, C.G., Liu, X.L., Connors, K., Matheis, N., Schuster, D.K., Menasco, D.J., Wagoner, W., Lightner, J. and Wagner, D.R. (2007) Activation tagging in tomato identifies a transcriptional regulator of anthocyanin biosynthesis, modification, and transport. *Plant Cell*, **15**, 1689–1703.

Michael, A.J., Furze, J.M., Rhodes, M.J.C. and Burtin, D. (1996) Molecular cloning and functional identification of a plant ornithine decarboxylase cDNA. *Biochem. J.*, **314**, 241–248.

Mizusaki, S., Tanabe, Y., Noguchi, M. and Tamaki, E. (1971) Phytochemical studies on tobacco alkaloids XIV. The occurrence and properties of putrescine *N*-methyltransferase in tobacco roots. *Plant Cell Physiol.*, **12**, 633–640.

Mizusaki, S., Tanaka, Y., Noguchi, M. and Tamaki, E. (1972) *N*-methylputrscine oxidase from tobacco roots. *Phytochemistry*, **11**, 2757–2762.

Mizusaki, S., Tanaka, Y., Noguchi, M. and Tamaki, E. (1973) Changes in the activities of ornithine decarboxylase, putrescine *N*-methyltransferase, and *N*-methylputrescine oxidase in tobacco roots in relation to nicotine biosynthesis. *Plant Cell Physiol.*, **14**, 103–110.

Morita, M., Shitan N., Sawada K., Van Montagu M.C., Inzé D., Rischer H., Goossens A., Oksman-Caldentey K.M., Moriyama Y. and Yazaki K. (2009) Vacuolar transport of nicotine is mediated by a multidrug and toxic compound extrusion (MATE) transporter in *Nicotiana tabacum. Proc. Natl. Acad. Sci. USA*, **106**, 2447–2452.

Murakami, Y., Matsufuji, S., Hayashi, S., Tanahashi, N. and Tanaka, K. (2000) Degradation of ornithine decarboxylase by the 26S proteasome. *Biochem. Biophys. Res. Commun.*, **267**, 1–6.

Oki, H. and Hashimoto, T. (2004) Jasmonate-responsive regions in a *Nicotiana sylvestris PMT* gene involved in nicotine biosynthesis. *Plant Biotech.*, **21**, 269–274.

Omote, H., Hiasa, M., Matsumoto, T., Otsuka, M. and Moriyama, Y. (2006) The MATE proteins as fundamental transporters of metabolic and xenobiotic organic cations. *Trends Pharmacol. Sci.*, **27**, 587–593.

Paschold, A., Halitschke, R. and Baldwin, I.T. (2007) Co(i)-ordinating defenses: NaCOI1 mediates herbivore-induced resistance in *Nicotiana attenuate* and reveals the role of herbivore movement in avoiding defenses. *Plant J.*, **51**, 79–91.

Riechers, D.E. and Timko, M.P. (1999) Structure and expression of the gene family encoding putrescine *N*-methyltransferase in *Nicotiana tabacum*: new clues to the evolutionary origin of cultivated tobacco. *Plant Mol. Biol.*, **41**, 387–401.

Rischer, H., Oresic, M., Seppanen-Laakso, T., Katajamaa, M., Lammertyn, F. Ardiles-Diaz, W., Van Montagu, M.C., Inze, D., Oksman-Caldentey, K.M. and Goossens, A. (2006) Gene-to-metabolite networks for terpenoid indole alkaloid biosynthesis in *Catharanthus roseus* cells. *Proc. Natl. Acad. Sci. USA*, **103**, 5614–5619.

Roberts, M.F. and Wink, M. (1998) *Alkaloids: Biochemistry, Ecology, and Medicinal Applications*, Plenum Press, New York.

Robinson, T. (1974) Metabolism and function of alkaloids in plants. *Science*, **184**, 430–435.

Rushton, P.J., Bokowiec, M.T., Laudeman, T.W., Brannock, J.F., Chen, X. and Timko, M.P. (2008) TOBFAC: the database of tobacco transcription factors. *BMC Bioinformatics*, **9**, 53.

Saitoh, F., Noma, M. and Kawashima, N. (1985) The alkaloid contents of sixty *Nicotiana* species. *Phytochemistry*, **24**, 477–480.

Sandemeier, E., Hale, T.I. and Christen, P. (1994) Multiple evolutionary origin of pyridoxal-5′-phosphate-dependent amino acid decarboxylases. *Eur. J. Biochem.*, **221**, 997–1002.

Sato, F., Hashimoto, T., Hachiya, A., Tamura, K.I., Chio, K.B., Morishige, T., Fujimoto, H. and Yamada, Y. (2001) Metabolic engineering of plant alkaloid biosynthesis. *Proc. Natl. Acad. Sci. USA*, **98**, 367–372.

Sato, F., Inai, K. and Hashimoto, T. (2007) Metabolic engineering in alkaloid biosynthesis: case studies in tyrosine- and putrescine-derived alkaloids, in *Applications of Plant Metabolic Engineering* (eds R. Verpoorte, A.W. Alfermann and T.S. Johnson), Springer, Dordrecht, pp. 145–173.

Saunders, J.Q. and Bush, L.P. (1979) Nicotine biosynthetic enzyme activities in *Nicotiana tabacum* L. genotypes with different alkaloid levels. *Plant Physiol.*, **64**, 236–240.

Scott, T.A. and Glynn, J.P. (1967) The incorporation of [2,3,7-^{14}C]-nicotinic acid into nicotine by *Nicotiana tabacum. Phytochemistry*, **6**, 505–510.

Shoji, T., Nakajima, K. and Hashimoto, T. (2000a) Ethylene suppresses jasmonate-induced gene expression in nicotine biosynthesis. *Plant Cell Physiol.*, **41**, 1072–1076.

Shoji, T., Yamada, Y. and Hashimoto, T. (2000b) Jasmonate induction of putrescine *N*-methyltransferase genes in roots of *Nicotiana sylvestris*. *Plant Cell Physiol.*, **41**, 831–839.

Shoji, T., Winz, R., Iwase, T., Nakajima, K., Yamada, Y. and Hashimoto, T. (2002) Expression patterns of two tobacco isoflavone reductase-like genes and their possible roles in secondary metabolism in tobacco. *Plant Mol. Biol.*, **50**, 427–440.

Shoji, T. and Hashimoto, T. (2008) Why does anatabine, but not nicotine, accumulate in jasmonate-elicited cultured tobacco BY-2 cells? *Plant Cell Physiol.*, **49**, 1209–1216.

Shoji, T., Ogawa, T. and Hashimoto, T. (2008) Jasmonate-induced nicotine formation in tobacco is mediated by tobacco *COI1* and *JAZ* genes. *Plant Cell Physiol.*, **49**, 1003–1012.

Shoji, T., Inai, K., Yazaki, Y., Sato, Y., Takase, H., Shitan, N., Yazaki, K., Goto, Y., Toyooka, K., Matsuoka, K. and Hashimoto, T. (2009) Multidrug and toxic compound extrusion-type transporters implicated in vacuolar sequestration of nicotine in tobacco roots. *Plant Physiol.*, **149**, 708–718.

Shoji, T., Kajikawa, M. and Hashimoto, T. (2010) Clustered transcription factor genes regulate nicotine biosynthesis in tobacco. *Plant Cell*, **22**, 3390–3409.

Siminszky, B., Gavilano, L., Bowen, S.W. and Dewey, R.E. (2005) Conversion of nicotine to nornicotine in *Nicotiana tabacum* is mediated by CYP82E4, a cytochrome P450 monooxygenase. *Proc. Natl. Acad. Sci. USA*, **102**, 14919–14924.

Sinclair, S.J., Murphy, K.J., Birch, C.D. and Hamill, D. (2000) Molecular characterization of quinolinate phosphoribosyltranferase (QPRTase) in *Nicotiana*. *Plant Mol. Biol.*, **44**, 603–617.

Staswick, P.E. and Tiryaki, I. (2004) The oxylipin signal jasmonic acid is activated by an enzyme that conjugates it to isoleucine in *Arabidopsis*. *Plant Cell* **16**, 2117–2127.

Stead, L.F. and Lancaster, T. (2007) Interventions to reduce harm from continued tobacco use. *Cochrane Database Syst. Rev.*, **18**, CD005231.

Stenzel, O., Teuber, M. and Dräger, B. (2006) Putrescine *N*-methyltransferase in *Solanum tuberosum* L., a calystegine-forming plant. *Planta*, **223**, 200–212.

Steppuhn, A., Gase, K., Krock, B., Halitschke, R. and Baldwin, I.T. (2004) Nicotine's defensive function in nature. *PLoS Biol.*, **2**, 1074–1080.

Suzuki, K., Yamada, Y. and Hashimoto, T. (1999) Expression of *Atropa belladonna* putrscine *N*-methyltransferase gene in root pericycle. *Plant Cell Physiol.*, **40**, 289–297.

Teuber, M., Azemi, M.E., Namjoyan, F., Meier, A.C., Wodak, A., Brandt, W. and Dräger, B. (2007) Putrescine *N*-methyltransferase – a structure-function analysis. *Plant Mol Biol.*, **63**, 787–801.

van der Fits, L. and Memelink, J. (2000) ORCA3, a jasmonate-responsive transcriptional regulator of plant primary and secondary metabolism. *Science*, **289**, 295–297.

von Dahl, C.C., Winz, R.A., Halitschke, R., Kühnemann, F., Gase, K. and Baldwin, I.T. (2007) Tuning the herbivore-induced ethylene burst: the role of transcript accumulation and ethylene perception in *Nicotiana attenuata*. *Plant J.*, **51**, 293–307.

Walton, N.J. and McLauchlan, W.R. (1990) Diamine oxidase and alkaloid production in transformed root cultures of *Nicotiana tabacum*. *Phytochemistry*, **29**, 1455–1457.

Wang, L., Allmann, S., Wu, J. and Baldwin, I.T. (2008) Comparisons of LIPOXYGENASE3- and JASMONATE-RESISTANT4/6-silenced plants reveal that jasmonic acid and jasmonic acid-amino acid conjugates play different roles in herbivore resistance of *Nicotiana attenuate*. *Plant Physiol.*, **146**, 904–915.

Wasternack, C. (2007) Jasmonates: an update on biosynthesis, signal transduction and action in plant stress response, growth and development. *Ann. Bot.*, **100**, 681–697.

Wernsman, E.A. and Matzinger, D.F. (1970) Relative stability of alleles at the nicotine conversion locus of tobacco. *Tob. Sci.*, **14**, 34–36.

Winz, R.A. and Baldwin, I.T. (2001) Molecular interactions between the specialist herbivore *Manduca sexta* (Lepidoptera, Sphingidea) and its natural host *Nicotiana attenuate*. IV. Insect-induced ethylene reduces jamonate-induced nicotine accumulation by regulating putrescine *N*-methyltransferase transcripts. *Plant Physiol.*, **125**, 2189–2202.

Xu, B. and Timko, M.P. (2004) Methyl jasmonate induced expression of the tobacco putrescine *N*-methyltransferase genes requires both G-box and GCC-motif elements. *Plant Mol. Biol.*, **55**, 743–761.

Xu, D., Shen, Y., Chappell, J., Cui, M. and Nielsen, M. (2007) Biochemical and molecular characterization of nicotine demethylase in tobacco. *Physiol. Plant.*, **129**, 307–319.

Yang, K.S., Gholson, R.K. and Waller, G.R. (1965) Studies on nicotine biosynthesis. *J. Am. Chem. Soc.*, **87**, 4184–4188.

Yazaki, K. (2005) Transporters of secondary metabolites. *Curr. Opin. Plant Biol.*, **8**, 301–307.

Yoshida, D. and Mitake, T. (1966) Agmatine and *N*-carbamylputrescine as intermediates in the formation of nicotine by tobacco plants. *Plant Cell Physiol.*, **7**, 301–305.

Zhang, Z.-P. and Baldwin, I.T. (1997) Transport of [2-^{14}C] jasmonic acid from leaves to roots mimics wound-induced changes in endogenous jasmonic acid pools in *Nicotiana sylvestris*. *Planta*, **203**, 436–441.

Ziegler, J., Voigtländer, S., Schmidt, J., Kramell, R., Miersch, O., Ammer, C., Gesell, A. and Kutchan, T.M. (2006) Comparative transcript and alkaloid profiling in *Papaver* species identifies a short chain dehydrogenase/reductase involved in morphine biosynthesis. *Plant J.*, **48**, 177–192.

Ziegler, J. and Facchini, P.J. (2008) Alkaloid biosynthesis: metabolism and trafficking. *Ann. Rev. Plant Biol.*, **59**, 735–769.

8

Terpenoid Biosynthesis

Dae-Kyun Ro

8.1 Introduction

Terpenoids were historically named after 'turpentine', a collection of lipidic natural products distilled from resinous woody plants such as pine and balsam trees. Turpentine was used as a solvent for paintings and raw materials for organic chemical synthesis in the pre-petroleum era. At the advent of petroleum-derived chemicals in the early 1900s, the petrochemicals were initially called 'turpentine substitutes', but being much cheaper they rapidly replaced turpentine. Currently, increasing numbers of scientists are investigating the possibility of adopting terpenoid natural products as 'petrochemical substitutes', thereby reducing our dependency on petroleum and mitigating greenhouse gas emissions. In the late 1970s, Melvin Calvin, the pioneering photosynthesis researcher, demonstrated that C15 sesquiterpenes collected from the trunk of the Amazon Copaiba (*Copaifera* sp.) tree could be directly used as fuel for diesel engines (Calvin, 1980). In earlier times, Thomas Edison's Botanic Research Company engaged in a breeding programme for goldenrod (*Solidago canadensis*) in an attempt to produce the terpenoid polymer, natural rubber (van Beilen and Poirier, 2007). In the future, we can anticipate an increasing use of renewable terpenoids in our daily lives (Bohlmann and Keeling, 2008). However, as will be discussed in this article, future commercial production of terpenoids will almost certainly not be based on the traditional distillation methods of the past. This article focuses on recent developments, 2005–2010, in the area of the biochemistry, cellular and molecular biology, and biotechnology of plant terpenoid metabolism. A concise updated biochemical overview of terpenoid metabolism will be given, with examples of biotechnological applications.

Plant Metabolism and Biotechnology, First Edition. Edited by Hiroshi Ashihara, Alan Crozier, and Atsushi Komamine.

8.2 Terpenoid Diversity

Among several classes of natural products (e.g. lipids, phenylpropanoids, polyketides, alkaloids and glucosinolates), terpenoids are by far the most diverse group with >30,000 structurally known compounds (Buckingham, 2007; Dewick, 2009). Some terpenoids are essential components in fundamental physiological processes. Examples include the plastoquinone and ubiquinone for electron transport chains, steroids for membrane fluidity and hormonal regulations, and protein prenylation for subcellular localisation. However, a majority of terpenoids are synthesized in plants as specialised or secondary metabolites. These are the non-essential compounds for survival, but which through millions of years of adaptive evolution now collectively influence the overall fitness and wellbeing of plants. Some terpenoids and their precursors, of medicinal and industrial importance, are illustrated in Figure 8.1.

The representative terpenoids depicted in Figure 8.1 have enriched the lives of mankind as pharmaceuticals, nutraceuticals, aroma and flavour components, and industrial materials. For example, artemisinin (sesquiterpene lactone endoperoxide) is currently the only antimalarial drug effective against drug-resistant malarial parasites. Nootkatone is an expensive grapefruit aroma, and astaxanthin has found widespread use as a pigment additive in aquaculture. For some terpenoids, specific *in vivo* molecular targets have been identified. Thapsigargin binds to sarco/endoplasmic reticulum calcium ATPase (SERCA) (Lytton *et al.*, 1991), and salvinorin A functions as a potent agonist of the kappa opioid receptor (Roth *et al.*, 2002). Detailed knowledge of terpenoid ligand and protein receptor interaction has led to pharmaceutical developments based on the frameworks of these terpenoids. However, it should be noted that terpenoids and their precursors used for semi-synthetic purposes are still purified from plants by traditional extraction methods.

Despite their immense structural diversity, all terpenoids can be reconstructed using a simple building block, the C5 isoprene unit (Figure 8.2A). In most cases, terpenoids are formed by head-to-tail condensations, but occasionally head-to-head (e.g. squalene) or head-to-middle (e.g. chrysanthemene) condensations can occur (Figure 8.2B). However, cleavage of C-C bonds can result in terpenoids which cannot be easily explained by the C5 isoprene rule (e.g. C11 or C16 homoterpene). Recently, a novel plant hormone strigolactone was discovered that *in planta* functions as a branching inhibitor, a recruiter of arbuscular mycorrhizal fungi and a germination stimulator for parasitic plants (Akiyama *et al.*, 2005; Gomez-Roldan *et al.*, 2008; Umehara *et al.*, 2008). Strigolactone is synthesized from the break-down product of carotenoids (Schwartz *et al.*, 2004). For such unusual molecules, feeding assays are required to validate their classification as terpenoids.

Isoprene, however, is not the true biochemical precursor in the terpenoid biosynthetic pathway. The diphosphate ester of isoprene, isopentenyl diphosphate (IPP), is the activated, biologically relevant precursor which can be readily incorporated into diverse terpenoid metabolites (Figure 8.2A). The central dogma in terpenoid metabolism is that IPP and its isomer dimethylallyl diphosphate (DMAPP) are the universal precursors for tens of thousands of distinct terpenoid natural products, whether they are synthesized from bacteria, fungi, plant or human. Although DMAPP alone is a direct precursor of the volatile product, isoprene, in some plants (Miller *et al.*, 2001), the more important biochemical role of DMAPP is to serve as a priming molecule for the synthesis of longer linear prenyl diphosphates. Catalyzed by a family of enzymes called *trans*-prenyl transferase, DMAPP is

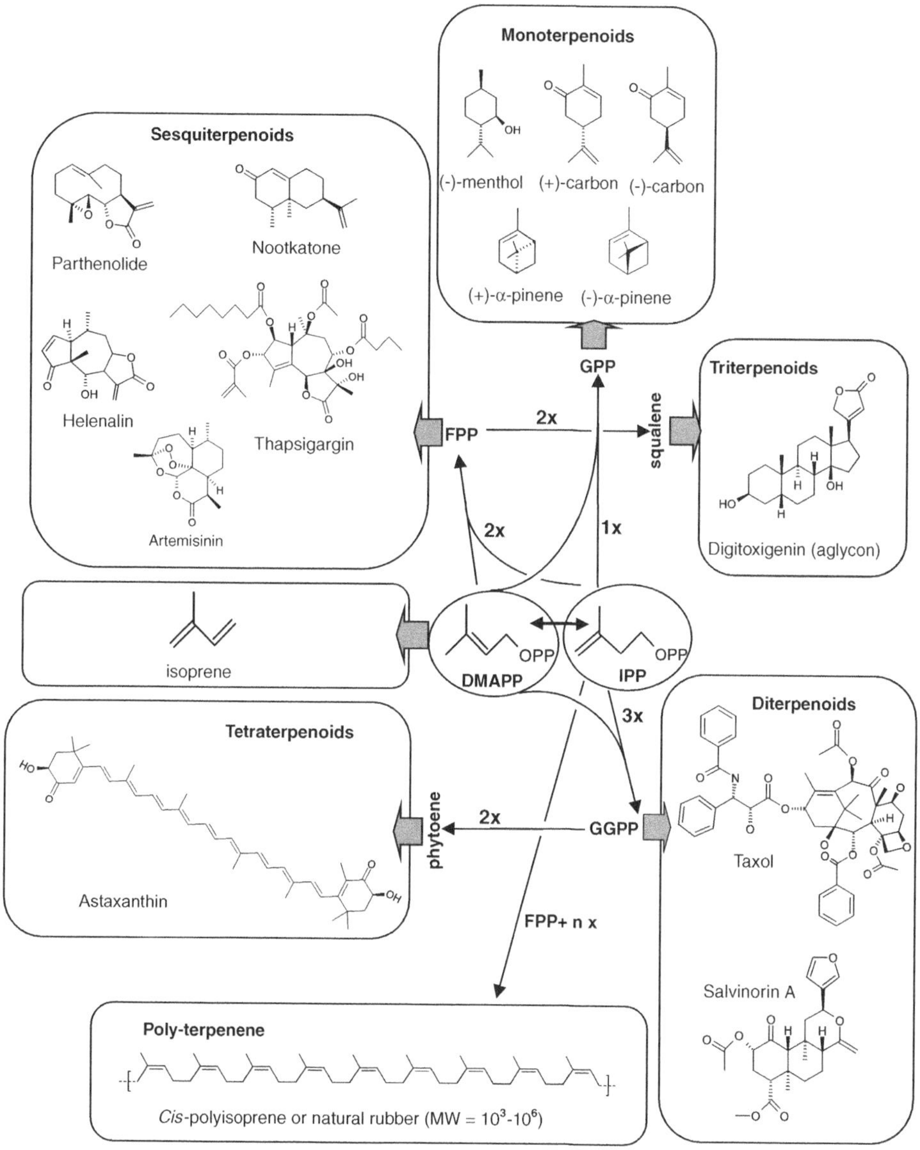

Figure 8.1 *Examples of diverse terpenoids used as aroma, flavours, medicine, and industrial chemicals. Condensation rules for mono-, sesqui-, di-, tri-, tetra- and poly-terpenes are depicted from the basic building blocks, IPP and DMAPP*

further extended by sequential condensations with IPP to yield C10 (1× IPP), C15 (2× IPP) and C20 (3× IPP) diphosphates, referred to as geranyl diphosphate (GPP), farnesyl diphosphate (FPP), and gerenylgeranyl diphosphate (GGPP), respectively (Figure 8.3A). These linear prenyl diphosphate molecules are the direct substrates of terpene synthases (TPSs), which constitute a superfamily of enzymes responsible for the synthesis of a plethora of

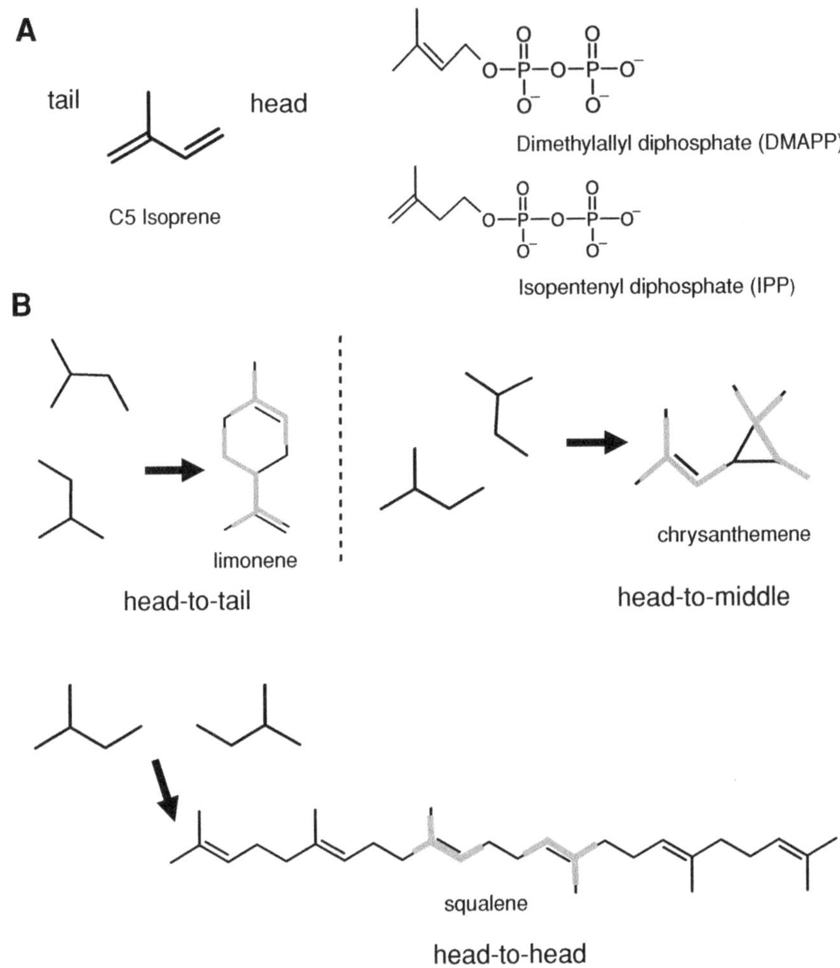

Figure 8.2 *Carbon backbones of terpenoids. (A) The structure of the basic building block, isoprene (left), and structures of DMAPP and IPP (the biological precursors of terpenoids, right) are illustrated. (B) Different types of isoprene condensation in plant. Grey coloured lines indicate the C5 isoprene units in the end-products*

natural terpene hydrocarbons. By a distinct set of enzymes (squalene synthase and phytoene synthase) that do not belong to the TPS superfamily, FPP and GGPP can be condensed to form C30 squalene and C40 phytoene, respectively, from which a diverse array of triterpenoid and tetraterpenoid derivatives (carotenoids) are synthesized. In addition to IPP's essential role as a base substrate for small molecules, IPP is the building block of isoprene polymers such as rubber (*cis*-polyisoprene). Approximately 2500 plant species, including the Brazilian rubber tree (*Hevea brasiliensis*), are able to synthesize terpene polymers. In these plants, thousands of IPP molecules are conjugated onto the FPP initiator in a unique *cis*-configuration. This *cis*-form accounts for the formation of spring-like structures that

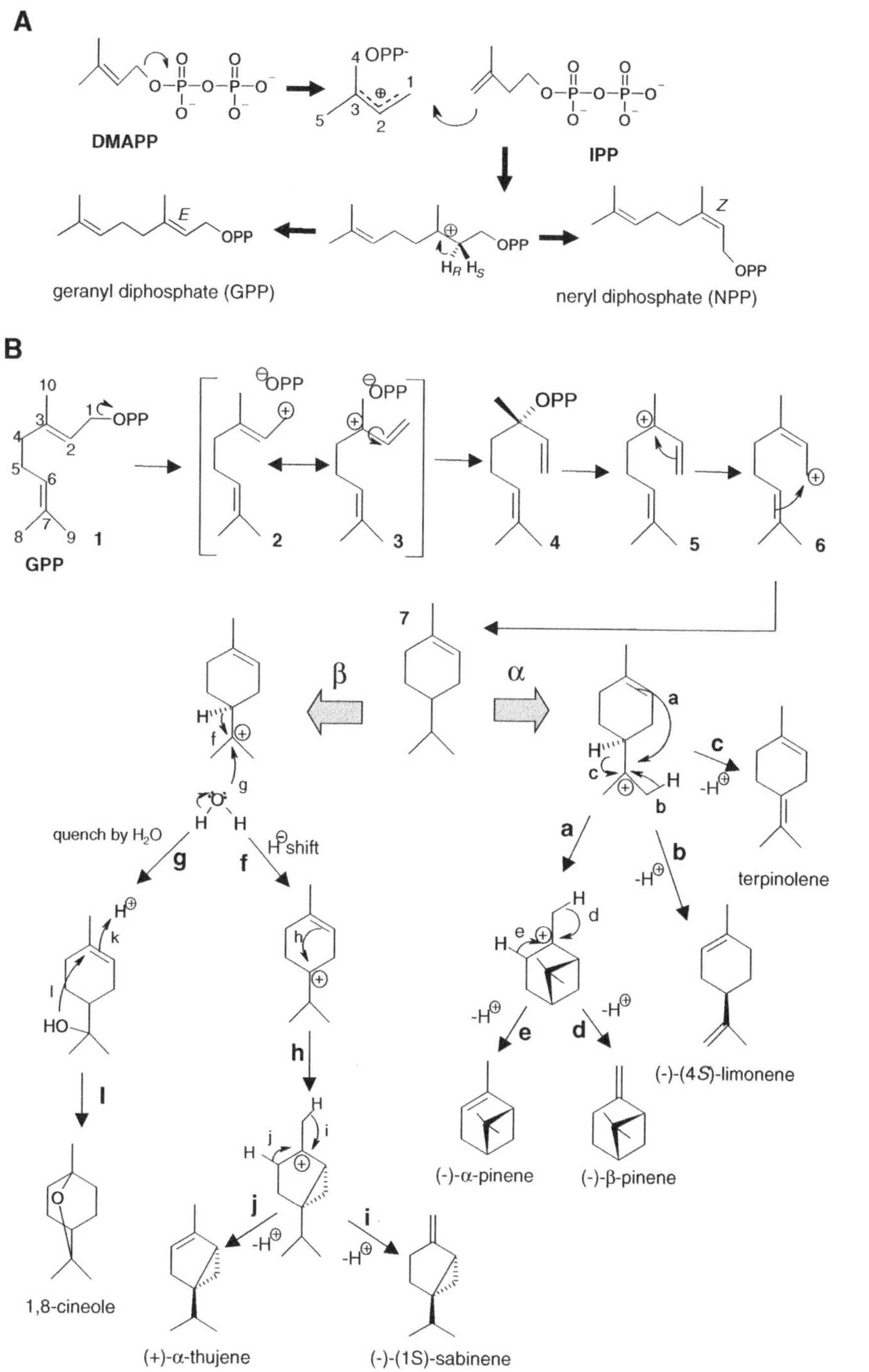

Figure 8.3 *Illustration of chemical mechanisms for prenyl diphosphate and terpene synthesis. (A) Mechanisms for* trans-*GPP and* cis-*GPP (neryl diphosphate). (B) Mechanisms for diverse terpene synthesis via carbocation intermediates*

confer unique elasticity to naturally produced rubber. It is amazing to notice the enormous chemical diversity of terpenoids that derive from the simple molecule, IPP.

8.3 Mechanistic Aspects of Terpenoid Biogenesis

The formation of an unstable carbocation intermediate is the central reaction mechanism that drives the biosynthesis of various allylic prenyl diphosphates (GPP, FPP and GGPP) as well as hydrocarbon terpenes. Prenyltransferase enzymes can cleave the diphosphate group from DMAPP to create a reactive carbocation isoprene intermediate whose electrons delocalise over C1 to C3 in DMAPP (Figure 8.3A). The prenyltransferase enzyme catalyzes the nucleophilic attack of the IPP double-bond (π-bond) on the DMAPP carbocation to synthesize C10 diphosphate carbocation. Stereo-specific deprotonation of H_R proton in IPP yields the C10 diphosphate GPP with a *trans*-double-bond by *trans*-prenyl transferase. Essentially, the same mechanism is used to produce C15 FPP and C20 GGPP by condensing two or three IPPs onto a single DMAPP.

Some TPS enzymes can isomerise GPP to neryl diphosphate (NPP) by their built-in isomerase activity before the cyclisation step (see below). However, a unique *cis*-prenyl transferase, belonging to a family of enzymes entirely distinct from the *trans*-prenyltransferase, was recently shown to synthesize and release NPP as a final product from DMAPP and IPP (Schilmiller *et al.*, 2009). This reaction is catalyzed by the stereo-specific loss of a H_S proton as observed in the biosynthesis of rubber (Figure 8.3A). In addition, a novel FPP composed exclusively of *cis*-double-bonds has been identified in tomato, and a unique *cis*-prenyltransferase was found to be responsible for this *cis,cis*-FPP synthesis (Sallaud *et al.*, 2009). These findings and biochemical kinetic data using the *cisoid* substrates suggest that small C10 and C15 *cisoids* are *bona fide* substrates for TPS enzymes in nature. The presence of distinct types of prenyl diphosphates in both *E*- and *Z*-isomers provides the first layer of structural diversity in terpenoids.

From the prenyl diphosphates, extremely diverse terpenoids with stereo- and regio-specificities arise, but the core principle of reaction mechanism is similar to that for the prenyl diphosphate synthesis – the formation of an unstable carbocation promoting a cascade of reactions to stabilise the reactive carbocation. Three types of similar chemical reactions, which drive the intramolecular skeletal rearrangements, are double-bond (π-bond) rearrangement, hydride (or methyl) shift, and single C-C bond (σ bond) rearrangement from the existing terpene carbon backbone. These reactions, however, keep relaying the reactive carbocation from one carbon to the next in a cascade, which can only be terminated by either a deprotonation or quenching by a nucleophile (e.g. a water molecule). The combination of these reactions ultimately results in the observed structural diversity of terpenoids.

The carbocation chemistry involved in terpene formation is illustrated in Figure 8.3B using GPP transformation (monoterpene formation) as a case study. With GPP as the substrate (1), TPS catalyses the formation of C10 carbocation. The resulting intermediate is resonance stabilised between 2 and 3 and, at this stage, the dissociated diphosphate anion is believed to form a complex with the carbocation intermediate. The single bond between C2 and C3 in one of the resonance forms (3) can rotate freely, and linalyl diphosphate (LPP, 4) can be generated. The LPP is thought to be an intermediate in the TPS-catalyzed synthesis of cyclic monoterpenes, which led to the idea that the initial isomerase activity is embedded in the TPS. The dissociation of diphosphate from LPP yields a reactive carbocation intermediate (5), and the nearby double bond is sequentially rearranged to form a cyclic carbocation intermediate (7). From carbocation 7, two reaction paths can be postulated. In the first path (α), the carbocation captures the C2/C3 double-bond and passes the carbocation onto C3 (a). This reaction is terminated by a deprotonation at C10 or C4 position (d or e), yielding two double-bond positional isomers of pinene. Similarly, distinct deprotonations of 7 can result in limonene (b) and terpinolene (c). In the second path (β), carbocation of 7 is neutralised by capturing a hydroxyl group from a water molecule (g), and the second round of carbocation is initiated by a protonation of C2 (k), and subsequently the C3 carbocation is stabilised by forming an ether linkage between C7 and C3, resulting in cineol formation (l). Alternatively, the carbocation at C7 can be transformed to carbocation C8 through hydride (H^-) shift (f). A combination of double-bond rearrangement and deprotonation results in the formation of thujene and sabinene (h, i and j). For the formation of C15 sesquiterpenes and C20 diterpenes, similar reaction mechanisms are utilised.

The resulting end products often display multiple chiral centres and double-bond positional variants, leading to the creation of regio- and stereo-isomers. Despite the subtle structural differences in terpenoid structures, animals and insects can respond to the isomers differently. For example, humans perceive the (+)-carbon as caraway aroma but (−)-carbon as mint (Figure 8.1). It is reasonable to believe that chemical sensing mediated by terpenoids is an important part of plant–plant and plant–insect interactions. Emerging data indeed support the view that terpenoids are the chemicals possessing information that allows plants to communicate with their environment (see Section 8.8.2). For these reasons, terpenoids are called 'info-chemicals' whereby stereo- and regio-chemistry generated by TPS reactions serve as unique chemical vocabularies.

8.4 Terpene Synthase – Structure, Evolution and Engineering

The countless folding patterns of GPP, FPP and GGPP via carbocation intermediates account for the terpenoid diversity that occurs in numerous plant species. However, a specific plant species synthesizes only a subset of terpenoid products which display unique stereo- and regio-specificity. Since TPS governs the pathway for prenyl diphosphate molecular folding, we should be able to redirect the folding patterns to produce different molecules by altering TPS structure. In order to achieve TPS molecular engineering, it is essential to understand the three-dimensional (3D) structure of the enzyme by which the key amino-acid residues in close proximity to the substrate can be mapped.

The 3D structure of a sesquiterpene synthase, tobacco 5-*epi*-aristolochene synthase (TEAS), with its substrate analogues, provides an invaluable insight into the TPS-guided folding mechanism (Starks *et al.*, 1997). In brief, the substrate-binding pocket of TPS has a hydrophilic entry domain to interact with the diphosphate ions, and a hydrophobic inner environment to accommodate the isoprene unit. At the entry point, the highly conserved Asp residues directly coordinate Mg^{2+} metals. These divalent metal ions and the adjacent Arg residues jointly create a focused area of positive charges which serve to cleave and pull the negatively charge diphosphate ions away from the active site. At the same time, the partial negative charges inside the pocket control the proper positioning of the carbocation intermediate. After that, a cascade of deprotonation, reprotonation, double-bond rearrangement, and hydride- and methyl-shift are proposed to occur by interactions of the residues surrounding the substrate.

In addition to the TEAS structure, three other TPS 3D structures became available – pentalenene synthase from bacteria, and aristolochene synthase and trichodiene synthase from fungi (Lesburg *et al.*, 1997; Caruthers *et al.*, 2000; Rynkiewicz *et al.*, 2001). The structures of FPP synthase, which catalyzes the condensation of DMAPP and IPP, and the squalene synthase for the condensation of two FPP molecules, have been determined (Tarshis *et al.*, 1994; Pandit *et al.*, 2000). Comparison of the crystal structures of these different enzymes, catalyzing closely related reactions, revealed that they all share the core five helixes of the TPS catalytic site, despite their unrelated primary sequences (Greenhagen and Chappell, 2001). From the comparative structural information, plant TPS has probably evolved from the primary metabolic enzyme FPP synthase and squalene synthase.

The evolutionary lineage of plant secondary metabolism to the primary metabolism can also be found in other pathways. Histidine ammonia lyase (HAL), which catalyses histidine degradation, is widely distributed throughout the plant kingdom. In contrast, phenylalanine ammonia lyase (PAL) is restricted to plants that carry out phenolic biosynthesis. Although HAL and PAL share less than 25% sequence identity, the residues positioned in their catalytic sites are superimposable when their crystal structures are compared (Ritter and Schulz, 2004). This demonstrates that the conservation of 3D structure at the active site is critical in enzyme evolution.

Several plant TPSs were modelled against the TEAS 3D structure, providing the basis for altering TPS product specificity by site-directed mutagenesis. One hypothesis is that an ancient enzyme with promiscuous functions evolved for the specific functionality. In order to evaluate enzyme evolution in an *in vitro* condition, promiscuous γ-humulene synthase which produces 52 different types of sesquiterpenes was modelled, and 19 amino-acid residues constituting the active site contour were identified (Yoshikuni *et al.*, 2006). Saturation mutagenesis on all 19 residues was carried out individually and the product profiles determined. The profiles were altered with some mutations, and a simple mathematical model, formulated from the activity charts, was devised to predict and design specific TPS enzymes. Consequently, several new TPSs, including some that are not found in nature, were created in a predictable manner *in vitro*.

Similarly, structural modelling of henbane (*Hyoscyamus niger*) premnaspirodiene synthase (HPS) identified nine residues at the active site. Full substitutions of these residues resulted in reciprocal interconversion of activities between TEAS and HPS (Greenhagen *et al.*, 2006). These nine different residues in TEAS were substituted with all possible combinations (512 mutants) to trace the mutational path that TEAS may have taken through evolution (O'Maille *et al.*, 2008). In a number of mutants, converged evolution appeared to occur because multiple additive and punctuated alterations of the TPS specificity resulted in an identical chemical phenotype. In an independent study, the interconversion of four key residues identified from the two paralogous diterpene synthases in conifers were sufficient to reciprocally interchange the product specificity between isopimaradiene and levopimaradiene synthase (Keeling *et al.*, 2008). Three of these residues were mapped to one side of the substrate-binding pocket, while the other residue occupied the secondary layer of the active site.

These mutation analyses of various TPSs demonstrated, in principle, that the TPS product specificity can be engineered to produce different types of terpenoids. These results confirmed that the catalytic plasticity of TPSs is the major driver for the terpenoid diversity in nature.

8.5 Two Distinct Pathways for Isopentenyl Diphosphate (IPP) Biosynthesis

The key terpenoid precursor IPP is synthesized by two independent pathways – mevalonate (MVA) pathway and methyl erythritol phosphate (MEP) pathway (Figure 8.4). MVA pathway has been studied for more than 50 years, but MEP pathway was discovered only a decade ago and has many unsolved questions. Plants possess both the MVA and MEP pathways often in the same cell type, and the biological significance of the MEP route in the context of the traditional MVA pathway has posed unique challenges to plant biologists. Characteristics of these two pathways with recent findings will be outlined in this section.

8.5.1 Mevalonate (MVA) Pathway

Intensive biochemical studies of sterols in 1950s firmly established the MVA pathway as the biosynthetic route for IPP and DMAPP in yeast and animals. In the MVA pathway, three acetyl CoA molecules are condensed to give 3-hydroxy-3-methylglutaryl CoA (HMGCoA) by acetoacetyl CoA synthase and HMGCoA synthase. The thio-ester group of HMGCoA is reduced to mevalonate by HMGCoA reductase that catalyzes a critical rate-determining reaction in the MVA pathway. Due to the central regulatory position of HMGCoA reductase, this enzyme has been an important molecular target for drugs, aiming at lowering cholesterol levels. Lovastatin is a polyketide isolated from the fungus *Aspergillus terreus*. This compound and its derivatives (e.g. simvastatin) are potent competitive inhibitors of HMGCoA reductase, and act as efficient drugs regulating hypercholesterolemia in humans. On the other hand, HMGCoA reductase is a primary target enzyme for metabolic flux engineering. Overexpression of HMGCoA reductase without its feedback-sensing *N*-terminal domain is known to dramatically increase the MVA metabolic flux in engineered yeast. In the subsequent steps, mevalonate is phosphorylated twice to yield mevalonate diphosphate, and finally decarboxylation and dehydration give rise to the key intermediate IPP which is further converted to DMAPP by IPP isomerase. In general, all enzymes in the MVA pathway are believed to operate in the cytosol, but HMGCoA reductase is localised to the endoplasmic reticulum (ER). It should be noted, however, that there has been recent debate about the possible peroxisomal localisation of the MVA pathway in animals and plants (see Section 8.6.1).

The regulation of MVA pathway in plants is less well understood than in animals, but recent research has provided evidence for a unique regulatory role of light in the MVA pathway in plants. In order to identify the mechanisms for IPP or other precursor exchanges between plastid and cytosol, activation-tagged arabidopsis lines were screened for simultaneous lovastatin- and fosmidomycin (MEP pathway inhibitor)-resistant phenotype (Rodriguez-Concepcion *et al.*, 2004). Instead of finding an activated gene, the identified mutant was a phytochrome knock-out allele. The MVA pathway was found to be repressed

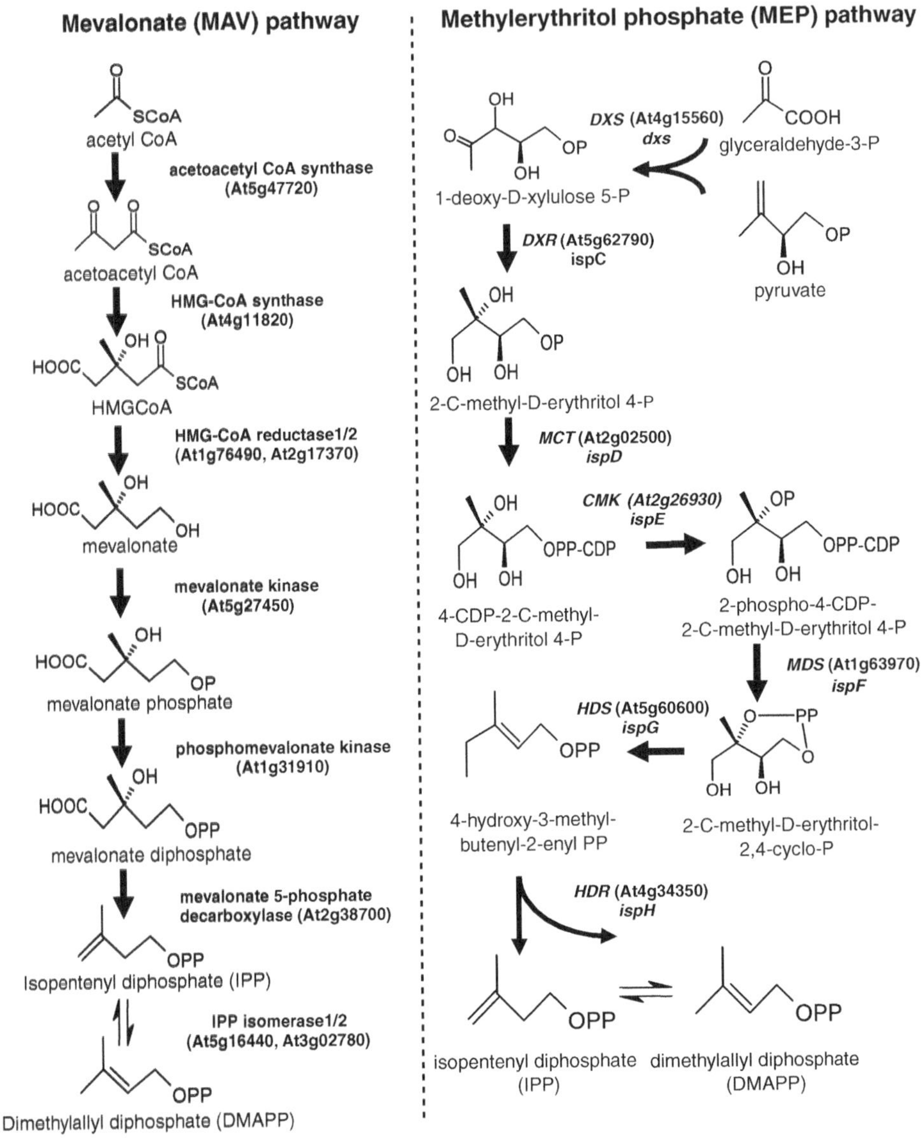

Figure 8.4 *Mevalonate and methyl erythritol phosphate pathway for IPP biosynthesis. Abbreviations used for MEP pathway genes were from the unified nomenclature proposed by Phillips* et al. *(2008b). Orthologous genes from* E. coli *and* Arabidopsis *gene numbers were also given. DXS, 1-deoxy-D-xylulose 5-phosphate synthase; DXR, 1-deoxy-D-xylylose 5-phosphate reductoisomerase; MCT, 2-C-methyl-D-erythritol 4-phosphate cytidylyltransferase; CMK, 4-(cytidine 5′-diphospho)-2-C-methyl-D-erythritol kinase; MDS, 2-C-methyl-D-erythritol 2,4-cyclodiphosphate synthase; HDS, 4-hydroxy-3-methylbut-2-enyl diphosphate synthase; HDR, 4-hydroxy-3-methylbut-2-enyl diphosphate reductase*

by light through downregulation of HMGCoA reductase from the phytochrome-mediated light signal in the seedling stage of development. Hence, de-repression (overexpression) of the *HMGCoA* in the phytochrome-mutant can overcome the lovastain-induced developmental arrest. The fact that this mutant is also resistant to fosmidomycin suggests that metabolite overflow occurs from cytosol to plastid. Since the MEP pathway is also regulated by light, fine-tuning of MVA and MEP metabolic flux by environmental factors such as light appears to be a critical element for plant physiology and development.

8.5.2 Methyl Erythritol Phosphate (MEP) Pathway

The MEP pathway is entirely independent of the MVA pathway, and it is perhaps the last hidden metabolic route evolutionarily conserved across the plant kingdom. Until the early 1990s, the MVA pathway was accepted as the sole biosynthetic pathway supplying IPP in all living organisms. However, labelled precursor-feeding experiments in bacteria showed that the labelling pattern of hopanoid (bacterial terpenoid equivalent to cholesterol in animals) deviated from the pattern predicted by the MVA pathway (Rohmer *et al.*, 1993). Similar inconsistency of labelling patterns was also reported by other groups, and subsequent experiments identified pyruvate and glyceraldehyde 3-phosphate (G3P) as novel precursors of IPP in bacteria (Rohmer *et al.*, 1996). Shortly after the initial identification of pyruvate and G3P as precursors for IPP, the entire biosynthetic pathway for IPP synthesis was elucidated in *E. coli* at the molecular level by combinatory approaches using biochemical purification, mutant isolation and bacterial genomics.

The name of this novel pathway is confusing for historical reasons. The first intermediate, 1-deoxy-D-xylulose-5-phosphate (DXP), is channelled to both IPP and pyridoxal phosphate (vitamin B6) synthesis in *E. coli*. However, the second intermediate, 2-*C*-methyl-D-erythritol-4-phosphate (MEP), is a committed molecule only for IPP biosynthesis. Therefore, the new pathway was referred to as the MEP pathway. Nonetheless, it is also called the non-mevalonate pathway, the DXP pathway, and the Rohmer pathway following the key scientific contributor's name. The nomenclature of the genes in MEP pathway is equally confusing due to the simultaneous publications by several groups. Here the genes in the MEP pathway are designated using the unified nomenclature system proposed by Rodriguez-Concepcion and co-workers (Figure 8.4) (Phillips *et al.*, 2008b).

Immediately after the characterisations of genes in the bacterial MEP pathway, homologous genes in *Arabidopsis* were identified by comparative genomics, followed by detailed characterisations using knock-out mutants and biochemical analysis. In *Arabidopsis*, all genes in the MEP pathway from *DXS* to *HDR* are encoded in single copies, and null mutants of these genes resulted in a lethal albino phenotype in seedlings. These results demonstrate the essential role of the MEP pathway in the synthesis of molecules necessary for photosynthesis.

The partial loss-of-function mutants in *Arabidopsis DXS* and *HDS* allowed physiological studies of the reduced MEP pathway flux. In one study, a leaky *HDS* mutant surprisingly gained enhanced pathogen resistance by elevating the endogenous level of salicylic acid (Gil *et al.*, 2005). Although it is not conclusive whether this unexpected link is a direct consequence of an impaired MEP pathway or an indirect result, the MEP pathway appears to modulate salicylic acid biosynthesis. A leaky mutant of *DXP* was also identified through lovastain insensitive screening, suggesting possible cross-talk between the MVA and MEP pathways in plants (Crowell *et al.*, 2003). In addition, the temperature-sensitive mutant for *DXP*, which permits normal growth at 22°C but restricts growth at 15°C, has been a useful research tool to investigate the physiological roles of the MEP pathway (Araki *et al.*, 2000). Considering the central position occupied by the MEP pathway in plants, the MEP pathway and the intermediates derived from it should be interlocked with many other physiological processes. Careful studies of those partial loss-of-function mutants will help us to understand the physiological roles of the MEP pathway in plants.

The presence of the two independent IPP supplying pathways is characteristic in different kingdoms. Animals and fungi exclusively use the MVA pathway, while most bacteria utilise the MEP pathway. In contrast, plants operate both the MVA and MEP pathways in the cytosol/ER and plastids, respectively. Intriguingly, apicomplexan protozoa operate the MEP pathway inside their unique organelle, the apicoplast (equivalent to plastid in plants). It is apparent that the MEP pathway is not present in humans but is unique to pathogenic bacteria and apicomplexan protozoa, such as the malaria parasite, *Plasmodium falciparum*. This finding opens up an exciting opportunity for a new area of antibacterial and antimalarial drugs targeting the enzymes of the MEP pathway.

8.6 Subcellular and Cellular Compartmentalisations of Terpenoid Metabolism

8.6.1 Subcellular Localisation and Metabolic Cross-Talk between MVA and MEP Pathways

The classical view of terpenoid biogenesis is that FPP is synthesized in the cytosol via the MVA pathway, while GPP and GGPP are synthesized in plastids via the MEP pathway (Chappell, 2002). Cytosol IPP is transported to mitochondria where ubiquinone is synthesized via FPP synthase. Due to this compartmentalisation, sesqui- and triterpenoids are exclusively composed of isoprenes derived from the MVA pathway, whereas mono- and diterpenoids are synthesized exclusively with isoprenes from the MEP pathway. However, mitochondrial terpenoid synthesis depends upon an external IPP supply. This traditional view requires a revision from recent new findings. In a labelling experiment using snapdragon (*Antirrhinum majus*) flowers, the MEP pathway was found to be responsible for the biosynthesis of the sesquiterpene nerolidol as well as the monoterpene myrcene, implying that unidirectional trafficking of IPP from plastid to cytosol occurs (Dudareva *et al.*, 2005). An even more complicated scenario came from a ^{13}C-CO_2 feeding experiment in *Artemisia annua*. In this experiment, the labelling patterns of the sesquiterpenoid artemisinin can best be explained by considering the incorporation of the middle isoprene unit derived from the MEP pathway, while the first and final isoprene units are produced by the MVA pathway (Schramek *et al.*, 2010).

Also, the textbook view of subcellular compartmentalisation might require revision in view of the recent finding that *Arabidopsis* isopentenyl diphosphate isomerase (IDI) is localised in peroxisomes (Sapir-Mir *et al.*, 2008). *Arabidopsis* encodes two homologous IDIs (IDI1 and IDI2), which possess two alternative translation start sites in their messages. Subcellular targeting experiments using GFP fusion internal to IDI1 and IDI2 showed that the long form (utilising the first ATG) of IDI1 was localised primarily in plastids, while the long form of IDI2 was localised in mitochondria. These data are in agreement with previous results (Phillips *et al.*, 2008a). However, both IDI1/2 short forms (using the second ATG) were targeted to peroxisomes. The peroxisomal localisation data are inconsistent with the previous result that showed the cytosolic localisations of the same proteins. However, Sapir-Mir *et al.* (2008) demonstrated that the previous results were caused by an artefact in the GFP-fusion construct. In line with the IDI localisation data, *Arabidopsis* peroxisome proteomics studies identified acetoacetyl CoA synthase (Reumann *et al.*, 2007), and in the subsequent *in silico* analyses identified the putative peroxisomal targeting motives from *Arabidopsis* HMGCoA synthase, mevalonate kinase, and mevalonate diphosphate decarboxylase (Sapir-Mir *et al.*, 2008). These results and the experimental evidence from mammalian systems reviewed by Kovacs *et al.* (2002) suggest that peroxisomes could be a subcellular site for the MVA pathway in plants (Figure 8.5). This new model requires more complicated metabolic cross-talks between subcellular organelles. Certainly, the metabolic contribution of peroxisomes to the MVA pathway requires further investigation.

8.6.2 Cellular Compartmentalisation of Terpenoid Metabolism

A majority of plant-specialised metabolites are synthesized for defensive purposes. In order to avoid self-toxicity, sophisticated sequestering mechanisms of the intermediates and/or final end-products have evolved. Extensive localisation studies in alkaloid metabolism revealed that multiple cell types coordinate the synthesis of complex alkaloids in opium poppy (*Papaver sominferum*) and Madagascar periwinkle (*Catharanthus roseus*) (Bird *et al.*, 2003; Weid *et al.*, 2004; Murata and De Luca, 2005; Murata *et al.*, 2008). Although multicellular coordination of terpenoid metabolism is poorly understood, recent data from *Artemisia annua* support the idea that different types of cells are involved in terpenoid biosynthesis (Olsson *et al.*, 2009). Trichomes of *Artemisia annua* are composed of ten cells forming a biseriate structure in which the apical cell layer is chloroplast-free and the two subapical middle cell layers are enriched with chloroplasts. Laser micro-dissection was used to separate apical cells from chloroplast-rich subapical cells. When the two cell types were subjected to reverse transcriptase PCR analysis, three transcripts responsible for the artemisinin biosynthesis were clearly detected, but no detectable signals were found in the subapical cells. The results demonstrated that the apical cells were enriched with transcripts for the artemisinin biosynthesis.

Taking the ^{13}C-labelling results of artemisinin (discussed in Section 9.6.1) and these cellular localisation data together, the artemisinin biosynthesis appears to involve metabolic cross-talk at both the subcellular and cellular levels (Schramek *et al.*, 2010). The condensation of the first DMAPP and second IPP appears to occur in the chloroplast of the subapical cells because the middle (second) isoprene unit originates from the MEP pathway in the chloroplast. However, the resulting GPP or subsequent reactant FPP must transfer from subapical cells to the apical cells where the transcripts for terpene synthase, P450 and reductase are present. After the completion of the synthesis, artemisinin again transfers to the apoplastic sac between the membrane and cuticular layer for sequestration. This model shows that the specialised metabolites are synthesized through delicate developmental programmes in plants.

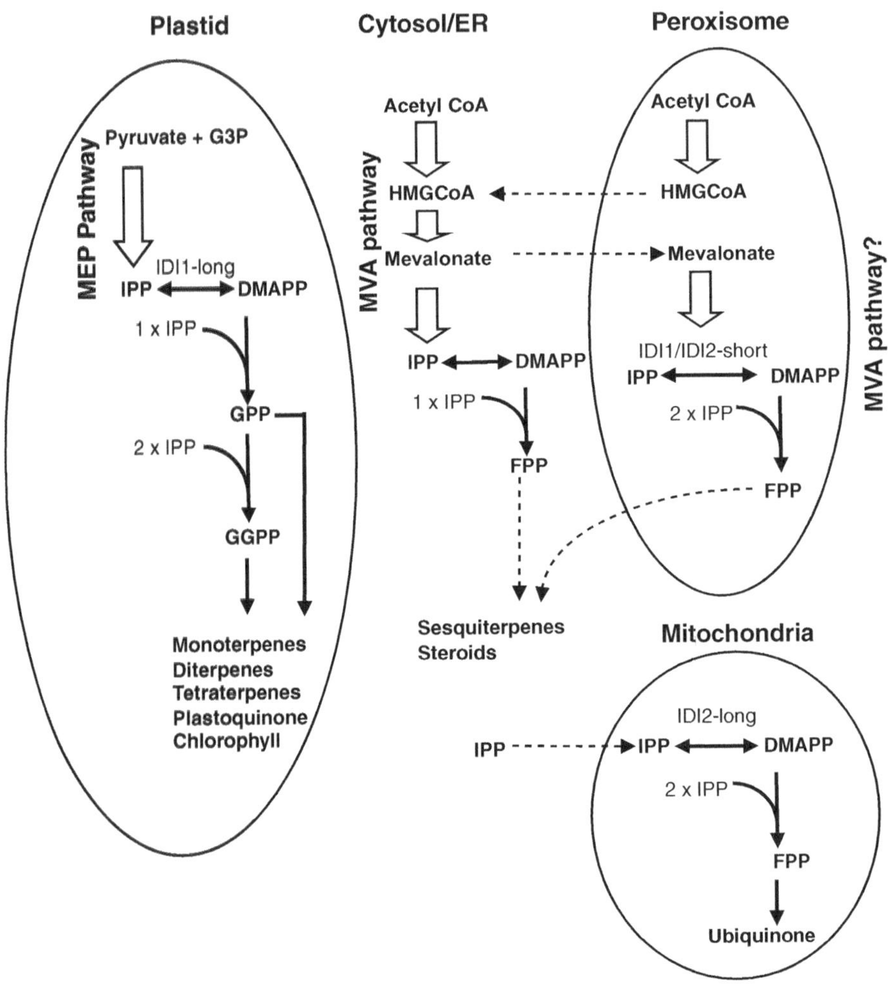

Figure 8.5 *Subcellular compartmentalisation of terpenoid metabolism in plants. A subcellular localisation study of IPP isomerase (IDI) 1 and 2 short forms in peroxisome by Sapir-Mir* et al. *(2008) suggested potential roles of peroxisome in terpenoid metabolism*

8.7 Gene Clusters in Terpenoid Metabolism

In prokaryotes, a group of genes in the same metabolic pathway is often expressed in a single polycistronic message, ensuring concerted transcription of functionally interdependent genes. In fungi, gene clusters for the synthesis of complex, specialised metabolites such as polyketides are common. In both instances, the existence of clustered metabolic genes markedly expedites the biochemical and molecular characterisation of the biochemical pathway. In contrast, metabolic genes in higher eukaryotes are known to be scattered in the large genome.

Intriguingly, the recent discovery of terpenoid gene clusters in *Arabidopsis* and rice (*Oryza sativa*) altered our view of the terpenoid metabolism at the perspective of genome evolution. In *Arabidopsis*, biochemical characterisation of four genes in a single locus revealed that these closely clustered genes are fully responsible for the synthesis of the triterpenoid thalianol (Field and Osbourn, 2008). Similarly, the existence of gene clusters for diterpenoid biosynthesis in rice was used as a basis to functionally identify one novel P450 gene (*CYP76M7*) (Swaminathan *et al.*, 2009). These two recent results strongly indicate that the early discovery of clustering of tryptophan synthase α subunit and multiple P450s in phytoalexin DIMBOA (2,4-dihydroxy-7-methoxy-1,4-benzoxazin-3-one) biosynthesis is not an isolated genomic event (Frey *et al.*, 1997), but the clustering of genes in the specialised metabolism can be much more common in plants.

Ample evidence suggests that gene expression in higher eukaryotes is not only regulated at the individual transcriptional level, but also at the status of chromatin condensation, thereby influencing the accessibility of transcription factors to specific chromosomal loci. In genome-wide analyses of *Caenorhabditis elegans*, co-expressed genes are grouped in 2–5 members along the chromosome with strong statistical significance (Roy *et al.*, 2002). Clustering genes in the terpenoid metabolism might have been under selection pressure during evolution due to the advantage of co-regulation at the chromatin level. Mechanistically, it is not difficult to imagine the clusters of gene families such as P450s in plants. Gene duplication by non-equal crossover, subsequent neofunctionalisation, and maintenance of clusters by evolutionary selection form the most simplified rationale for the creation of clustered genes of the same gene family. However, clustering of genes that originated from different gene families may have required a higher level of genome reorganisation mechanisms or genome evolution.

Whole genome sequencing is expanding to non-model plant species with a notably large genome. Future genome sequence data will reveal whether gene clustering is a common genomic feature in terpenoid and other specialised metabolisms in plants. The existence of gene clusters in terpenoid metabolism provides the hope that the characterisation of metabolic pathway can be easily achieved once we are able to sequence the entire genome of different medicinal plant species.

8.8 Metabolic Engineering of Terpenoid Metabolism

Due to the immediate applicability of terpene-based pharmaceuticals, nutraceuticals and industrial chemicals, metabolic engineering of microbes and plants to acquire desired traits through modulating terpenoid levels or altering the terpenoid profiles has been an intensive area of research. In recent years, chemical engineers have been motivated by the discoveries of key genes essential for the synthesis of valuable pharmaceuticals, and signalling and defence molecules important for agronomic traits. For example, in the 19-step taxol biosynthetic pathway, 14 molecular clones have been identified from the bark of Pacific yew (*Taxus brevifolia*) (Croteau *et al.*, 2006, and references therein). Also, three essential genes required for the biosynthesis of the antimalarial drug artemisinin, namely amorphadiene synthase, amorphadiene oxidase and artemisinic aldehyde reductase, have been identified from *Artemisia annua* (Covello, 2008, and references therein). With those clones in hands, the next step is to produce high-value molecules from innovative biological platforms.

Quantitative increases or decreases in specific terpenoids are the central theme in metabolic engineering. On the other hand, qualitative modulation of the terpenoid profile to improve the nutritional value of crops or to alter the interactions among organisms (e.g. plant-plant or plant-insect) is another important aspect. Here, the latter will be discussed in the context of plant metabolic engineering, and yield improvement will be discussed in the context of microbial engineering and emerging synthetic biology.

8.8.1 Microbial Metabolic Engineering

The microbial production of valuable terpenoids through reliable fermentation technologies has been an ultimate goal for metabolic engineers. The infrastructures for fermentation are already in place, and many chemicals including bio-ethanol and amino acids have been successfully produced on an industrial scale by fermentations. The development of microbial terpenoid production is a rational approach when considering the availability of inexpensive renewable carbon sources for fermentation. The progress of genetic and metabolic engineering in recent years has proven to be realistic for the manufacture of customised terpenoids by microbes.

Terpenoid microbial engineering has been focused on increasing the yield of amorphadiene and artimisinic acid, the central precursors of antimalarial drug artemisinin, in two model microbes, *E. coli* and yeast (*Saccharomyces cerevisiae*). One breakthrough in *E. coli* metabolic engineering came from the complete reconstitution of the *S. cerevisiae* MVA pathway in *E. coli*, which does not operate the MVA pathway and whose IPP supply is solely dependent on the MEP pathway (Martin *et al.*, 2003). A series of improvements in *E. coli* MVA pathway fermentation procedures optimised amorphadiene production titre at 27 g/l (Tsuruta *et al.*, 2009). Such a high titre of hydrocarbon production, however, was not fully coupled with the subsequent oxidation reactions by the cytochrome P450 enzyme, amorphadiene oxidase (AMO), although marginal success of oxidative conversion with 100 mg/l of artemisinic acid was achieved by engineering the AMO membrane domain (Chang *et al.*, 2007). Engineering of AMO is necessary to properly express membrane-bound P450 enzyme in the prokaryotic platform, which lacks an ER system. Although bacteria have P450s, they are soluble and often receive electrons from a two-component system analogous to ferredoxin and ferredoxin reductase, while plant P450s interact with another membrane-bound enzyme, cytochrome P450 reductase. Interestingly, one bacterial P450 BM3 from *Bacillus megaterium* forms natural fusion with reductase (Narhi and Fulco, 1986), and the crystal structure of this P450-reductase hybrid enzyme has been determined (Ravichandran *et al.*, 1993). The BM3 scaffold may be a good reference for engineering AMO that is soluble and fused with reductase without losing its native catalytic property.

Comparable efforts by two independent groups have been made to enhance FPP flux in yeast. The key difference is that one group employed a conditional downregulation of squalene synthase, a competing pathway for the FPP precursor, and the other group isolated a mutant that can take up steroids from media in the squalene synthase knock-out background (Ro *et al.*, 2006; Takahashi *et al.*, 2007). The production titre of amorphadiene was an order of magnitude lower in yeast in comparison with the *E. coli* system with engineered MVA pathway. However, the *in vivo* catalytic coupling of ADS and AMO was superior in yeast, with 2.5 g/l of artemisinic acid production (Lenihan *et al.*, 2008; Ro *et al.*, 2008).

One key lesson from all these microbial pathway optimisations is that the catalytic balance, or 'just enough' enzyme, is far more important than simply a high abundance of enzymes in cells. In the MVA pathway comprising seven enzymes, accumulation of HMGCoA caused cytotoxicity in *E. coli* (Pitera *et al.*, 2007). Moreover, unnecessary over-production of recombinant enzymes can slow down the biomass increase of microbes. In order to balance enzyme activities, a number of elegant methods for *in vivo* enzyme titrations have been designed using synthetic biology approaches. Synthetic biology, in essence, focuses on the creation of a tunable and predictable biological system for a specific purpose. Systematic modulations of translation efficiency of multiple genes in MVA pathways resulted in a marked increase in the final product titre (Pfleger *et al.*, 2006). Also, a genome-wide optimisation of the ribosomal binding sites in the MEP pathway genes in *E. coli* improved the metabolic flux by multiplex automated genome engineering (MAGE), where billions of combinatory genome variants are evaluated (Wang *et al.*, 2009). At the post-translational level, it has been believed that multi-enzyme complexes (or metabolons) play key roles in reducing the release of cytotoxic intermediates and enhancing the overall flux and cell viability. Recently, synthetic protein scaffolding was devised to anchor three enzymes (acetoacetyl CoA synthase, HMGS and HMGR) in the MVA pathway, thereby creating a synthetic enzyme complex in *E. coli* (Dueber *et al.*, 2009). The synthetic enzyme complex increased the product yield by 77-fold.

The ultimate goal of microbial engineering is to create a strain that can channel most of its energy and carbon sources to the synthesis of custom terpenoid compounds while they maintain a healthy cell physiology necessary for survival and cell division. Some recent studies demonstrate that semi-synthetic microbes for custom terpenoid production will be achieved in the near future. For example, microbial genomes become more manipulative as demonstrated by the artificial reduction in the *E. coli* genome size by 8% without perturbing its growth rate (Kolisnychenko *et al.*, 2002). In addition, a 582-Kb synthetic genome of *Mycoplasma genitalium* was chemically synthesized *in vitro* with the removal of its pathogenicity loci (Gibson *et al.*, 2008). Examples of these top-down (genome reduction) and bottom-up (genome synthesis) approaches suggest that synthetic *E. coli* with a small genome (native genome size, 4.6 Mbp) designed for dedicated terpenoid production could indeed be feasible.

8.8.2 Plant Metabolic Engineering

The stunning achievements in microbial engineering has resulted in the perception that plant metabolic engineering is unsatisfactory. However, it should be emphasised that the photosynthetic capability programmed into plant metabolism is the only biochemical process to fix atmospheric carbon dioxide. Therefore, fine chemical and feedstock production using plants has fundamental advantages over microbial fermentation which depends upon external energy input.

Since tobacco (*Nicotiana tobacum*) plants grow quickly and can be transformed easily by *Agrobacterium*, the feasibility of using tobacco as a plant platform to produce terpenoids has been evaluated. In initial trials, transgenic plants transformed with TPS did not result in detectable levels of targeted terpenenoids due to limited FPP substrate availability in cells. This limitation was overcome by simultaneous targeting of FPP synthase and sesquiterepene synthases (*ADS* and patchoulol synthase) to the IPP-rich chloroplasts in tobacco. The

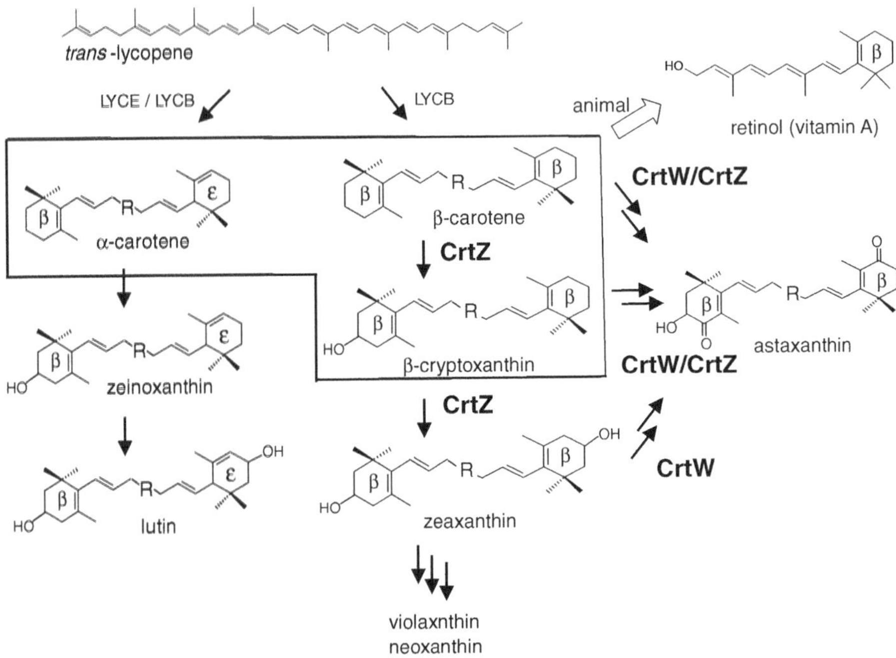

Figure 8.6 A biosynthetic pathway for various oxygenated carotenoids including pro-vitamin A (boxed area) and commercially important astaxanthin. Bacterial genes (CrtZ and CrtW) for the conversion of β-carotene to astaxanthin are shown in bold. Detailed information about carotenoid metabolism is given in Chapter 13

result was a more than 1000-fold increase in amorphadiene and patchoulol synthesis (~30 mg/g fresh weight) (Wu *et al.*, 2006). In another case, the novel marine bacteria genes, *CrtZ* and *CrtW*, which can convert β-carotene to astaxanthin (Figure 8.6), were expressed in tobacco chloroplast by plastid transformation (Hasunuma *et al.*, 2008). This transplastomics engineering mimicked natural conditions because bacteria genes were expressed in prokaryote-origin chloroplasts and hence artificial chloroplast-targeting was not necessary. Remarkably, 5 mg/g dry weight of astaxanthin was synthesized while the photosynthesis and growth rates of the transgenic plants were unaffected. This result is surprising because β-carotene is the precursor for photosynthetic pigments. Perhaps the β-carotene pool in the tobacco chloroplast is sufficiently large to support both the β-xanthophyll and foreign astaxanthin biosynthetic pathways. These data demonstrate that tobacco plants can be used to produce high-value small molecules.

The most significant progress in agro-biotechnology of plant terpenoid in the last five years is the engineering of *Arabidopsis* to acquire a novel chemical trait to attract natural parasitoids or carnivorous mites. Compelling experimental evidence for eco-physiological roles of sesquiterpenes was provided by three independent experiments involving transgenic *Arabidopsis*. Nerolidol synthase was targeted to IPP-rich mitochondria to allow synthesis of two new sesquiterpenes by which carnivorous mites, a natural predator of aphids, can be attracted, thereby defending plants against herbivores (Kappers *et al.*, 2005).

Similarly, *Arabidopsis* lines transgenic for either peppermint (*E*)-β-farnesene or maize (*E*)-α-bergamotene/(*E*)-β-farnesene were able to recruit natural parasitoids of herbivores (Beale *et al.*, 2006; Schnee *et al.*, 2006). Although the level of sesquiterpene production was not sufficiently high in all three cases, insects and mites can sense minute amounts of signalling chemicals. Therefore, the subtle qualitative alteration of the terpenoid profile in *Arabidopsis* was sufficient to modulate plant–insect interactions.

In all of the above cases, constitutive CaMV 35S promoters were used to drive the ectopic TPS gene expressions in plants. However, plants have always co-evolved specialised cellular or non-cellular compartments, such as trichome sacs, to sequester toxic terpenoids. Considering the natural strategies that plants adopt, future directions should focus on cell-specific expression of foreign genes.

8.9 Concluding Remarks

There can be no doubt that progress in terpenoid research in the last five years has been impressive. Yet many unsolved questions still remain to be answered. The cellular and subcellular compartmentalisation of terpenoid metabolism appears to be more complicated than originally thought. The discovery of *cisoid* substrates for TPS raises a question as to the diversity and functions of *cisoid* terpenoids. What are the molecular mechanisms enabling the evolution and maintenance of the terpenoid gene clusters, and how commonly do the clusters occur in various plant genomes? Can we create a synthetic microbe for dedicated terpenoid production reaching 100% of the theoretical metabolic yield? Although these are all difficult problems, there is every indication that they can be overcome by emerging technologies. Sequencing costs have become dramatically cheaper due to the invention of massive parallel sequencers (e.g. 454 and illumina), and cost-effective synthesis of large pieces of DNA has become realistic. Various reverse genetic tools (RNAi and artificial miRNA) have been developed for multiple plant species for functional genomics. In particular, targeted genome editing has become feasible due to the development of zinc finger nuclease (Shukla *et al.*, 2009; Townsend *et al.*, 2009). Creative integration of these leading technologies and classical terpenoid biochemistry promises a bright and exciting future.

References

Akiyama, K., Matsuzaki, K. and Hayashi, H. (2005) Plant sesquiterpenes induce hyphal branching in arbuscular mycorrhizal fungi. *Nature*, **435**, 824–827.

Araki, N., Kusumi, K., Masamoto, K., Niwa, Y. and K., I. (2000) Temperature-sensitive Arabidopsis mutant defective in 1-deoxy-D-xylulose 5-phosphate synthase within the plastid non-mevalonate pathway of isoprenoid biosynthesis. *Physiol. Plant.*, **108**, 19–24.

Beale, M.H., Birkett, M.A., Bruce, T.J., Chamberlain, K., Field, L.M., Huttly, A.K., Martin, J.L., Parker, R., Phillips, A.L., Pickett, J.A., Prosser, I.M., Shewry, P.R., Smart, L.E., Wadhams, L.J., Woodcock, C.M. and Zhang, Y. (2006) Aphid alarm pheromone produced by transgenic plants affects aphid and parasitoid behavior. *Proc. Natl. Acad. Sci. USA*, **103**, 10509–10513.

Bird, D.A., Franceschi, V.R. and Facchini, P.J. (2003) A tale of three cell types: alkaloid biosynthesis is localized to sieve elements in opium poppy. *Plant Cell*, **15**, 2626–2635.

Bohlmann, J. and Keeling, C.I. (2008) Terpenoid biomaterials. *Plant J.*, **54**, 656–669.

Buckingham, J. (2007) *Dictionary of Natural Products on DVD*. CRC Press, London.

Calvin, M. (1980) Hydrocarbons from plants: analytical methods and observation. *Naturwissenschaften*, **67**, 525–533.

Caruthers, J.M., Kang, I., Rynkiewicz, M.J., Cane, D.E. and Christianson, D.W. (2000) Crystal structure determination of aristolochene synthase from the blue cheese mold, Penicillium roqueforti. *J. Biol. Chem*, **275**, 25533–25539.

Chang, M.C., Eachus, R.A., Trieu, W., R Ro, D.K. and Keasling, J.D. (2007) Engineering Escherichia coli for production of functionalized terpenoids using plant P450s. *Nat. Chem. Biol.*, **3**, 274–277.

Chappell, J. (2002) The genetics and molecular genetics of terpene and sterol origami. *Curr. Opin. Plant Biol.*, **5**, 151–157.

Covello, P.S. (2008) Making artemisinin. *Phytochemistry*, **69**, 2881–2885.

Croteau, R., Ketchum, R.E.B., Long, R.M., Kaspera, R. and Wildung, M.R. (2006) Taxol biosynthesis and molecular genetics. *Phytochem. Rev.*, **5**, 75–97.

Crowell, D.N., Packard, C.E., Pierson, C.A., Giner, J.L., Downes, B.P. and Chary, S.N. (2003) Identification of an allele of CLA1 associated with variegation in *Arabidopsis thaliana*. *Physiol. Plant.*, **118**, 29–37.

Dewick, P.M. (2009) *Medicinal Natural Products: A Biosynthetic Approach* (3rd edn). John Wiley and Sons Ltd, Chichester, UK.

Dudareva, N., Andersson, S., Orlova, I., Gatto, N., Reichelt, M., Rhodes, D., Boland, W. and Gershenzon, J. (2005) The nonmevalonate pathway supports both monoterpene and sesquiterpene formation in snapdragon flowers. *Proc. Natl. Acad. Sci. USA*, **102**, 933–938.

Dueber, J.E., Wu, G.C., Malmirchegini, G.R., Moon, T.S., Petzold, C.J., Ullal, A.V., Prather, K.L. and Keasling, J.D. (2009) Synthetic protein scaffolds provide modular control over metabolic flux. *Nat. Biotechnol.*, **27**, 753–759.

Field, B. and Osbourn, A.E. (2008) Metabolic diversification – independent assembly of operon-like gene clusters in different plants. *Science*, **320**, 543–547.

Frey, M., Chomet, P., Glawischnig, E., Stettner, C., Grun, S., Winklmair, A., Eisenreich, W., Bacher, A., Meeley, R.B., Briggs, S.P., Simcox, K. and Gierl, A. (1997) Analysis of a chemical plant defense mechanism in grasses. *Science*, **277**, 696–699.

Gibson, D.G., Benders, G.A., Andrews-Pfannkoch, C., Denisova, E.A., Baden-Tillson, H., Zaveri, J., Stockwell, T.B., Brownley, A., Thomas, D.W., Algire, M.A., Merryman, C., Young, L., Noskov, V.N., Glass, J.I., Venter, J.C., Hutchison, C.A., 3rd and Smith, H.O. (2008) Complete chemical synthesis, assembly, and cloning of a *Mycoplasma genitalium* genome. *Science*, **319**, 1215–1220.

Gil, M.J., Coego, A., Mauch-Mani, B., Jorda, L. and Vera, P. (2005) The Arabidopsis csb3 mutant reveals a regulatory link between salicylic acid-mediated disease resistance and the methyl-erythritol 4-phosphate pathway. *Plant J.*, **44**, 155–166.

Gomez-Roldan, V., Fermas, S., Brewer, P.B., Puech-Pages, V., Dun, E.A., Pillot, J.P., Letisse, F., Matusova, R., Danoun, S., Portais, J.C., Bouwmeester, H., Becard, G., Beveridge, C.A., Rameau, C. and Rochange, S.F. (2008) Strigolactone inhibition of shoot branching. *Nature*, **455**, 189–194.

Greenhagen, B. and Chappell, J. (2001) Molecular scaffolds for chemical wizardry: learning nature's rules for terpene cyclases. *Proc. Natl. Acad. Sci. USA*, **98**, 13479–13481.

Greenhagen, B.T., O'Maille, P.E., Noel, J.P. and Chappell, J. (2006) Identifying and manipulating structural determinates linking catalytic specificities in terpene synthases. *Proc. Natl. Acad. Sci. USA*, **103**, 9826–9831.

Hasunuma, T., Miyazawa, S., Yoshimura, S., Shinzaki, Y., Tomizawa, K., Shindo, K., Choi, S.K., Misawa, N. and Miyake, C. (2008) Biosynthesis of astaxanthin in tobacco leaves by transplastomic engineering. *Plant J.*, **55**, 857–868.

Kappers, I.F., Aharoni, A., van Herpen, T.W., Luckerhoff, L.L., Dicke, M. and Bouwmeester, H.J. (2005) Genetic engineering of terpenoid metabolism attracts bodyguards to Arabidopsis. *Science*, **309**, 2070–2072.

Keeling, C.I., Weisshaar, S., Lin, R.P. and Bohlmann, J. (2008) Functional plasticity of paralogous diterpene synthases involved in conifer defense. *Proc. Natl. Acad. Sci. USA*, **105**, 1085–1090.

Kolisnychenko, V., Plunkett, G., Herring, C.D., Feher, T., Posfai, J., Blattner, F.R. and Posfai, G. (2002) Engineering a reduced *Escherichia coli* genome. *Genome Res.*, **12**, 640–647.

Kovacs, W.J., Olivier, L.M. and Krisans, S.K. (2002) Central role of peroxisomes in isoprenoid biosynthesis. *Prog. Lipid Res.*, **41**, 369–391.

Lenihan, J.R., Tsuruta, H., Diola, D., Renninger, N.S. and Regentin, R. (2008) Developing an industrial artemisinic acid fermentation process to support the cost-effective production of antimalarial artemisinin-based combination therapies. *Biotechnol. Prog.*, **24**, 1026–1032.

Lesburg, C.A., Zhai, G., Cane, D.E. and Christianson, D.W. (1997) Crystal structure of pentalenene synthase: mechanistic insights on terpenoid cyclization reactions in biology. *Science*, **277**, 1820–1824.

Lytton, J., Westlin, M. and Hanley, M.R. (1991) Thapsigargin inhibits the sarcoplasmic or endoplasmic reticulum Ca-ATPase family of calcium pumps. *J. Biol. Chem.*, **266**, 17067–17071.

Martin, V.J., Pitera, D.J., Withers, S.T., Newman, J.D. and Keasling, J.D. (2003) Engineering a mevalonate pathway in *Escherichia coli* for production of terpenoids. *Nat. Biotechnol.*, **21**, 796–802.

Miller, B., Oschinski, C. and Zimmer, W. (2001) First isolation of an isoprene synthase gene from poplar and successful expression of the gene in *Escherichia coli*. *Planta*, **213**, 483–487.

Murata, J. and De Luca, V. (2005) Localization of tabersonine 16-hydroxylase and 16-OH-tabersonine-16-*O*-methyltransferase to leaf epidermal cells defines them as a major site of precursor biosynthesis in the vindoline pathway in *Catharanthus roseus*. *Plant J.*, **44**, 581–594.

Murata, J., Roepke, J., Gordon, H. and De Luca, V. (2008) The leaf epidermome of *Catharanthus roseus* reveals its biochemical specialization. *Plant Cell*, **20**, 524–542.

Narhi, L.O. and Fulco, A.J. (1986) Characterization of a catalytically self-sufficient 119,000-dalton cytochrome P-450 monooxygenase induced by barbiturates in *Bacillus megaterium*. *J. Biol. Chem.*, **261**, 7160–7169.

Olsson, M.E., Olofsson, L.M., Lindahl, A.L., Lundgren, A., Brodelius, M. and Brodelius, P.E. (2009) Localization of enzymes of artemisinin biosynthesis to the apical cells of glandular secretory trichomes of *Artemisia annua* L. *Phytochemistry*, **70**, 1123–1128.

O'Maille, P.E., Malone, A., Dellas, N., Andes Hess, B., Jr., Smentek, L., Sheehan, I., Greenhagen, B.T., Chappell, J., Manning, G. and Noel, J.P. (2008) Quantitative exploration of the catalytic landscape separating divergent plant sesquiterpene synthases. *Nat. Chem. Biol.*, **4**, 617–623.

Pandit, J., Danley, D.E., Schulte, G.K., Mazzalupo, S., Pauly, T.A., Hayward, C.M., Hamanaka, E.S., Thompson, J.F. and Harwood, H.J., Jr. (2000) Crystal structure of human squalene synthase. A key enzyme in cholesterol biosynthesis. *J. Biol. Chem.*, **275**, 30610–30617.

Pfleger, B.F., Pitera, D.J., Smolke, C.D. and Keasling, J.D. (2006) Combinatorial engineering of intergenic regions in operons tunes expression of multiple genes. *Nat. Biotechnol.*, **24**, 1027–1032.

Phillips, M.A., D'Auria, J.C., Gershenzon, J. and Pichersky, E. (2008a) The Arabidopsis thaliana type I isopentenyl diphosphate isomerases are targeted to multiple subcellular compartments and have overlapping functions in isoprenoid biosynthesis. *Plant Cell*, **20**, 677–696.

Phillips, M.A., Leon, P., Boronat, A. and Rodriguez-Concepcion, M. (2008b) The plastidial MEP pathway: unified nomenclature and resources. *Trends Plant Sci.*, **13**, 619–623.

Pitera, D.J., Paddon, C.J., Newman, J.D. and Keasling, J.D. (2007) Balancing a heterologous mevalonate pathway for improved isoprenoid production in *Escherichia coli*. *Metab. Eng.*, **9**, 193–207.

Ravichandran, K.G., Boddupalli, S.S., Hasermann, C.A., Peterson, J.A. and Deisenhofer, J. (1993) Crystal structure of hemoprotein domain of P450BM-3, a prototype for microsomal P450s. *Science*, **261**, 731–736.

Reumann, S., Babujee, L., Ma, C., Wienkoop, S., Siemsen, T., Antonicelli, G.E., Rasche, N., Luder, F., Weckwerth, W. and Jahn, O. (2007) Proteome analysis of arabidopsis leaf peroxisomes reveals novel targeting peptides, metabolic pathways, and defense mechanisms. *Plant Cell*, **19**, 3170–3193.

Ritter, H. and Schulz, G.E. (2004) Structural basis for the entrance into the phenylpropanoid metabolism catalyzed by phenylalanine ammonia-lyase. *Plant Cell*, **16**, 3426–3436.

Ro, D.K., Paradise, E.M., Ouellet, M., Fisher, K.J., Newman, K.L., Ndungu, J.M., Ho, K.A., Eachus, R.A., Ham, T.S., Kirby, J., Chang, M.C., Withers, S.T., Shiba, Y., Sarpong, R. and Keasling, J.D. (2006) Production of the antimalarial drug precursor artemisinic acid in engineered yeast. *Nature*, **440**, 940–943.

Ro, D.K., Ouellet, M., Paradise, E.M., Burd, H., Eng, D., Paddon, C.J., Newman, J.D. and Keasling, J.D. (2008) Induction of multiple pleiotropic drug resistance genes in yeast engineered to produce an increased level of anti-malarial drug precursor, artemisinic acid. *BMC Biotechnol.*, **8**, 83.

Rodriguez-Concepcion, M., Fores, O., Martinez-Garcia, J.F., Gonzalez, V., Phillips, M.A., Ferrer, A. and Boronat, A. (2004) Distinct light-mediated pathways regulate the biosynthesis and exchange of isoprenoid precursors during arabidopsis seedling development. *Plant Cell*, **16**, 144–156.

Rohmer, M., Knani, M., Simonin, P., Sutter, B. and Sahm, H. (1993) Isoprenoid biosynthesis in bacteria: a novel pathway for the early steps leading to isopentenyl diphosphate. *Biochem J.*, **295** (Pt 2), 517–524.

Rohmer, M., Seemann, M., Horbach, S., Bringer-Meyer, S. and Sahm, H. (1996) Glycealdehyde 3-phosphate and pyruvate as precursors of isoprenic units in an alternative non-mevalonate pathway for terpenoid biosynthesis. *J. Am. Chem. Soc.*, **118**, 2564–2566.

Roth, B.L., Baner, K., Westkaemper, R., Siebert, D., Rice, K.C., Steinberg, S., Ernsberger, P. and Rothman, R.B. (2002) Salvinorin A: a potent naturally occurring non-nitrogenous k-opioid selective agonist. *Proc. Natl. Acad. Sci. USA*, **99**, 11934–11939.

Roy, P.J., Stuart, J.M., Lund, J. and Kim, S.K. (2002) Chromosomal clustering of muscle-expressed genes in *Caenorhabditis elegans*. *Nature*, **418**, 975–979.

Rynkiewicz, M.J., Cane, D.E. and Christianson, D.W. (2001) Structure of trichodiene synthase from *Fusarium sporotrichioides* provides mechanistic inferences on the terpene cyclization cascade. *Proc. Natl. Acad. Sci. USA*, **98**, 13543–13548.

Sallaud, C., Rontein, D., Onillon, S., Jabes, F., Duffe, P., Giacalone, C., Thoraval, S., Escoffier, C., Herbette, G., Leonhardt, N., Causse, M. and Tissier, A. (2009) A novel pathway for sesquiterpene biosynthesis from *Z*, *Z*-farnesyl pyrophosphate in the wild tomato *Solanum habrochaites*. *Plant Cell*, **21**, 301–317.

Sapir-Mir, M., Mett, A., Belausov, E., Tal-Meshulam, S., Frydman, A., Gidoni, D. and Eyal, Y. (2008) Peroxisomal localization of arabidopsis isopentenyl diphosphate isomerases suggests that part of the plant isoprenoid mevalonic acid pathway is compartmentalized to peroxisomes. *Plant Physiol.*, **148**, 1219–1228.

Schilmiller, A.L., Schauvinhold, I., Larson, M., Xu, R., Charbonneau, A.L., Schmidt, A., Wilkerson, C., Last, R.L. and Pichersky, E. (2009) Monoterpenes in the glandular trichomes of tomato are synthesized from a neryl diphosphate precursor rather than geranyl diphosphate. *Proc. Natl. Acad. Sci. USA*, **106**, 10865–10870.

Schnee, C., Kollner, T.G., Held, M., Turlings, T.C., Gershenzon, J. and Degenhardt, J. (2006) The products of a single maize sesquiterpene synthase form a volatile defense signal that attracts natural enemies of maize herbivores. *Proc. Natl. Acad. Sci. USA*, **103**, 1129–1134.

Schramek, N., Wang, H., Romisch-Margl, W., Keil, B., Radykewicz, T., Winzenhorlein, B., Beerhues, L., Bacher, A., Rohdich, F., Gershenzon, J., Liu, B. and Eisenreich, W. (2010) Artemisinin biosynthesis in growing plants of *Artemisia annua*. A $^{13}CO_2$ study. *Phytochemistry*, **71**, 179–187.

Schwartz, S.H., Qin, X. and Loewen, M.C. (2004) The biochemical characterization of two carotenoid cleavage enzymes from arabidopsis indicates that a carotenoid-derived compound inhibits lateral branching. *J. Biol. Chem.*, **279**, 46940–46945.

Shukla, V.K., Doyon, Y., Miller, J.C., DeKelver, R.C., Moehle, E.A., Worden, S.E., Mitchell, J.C., Arnold, N.L., Gopalan, S., Meng, X., Choi, V.M., Rock, J.M., Wu, Y.Y., Katibah, G.E., Zhifang, G., McCaskill, D., Simpson, M.A., Blakeslee, B., Greenwalt, S.A., Butler, H.J., Hinkley, S.J., Zhang, L., Rebar, E.J., Gregory, P.D. and Urnov, F.D. (2009) Precise genome modification in the crop species *Zea mays* using zinc-finger nucleases. *Nature*, **459**, 437–441.

Starks, C.M., Back, K., Chappell, J. and Noel, J.P. (1997) Structural basis for cyclic terpene biosynthesis by tobacco 5-*epi*-aristolochene synthase. *Science*, **277**, 1815–1820.

Swaminathan, S., Morrone, D., Wang, Q., Fulton, D.B. and Peters, R.J. (2009) CYP76M7 is an *ent*-cassadiene C11a-hydroxylase defining a second multifunctional diterpenoid biosynthetic gene cluster in rice. *Plant Cell*, **21**, 3315–3325.

Takahashi, S., Yeo, Y., Greenhagen, B.T., McMullin, T., Song, L., Maurina-Brunker, J., Rosson, R., Noel, J.P. and Chappell, J. (2007) Metabolic engineering of sesquiterpene metabolism in yeast. *Biotechnol. Bioeng.*, **97**, 170–181.

Tarshis, L.C., Yan, M., Poulter, C.D. and Sacchettini, J.C. (1994) Crystal structure of recombinant farnesyl diphosphate synthase at 2.6-A resolution. *Biochemistry*, **33**, 10871–10877.

Townsend, J.A., Wright, D.A., Winfrey, R.J., Fu, F., Maeder, M.L., Joung, J.K. and Voytas, D.F. (2009) High-frequency modification of plant genes using engineered zinc-finger nucleases. *Nature*, **459**, 442–445.

Tsuruta, H., Paddon, C.J., Eng, D., Lenihan, J.R., Horning, T., Anthony, L.C., Regentin, R., Keasling, J.D., Renninger, N.S. and Newman, J.D. (2009) High-level production of amorpha-4,11-diene, a precursor of the antimalarial agent artemisinin, in *Escherichia coli*. *PLoS One*, **4**, e4489.

Umehara, M., Hanada, A., Yoshida, S., Akiyama, K., Arite, T., Takeda-Kamiya, N., Magome, H., Kamiya, Y., Shirasu, K., Yoneyama, K., Kyozuka, J. and Yamaguchi, S. (2008) Inhibition of shoot branching by new terpenoid plant hormones. *Nature*, **455**, 195–200.

van Beilen, J.B. and Poirier, Y. (2007) Establishment of new crops for the production of natural rubber. *Trends Biotechnol.*, **25**, 522–529.

Wang, H.H., Isaacs, F.J., Carr, P.A., Sun, Z.Z., Xu, G., Forest, C.R. and Church, G.M. (2009) Programming cells by multiplex genome engineering and accelerated evolution. *Nature*, **460**, 894–898.

Weid, M., Ziegler, J. and Kutchan, T.M. (2004) The roles of latex and the vascular bundle in morphine biosynthesis in the opium poppy, *Papaver somniferum*. *Proc. Natl. Acad. Sci. USA*, **101**, 13957–13962.

Wu, S., Schalk, M., Clark, A., Miles, R.B., Coates, R. and Chappell, J. (2006) Redirection of cytosolic or plastidic isoprenoid precursors elevates terpene production in plants. *Nat Biotechnol.*, **24**, 1441–1447.

Yoshikuni, Y., Ferrin, T.E. and Keasling, J.D. (2006) Designed divergent evolution of enzyme function. *Nature*, **440**, 1078–1082.

9

Benzylisoquinoline Alkaloid Biosynthesis

Isabel Desgagné-Penix and Peter J Facchini

9.1 Introduction

Alkaloids are a large and diverse group of low-molecular weight, nitrogenous compounds found in about 20% of flowering plant species. Most plant alkaloids are derived from amino acids, although other compounds such as purine nucleotides also serve as precursors. Alkaloid classification is based upon the skeletal configuration of their carbon–nitrogen ring systems, which include isoquinoline, indole, pyrrole, pyridine and piperidine structures (Petterson *et al.*, 1991). The toxicity and bitter taste of alkaloids are thought to play a role in the defence of plants against herbivory and pathogen challenge.

Benzylisoquinoline alkaloids (BIAs) are derived from the aromatic amino acid tyrosine and consist of approximately 2500 known compounds. Over 90% of the plants that produce BIAs are found in members of the basal angiosperm families Papaveraceae, Berberidaceae, Menispermaceae, Ranunculaceae and Magnoliaceae (Ziegler and Facchini, 2008; Facchini and DeLuca, 2008). Several BIAs are used for medicinal purposes owing to their potent pharmacological activity (Table 9.1), which is a putative indication of their biological function. For example, the effectiveness of morphine as an analgesic or (+)-tubocurarine as a muscle relaxant suggests a role as herbivore deterrents. Similarly, the antimicrobial properties of sanguinarine and berberine suggest that they confer protection against pathogens. Many ancient civilizations were aware of the pharmacological properties of BIA-producing plants. Opium, the dried latex of opium poppy (*Papaver somniferum*) containing morphine, codeine and several other alkaloids, was among the first drugs used by humankind. The

Plant Metabolism and Biotechnology, First Edition. Edited by Hiroshi Ashihara, Alan Crozier, and Atsushi Komamine.

Table 9.1 Important benzylisoquinoline alkaloids

Compound	Structural type	Plant source (or example)	Application
Berberine	Protoberberine	*Berberis vulgaris*	Antimicrobial
Codeine	Morphinan	*Papaver somniferum*	Antitussive, analgesic
Jatrorrhizine	Protoberberine	*Tinospora cordifolia*	Antimalarial
Laudanine	Simple	*Papaver somniferum*	Poison
Magnoflorine	Aporphine	*Coptidis rhizoma*	Antioxidant
Morphine	Morphinan	*Papaver somniferum*	Analgesic
Noscapine	Phthalideisoquinoline	*Papaver somniferum*	Antitussive, anticancer
Palmatine	Protoberberine	*Berberis aristata*	Hypotensive
Papaverine	Simple	*Papaver somniferum*	Vasodilator
Sanguinarine	Benzophenanthridine	*Sanguinaria canadensis*	Anti-microbial
Tubocurarine	Bisbenzylisoquinoline	*Chondrodendron tomentosum*	Muscle relaxant

Sumerians described the medicinal properties of opium as early as 4000 BCE, and the ancient Assyrians, Greeks and Romans all used opium for relieving pain and inducing sleep (Petterson *et al.*, 1991).

Different structural categories of BIAs are generally associated with specific plant taxa, leading to the use of various species in the study of BIA metabolism. Opium poppy has emerged as the premier model system especially with respect to investigations of morphinan alkaloid biosynthesis. Other widely used plants include *Eschscholzia californica* (California poppy) and *Sanguinaria canadensis* (Bloodroot), which produce mainly benzophenanthridine alkaloids such as sanguinarine, and *Thalictrum flavum* (Meadow-rue), *Berberis stolonifera* (Barberry) and *Coptis japonica* (Japanese goldthread), which accumulate protoberberine and aporphine alkaloids (Figure 9.1). Overall, molecular clones encoding more than 20 enzymes involved in BIA biosynthesis have been isolated, and the pace of gene discovery is increasing with the application of new technologies. In this chapter, we review BIA biosynthesis in plants with a focus on localisation, regulation and biotechnological applications.

9.2 Biosynthesis

9.2.1 (*S*)-Norcoclaurine

The benzylisoquinoline skeleton is the building block in the formation of several structural categories of BIAs including aporphines, benzophenanthridines, bisbenzylisoquinolines, protopines, protoberberines and morphinans (Figure 9.1). BIA biosynthesis begins with a combination of decarboxylations, *meta*-hydroxylations and deaminations that convert tyrosine to both dopamine and 4-hydroxyphenylacetaldehyde (4-HPAA) (Figure 9.2). Cognate cDNA encoding tyrosine decarboxylase (TYDC), which converts L-tyrosine and L-DOPA to tyramine and dopamine, respectively, were among the first isolated BIA biosynthetic

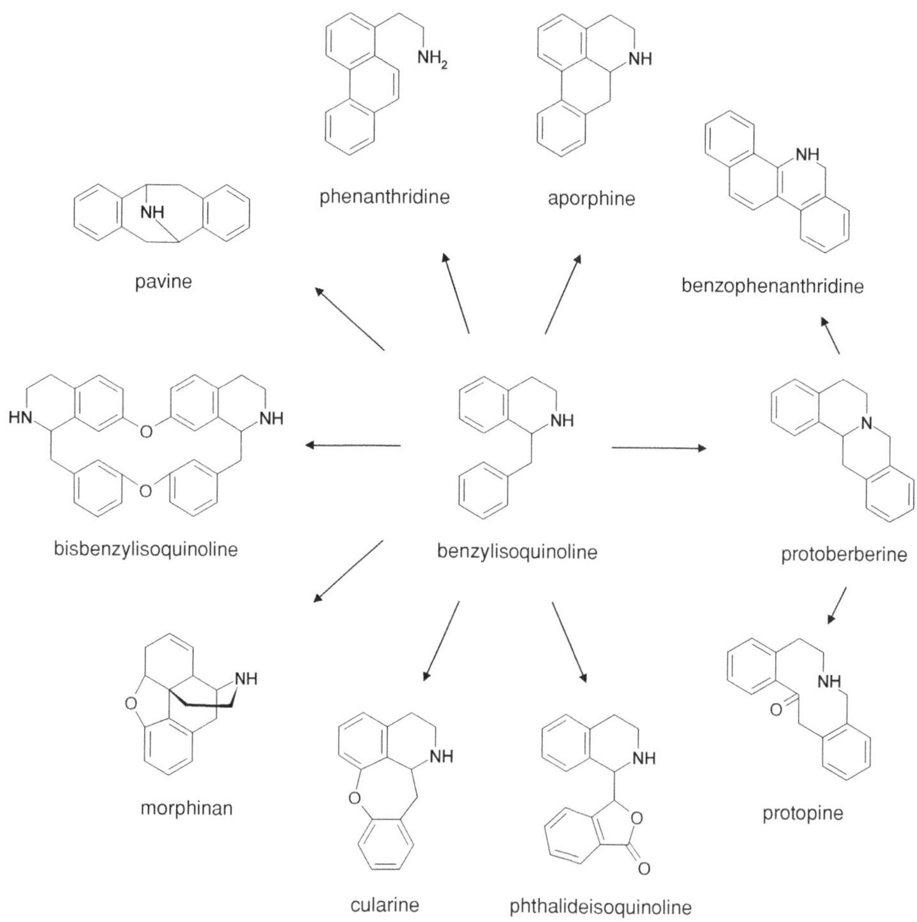

Figure 9.1 *Examples of diverse chemical structures of BIA categories derived from the simple benzylisoquinoline skeleton*

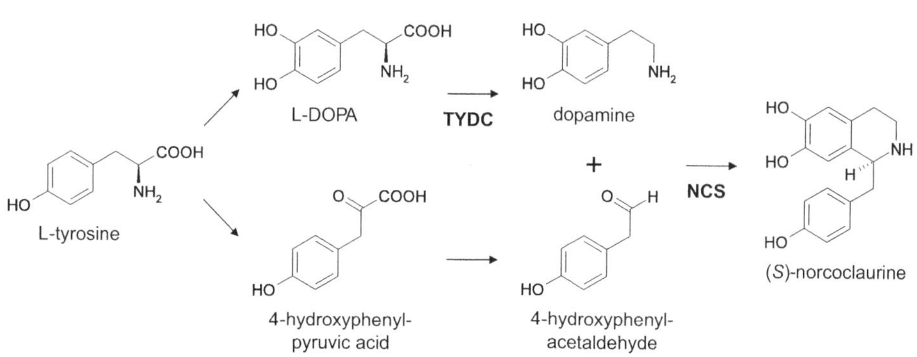

Figure 9.2 *Biosynthesis of (S)-norcoclaurine. Enzymes with cognate cDNAs are in bold. Abbreviations: TYDC, tyrosine/dopa decarboxylase; NCS, norcoclaurine synthase*

genes (Facchini and De Luca, 1994; Facchini *et al.*, 2000). The first committed step in the biosynthesis of BIA involves the coupling of dopamine and 4-HPAA via a Pictet-Spengler condensation catalyzed by norcoclaurine synthase (NCS) to stereoselectively produce the trihydroxylated alkaloid (*S*)-norcoclaurine (Figure 9.2). NCS is a member of the pathogenesis-related (PR)10/Betv1 protein family, and corresponding cDNAs have been isolated from *P. somniferum* and *T. flavum* (Samanani and Facchini,, 2002; Samanani *et al.*, 2004; Liscombe *et al.*, 2005). A cDNA isolated from *C. japonica* and encoding a protein with similarity to the 2-oxoglutarate-dependent dioxygenase family has also been reported to possess NCS activity (Minami *et al.*, 2007), although the catalytic mechanism and role of the enzyme in alkaloid metabolism are not known. (*S*)-norcoclaurine is the central precursor to more than 2500 known BIAs. The structural diversity of these compounds primarily results from: (1) modification of the benzylisoquinoline backbone via region-specific carbon–carbon and carbon–oxygen phenol coupling of (*S*)-reticuline or, in some cases, upstream intermediates (Figure 9.1); and (2) functional group transformation including *O*- and *N*-methylation, *O*-demethylation, hydroxylation, *O*-acetylation, reduction or oxidation (Ziegler and Facchini, 2008).

9.2.2 (*S*)-Reticuline

The conversion of (*S*)-norcoclaurine to (*S*)-reticuline involves *O*-methylation at position 6 by the *S*-adenosyl-L-methionine (SAM)-dependent (*R,S*)-norcoclaurine 6-*O*-methyltransferase (6OMT) to produce (*S*)-coclaurine (Figure 9.3) (Stadler and Zenk, 1990; Sato *et al.*, 1994; Ounaroon *et al.*, 2003; Zeigler *et al.*, 2005). The *N*-methylation of (*S*)-coclaurine catalyzed by the SAM-dependent (*S*)-coclaurine-*N*-methyltransferase (CNMT) yields *N*-methylcoclaurine (Loeffler *et al.*, 1995; Choi *et al.*, 2002; Facchini and Park, 2003; Morishige *et al.*, 2004). Hydroxylation at the 3′ position of *N*-methylcoclaurine by the cytochrome P450-dependent monooxygenase, (*S*)-*N*-methylcoclaurine 3′-hydroxylase (NMCH), results in the formation of (*S*)-3′-hydroxy-*N*-methylcoclaurine (Figure 9.3) (Pauli and Kutchan, 1998; Huang and Kutchan, 2000; Samanani *et al.*, 2005). Finally, the stereoselective 4′-*O*-methylation of (*S*)-3′-hydroxy-*N*-methylcoclaurine by (*S*)-3′-hydroxy-*N*-methylcoclaurine 4′-*O*-methyltransferase (4′OMT) produces (*S*)-reticuline (Frenzel and Zenk, 1990; Morishige *et al.*, 2000; Facchini and Park, 2003; Ziegler *et al.*, 2005). Cognate cDNAs encoding each enzyme have been isolated from several species.

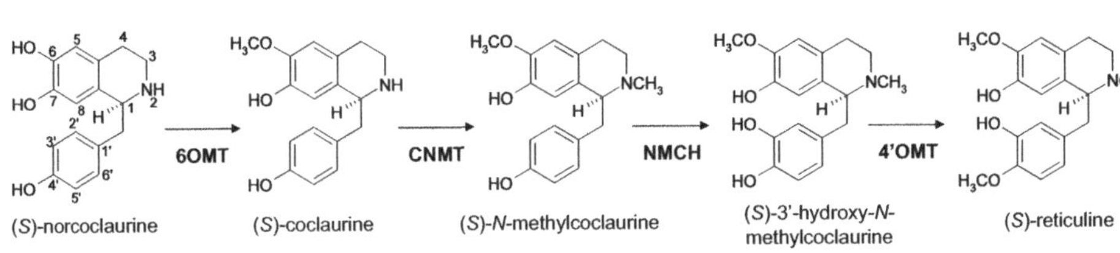

Figure 9.3 *Biosynthesis of (S)-reticuline. Enzymes with cognate cDNAs are in bold. Abbreviations: 6OMT, (R,S)-norcoclaurine 6-O-methyltransferase; CNMT, (S)-coclaurine-N-methyltransferase; NMCH, (S)-N-methylcoclaurine 3′-hydroxylase; 4′OMT, (S)-3′-hydroxy N-methylcoclaurine 4′-O-methyltransferase*

9.2.3 Morphinan Alkaloids

The first steps in the biosynthesis of morphinan alkaloids, catalyzed by 1,2-dehydroreticuline synthase (DRS) and 1,2-dehydroreticuline reductase (DRR), epimerise (*S*)-reticuline to (*R*)-reticuline (Figure 9.4) (De-Eknamkul and Zenk, 1990, 1992; Hirata *et al.*, 2004). Intramolecular carbon–carbon phenol coupling of (*R*)-reticuline by salutaridine synthase (SalSyn) results in the formation of salutaridine (Gerardy and Zenk, 1993a; Gesell *et al.*, 2009). Unlike most cytochrome P450-dependent enzymes, SalSyn is an oxidase rather than a monooxygenase (i.e. oxygen is not incorporated in the substrate during the reaction (Novak *et al.*, 2000). Subsequently, the stereospecific reduction of salutaridine to salutaridinol is catalyzed by the NADPH-dependent salutaridine reductase (SalR), a unique member of the short chain dehydrogenase/reductase (SDR) family (Gerardy and Zenk, 1993b; Ziegler *et al.*, 2006). Salutaridinol acetyltransferase (SalAT) converts salutaridinol to salutaridinol-7-*O*-acetate (Figure 9.4) (Lenz and Zenk, 1994, 1995a; Grothe *et al.*, 2001; Facchini and Park, 2003), but the acetyl group is eliminated in the formation of thebaine either spontaneously or enzymatically by thebaine synthase (THS) (Lenz and Zenk, 1995a; Fisinger *et al.*, 2007). A 2-oxoglutarate-dependent dioxygenase (T6ODM) catalyzing the 6-*O*-demethylation of thebaine to codeinone and oripavine to morphinone has recently been reported (Hagel and Facchini, 2010). Codeinone and morphinone are converted to codeine and morphine by the NADPH-dependent codeinone reductase (COR), a member of the aldo-keto reductase family (Lenz and Zenk, 1995b; Unterlinner *et al.*, 1999). A second 2-oxoglutarate-dependent dioxygenase (CODM) with extensive similarity to T6ODM catalyzes the 3-*O*-demethylation of codeine to morphine and thebaine to oripavine (Hagel and Facchini, 2010). Corresponding cDNAs encoding all enzymes of the morphinan branch pathway have been isolated except DRS, DRR and THS.

9.2.4 Sanguinarine

The first committed step in benzophenanthridine and protoberberine alkaloid biosynthesis is catalyzed by the FAD-dependent oxidoreductase berberine bridge enzyme (BBE), which catalyzes stereospecific oxidation and methylene bridge formation of (*S*)-reticuline to yield (*S*)-scoulerine (Figure 9.5) (Steffens *et al.*, 1985; Dittrich and Kutchan, 1991; Facchini *et al.*, 1996; Huang and Kutchan, 2000; Samanani *et al.*, 2005; Winkler *et al.*, 2006, 2007, 2008, 2009). The biosynthesis of benzophenanthridines such as sanguinarine begins with the consecutive formation of two methylenedioxy bridges in (*S*)-scoulerine by the cytochromes P450 (*S*)-cheilanthifoline synthase (CFS) and (*S*)-stylopine synthase (STS) (Figure 9.5) (Bauer and Zenk, 1991; Ikezawa *et al.*, 2007, 2009). (*S*)-stylopine is then *N*-methylated by (*S*)-tetrahydroprotoberberine *cis*-*N*-methyltransferase (TNMT) to (*S*)-*cis*-*N*-methylstylopine (O'Keefe and Beecher, 1994; Rueffer and Zenk, 1986, 1990; Liscombe and Facchini, 2007). The putative cytochromes P450, (*S*)-*cis*-*N*-Methyltetrahydroprotoberberine 14-hydroxylase (MSH) and protopine 6-hydroxylase (P6H), lead to protopine and 6-hydroxyprotopine, respectively (Rueffer and Zenk, 1987; Tanahashi and Zenk, 1990). 6-hydroxyprotopine is spontaneously rearranged to the benzophenanthridine alkaloid dihydrosanguinarine (Tanahashi and Zenk, 1990). Subsequently, dihydrosanguinarine is converted to sanguinarine by dihydrobenzophenanthridine oxidase (DBOX) (Schumacher and Zenk, 1988; Ignatov *et al.*, 1996). Sanguinarine reductase

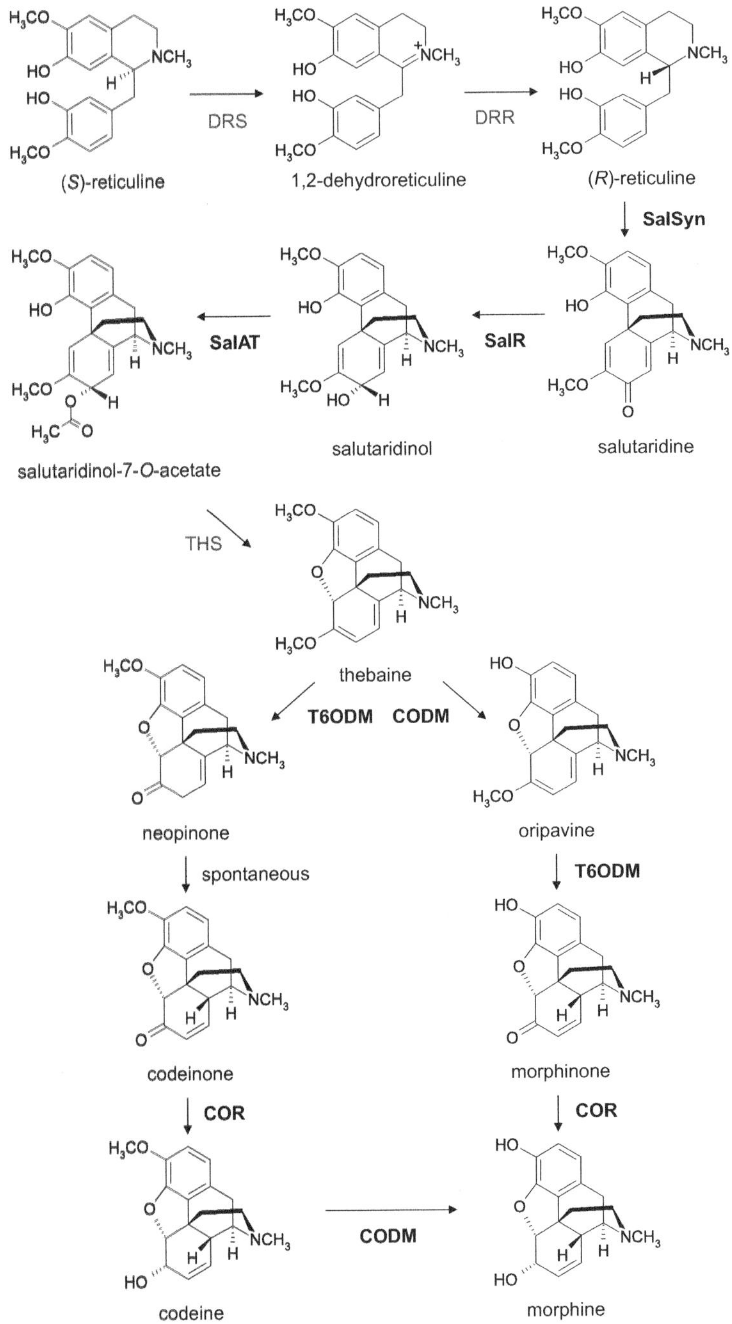

Figure 9.4 *Biosynthesis of morphine and codeine from (S)-reticuline. Enzymes with cognate cDNAs are in bold. Abbreviations: DRS, 1,2-dehydroreticule synthase; DRR, 1,2-dehydroreticuline reductase; SalSyn, salutaridine synthase; SalR, salutaridine reductase; SalAT, salutaridinol 7-O-acetyltransferase; THS, thebaine synthase;* T6ODM, *thebaine 6-O-demethylase;* COR, *codeinone reductase;* CODM, *codeine demethylase*

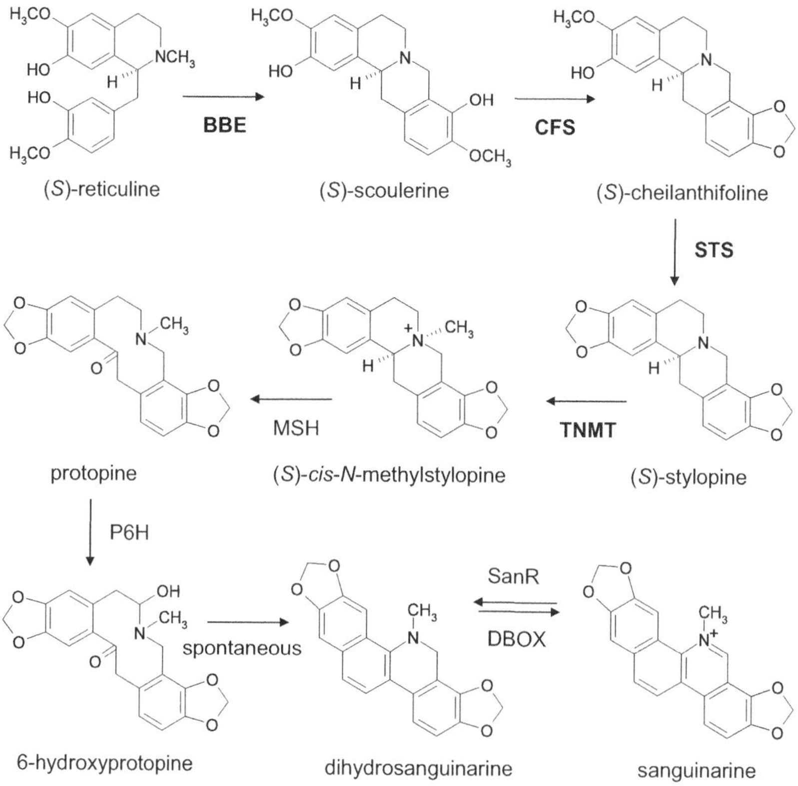

Figure 9.5 *Biosynthesis of sanguinarine from (S)-reticuline. Enzymes with cognate cDNAs are in bold. Abbreviations: BBE, berberine bridge enzyme; CFS, cheilanthifoline synthase; STS, stylopine synthase; TNMT, tetrahydroprotoberberine-N-methyltransferase; MSH, methylstylopine hydroxylase; P6H, protopine 6-hydroxylase; DBOX, dihydrobenzophenanthridine oxidase; SanR, sanguinarine reductase*

(SanR), putatively involved in the detoxification of sanguinarine, was reported in *E. californica* cell cultures (Weiss *et al.*, 2006). Cognate cDNAs have been reported for BBE, CFS, STS and TNMT from various species.

9.2.5 Aporphine and Protoberberine Alkaloids

(*S*)-reticuline is converted to the aporphine alkaloid (*S*)-corytuberine by carbon–carbon phenol coupling catalyzed by (*S*)-corytuberine synthase (CYP80G2) (Figure 9.6) (Ikezawa *et al.*, 2008). Magnoflorine probably results from the *N*-methylation of (*S*)-corytuberine by CNMT. In contrast, berberine biosynthesis begins with the 9-*O*-methylation of (*S*)-scoulerine by the SAM-dependent (*S*)-scoulerine 9-*O*-methyltransferase (SOMT) to yield (*S*)-tetrahydrocolumbamine (Figure 9.6) (Muemmler *et al.*, 1985; Fujiwara *et al.*, 1993; Takeshita *et al.*, 1995). (*S*)-canadine is formed via region-specific methylenedioxy bridge formation in (*S*)-tetrahydrocolumbamine by the cytochrome P450 (*S*)-canadine synthase (CAS) (Rueffer and Zenk, 1994; Ikezawa *et al.*, 2003), and is converted to berberine

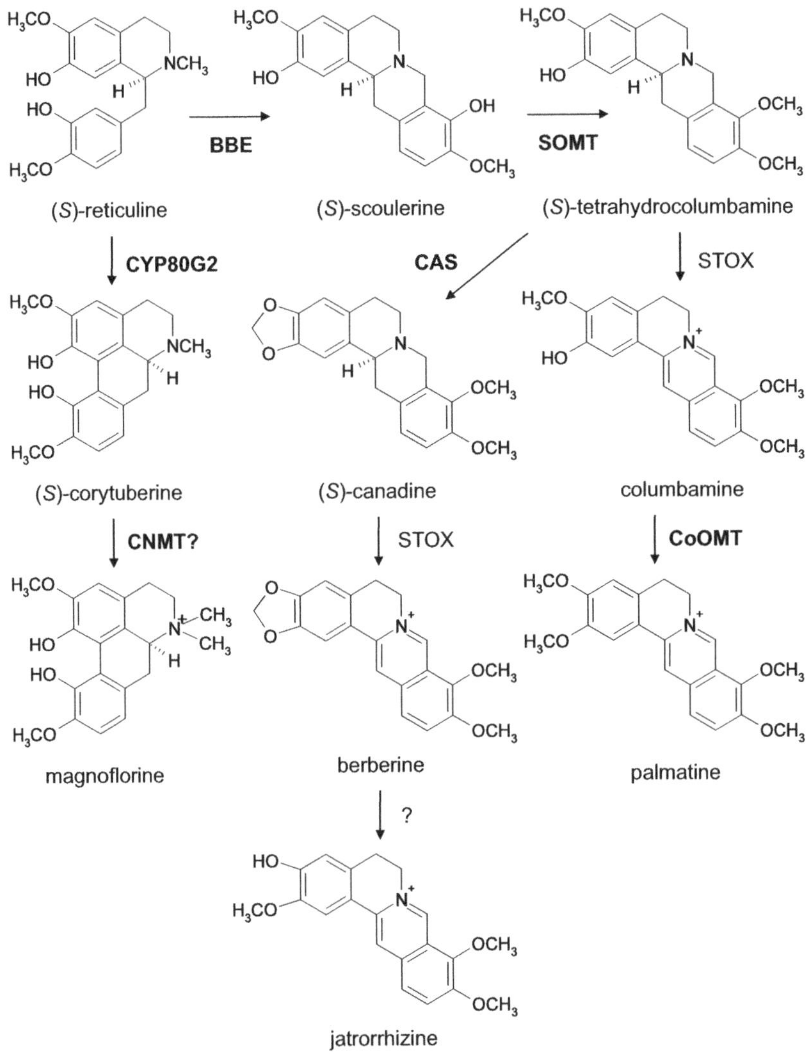

Figure 9.6 Biosynthesis of magnoflorine, berberine, palmatine and jatrorrhizine from (S)-reticuline. Enzymes with cognate cDNAs are in bold. Abbreviations: CYP80G2, (S)-corytuberine synthase; CNMT, coclaurine-N-methyltransferase; BBE, berberine bridge enzyme; SOMT, (S)-scoulerine 9-O-methyltransferase; CAS, canadine synthase; STOX, (S)-tetrahydroberberine oxidase; CoOMT, columbamine O-methyltransferase

by the FAD-dependent enzyme (*S*)-tetrahydroberberine oxidase (STOX) (Amann *et al.*, 1984; Okada *et al.*, 1988, 1989). In another pathway branch, STOX converts tetrahydrocolumbamine to columbamine, which is *O*-methylated by SAM-dependent columbamine *O*-methyltransferase (CoOMT) to palmatine (Rueffer *et al.*, 1986; Morishige *et al.*, 2002). Jatrorrhizine, the major protoberberine alkaloid in cell cultures of *Berberis spp.*, possesses an *O*-methylation pattern different from reticuline. Labelling studies led to the suggestion

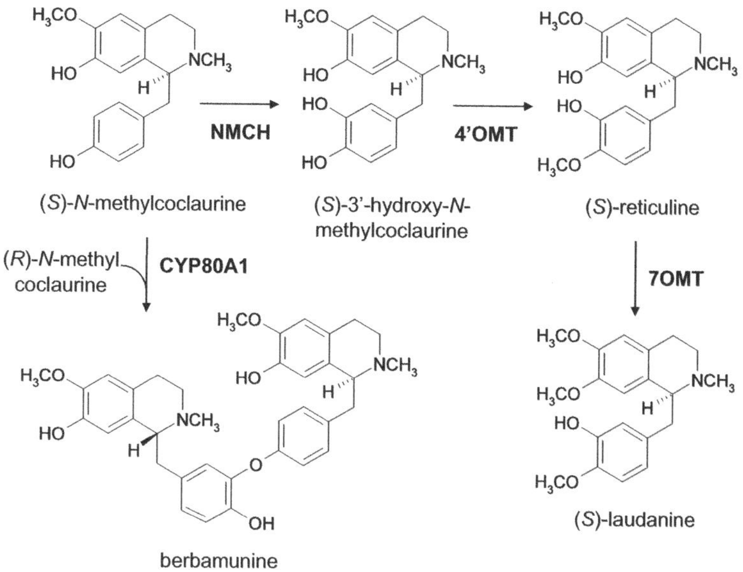

Figure 9.7 *Biosynthesis of berbamunine and laudanine from (S)-N-methylcoclaurine. Enzymes with cognate cDNAs are in bold. Abbreviations: CYP80A1, berbamunine synthase; NMCH, (S)-N-methylcoclaurine 3'-hydroxylase; 4'OMT, (S)-3'-hydroxy N-methylcoclaurine 4'-O-methyltransferase; 7OMT, (R,S)-reticuline 7-O-methyltransferase*

that jatrorrhizine is formed from berberine by reopening of the methylenedioxy group (Rueffer *et al.*, 1983), but the 3-*O*-demethylation of scoulerine by T6ODM and related dioxygenases in opium poppy supports an alternative route (Hagel and Facchini, 2010). With the exception of STOX, cDNAs encoding all of these enzymes have been reported.

9.2.6 Bisbenzylisoquinoline Alkaloids and Laudanine

The formation of dimeric bisbenzylisoquinoline alkaloids (bisBIAs) begins with the carbon–oxygen phenol coupling of *N*-methylcoclaurine by the cytochrome P450 berbamunine synthase (CYP80A1) (Figures 9.3 and 9.7) *B. stolonifera* (Stadler and Zenk, 1993; Kraus and Kutchan, 1995). CYP80A1 couples two molecules of (*R*)-*N*-methylcoclaurine or one (*R*)- and one (*S*)-*N*-methylcoclaurine to yield (*R*,*R*)-guattegaumerine or (*R*,*S*)-berbamunine, respectively. Finally, (*R*,*S*)-reticuline can be converted to laudanine by the (*R*,*S*)-reticuline 7-*O*-methyltransferase (7OMT) (Figure 9.7) (Ounaroon *et al.*, 2003). Both CYP80A1 and 7OMT cDNAs have been reported.

9.3 Localisation and Transport of Benzylisoquinoline Alkaloids and their Biosynthetic Enzymes

9.3.1 Cellular and Subcellular Localisation

Benzylisoquinoline alkaloid-producing plants contain specialised cells or structures to sequester their cytotoxic constituents. In opium poppy, BIAs accumulate in the vesicles

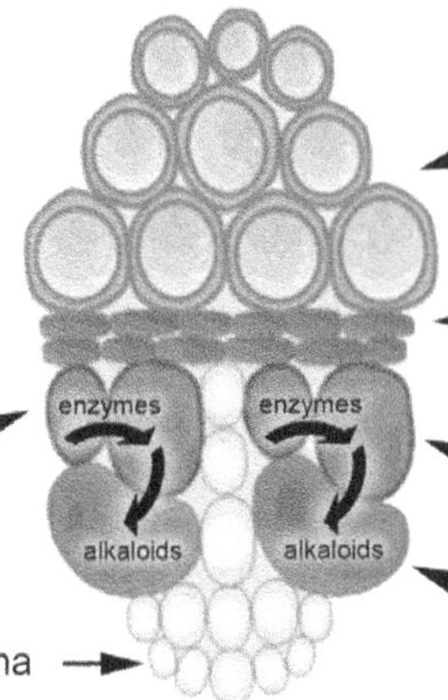

Figure 9.8 *Localisation of benzylisoquinoline alkaloid biosynthesis in the vascular bundles of opium poppy aerial organs. Biosynthetic genes are expressed, and corresponding enzymes are produced in the companion cells of the phloem. Biosynthetic enzymes are transported to the accompanying sieve elements in which alkaloid biosynthesis occurs. Subsequently, alkaloids are sequestered to the adjacent laticifers*

of laticifers that are proximal to sieve elements of the phloem. *In situ* hybridisation and immunolocalisation studies showed that BIA biosynthetic enzymes are located in sieve elements and the corresponding transcripts are found in adjacent companion cells (Figure 9.8). These results suggest that BIA biosynthetic genes are expressed and corresponding transcripts are translated in companion cells. Cognate enzymes are then transported to sieve elements where BIA biosynthesis occurs (Bird *et al.*, 2003; Samanani *et al.*, 2006). Alkaloids are then translocated to laticifers for storage. This complex spatial distribution of gene transcripts, enzymes and metabolites implicates multiple levels of regulation in the biosynthesis of BIAs in opium poppy.

In contrast, different cell types are involved in BIA biosynthesis in *T. flavum* (Samanani *et al.*, 2002, 2005). Protoberberine alkaloids accumulate in the root endodermis, and in the rhizome pith and cortex (Samanani *et al.*, 2002). Transcripts from nine genes involved in protoberberine alkaloid metabolism were co-localised to the immature root endodermis and the protoderm of leaf primordia in the rhizome, suggesting that the biosynthesis and accumulation of BIAs are both temporally and spatially uncoupled in *T. flavum* roots and rhizomes, respectively (Samanani *et al.*, 2002, 2005). Berberine biosynthetic enzymes have been reported specifically in root tissues of *C. japonica*, although the alkaloid also accumulates in the rhizome, suggesting a role for intercellular transport (Fujiwara *et al.*, 1993).

Most BIA biosynthetic enzymes have been associated with the cytosol with the exception of the cytochromes P450, BBE, STOX, and possibly NCS, which are associated with the endoplasmic reticulum (ER). Cytochrome P450 is integral to membrane proteins whose catalytic domains are typically on the cytosolic face of the ER, whereas BBE – and probably STOX and NCS – contain an ER signal peptide and additional cellular sorting information. The subcellular localisation of enzymes on both the cytosolic and luminal sides of the ER suggests a key role for transport processes in regulating BIA metabolism.

9.3.2 Transport

Membrane-bound, multidrug resistance (MDR) pumps are proteins that mediate the efflux of various and frequently cytotoxic molecules. MDR pumps have been shown to transport berberine and palmatine (Lewis, 2001), and thus are potential candidates for the intra- and intercellular translocation of BIAs. An MDR-type ATP-binding cassette (ABC) transporter from cultured *C. japonica* cells (*CjMDR1*) was suggested to participate in the translocation of berberine from its site of biosynthesis in the root to its site of accumulation in xylem cells of the rhizome (Shitan *et al.*, 2003).

9.4 Regulation

9.4.1 Gene Regulation

The control of biosynthetic gene expression is a primary determinant for the tissue- and cell type-specific localisation of BIA metabolism, and is a key contributor to the developmental and environmental regulation of product accumulation both qualitatively and quantitatively. The transcriptional regulation of BIA biosynthetic genes has not been well studied. Recently, however, a transcription factor belonging to the WRKY family and linked to the regulation of BIA metabolism was isolated from *C. japonica* cell cultures (Kato *et al.*, 2007). Silencing of the *CjWRKY1* gene in transformed protoplasts of cultured *C. japonica* cells was associated with a substantial reduction in the abundance of berberine biosynthetic gene transcripts. In contrast, overexpression of *CjWRKY1* increased the abundance of these same gene transcripts (Kato *et al.*, 2007).

Transactivation screens using BIA biosynthetic gene promoters fused to the luciferase reporter gene identified transcription factors from plant species that do not accumulate BIAs, such as *Arabidopsis*, maize and soybean. The heterologous transcription factors belonged to a variety of families (e.g. WRKY, AP2/ERF and MYB) and activated the *TYDC*, *6OMT*, *4'OMT*, *SalAT* and *COR* promoters in transgenic *P. somniferum*, and the *NMCH* and *BBE* promoters in transformed *E. californica* (Apuya *et al.*, 2008). The accumulation of some BIAs increased by 30-fold in some cases.

BIA biosynthetic genes and pathogen-responsive (PR) genes display expression characteristics suggesting their potential co-regulation (Vom Endt *et al.*, 2002), and the role of BIA metabolism as a plant defence response. All known genes involved in sanguinarine biosynthesis in cell cultures of opium poppy and other investigated members of the Papaveraceae are induced in response to treatment with a fungal elicitor (Zulak *et al.*, 2007). It is also interesting to note that NCS is a member of the PR10 family (Samanani *et al.*, 2004; Liscombe *et al.*, 2005). Elicitor treatment of opium poppy cell cultures also resulted in the induction of several genes encoding enzymes coupled to the production of tyrosine and other metabolites relevant to BIA biosynthesis, such as *S*-adenosylmethionine, providing evidence for the coordination of primary and secondary metabolism (Zulak *et al.*, 2007).

9.4.2 Signal Transduction

Two signal transduction pathways have been implicated in the induction of BIA metabolism in cultured *E. californica* cells treated with a fungal elicitor: (1) a jasmonate-dependent cascade responding to high concentrations of elicitor; and (2) a jasmonate-independent pathway active at low elicitor concentrations (Roos *et al.*, 2006). The jasmonate-independent cascade involves Gα proteins that activate phospholipase A2, which leads to a transient proton efflux from the vacuole and a subsequent activation of other cytoplasmic components (Viehweger *et al.*, 2006; Heinze *et al.*, 2007).

9.5 Application to Biotechnology

Altering the accumulation of specific BIA in plants has proven difficult using conventional breeding techniques. Recently, the application of molecular biotechnology to the engineering of BIA metabolism has produced examples showing the effectiveness of mutagenic and transgenic approaches in the modulation of specific BIA pathways, especially in opium poppy. In general, metabolic engineering is used to either increase the level of a valuable compound or reduce the accumulation of undesirable products.

9.5.1 Mutagenesis

Mutagenesis is a proven and powerful approach to creating advantageous new traits in plants. Standard techniques to establish mutant seed stocks can be applied to many BIA-producing plants. However, the genomics resources required to identify mutated genes are not yet available for most species. The recent development of expressed sequence tag (EST) databases and DNA microarrays in opium poppy is providing unprecedented opportunities to isolate new genes using comparative genomics of mutant and wild-type plants. The opium poppy mutant *top1* was isolated by screening plants generated from a chemically mutagenised seed stock for alterations in their BIA profile (Millgate *et al.*, 2004). The *top1* line is rich in thebaine and oripavine, but does not accumulate codeine or morphine. Although a mutation affecting the *T6ODM* gene (Figure 9.4) was proposed, the genetic basis of the *top1* mutation was not identified (Millgate *et al.*, 2004). The recent discovery of this gene using a custom-made DNA microarray and the comparative genomics of a similar opium poppy line displaying the high-thebaine/oripavine, codeine/morphine-free phenotype compared with wild-type plants demonstrates the power of these approaches for the isolation of elusive genes (Hagel and Facchini, 2010). Interestingly, the *top1* phenotype has been reported in natural populations (Nyman, 1978, 1980), suggesting that the mutation in not uncommon. In contrast with *top1*, the opium poppy mutant Przemko contains only trace BIA levels (Hagel *et al.*, 2008), which could implicate a non-functional biosynthetic enzyme, transporter or transcriptional regulator.

9.5.2 Genetic Transformation and Metabolic Engineering

Genetic transformation provides an opportunity to alter the expression of genes involved in BIA biosynthesis, which can potentially modulate the accumulation of valuable intermediates or end products. Stable genetic transformation methods for BIA-producing species are generally inefficient, especially with respect to plant regeneration. Nevertheless, transformation protocols for intact plants and cultured cells have been developed for several species including *P. somniferum* (Park and Facchini, 2000a; Chitty *et al.*, 2003; Facchini *et al.*, 2008) and *E. californica* (Park and Facchini, 2000b).

Post-transcriptional gene silencing methods involving RNA interference (RNAi) or virus-induced gene silencing (VIGS) have been used as effective tools to silence specific BIA biosynthetic genes and alter alkaloid profiles in plants and cell cultures, especially in opium poppy (Allen *et al.*, 2004; Hileman *et al.*, 2005). Both approaches rely on plant defence mechanisms involving the degradation of the RNA transcripts encoded by the target gene (Baulcombe, 2004; Eamens *et al.*, 2008). Briefly, target gene sequence-specific double-stranded dsRNAs are produced, recognised and processed into small-interfering siRNAs by the RNase-III enzyme Dicer. These siRNAs are detected, bound and loaded onto the RNA-induced silencing complexes (RISCs) where they direct the targeted degradation of complementary RNA transcripts leading to a knockdown in transcript levels of the target gene. RNAi and VIGS have been used to silence several BIA biosynthetic genes often with unexpected results. For example, RNAi-mediated silencing of *COR* genes in transgenic opium poppy plants resulted in the accumulation of (*S*)-reticuline (Figure 9.4) (Allen *et al.*, 2004). The accumulation of a pathway intermediate seven enzymatic steps upstream of COR was suggested to result either from negative feedback regulation, or from disruption of a putative enzyme complex possibly composed of all enzymes in the morphinan alkaloid branch pathway (Allen *et al.*, 2004). Transgenic opium poppy plants overexpressing the *COR1* gene were reported to show a 15–30% increase in morphinan alkaloid content (Larkin *et al.*, 2007). Similarly, overexpression of *NMCH* resulted in a 450% increase in total alkaloid accumulation in opium poppy latex (Frick *et al.*, 2007). In contrast, an 84% reduction in total alkaloid content was reported in opium poppy plants expressing an antisense-*NMCH* gene (Frick *et al.*, 2007). Overexpression of *SalAT* increased BIA levels by 40%, whereas RNAi silencing of *SalAT* resulted in the accumulation of salutaridine rather than salutaridinol (Figure 9.4) (Allen *et al.*, 2008). The accumulation of a pathway intermediate an enzymatic step upstream of SalAT was suggested to result from the disruption of a complex between SalR and SalAT (Allen *et al.*, 2008).

Efforts to increase the BIA content of plant cell cultures, which have long been regarded as a potential alternative for the commercial production of valuable natural products, have also been reported. The transformation of *E. californica* cell cultures using antisense-*BBE* or antisense-*NCMH* genes reduced the overall benzophenanthridine alkaloid content and increased the pool size of several amino acids, but did not result in the expected accumulation of pathway intermediates (Park *et al.*, 2002). In contrast, overexpression of *BBE* in transformed *E. californica* root cultures increased benzophenanthridine alkaloid levels and reduced the pool size of several amino acids. Roots transformed with an antisense-*BBE* gene showed lower BIA levels and larger amino acid pools, but did not accumulate reticuline as expected (Park *et al.*, 2003). However, the RNAi-mediated silencing of the *BBE* gene in cultured *E. californica* cells reduced sanguinarine accumulation and increased reticuline levels (Fujii *et al.*, 2007). The overexpression of the *6OMT* gene from *C. japonica* in cultured *E. californica* cells increased BIA content by 7.5-fold, whereas overexpression of the *C. japonica 4′OMT* gene had only a marginal effect on BIA content (Inui *et al.*, 2007).

BIA biosynthetic enzymes catalyzing reactions at pathway junctions are particularly interesting candidates for the engineering of BIA metabolism. For example, SOMT is involved in the biosynthesis of berberine and columbamine (Figure 9.7). Expression of the *C. japonica SOMT* gene in cultured *E. californica* created a new branch pathway leading to the accumulation of berberine and columbamine, which are not normally produced in this system (Sato *et al.*, 2001).

9.5.3 Metabolic Engineering

The reconstitution of BIA biosynthetic pathways in microorganisms is a potential alternative to metabolic engineering in plants for the commercial production of valuable compounds (Chemler and Koffas, 2008; Lee *et al.*, 2009; Leonard *et al.*, 2009). The introduction into *Escherichia coli* of a combination of microbial and plant biosynthetic genes facilitated the microbial production of (*S*)-reticuline (Minami *et al.*, 2008). In the engineered pathway, monoamine oxidase (MAO) was used to produce 3,4-DHPAA (3,4-dihydroxyphenylacetaldehyde) from exogenous dopamine, and the two compounds were then condensed by norcoclaurine synthase (NCS) to yield (*S*)-norlaudanosoline. It is noteworthy that the *C. japonica* NCS (i.e. CjPR10A) used for this study is a member of the PR10 protein family. The purported *C. japonica* NCS belonging to the 2-oxoglutarate-dependent family (i.e. CjNCS1) showed considerably less activity in *E. coli*. Subsequently, 6OMT, CNMT and 4′OMT from *C. japonica* were used to convert (*S*)-norlaudanosoline to (*S*)-reticuline. Ultimately, magnoflorine or (*S*)-scoulerine (Figure 9.6) were produced by co-cultivation of the engineered *E. coli* with yeast (*Saccharomyces cerevisiae*) expressing different BIA biosynthetic enzymes operating downstream of (*S*)-reticuline in *C. japonica*. Magnoflorine was produced when the engineered *S. cerevisiae* harboured the genes encoding corytuberine synthase (CYP80G2) and CNMT (Figure 9.6), whereas *S. cerevisiae* expressing the *BBE* gene produced (*S*)-scoulerine (Minami *et al.*, 2008). Using a similar approach, the metabolic engineering of yeast for the production of promorphinan (Figure 9.4), benzophenanthridine (Figure 9.5) and protoberberine (Figure 9.6) alkaloids has also been reported (Hawkins and Smolke, 2008). A noteworthy achievement involved the use of a human P450-dependent enzyme to convert (*R*)-reticuline to salutaridine in the engineered yeast.

9.6 Conclusions

Plants remain the only source for several pharmaceutically important BIAs including the widely used analgesics morphine and codeine. Until recently, the biotechnological development of commercial production systems for valuable compounds has been impaired by our limited knowledge of BIA metabolism. However, recent applications of genomics technologies combined with several decades of impressive advances using well-established biochemical and molecular biological approaches has resulted in the identification of more than 20 BIA biosynthetic genes. Our rapidly expanding knowledge of the genetic and biochemical mechanisms underlying BIA metabolism now provides unprecedented opportunities to engineer BIA biosynthetic pathways in plants and microorganisms

References

Allen, R.S., Millgate, S.G., Chitty, J.A., Thisleton, J., Miller, J.A., Fist, A.J. Gerlach, W.L. and Larkin, P.J. (2004) RNAi-mediated replacement of morphine with the nonnarcotic alkaloid reticuline in opium poppy. *Nat. Biotechnol.*, **12**, 1559–1566.

Allen, R.S., Miller, J.A.C., Chitty, J.A., Fist, A.J., Gerlach, W.L. and Larkin, P.J. (2008) Metabolic engineering of morphinan alkaloids by over-expression and RNAi suppression of salutaridinol 7-*O*-acetyltransferase in opium poppy. *Plant Biotechnol. J.*, **6**, 22–30.

Amann, M., Nagakura, N. and Zenk, M.H. (1984) (*S*)-Tetrahydroprotoberberine oxidase the final step in protoberberine biosynthesis. *Tetrahedron Lett.*, **25**, 953–954.

Apuya, N.R., Park, J-H, Liping, Z., Ahyow, M., Davidow, P., Van Fleet, J., Rarang, J.C., Hiplley, M., Johnson, T.W., Yoo, H-D., Trieu, A., Krueger, S., Wu, C-y., Lu, Y-p., Flavell, R.B. and Bobzin, S.C. (2008) Enhancement of alkaloid production in opium and California poppy by transactivation using heterologous regulatory factors. *Plant Biotech. J.*, **6**, 160–175.

Bauer, W. and Zenk, M.H. (1991) Two methylenedioxy bridge forming cytochrome P-450 dependent enzymes are involved in (*S*)-stylopine biosynthesis. *Phytochemistry*, **30**, 2953–2961.

Baulcombe, D. (2004) RNA silencing in plants. *Nature*, **431**, 356–363.

Bird, D.A., Franceschi, V.R. and Facchini, P.J. (2003) A tale of three cell types: alkaloid biosynthesis is localized to sieve elements in opium poppy. *Plant Cell*, **15**, 2626–2635.

Chemler, J.A. and Koffas, M.A.G. (2008) Metabolic engineering for plant natural product biosynthesis in microbes. *Curr. Opin. Biotech.*, **19**, 597–605.

Chitty, J.A., Allen, R.S., Fist, A.J. and Larkin, P.J. (2003) Genetic transformation in commercial Tasmanian cultivars of opium poppy, *Papaver somniferum*, and movement of transgenic pollen in the field. *Funct. Plant Biol.*, **30**, 1045–1058.

Choi, K.B., Morihige, T., Shitan, N., Yazaki, K. and Sato, F. (2002) Molecular cloning and characterization of coclaurine *N*-methyltransferase from cultured cells of *Coptis japonica*. *J. Biol. Chem.*, **277**, 830–835.

De-Eknamkul, W. and Zenk, M.H. (1990) Enzymatic formation of (*R*)-reticuline from 1,2-dehydroreticuline in the opium poppy plant. *Tetrahedron Lett.*, **31**, 4855–4858.

De-Eknamkul, W. and Zenk, M.H. (1992) Purification and properties of 1,2-dehydroreticuline reductase from *Papaver somniferum* seedlings. *Phytochemistry*, **31**, 813–821.

Dittrich, H. and Kutchan T.M. (1991) Molecular cloning, expression, and induction of berberine bridge enzyme, an enzyme essential to the formation of benzophenanthridine alkaloids in the response of plants to pathogenic attack. *Proc. Natl Acad. Sci. USA*, **88**, 9969–9973.

Eamens, A., Wang, M.-B., Smith, N.A. and Waterhouse, P.M. (2008) RNA silencing in plants: yesterday, today and tomorrow. *Plant Physiol.*, **147**, 456–468.

Facchini P.J. and De Luca, V. (1994) Differential and tissue-specific expression of a gene family for tyrosine/dopa decarboxylase in opium poppy. *J. Biol. Chem.*, **269**, 16684–26690.

Facchini, P.J., Penzes, C., Johnson, A.G. and Bull, D. (1996) Molecular characterization of berberine bridge enzyme genes from opium poppy. *Plant Physiol.*, **112**, 1669–1677.

Facchini, P.J., Huber-Allanach, K.L. and Tari, L.W. (2000) Plant aromatic L-amino acid decarboxylases: evolution, biochemistry, regulation, and metabolic engineering applications. *Phytochemistry*, **54**, 121–138.

Facchini, P.J. and Park, S.U. (2003) Developmental and inducible accumulation of gene transcripts involved in alkaloid biosynthesis in opium poppy. *Phytochemistry*, **64**, 177–186.

Facchini, P.J. and De Luca, V. (2008) Opium poppy and Madagascar periwinkle: model non-model systems to investigate alkaloid biosynthesis in plants. *Plant J.*, **54**, 763–784.

Facchini, P.J., Loukanina, N. and Blanche, V. (2008) Genetic transformation via somatic embryogenesis to establish herbicide-resistant opium poppy. *Plant Cell Rep.*, **27**, 719–727.

Fisinger, U., Grobe, N. and Zenk, M.H. (2007) Thebaine synthase: a new enzyme in the morphine pathway in *Papaver somniferum. Nat. Prod. Commun.*, **2**, 249–253.

Frenzel, T. and Zenk, M.H. (1990) *S*-adenosyl-L-methionine: 3′-hydroxyl-*N*-methyl-(*S*)-coclaurine-4′-*O*-methyltransferase, a region- and stereoselective enzyme of the (*S*)-reticuline pathway. *Phytochemistry*, **29**, 3505–3511.

Frick, S., Krammell, R. and Kutchan, T.M. (2007) Metabolic engineering with a morphine biosynthetic P450 in opium poppy surpasses breeding. *Metab. Eng.*, **9**, 169–176.

Fujiwara, H., Takeshita, N., Terano, Y., Fitchen, J.H., Tsujita, T., Katagiri, Y., Sato, F. and Yamada, Y. (1993) Expression of (*S*)-scoulerine 9-*O*-methyltransferase in *Coptis japonica* plants. *Phytochemistry*, **34**, 949–954.

Fujii, N., Inui, T., Iwasa, K., Morishige, T. and Sato, F. (2007) Knockdown of berberine bridge enzyme by RNAi accumulates (*S*)-reticuline and activates a silent pathway in cultures California poppy cells. *Transgenic Res.*, **16**, 363–375.

Gerardy, R. and Zenk, M.H. (1993a) Formation of salutaridine from (*R*)-reticuline by a membrane-bound cytochrome P-450 enzyme from *Papaver somniferum. Phytochemistry*, **32**, 79–86.

Gerardy, R. and Zenk, M.H. (1993b) Purification and characterization of salutaridine: NADPH 7-oxidoreductase form *Papaver somniferum. Phytochemistry*, **34**, 125–132.

Gesell, A., Rolf, M., Ziegler, J. and Kutchan, T.M. (2009) CYP719B1 is salutaridine synthase, the C-C phenol-coupling enzyme of morphine biosynthesis in opium poppy. *J. Biol. Chem.*, **284**, 24432–24442.

Grothe, T., Lenz, R. and Kutchan, T.M. (2001) Molecular characterization of the salutaridinol 7-*O*-acetyltransferase involved in morphine biosynthesis in opium poppy *Papaver somniferum. J. Biol. Chem.*, **276**, 30717–30723.

Hagel, J.M., Weljie, A., Vogel, H.J. and Facchini, P.J. (2008) Quantitative ^{1}H-NMR metabolomics as a biochemical genomics platform to study alkaloid biosynthesis in opium poppy. *Plant Physiol.*, **147**, 1805–1821.

Hagel, J.M. and Facchini, P.J. (2010) Novel *O*-demethylases of morphine biosynthesis in opium poppy. *Nat. Chem. Biol.*, **4**, 273–275.

Hawkins, K.M. and Smolke, C.D. (2008) Production of benzylisoquinoline alkaloids in *Saccharomyces cerevisiae. Nat. Chem. Biol.*, **4**, 564–573.

Heinze, M., Steighardt, J., Gesell, A., Schwartze, W. and Roos, W. (2007) Regulatory interaction of the Gα protein with phospholipase A(2) in the plasma membrane of *Eschscholtzia californica. Plant J.*, **52**, 1041–1051.

Hileman, L.C., Drea, S., de Martino, G., Litt, A. and Irish, V.F. (2005) Virus-induced gene silencing is an effective tool for assaying gene function in the basal eudicot species *Papaver somniferum* (opium poppy). *Plant J.*, **44**, 334–341.

Hirata, K., Poeaknapo, C., Schmidt, J. and Zenk, M.H. (2004) 1,2-Dehydroreticuline synthase, the branch point enzyme opening the morphinan biosynthetic pathway. *Phytochemistry*, **65**, 1039–1046.

Huang, F.C. and Kutchan, T.M. (2000) Distribution of morphinan and benzo[c] phenanthridine alkaloid gene transcript accumulation in *Papaver somniferum. Phytochemistry*, **53**, 555–564.

Ignatov, A., Clark, W.G., Cline, S.D., Psenak, M., Krueger, J. and Coscia, C.J. (1996) Elicitation of dihydrobenzophenanthridine oxidase in *Sanguinaria canadensis* cell cultures. *Phytochemistry*, **43**, 1141–1144.

Ikezawa, N., Tanaka, M., Nagayoshi, M., Shinkyo, R., Sakaki, T., Inouye, K. and Sato, F. (2003) Molecular cloning and characterization of CYP719, a methylenedioxy bridge-forming enzyme that belongs to a novel P450 family, from cultured *Coptis japonica* cells. *J. Biol. Chem.*, **278**, 38557–38565.

Ikezawa, N., Iwasa, K. and Sato, F. (2007) Molecular cloning and characterization of methylenedioxy bridge-forming enzymes involved in stylopine biosynthesis in Eschscholzia californica. *FEBS J.*, **274**, 1019–1035.

Ikezawa, N., Iwasa, K. and Sato, F. (2008) Molecular cloning and characterization of CYP80G2, a cytochrome P450 that catalyzes an intramolecular C-C phenol coupling of (*S*)-reticuline in magnoflorine biosynthesis, from cultured *Coptis japonica* cells. *J. Biol. Chem.*, **283**, 8810–8821.

Ikezawa, N., Iwasa, K. and Sato, F. (2009) CYP719A subfamily of cytochrome P450 oxygenases and isoquinoline alkaloid biosynthesis in *Eschscholzia californica. Plant Cell Rep.*, **28**, 123–133.

Inui, T., Tamura, K-I., Fujii, N., Morishige, T. and Sato, F. (2007) Overexpression of *Coptis japonica* norcoclaurine 6-*O*-methyltransferase overcomes the rate-limiting step in benzylisoquinoline alkaloid biosynthesis in cultured *Eschscholzia californica. Plant Cell Physiol.*, **48**, 252–262.

Kato, N., Dubouzet, E., Kokabu, Y., Yoshida, S., Taniguchi, Y., Dubouzet, J.G., Yazaki, K. and Sato, F. (2007) Identification of a WRKY protein as a transcriptional regulator of benzylisoquinoline alkaloid biosynthesis in *Coptis japonica. Plant Cell. Physiol.*, **48**, 8–18.

Kraus, P.F. and Kutchan, T.M. (1995) Molecular cloning and heterologous expression of a cDNA encoding berbamunine synthase, a C-O phenol-coupling cytochrome P450 from the higher plant *Berberis stolonifera. Proc. Natl Acad. Sci. USA*, **92**, 2071–2075.

Larkin, P.J., Miller, J.A., Allen, R.S., Chitty, J.A., Gerlach, W.L., Frick, S., Kutchan, T.M. and Fist, A.J. (2007) Increasing morphinan alkaloid production by over-expressing codeinone reductase in transgenic *Papaver somniferum. Plant Biotechnol. J.*, **5**, 26–37.

Lee, S.Y., Kim, H.U., Park, J.H., Park, J.M. and Kim, T.Y. (2009) Metabolic engineering of microorganisms: general strategies and drug production. *Drug Discovery Today*, **14**, 78–88.

Lenz, R. and Zenk, M.H. (1994) Closure of the oxide bridge in morphine biosynthesis. *Tetrahedron Lett.*, **35**, 3897–3900.

Lenz, R. and Zenk, M.H. (1995a) Acetyl coenzyme A: salutaridinol-7-*O*-acetyltransferase from *Papaver somniferum* plant cell cultures. The enzyme catalyzing the formation of thebaine in morphine biosynthesis. *J. Biol. Chem.*, **270**, 31091–31096.

Lenz, R. and Zenk, M.H. (1995b) Purification and properties of codeinone reductase (NADPH) from *Papaver somniferum* cell cultures and differentiated plants. *Eur. J. Biochem.*, **233**, 132–139.

Leonard, E., Runguphan, W., O'Connor, S. and Prather, K.J. (2009) Opportunities in metabolic engineering to facilitate scalable alkaloid production. *Nature Chem. Biol.*, **5**, 292–300.

Lewis, K (2001) In search of natural substrates and inhibitors of MDR pumps. *J. Mol. Microbiol. Biotechnol.*, **3**, 247–254.

Liscombe, D.K., MacLeod, B.P., Loukanina, N., Nandi, O.I. and Facchini, P.J. (2005) Evidence for the monophyletic evolution of benzylisoquinoline alkaloid biosynthesis in angiosperms. *Phytochemistry*, **66**, 2501–2520.

Liscombe, D.K. and Facchini, P.J. (2007) Molecular cloning and characterization of tetrahydroprotoberberine cis-*N*-methyltransferase, an enzyme involved in alkaloid biosynthesis in opium poppy. *J. Biol. Chem.*, **282**, 14741–14751.

Loeffler, S., Deus-Neumann, B. and Zenk, M.H. (1995) *S*-adenosyl-L-methionine: (*S*)-coclaurine-*N*-methyltransferase from *Tinospora cordifolia*. *Phytochemistry*, **38**, 1387–1395.

Millgate, A.G., Pogson, B.J., Wilson, I.W., Kutchan, T.M., Zenk, M.H., Gerlach, W.L., Fist, A.J. and Larkin, P.J. (2004) Analgesia: morphine pathway block in *top1* poppies. *Nature*, **431**, 413–414.

Minami, H., Dubouzet, E., Iwasa, K. and Sato, F. (2007) Functional analysis of norcoclaurine synthase in *Coptis japonica*. *J. Biol. Chem.*, **282**, 6274–6282.

Minami, H., Kim, J-S., Ikezawa, N., Takemura, T., Katayama, T., Kumagai, H. and Sato, F. (2008) Microbial production of plant benzylisoquinoline alkaloids. *Proc. Natl. Acad. Sci. USA*, **105**, 7393–7398.

Morishige, T., Tsujita, T., Yamada, Y. and Sato, F. (2000) Molecular characterization of the *S*-adenosyl-L-methionine: 3′-hydroxy-*N*-methylcoclaurine 4′-*O*-methyltransferase involved in isoquinoline alkaloid biosynthesis in *Coptis japonica*. *J. Biol. Chem.*, **275**, 23398–23405.

Morishige, T., Dubouzet, E., Choi, K., Yazaki, K. and Sato, F. (2002) Molecular cloning of columbamine *O*-methyltransferase from cultured *Coptis japonica* cells. *Eur. J. Biochem.*, **269**, 5659–5667.

Morishige, T., Choi, K-B. and Sato, F. (2004) *In vivo* bioconversion of tetrahydroisoquinoline by recombinant coclaurine *N*-methyltransferase. *Biosci. Biotechnol. Biochem.*, **68**, 939–941.

Muemmler, S., Rueffer, M., Nagakura, N. and Zenk, M.H. (1985) *S*-adenosyl-L-methionine: (*S*)-scoulerine 9-*O*-methyltransferase, a highly stereo- and region-specific enzyme in tetrahydroportoberberine biosynthesis. *Plant Cell Reports*, **4**, 36–39.

Novak, B.H., Hudlicky, T., Reed, J.W., Mulzer, J. and Trauner, D. (2000) Morphine synthesis and biosynthesis – an update. *Curr. Org. Chem.* **4**, 343–362.

Nyman, U. (1978) Selection for high thebaine/low morphine content (cpv. *Morph:The*) in *Papaver somniferum* L. *Hereditas*, **89**, 43–50.

Nyman, U. (1980) Selection for high thebaine/low morphine content in *Papaver somniferum* L. II. Studies of the I_4 and I_5 generations from a spontaneous mutant. *Hereditas*, **93**, 121–124.

Okada, N., Shinmyo, A., Okada, H. and Yamada, Y. (1988) Purification and characterization of (*S*)-tetrahydroprotoberberine oxidase from cultured *Coptis japonica* cells. *Phytochemistry*, **27**, 979–982.

Okada, N., Koizumi, N., Tanaka, T., Ohkubo, H. and Nakanishi, S. (1989) Isolation, sequence, and bacterial expression of a cDNA for (*S*)-tetrahydroprotoberberine oxidase from cultured berberine-producing *Coptis japonica* cells. *Proc. Natl Acad. Sci. USA*, **86**, 534–538. Published erratum in *Proc. Natl. Acad. Sci. USA*, **87**, 6928.

O'Keefe, B.R. and Beecher, C.W.W. (1994) Isolation and characterization of *S*-adenosyl-L-methionine:tetrahydroberberine-*cis*-*N*-methyltransferase from suspension cultures of *Sanguinaria canadensis* L. *Plant Physiol.*, **105**, 395–403.

Ounaroon, A., Decker, G., Schmidt, J., Lottspeich, F. and Kutchan, T.M. (2003) (*R,S*)-Reticuline 7-*O*-methyltransferase and (*R,S*)-norcoclaurine 6-*O*-methyltransferase of *Papaver somniferum* – cDNA cloning and characterization of methyl transfer enzymes of alkaloid biosynthesis in opium poppy. *Plant J.*, **36**, 808–819.

Park, S-U. and Facchini, P.J. (2000a) *Agrobacterium*-mediated transformation of opium poppy, *Papaver somniferum* L., via shoot organogenesis. *J. Plant Physiol.*, **157**, 207–214.

Park, S-U. and Facchini, P.J. (2000b) *Agrobacterium*-mediated genetic transformation of California poppy, *Eschscholzia californica* Cham., via somatic embryogenesis. *Plant Cell Rep.*, **19**, 1006–1012.

Park, S-U., Yu, M. and Facchini, P.J. (2002) Antisense RNA-mediated suppression of benzophenanthridine alkaloid biosynthesis in transgenic cell cultures of California poppy. *Plant Physiol.*, **128**, 696–706.

Park, S-U., Yu, M. and Facchini, P.J. (2003) Modulation of berberine bridge enzyme levels in transgenic root cultures of California poppy alters the accumulation of benzophenanthridine alkaloids. *Plant Mol. Biol.*, **51**, 153–164.

Pauli, H.H. and Kutchan T.M. (1998) Molecular cloning and functional heterologous expression of two alleles encoding (*S*)-*N*-methylcoclaurine 3′-hydroxylase (CYP80b1), a new methyl jasmonate-inducible cytochrome P-450-dependent monooxygenase of benzylisoquinoline alkaloid biosynthesis. *Plant J.*, **13**, 793–801.

Petterson, D.S., Harris, D.J. and Allen, D.G. (1991) Alkaloids, in *Toxic Substance in Crop Plants* (eds J.P.F. D'Mello, C.M. Duffus and J.H. Duffus), Royal Society of Chemistry, Cambridge, pp. 148–179.

Roos, W., Viehweger, K., Dordschbal, B., Schumann, B., Evers, S., Steighardt, J. and Schwartze, W. (2006) Intracellular pH signals in the induction of secondary pathways-the case of *Eschscholtzia californica. J. Plant Physiol.*, **163**, 369–381.

Rueffer, M., Ekundayo, O., Nagakura, N. and Zenk, M.H. (1983) Biosynthesis of the protoberberine alkaloid jatrorrhizine. *Tetrahedron Lett.*, **24**, 2643–2644.

Rueffer, M. and Zenk, M.H. (1986) *S*-Adenosyl-L-methionine: (*S*)-7,8,13,14-tetrahydroberberine-*cis*-*N*-methyltransferase, a branch point enzyme in the biosynthesis of benzophenanthridine and protopine alkaloids. *Tetrahedron Lett.*, **27**, 5603–5604.

Rueffer, M., Amann, M. and Zenk, M.H. (1986) *S*-adenosyl-L-methionine: columbamine *O*-methyltransferase, a compartmentalized enzyme in protoberberine biosynthesis. *Plant Cell Rep.*, **3**, 182–185.

Rueffer, M. and Zenk, M.H. (1987) Enzymatic formation of protopines by a microsomal cytochrome P-450 system of *Corydalis vaginans. Tetrahedron Lett.*, **28**, 5307–5310.

Rueffer, M. and Zenk, M.H. (1990) Partial purification and properties of *S*-adenosyl-L-methionine:(*S*)-tetrahydroprotoberberine-*cis*-*N*-methyltransferase from suspension-cultured cells of Eschscholtzia and Corydalis. *Phytochemistry*, **29**, 3727–3733.

Rueffer, M. and Zenk, M.H. (1994) Canadine synthase from *Thalictrum tuberosum* cell cultures catalyses the formation of the methylenedioxy bridge in berberine synthesis. *Phytochemistry*, **36**, 1219–1223.

Samanani, N. and Facchini, P.J. (2002) Purification and characterization of norcoclaurine synthase. The first committed enzyme in benzylisoquinoline alkaloid biosynthesis in plants. *J. Biol. Chem.*, **277**, 33878–33883.

Samanani, N., Yeung, E.C. and Facchini, P.J. (2002) Cell type-specific protoberberine accumulation in *Thalictrum flavum*. *J. Plant Physiol.*, **159**, 1189–1196.

Samanani, N., Liscombe, D.K. and Facchini, P.J. (2004) Molecular cloning and characterization of norcoclaurine synthase, an enzyme catalyzing the first committed step in benzylisoquinoline alkaloid biosynthesis. *Plant J.*, **40**, 302–313.

Samanani, N., Park, S.U. and Facchini, P.J. (2005) Cell type-specific localization of transcripts encoding nine consecutive enzymes involved in protoberberine alkaloid biosynthesis. *Plant Cell*, **17**, 915–926.

Samanani, N., Alcantara, J., Bourgault, R., Zulak, K.G. and Facchini, P.J. (2006) The role of phloem sieve elements and laticifers in the biosynthesis and accumulation of alkaloids in opium poppy. *Plant J.*, **47**, 547–563.

Sato, F., Tsujita, T., Katagiri, Y., Yoshida, S. and Yamada, Y. (1994) Purification and characterization of *S*-adenosyl-L-methionine: norcoclaurine 6-*O*-methyltransferase from cultured *Coptis japonica* cells. *Eur. J. Biochem.*, **225**, 125–131.

Sato, F., Hashimoto, T., Hachiya, A., Tamura, K., Choi, K., Morishige, T., Fijimoro, H. and Yamada, Y. (2001) Metabolic engineering of plant alkaloid biosynthesis. *Proc. Natl. Acad. Sci. USA*, **98**, 367–372.

Schumacher, H.-M. and Zenk, M.H. (1988) Partial purification and characterization of dihydrobenzophenanthridine oxidase from *Eschscholtzia californica* cell suspension cultures. *Plant Cell Reports*, **7**, 43–46.

Shitan, N., Bazin, I., Dan, K., Obata, K., Kigawa, K., Ueda, K., Sato, F., Forestier, C. and Yazaki, K. (2003) Involvement of CjMDR1, a plant multidrug-resistance-type ATP-binding cassette protein, in alkaloid transport in *Coptis japonica*. *Proc. Natl. Acad. Sci. USA*, **100**, 751–756.

Stadler, R. and Zenk, M.H. (1990) A revision of the generally accepted pathway for the biosynthesis of the benzyltetrahydroisoquinoline reticuline. *Liebigs Ann. Chem.*, 555–562.

Stadler, R. and Zenk, M.H. (1993) The purification and characterization of a unique cytochrome P-450 enzyme from *Berberis stolonifera* plant cell cultures. *J. Biol. Chem.*, **268**, 823–831.

Steffens, P., Nagakura, N. and Zenk, M.H. (1985) Purification and characterization of the berberine bridge enzyme genes from *Berberis beaniana* cell cultures. *Phytochemistry*, **24**, 2577–2583.

Takeshita, N., Fujiwara, H., Mimura, H., Fitchen, J., Yamada, Y. and Sato, F. (1995) Molecular cloning and characterization of *S*-adenosyl-L-methionine: scoulerine 9-*O*-methyltransferase from cultured cells of *Coptis japonica*. *Plant Cell Physiol.*, **36**, 29–36.

Tanahashi, T. and Zenk, M.H. (1990) Elicitor induction and characterization of microsomal protopine-6-hydroxylase, the central enzyme in benzophenanthridine alkaloid biosynthesis. *Phytochemistry*, **29**, 1113–1122.

Unterlinner, B., Lenz, R. and Kutchan, T.M. (1999) Molecular cloning and functional expression of codeinone reductase: the penultimate enzyme in morphine biosynthesis in the opium poppy *Papaver somniferum*. *Plant J.*, **18**, 465–475.

Viehweger, K., Schwartze, W., Schumann, B., Lein, W. and Roos, W. (2006) The Gα protein controls a pH-dependent signal path to the induction of phytoalexin biosynthesis in *Eschscholtzia californica*. *Plant Cell*, **18**, 1510–1523.

Vom Endt, D., Kijne, J.W. and Memelink, J. (2002) Transcription factors controlling plant secondary metabolism: what regulates the regulators? *Phytochemistry*, **61**, 107–114.

Weiss, D., Baumert, A., Vogel, M. and Roos, W. (2006) Sanguinarine reductase, a key enzyme of benzophenanthridine detoxification. *Plant Cell Environ.*, **29**, 291–302.

Winkler, A., Hartner, F., Kutchan, T.M., Glieder, A. and Macheroux, P. (2006) Biochemical evidence that berberine bridge enzyme belongs to a novel family of flavoproteins containing a bi-covalently attached FAD cofactor. *J. Biol. Chem.*, **281**, 21276–21285.

Winkler, A., Kutchan, T.M. and Macheroux, P. (2007) 6-*S*-Cysteinylation of bi-covalently attached FAD in berberine bridge enzyme tunes the redox potential for optimal activity. *J. Biol. Chem.*, **282**, 24437–24443.

Winkler, A., Lyskowski, A., Riedl, S., Puhl, M., Kutchan, T.M., Macheroux, P. and Gruber, K. (2008) A concerted mechanism for berberine bridge enzyme. *Nature Chem. Biol.*, **4**, 739–741.

Winkler, A., Puhl, M., Weber, H., Kutchan, T.M., Gruber, K. and Macheroux, P. (2009) Berberine bridge enzyme catalyzes the six electron oxidation of (*S*)-reticuline to dehydroscoulerine. *Phytochemistry*, **70**, 1092–1097.

Ziegler, J., Diaz-Chavez, M.L., Kramell, R., Ammer, C. and Kutchan, T.M. (2005) Comparative macroarray analysis of morphine containing *Papaver somniferum* and eight morphine free *Papaver* species identifies an *O*-methyltransferase involved in benzylisoquinoline biosynthesis. *Planta*, **222**, 458–471.

Ziegler, J., Voighlander, S., Schmidt, J., Kramell, R., Miersch, O., Ammer, C., Gesell, A. and Kutchan, T.M. (2006) Comparative transcript and alkaloid profiling in *Papaver* species identifies a short chain dehydrogenase/reductase involved in morphine biosynthesis. *Plant J.*, **48**, 177–192.

Ziegler, J. and Facchini, P.J. (2008) Alkaloid biosynthesis: metabolism and trafficking. *Ann. Rev. Plant Biol.*, **59**, 735–769.

Zulak, K.G., Cornish, A., Daskalchuk, T.E., Deyholos, M.K., Goodenowe, D.B., Gordon P.P., Klassen, D., Pelcher, L.E., Sensen, C.W. and Facchini, P.J. (2007) Gene transcript and metabolite profiling of elicitor-induced opium poppy cell cultures reveals the coordinate regulation of primary and secondary metabolism. *Planta*, **225**, 1085–1106.

10

Monoterpenoid Indole Alkaloid Biosynthesis

Vincenzo De Luca

10.1 Introduction

Approximately 20% of plant species contain alkaloids with a broad range of physiological properties that protect them from various types of herbivores and pathogens. Within this abundant group of nitrogen-containing secondary metabolites, the monoterpenoid indole alkaloids (MIAs) make up the largest and most diverse class of compounds that are characteristically found within the Apocynaceae, Loganiaceae and Rubiaceae plant families. The complexity of MIA chemistry is matched by their remarkably diverse effects on living organisms which has led to their use as drugs (Figure 10.1) for treating neurological disorders (reserpine), cancer (camptothecin, vinblastine and vincristine) and as vasodilators (yohimbine) in humans. Serpentine has recently been shown to be a powerful acetylcholinesterase inhibitor with potential use for treatment of Alzheimer's disease (Pereira *et al.*, 2010).

Catharanthus roseus is the best characterised MIA-producing plant, mostly due to the commercial value of its anticancer alkaloids, vinblastine and vincristine. Several recent reviews have been published on the analysis (Hisiger and Jolicoeur, 2007), chemistry (O'Connor and Maresh, 2006), biosynthesis (Facchini and De Luca, 2008; Oudin *et al.*, 2007a; El-Sayed and Verpoorte, 2007; Loyola-Vargas *et al.*, 2007; Hedhili *et al.*, 2007), regulation (Memelink and Gantet, 2007; Memelink, 2009), intracellular, intercellular and organ-specific compartmentation (Mahroug *et al.*, 2007), plant cell culture-based

Plant Metabolism and Biotechnology, First Edition. Edited by Hiroshi Ashihara, Alan Crozier, and Atsushi Komamine.

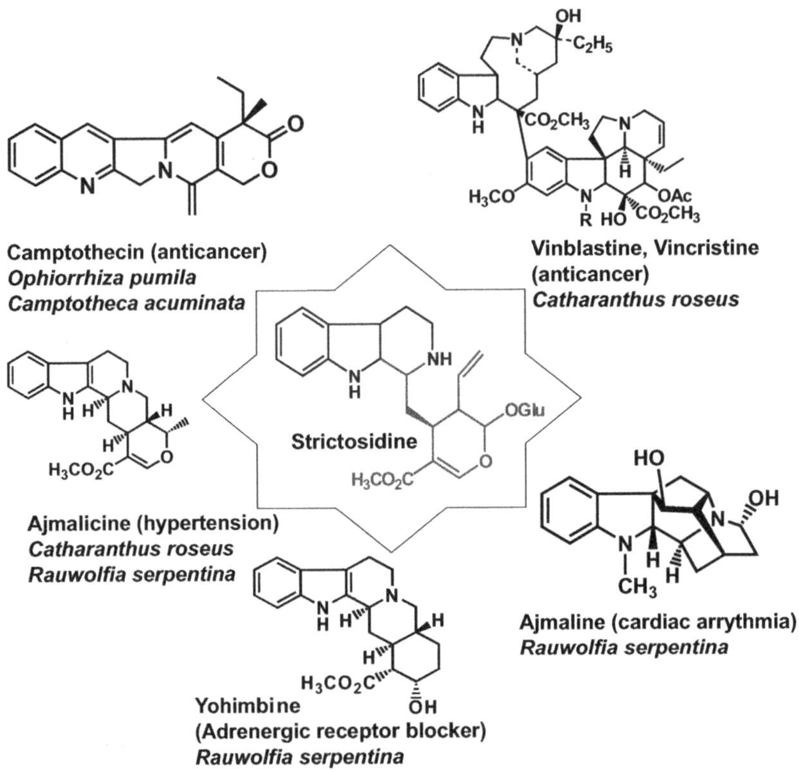

Figure 10.1 *The central intermediate, strictosidine, is the precursor for several thousand MIAs with powerful biological activities. These include the commercially valuable anticancer alkaloids vinblastine and camptothecin, as well as MIAs with other potential uses such as ajamaline, ajmalicine and yohimbine*

production (Zhao and Verpoorte, 2007), biotechnological production (Zárate and Verpoorte, 2007; Zhou *et al.*, 2009), functional genomics (Goossens and Rischer, 2007) and transport (Roytrakul and Verpoorte, 2007) of MIAs in *Catharanthus*. Although there have been fewer investigations, studies with *Rauvolfia serpentina* (ajmaline and reserpine) (Stöckigt *et al.*, 2007), *Camptotheca acuminata* (camptothecin) (Lorence and Nessler, 2004; Sirikantaramas *et al.*, 2007) and *Ophiorrhiza pumila* (camptothecin) (Sirikantaramas *et al.*, 2007) have also contributed significantly to our biochemical and molecular knowledge of MIA biosynthesis. This review describes some of the significant discoveries with *Catharanthus roseus*, *Rauvolfia serpentina*, *Camptotheca acuminata* and *Ophiorrhiza pumila* that have helped to highlight information on MIA biosynthesis, its regulation during plant growth and development, the contributions of multiple cell types to different sections of this complex pathway, and the involvement of different intracellular compartments in biosynthesis in the whole plant and/or in cell cultures.

10.2 Monoterpenoid Indole Alkaloid (MIA) Biosynthesis

10.2.1 Contributions of Two Separate Pathways in MIA Assembly

A unique feature of MIA assembly is the participation of separate pathways that distinguish tryptamine from the tryptophan branch of the shikimate pathway and secologanin from the 2-*C*-methyl-D-erythritol-4-phosphate (MEP) pathway rather than the mevalonic acid (MVA) pathway (Figure 10.2). Many biochemical studies on tryptophan decarboxylase (TDC) involved in tryptamine formation from tryptophan have been conducted with *Catharanthus roseus* cell cultures and in intact plants. The cDNA for this gene was the first to be identified and functionally characterised in *Catharanthus* (De Luca *et al.*, 1989), with the recombinant protein being able to decarboxylate tryptophan but not tyrosine or phenylalanine. Labelling studies using *Catharanthus roseus* (Contin *et al.*, 1998) and *Ophiorrhiza pumila* (Yamazaki *et al.*, 2004) cell cultures with [1-^{13}C]-glucose showed that the MEP pathway, rather than the mevalonic pathway, contributes to the formation of this key precursor. While many of the *Catharanthus* genes for the MEP pathway, including deoxyxylulose 5-phosphate synthase (DXS), deoxyxylulose 5-phosphate reductoisomerase (DXR),

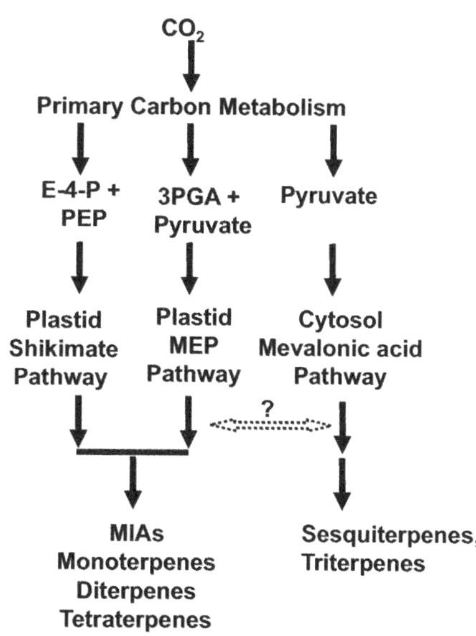

Figure 10.2 *Linking primary metabolism and MIA biosynthesis. Primary metabolism converts CO_2 fixed during photosynthesis to erythrose-4-phosphate (E-4-P), phospoenolpyruvate (PEP), 3-phosphoglycerate (3PGA) and pyruvate which are used to elaborate the plastid-localised shikimate and MEP pathways together with the cytosol localised mevalonic acid pathway for biosynthesis of MIAs, as well as different terpenes. The shikimate pathway provides the tryptophan that is then converted in the cytosol to tryptamine which is utilised in the biosynthesis of MIAs. It is not clear how much crosstalk takes place between the MEP and mevalonic acid pathways*

methylerythritol 2,4-diphosphate synthase (MECS) and hydroxymethylbutenyl diphosphate synthase (HDS) have been identified (Chahed *et al.*, 2000; Veau *et al.*, 2000: Oudin *et al.*, 2007b), most of their biochemical properties have not, as yet, been characterised in detail.

10.2.2 Genes for the Biosynthesis of Secologanin

The secologanin biosynthesis pathway from geraniol comprises at least eight steps involving geraniol-10-hydroxylase (G10H; Cyp76B6) (Collu *et al.*, 2001), 10-hydroxygeraniol oxidoreductase (10HGO), which is also known as acyclic monoterpene primary alcohol dehydrogenase (Ikeda *et al* 1991), loganic acid methyltransferase (LAMT) (Murata *et al.*, 2008) and secologanin synthase (SLS; Cyp72A1) (Irmler *et al.*, 2000) whose genes have been cloned and functionally characterised (Figure 10.3). While several more genes remain to be characterised, NADPH-dependent 10-oxogeranial cyclase from *Rauvolfia* cell cultures (Uesato *et al.*, 1986, 1987) and *Catharanthus* hairy root cultures (Sanchez-Iturbe *et al.*, 2005) have been partially purified and characterised. During the cloning and characterisation of the secologanin synthase gene, preliminary studies indicated that deoxyloganin hydroxylase (DLGT) was a cytochrome P450 enzyme (Irmler *et al.*, 2000) as previously suggested for the same enzyme in *Lonicera japonica* cell cultures (Katano *et al.*, 2001). Bacterial expression and biochemical characterisation of LAMT (Murata *et al.*, 2008) confirmed early substrate specificity studies with the partially purified enzyme (Madyastha *et al.*, 1973) demonstrating that *O*-methylation occurs after the hydroxylation of 7-deoxyloganic acid (Figure 10.3) in *Catharanthus.* It will be interesting to see whether the substrate specificity of LAMT is different in *Lonicera japonica* (Yamamoto *et al.*, 1998), where the possibility has been raised that hydroxylation may occur after methylation.

10.2.3 MIA Biosynthesis in *Catharanthus Roseus*

The MIA pathway of *Catharanthus roseus* has been extensively studied in both plant cell and in hairy root cultures as well as in intact plants. Tryptamine, derived from tryptophan by the action of tryptophan decarboxylase (TDC) (De Luca *et al.*, 1989), is coupled to secologanin by strictosidine synthase (STR) (Kutchan, 1989; McKnight *et al.*, 1990) to form strictosidine, the central intermediate to over 130 MIAs that have been detected in *Catharanthus* (van der Heijden *et al.*, 2004). Strictosidine β-glucosidase (SDG), a large molecular weight, putative membrane-associated enzyme (Geerlings *et al.*, 2000), shows strict specificity for strictosidine to produce a highly reactive ring-opened dialdehyde intermediate involved in the formation of corynantheine, iboga and aspidosperma classes of MIAs. The pathways leading to each class of MIA remains to be characterised and will require the availability of intermediates to allow discovery of the steps leading to the formation of molecules such as tabersonine and catharanthine.

The biochemistry for converting tabersonine to vindoline has been extensively characterised (Facchini and De Luca, 2008) (Figure 10.4). Tabersonine is converted to 16-hydroxytabersonine by the cytochrome P450-dependent tabersonine-16-hydroxylase (T16H; Cyp71D12) (Schröder *et al.*, 1999). The next step, conversion of 16-hydroxytabersonine to 16-methyoxytabersonine, is catalyzed by 16-hydroxytabersonine-16-*O*-methyltransferase (16OMT) (Levac *et al.*, 2008). The high affinity of 16OMT for

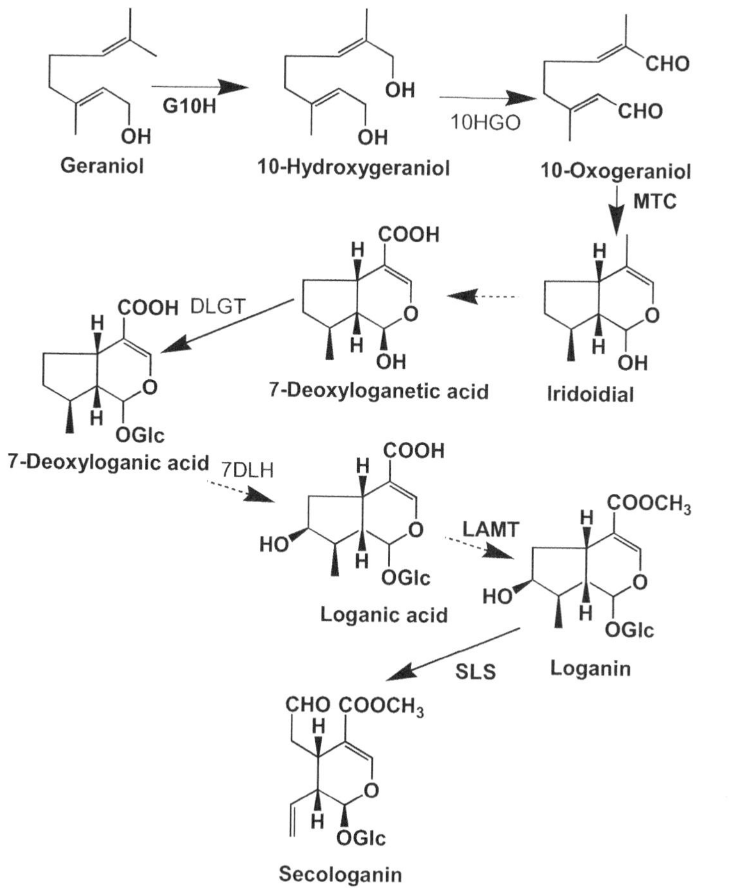

Figure 10.3 The pathway for secologanin biosynthesis from geraniol. Geraniol derived from the MEP pathway within IPAP chloroplasts is converted to 10-hydroxygeraniol by geraniol-10-hydroxylase (G10H) associated with the cytoplasmic face of the endoplasmic reticulum. Subsequent reactions include 10-hydroxygeraniol oxidoreductase (10HGO), monoterpene cyclase (MTC), deoxyloganetic acid glucosyltransferase (DLGT), 7-deoxyloganetic acid hydroxylase (7DLH), loganic acid carboxymethyltransferase (LAMT) and secologanin synthase (SLS). Biochemical reactions for all these reactions have been described and those highlighted in bold represent functionally characterised cloned genes. The oxidation of iridoidial to 7-deoxyloganetic acid requires additional steps that remain to be described

16-hydroxytabersonine, but not 16-hydroxy-2,3-dihydrotabersonine, suggests strongly that this reaction occurs before substitution of the 2,3-position on the molecule (Levac *et al.*, 2008). While the conversion of 16-methoxytabersonine to 16-methoxy-2,3-dihydrotabersonine remains to be described, a chloroplast associated-*N*-methyltransferase (NMT) catalyses metabolism of 16-methoxy-2,3-dihydrotabersonine to 4-desacetoxyvindoline, the third to last step in vindoline biosynthesis (De Luca and Cutler,

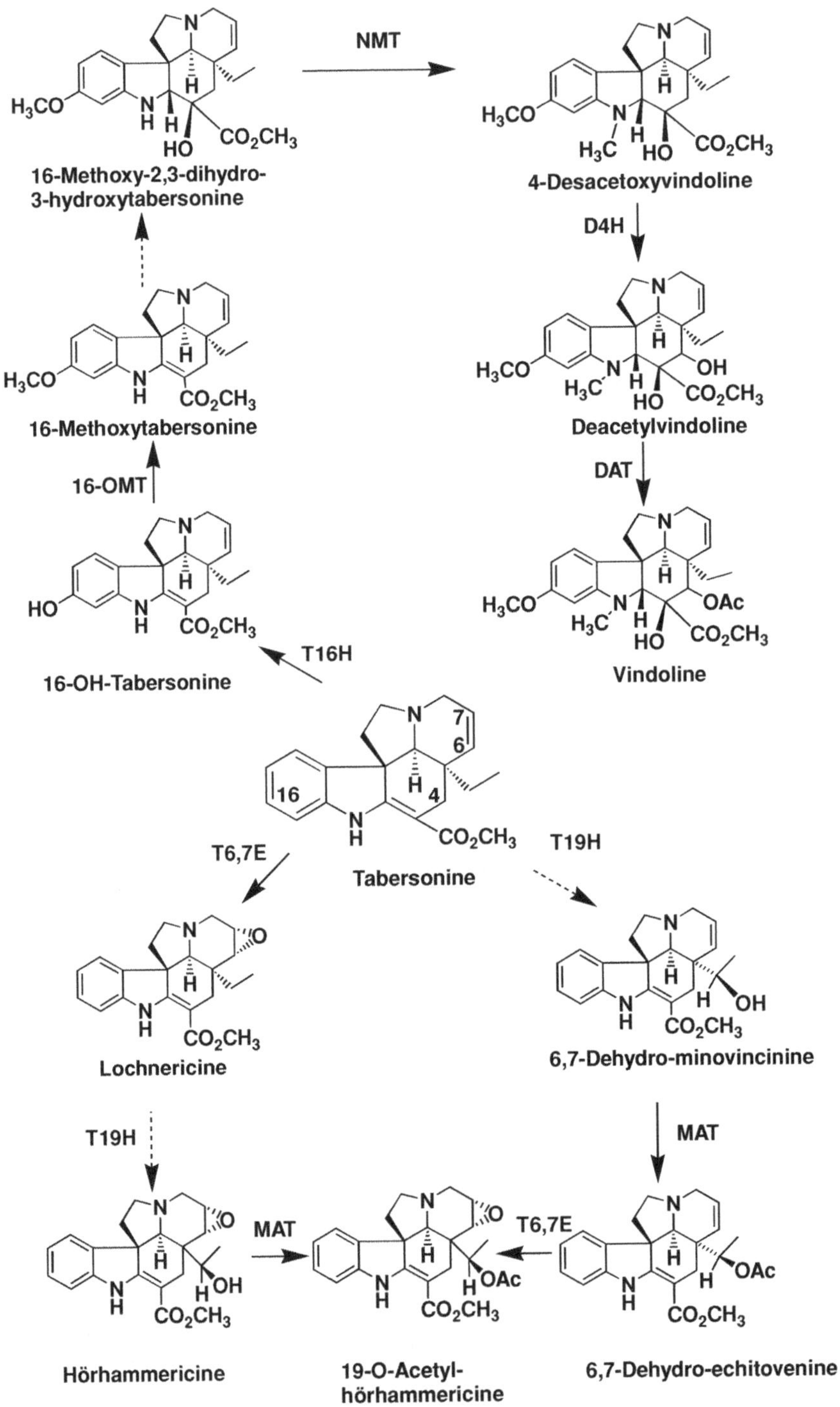

Figure 10.4 (Continued)

1987; Dethier and De Luca, 1993). Unlike other oxidases described for this pathway, desacetoxyvindoline 4-hydroxylase (D4H) (Vazquez-Flota *et al.*, 1997), which catalyzes the penultimate reaction in the vindoline pathway, is a 2-oxoglutarate dioxygenase. The final step in the pathway is catalyzed by deacetylvindoline 4-*O*-acetyltransferase (DAT) (Figure 10.4) (St-Pierre *et al.*, 1998).

Catharanthus roots accumulate lochnericine and hörhammericine, which are oxidised derivatives of tabersonine (Figure 10.4). The different fate of tabersonine in roots involves the formation of lochnericine and/or 19-*O*-acetylhörhammericine by a cytochrome P450-6,7-epoxidase (T6,7E) associated with microsomes in *Catharanthus* hairy root cultures (Rodriguez *et al.*, 2003). The production of hörhammericine requires an additional 19-hydroxylase (T19H) that remains to be characterised, while the terminal reaction to produce 19-*O*-acetylhörhammericine involves minovincinine-*O*-acetyltransferase (MAT) which has been cloned and functionally characterised (Laflamme *et al.*, 2001). Since deacetylvindoline can be *O*-acetylated by MAT, this activity in roots has been confused with DAT.

10.2.4 MIA Biosynthesis in *Rauvolfia Serpentina*

The medicinal MIAs of *Rauvolfia serpentina* include serpentine and ajmalicine which have been variously used as antipsychotics, antihypertensives and antiarrhythmics. The pathway for ajmaline biosynthesis has been extensively studied in cell cultures of *Rauvolfia serpentina* (Stöckigt *et al.*, 2007). The pathway beyond strictosidine involves a sequence of nine well-characterised enzymatic reactions that convert dehydrogeissoschizine → polyneuridine → epi-vellosimine → vinorine → vomilenine → 1,2-dihydrovomilenine → acetylnorajmaline → norajmaline → ajmaline (Figure 10.5). Compared with the 12 genes functionally characterised in *Catharanthus*, the functions of the six *Rauvolfia* MIA genes are known including those encoding STR, SGD, polyneuridine aldehyde esterase, vinorine synthase, cytochrome P450 reductase and acetylajmalan acetylesterase in *R. serpentina* (Ruppert *et al.*, 2005). The crystal structures for several of these enzymes have been elucidated, which has provided insights into the reaction mechanisms involved in MIA biosynthesis (Stöckigt *et al.*, 2007).

10.2.5 MIA Biosynthesis in *Camptotheca Acuminata* and *Ophiorrhiza Pumila*

Camptothecin is an inhibitor of DNA topoisomerase I that also induces extensive single strand breaks in DNA as a result of its covalent binding of topoisomerase I to the 3′ end of the broken DNA (Hsiang *et al.*, 1985). This property has led to the development of powerful

---

Figure 10.4 Biosynthesis of MIAs in Catharanthus roseus. *Enzymes for which corresponding cDNAs have been isolated are shown in bold. Abbreviations: T16H, tabersonine 16-hydroxylase; 16OMT, 16-hydroxytabersonine-16-O-methyltransferase; NMT, N-methyltransferase; D4H, desacetoxyvindoline 4-hydroxylase; DAT, deacetylvindoline 4-O-acetyltransferase; T19H, tabersonine-19-hydroxylase; MAT, minovincinine acetyltransferase; T6,7E, tabersonine 6,7 epoxidase*

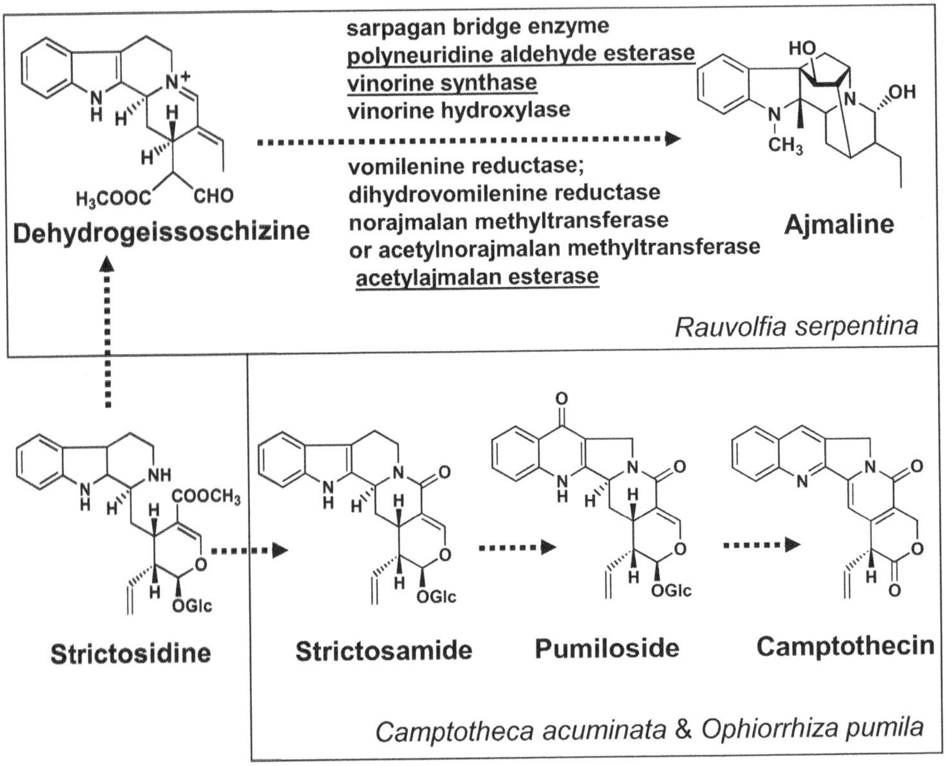

Figure 10.5 *Biosynthesis of MIAs in* Rauvolfia serpentina *(ajmaline),* Camptotheca acuminata *(camptothecin) and* Ophiorrhiza pumila *(camptothecin). Many of the biochemical steps involved in the biosynthesis of ajmaline have been characterised, and three clones (underlined) have been functionally expressed. Many challenges remain in the characterisation of the camptothecin pathway beyond strictosidine as illustrated by the putative intermediates involved and the biochemical steps that must be described*

camptothecin derivatives that are extensively used in cancer chemotherapy (Lorence and Nessler, 2004; Sirikantaramas *et al.*, 2007). The TDC, STR and NADPH:cytochrome P450 reductase involved in MIA biosynthesis have been characterised from *Ophiorrhiza pumila* (Yamazaki *et al.*, 2003), while only TDC has been characterised from *Camptotheca acuminata* (Lopez-Meyer and Nessler, 1997). Several biochemical steps remain to be characterised and genes cloned in order to fully elucidate the conversion of strictosidine to camptothecin (Sirikantaramas *et al.*, 2007) (Figure 10.5). Recently Sirikantamaras *et al.* (2008) reported that camptothecin-producing *Camptotheca acuminata, Ophiorrhiza pumila and Ophiorrhiza liukiuensis* express DNA topoisomerase I enzymes with point mutations that confer resistance to this alkaloid. One point mutation (Asn722Ser) was identical to the mutation found in human DNA topoisomerase I that confers resistance in camptothecin-resistant cancer cell lines, while the two other point mutations (Asn421Lys, Leu530Ile) also conferred camptothecin resistance.

10.3 MIA Pathway Gene Discovery will be Enhanced by Large-Scale Sequencing and Comparative Analyses

10.3.1 Pyrosequencing

The advent of inexpensive high-throughput sequencing methods promises to revolutionise the discovery of secondary metabolism pathways within non-model medicinal plant species (Facchini and De Luca, 2008). A number of projects have been initiated around the globe to apply 454 pyrosequencing in order to expand the biological sources of candidate genes from medicinal plants. In a recent example, such a strategy was applied in the discovery of candidate genes involved in the *de novo* biosynthesis of toxic cyanogenic glycosides in the moth *Zygaena filipendulae*, a property unique to this family of insects (Zagrobelny *et al.*, 2009). Since cyanogenic glucoside biosynthesis in plants involves members of the cytochrome P450 and glucosyltransferase gene families, candidate genes based on these families were selected from the transcriptome of *Zygaena filipendulae* obtained by pyrosequencing. In another study, *Artemisia annua*, which accumulates the antimalarial drug artemisin within glandular trichomes, was analysed by 454 pyrosequencing (Wang *et al.*, 2009). This glandular trichome transcriptome confirmed the deep coverage of transcripts provided by this technology by supplying putative functions to over 28,000 unigenes including genes involved in the assembly of artemisin.

10.3.2 Sequencing by Expressed Sequence Tag Approaches

Over the past ten years, many efforts to identify pathways for secondary metabolism have depended upon methods to isolate specialised cell types that preferentially express and accumulate particular secondary metabolites. Successful isolation of specialised cells followed by isolation of RNA could then be used for construction of cDNA libraries and submission of those libraries to random sequencing. While this is comparatively costly compared with pyrosequencing, this 'old'-generation sequencing technique has been highly effective for the characterisation of the specialised gland chemistry of plants that are often responsible for biosynthesis and accumulation of secondary metabolites. This functional genomic, computational and expression-based approach was first successfully used to harvest candidate genes for menthol biosynthesis in *Mentha piperita* (Lange *et al.*, 2000) in addition to pathways for highly methylated flavone biosynthesis, for transport processes and for secretion. The use of glandular trichomes from plants to isolate transcripts for the process of random sequencing has been used on many occasions to identify candidate genes in a variety of plant species (Schilmiller *et al.*, 2008).

The specialised nature of glandular trichomes can be extended to the other cell types that make up the epidermis of plants. This layer of cells both covers and protects the plant and its role is highly versatile during growth and development. In young tissues such as developing leaves, the epidermis has a secretory function to supply the surface with a hydrophobic and complex wax layer. At this stage of growth, epidermal cells will need to express pathways that will supply the secretory components for assembly of these protective surface layers. When transcriptomes from different *Catharanthus* organs, tissues and cell cultures with high levels of MIA biosynthesis (St-Pierre *et al.*, 1999; Laflamme *et al.*, 2001; Rischer *et al.*, 2006; Shukla *et al.*, 2006) were submitted to extensive random sequencing

(>25,000 ESTs) (Murata *et al.*, 2006; Rischer *et al.*, 2006; Shukla *et al.*, 2006), few known transcripts were obtained.

More recently, a new technique involving caborundum abrasion permitted the isolation of *Catharanthus* epidermis-enriched mRNA from young leaves (Murata *et al.*, 2008). Transcriptomic analysis of a representative cDNA library showed that all previously characterised MIA biosynthesis and regulatory genes were represented in the >10,000 unique ESTs that were obtained. These results also suggested that the remaining candidate genes for catharanthine and tabersonine biosynthesis were likely to be represented. One particular *O*-methyltransferase, represented 13 times in this EST population, encoded a full-length open reading frame that had a very high sequence similarity to salicylic acid *O*-carboxymethyltransferase. This led to its cloning and functional expression to identify LAMT, the enzyme catalyzing the second-to-last step in secologanin biosynthesis (Figure 10.3) (Murata *et al.*, 2008). As expected, LAMT activity was enriched in protein extracts from leaf epidermal cells compared with whole leaf extracts.

Another candidate *O*-methyltransferase represented four times in this database was suspected to be 16OMT, based on its sequence similarity to flavonoid methyltransferases. However, previous efforts to clone this gene with this type of information had been unsuccessful (Cacacea *et al.*, 2003, Schröder *et al.*, 2004). Cloning of *16OMT* was facilitated after it was realised that leaf epidermal proteins could be isolated on a large scale by carborundum abrasion (Levac *et al.*, 2008), leading to the purification of 16OMT to homogeneity and its sequencing. The use of carborundum abrasion also facilitated the biochemical identification of other enzymes localised in *Catharanthus* leaf epidermal cells. Both LAMT (Murata *et al.*, 2008) and *16OMT* (Levac *et al.*, 2008) were located primarily within this cell type, while the later steps in vindoline biosynthesis catalyzed by NMT and DAT were found in whole leaves.

The same epidermis-enriched cDNA library provided a candidate gene for the 4-diphosphocytidyl-2-C-methyl-D-erythritol synthase of the MEP pathway, three candidate genes common to the MEP/MVA pathways (isopentenyl pyrophosphate isomerase; farnesyl pyrophosphate synthase; geranyl pyrophosphate synthase) and four MVA pathway genes (3-hydroxy-3-methylglutaryl coenzyme A reductase, HMGR; 3-ketoacyl-CoA thiolase AACT1; acetoacetyl-CoA thiolase; HMG-CoA synthase HMGS) (Murata *et al.*, 2008). Apparently, enrichment of the MVA pathway in the epidermal cells of *Catharanthus* leaves is a consequence of them being a specialised site for the biosynthesis and accumulation of the triterpene ursolic acid (Usia *et al.*, 2005; Murata *et al.*, 2008). Furthermore, over 60 candidate genes involved in the biosynthesis of very long chain fatty acids were also identified.

10.4 Developmental and Environmental Regulation of MIA Biosynthesis

Early studies on MIA biosynthesis in intact plants provided important insights into developmental controls that regulate this pathway and the accumulation of its end products. Studies with developing seedlings of *Catharanthus roseus* revealed that expression of the MIA pathway was under strict development-, tissue-, and environment-specific controls that regulated the timing and accumulation rates of *Catharanthus* alkaloids. Fresh *Catharanthus* seeds have high and uniform germination rates, and the collection of seedlings at different stages of growth provides material that can be probed for gene expression, enzyme activities of individual MIA pathway steps. and accumulation of MIAs (De Luca *et al.*, 1986, 1988; Fernandez *et al.*, 1989). Seedlings grown in the absence of light transiently expressed TDC which peaked 4–6 days after germination commenced (Figure 10.6A). As TDC began to increase, catharanthine accumulation commenced and peaked after 8–10 days of growth (Figure 10.6B). Vindoline did not accumulate when seedlings were grown in darkness, which was a consequence of the etiolated seedlings lacking DAT (Figure 10.6A) and D4H which catalyse the last two steps in vindoline biosynthesis (Figure 10.4).

The large-scale extractions of etiolated seedlings yielded significant quantities of tabersonine and 16-methoxytabersonine together with low amounts of 16-hydroxytabersonine, 4-desacetoxyvindoline, deacetylvindoline and vindoline (Balsevich *et al.*, 1986). The tabersonine derivatives were converted quantitatively to vindoline when etiolated seedlings were transferred to light, indicating the operation of the six-step pathway illustrated in Figure 10.4 (Balsevich *et al.*, 1986). While seedlings grown in the presence of light showed the same developmental profile with respect to the transient expression of TDC and catharanthine accumulation, the last two steps in vindoline biosynthesis are activated by light treatment, peaking between 7 and 9 days of seedling development, with the accumulation of vindoline (Figure 10.6D) instead of tabersonine. These results specify important developmental and environmental differences in the regulation of the catharanthine/tabersonine component of these pathways and the last six steps that convert tabersonine into vindoline. Further studies have shown that the last two steps in vindoline biosynthesis can be activated by exposing etiolated seedlings to short periods of red light and the process can be reversed by far red light, implying the involvement of phytochrome in the activation process (Aerts and De Luca, 1992; Vazquez-Flota and De Luca, 1998).

10.4.1 Why is the Biosynthesis of MIAs in *Catharanthus* Compartmented in Different Cell Types and Within Different Organelles?

The cell-specific and organelle compartmentation of MIA biosynthesis in *Catharanthus roseus* remains something of a mystery. This plant has evolved the use of more than one cell type to accommodate the biosynthesis of vindoline, whereas the biosynthesis of tabersonine and catharanthine appear to be restricted to cortical root tip cells and the epidermis of the leaf (Figure 10.7). One possible explanation is that this enables the plant to sequester individual MIAs that acquire novel cytotoxicity when they form dimeric MIAs, as is the case with combining vindoline and catharanthine to form vinblastine. This unusually complex assembly of vindoline raises important issues concerning the organisation of MIA biosynthesis in other species of plants. Unfortunately, this is a little-investigated topic.

10.4.2 Why does MIA Biosynthesis Occur in at Least Five Subcellular Compartments?

The subcellular compartmentation of MIA biosynthesis in *Catharanthus roseus* has been shown to be very complex as different parts of the pathway have been discovered. The earliest biochemical localisation studies separated organelles by sucrose density gradient centrifugation and showed that early (TDC) and late steps (D4H and DAT) of vindoline

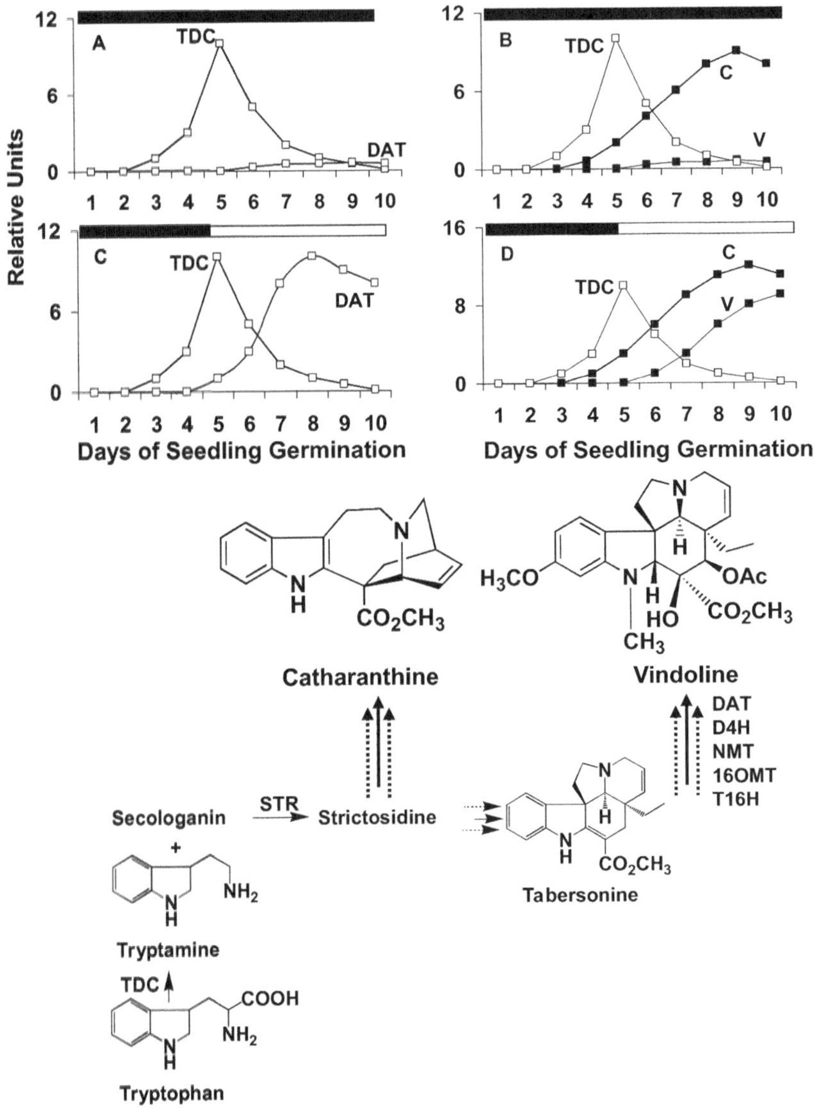

Figure 10.6 Expression of MIA biosynthesis is controlled by development-, tissue- and environment-specific controls. Seeds were germinated either in darkness (black bar), or darkness followed by light treatment after 4.5 days (black followed by white bar). During the course of their development, seedlings were analysed for TDC and DAT enzyme activity as well as catharanthine and vindoline accumulation. The data were assembled from different reports described in the review. Abbreviations: 16OMT, 16-hydroxytabersonine-16-O-methyltransferase; T16H, tabersonine 16-hydroxylase; D4H, desacetoxyvindoline 4-hydroxylase; DAT, deacetylvindoline 4-O-acetyltransferase; NMT, N-methyltransferase; STR, strictosidine synthase; TDC, tryptophan decarboxylase; C, catharanthine; V, vindoline

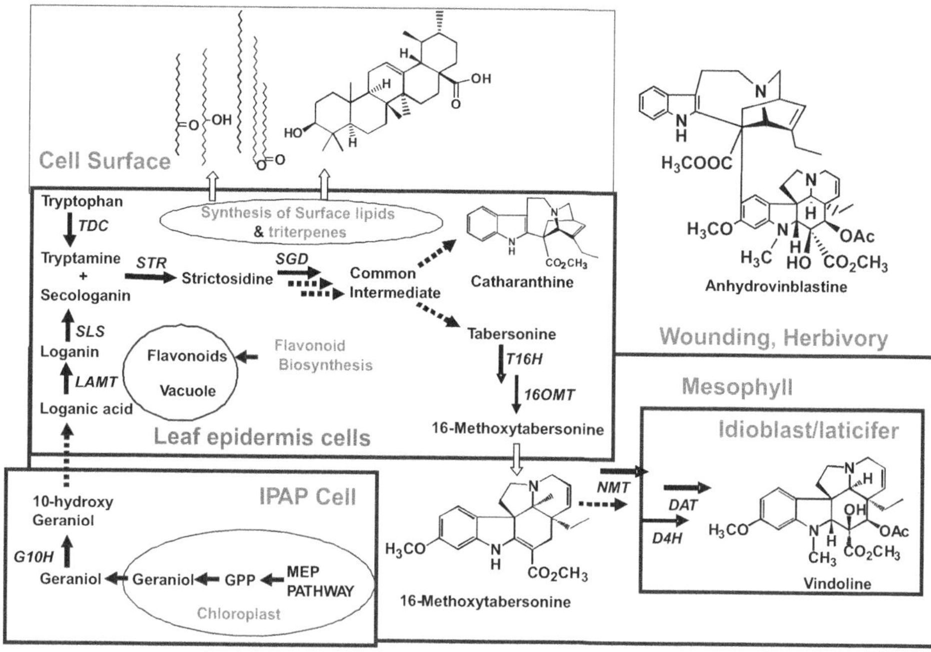

Figure 10.7 Model of MIA assembly in Catharanthus roseus *leaves. The leaf epidermis is specialised for biosynthesis of flavonoids, the fatty acid components of waxes, triterpenes and MIAs (Murata* et al.*, 2008). 10-hydroxygeraniol is made in specialised internal associated phloem parenchyma cells (IPAP) via the MEP pathway and geraniol-10-hydroxylase (G10H). This metabolite or another intermediate is transported to the leaf epidermis where loganic acid is converted to secologanin in reactions catalyzed by loganic acid carboxymethyltransferase (LAMT) and secologanin synthase (SLS). Within the leaf epidermis, tryptophan is decarboxylated in the cytoplasm to tryptamine by tryptophan decarboxylase (TDC), and strictosidine is formed by the coupling of tryptamine and secologanin in the leaf epidermal vacuole by strictosidine synthase (STR). Strictosidine is released into the cytoplasm as a reactive aglycone by cytosolic strictosidine β-glucosidase (SGD), where mostly unknown biochemical transformations lead to the production of the three major MIA bac, corynanthe, iboga and aspidosperma. A common intermediate is converted to catharanthine or to tabersonine by uncharacterised biochemical reactions. Catharanthine is synthesized in the leaf epidermis, as are the fatty acid components of waxes and the triterpene, ursolic acid, that are secreted into the leaf cell. Tabersonine is further metabolised to 16-methoxytabersonine in the leaf epidermis by tabersonine-16-hydroxylase (T16H) and 16OMT-catalysed conversions. 16-methoxytabersonine may be secreted into leaf mesophyll cells where a further uncharacterised oxidation takes place, followed by an* N*-methylation (NMT), located in chloroplast thylakoids, hydroxylation (D4H) and* O*-acetylation (DAT) to yield vindoline. The last two steps in vindoline biosynthesis appear to occur in specialised leaf idioblast and laticifer cells. It is postulated that wounding or herbivory may result in coupling of catharanthine and vindoline, forming dimeric anticancer MIAs such as anhydrovinblastine. The dotted lines represent uncharacterised reactions or unknown mechanisms that remain to be documented*

biosynthesis were located in the 'cytosol' (De Luca and Cutler, 1987). While this was true, later studies established that these reactions took place in the cytosol of different leaf cells. In a similar manner, strictosidine synthase (STR) (McKnight *et al.*, 1991) (Figure 10.7) and the peroxidase that couples vindoline to catharanthine for dimer production (Sottomayor *et al.*, 1998) both contain vacuole-targeting signals, but it remains to be established whether both proteins occur in the same vacuole, in vacuoles of different cells, or even whether their expression takes place at the same stage of growth and development. The next reaction in MIA biosynthesis involves strictosidine β-glucosidase (SGD), a soluble enzyme thought to be associated with the cytoplasmic face of the endoplasmic reticulum/vacuole (Stevens *et al.*, 1993). This association with these membranes presumably permits the conversion of strictosidine to a reactive aglycone that can be used by subsequent enzymes for producing different MIAs. A number of cytochrome P450 monooxygenases (CYPs) have also been characterised and all appear to be associated with the endoplasmic reticulum (St-Pierre and De Luca, 1995; Yamamoto *et al.*, 2000; Collu *et al.*, 2001). However, as described below, expression of different MIA pathway CYPs occurs in different leaf cells (Figure 10.7). Finally, the *N*-methyltransferase that catalyses the third to last step in vindoline biosynthesis is closely associated with thylakoid membranes within chloroplasts of leaf mesophyll/idioblast/laticifer cells (Figure 10.7) (Dethier and De Luca, 1993; Murata *et al.*, 2008).

10.4.3 Why is the MEP Pathway and Geraniol-10-Hydroxylase Expressed in Internal Phloem-Associated Parenchyma Cells?

With the identification of the MEP pathway as the primary supply of IPP for secologanin biosynthesis in *Catharanthus* cell cultures (Contin *et al.*, 1998) and *Ophiorrhiza pumila* hairy roots (Yamazaki *et al.*, 2004), four genes involved in this pathway were localised by *in situ* RNA hybridisation studies in cells surrounding the phloem vasculature in young leaves (Burlat *et al.*, 2004; Mahroug *et al.*, 2006; Oudin *et al.*, 2007b). Transcripts for G10H (Cyp76B6) that commits geraniol to the biosynthesis of secologanin were also localised within the same cell types, and led to the term 'internal phloem-associated parenchyma (IPAP) cells' for this biochemical specialisation (Figure 10.7). This discovery led to the hypothesis that chloroplasts within IPAP cells are enriched with the MEP pathway together with a putative geraniol synthase that supplies geraniol for the biosynthesis of 10-hydroxygeraniol mediated by G10H associated with the cytoplasmic face of the endoplasmic reticulum within the same cells. However, several studies suggest that chloroplasts can also be the site of certain cytochrome P450 reactions such as allene oxide synthase (Froehlich *et al.*, 2001), fatty acid hydroperoxide lyase (Froehlich *et al.*, 2001), *ent*-kaurene oxidase (Helliwell *et al.*, 2001) and carotenoid ε hydroxylase (Tian *et al.*, 2003). In recent studies using transient expression of the MEP pathway protein, hydroxymethylbutenyl-4-diphosphate synthase (HDS) tagged with green fluorescent protein targeted the gene in *Catharanthus* cell culture plastids, and also indicated enrichment within stromules closely associated with the endoplasmic reticulum (Guirimand *et al.*, 2009). These results confirm electron microscopy immunogold labelling studies that localised HDS in the stroma of chloroplasts of IPAP cells (Oudin *et al.*, 2007b). The same studies performed with G10H showed it was associated with the endoplasmic reticulum but not the plastid (Guirimand *et al.*, 2009). After biosynthesis of 10-hydroxygeraniol, it is not known how many more steps are carried out within IPAP cells, or whetehr this intermediate is transported to the leaf epidermis for conversion to secologanin and, ultimately, MIAs.

Chrysomelid beetles, *Phaedon cochleariae* and *Gastrophysa viridula*, synthesize iridoids such as chrysomelidial which accumulate, seemingly as defensive compounds, within nine pairs of glandular reservoirs on the back of the insects (Burse *et al.*, 2007). The *de novo* biosynthesis of iridoid precursors appears to involve the fat body that expresses the mevalonic acid pathway, G10H and 10-hydroxygeraniol glucosyltransferase (10HGGT) which produce 10-hydroxygeraniol-10-*O*-glucoside. This soluble glucoside may then be transferred from the fat body to the hemolymph of the insect for transport to the glandular reservoir where a β-glucosidase releases 10-hydroxygeraniol for conversion into beetle iridoids by monoterpene cyclase and other enzymes. It would be interesting to test whether a 10HGGT exists within IPAP cells and whether 10-hydroxygeraniol-10-*O*-glucoside is the intermediate being transported to the leaf epidermis. Similar to the insect model, an epidermal β-glucosidase would then be required to release the aglycone for use in subsequent enzyme reactions in the biosynthesis of secologanin. If this were correct, the entire pathway beyond 10-hydroxygeraniol would be restricted to the leaf epidermis.

The biochemical specialisation of IPAP cells raises questions about their possible general involvement in the biosynthesis of iridoids, since they occur widely in plants and have been used as important chemical markers in studies on plant classification, phylogeny and evolution (Sampaio-Santosa and Kaplan, 2001). Snapdragon (*Antirrhinum majus*) contains two major iridoids, antirrhide and antirrhinoside (Guiso and Scarpati, 1969). When photosynthesising leaves were fed with $^{14}CO_2$, the partitioning of ^{14}C between the leaf and petiole was monitored and it was shown that 47% of the phloem mobile ^{14}C-photoassimilate was antirrhinoside while the rest was sucrose (Beninger *et al.*, 2007) This suggests that in snapdragon, and perhaps in other plants, iridoids may partially replace sucrose in the translocation of photoassimilates. In this context iridoids can function as osmoregulators, playing a similar role to sucrose, but their toxicity will have the advantage of deterring herbivores. Do iridoid-translocating plants have IPAP cells? If they do, are the pathways for their biosynthesis present within these cells? The answers to these questions may establish the primary purpose of such biochemical specialisation and help to identify the ancestry of IPAP cells and broaden their biological significance.

10.4.4 Why is MIA Biosynthesis Regulated and Organised Differently in above- and below-Ground Organs in *Catharanthus* Roseus?

While catharanthine appears to be found within all organs of *Catharanthus roseus* plants, tabersonine is metabolised in a different manner in above- and below-ground organs. Within roots tabersonine undergoes two oxidations and one *O*-acylation to form 19-*O*-acetoxyhörhammericine (Morgan and Shanks, 1999; Magnotta *et al.*, 2007), while in above-ground parts it undergoes three oxidations, one *O*-methylation, one *N*-methylation and one *O*-acylation to form vindoline (Figures 10.4 and 10.6) (De Luca and St. Pierre, 2000). The biosynthesis of catharanthine and 19-*O*-acetoxy-hörhammericine appears to take place within the protoderm and cortical cells close to the root apical meristem as determined by *in situ* RNA hybridisation (St-Pierre *et al.*, 1999; Laflamme *et al.*, 2001). While it has yet to be determined, it would be important to ascertain whether the MEP pathway and G10H that supply intermediates for secologanin biosynthesis are expressed within the same cells as the rest of the MIA pathway in *Catharanthus* roots. This would suggest quite different intracellular compartmentation and gene regulation for supplying precursors for the root and the shoot.

Studies using immunocytochemistry and *in situ* RNA hybridisation provided clear evidence that the epidermis of *Catharanthus* leaves, stems and flower buds is responsible for the biosynthesis of catharanthine (unpublished results) and 16-methoxytabersonine (Murata and De Luca, 2005), while the last four steps (an oxidation, one *N*-methylation, an oxidation and one *O*-acylation) of vindoline biosynthesis occurs within mesophyll cells (St-Pierre *et al.*, 1999) (Figure 10.7). Furthermore, these studies confirm biochemical assays that showed that MIA biosynthesis and accumulation occurred specifically in the youngest tissues. *TDC* transcripts and antigens as well as *STR* transcripts were localised to the epidermis of leaves, stems and flower buds. The transcripts for the last two steps in vindoline biosynthesis (*D4H*, *DAT*) were associated with specialised laticifer and idioblast cells in these organs. This result provided evidence that an intermediate such as 16-methoxytabersonine must be exported from the leaf epidermis in order to continue the biosynthesis of vindoline within cells that express the last four steps in the pathway. In addition to MIA biosynthesis, the young leaf epidermis also expresses the pathways for triterpenoids that allow production and secretion of ursolic acid at the leaf surface (Murata *et al.*, 2008) and of flavonoids (Murata *et al.*, 2008; Mahroug *et al.*, 2006), which are transported and accumulate within the vacuole of the epidermal cells (Figure 10.7).

It is also important to note that *Catharanthus* hairy roots express neither *D4H* nor *DAT* transcripts nor related enzyme activities consistent with the accumulation of vindoline in aerial organs (St-Pierre *et al.*, 1999; Magnotta *et al.*, 2007). It has been suggested that, under certain conditions, cell cultures can be stimulated to produce and accumulate vindoline by treatment with UVB light (Ramani and Jayabaskaran, 2008), while another study showed that *Catharanthus* cell cultures do not accumulate vindoline, confirming the widely observed view that the late steps in vindoline biosynthesis are transcriptionally blocked within cell cultures (Vazquez-Flota *et al.*, 2002). Ramani and Jayabaskaran (2008) claimed that UV-treated cell cultures can accumulate up to 0.6 mg of vindoline per gram (dry weight), but this was based on a poorly described HPLC analysis. Evidence based on mass spectrometry and NMR is required to substantiate this claim which is currently not supported by extensive literature on the topic.

10.4.5 How is MIA Biosynthesis Regulated?

MIA biosynthesis can be activated in *Catharanthus roseus* cell cultures by removing auxin from the culture medium, while the addition of methyl-jasmonate has been shown to greatly increase MIA levels (Gantet *et al.*, 1998). Cultures growing in the absence of auxin will accumulate lower levels of MIAs when they are treated with inhibitors of the octadecanoid pathway. Together these results suggest that auxin depletion triggers jasmonate biosynthesis and signal transduction that leads to MIA pathway activation and accumulation (Figure 10.8). In contrast, addition of auxin to cell cultures suppress ajmalicine accumulation (Gantet *et al.*, 1998) but stimulates the accumulation of tabersonine and catharanthine (Rischer *et al.*, 2006). Other studies have shown that a farnesyltransferase (PFT) and a type-I protein geranylgeranyl transferase (PGGT-I) catalyze the resepctive C-14 and C-20 prenylations of unidentified proteins involved in activation of the MIA pathway

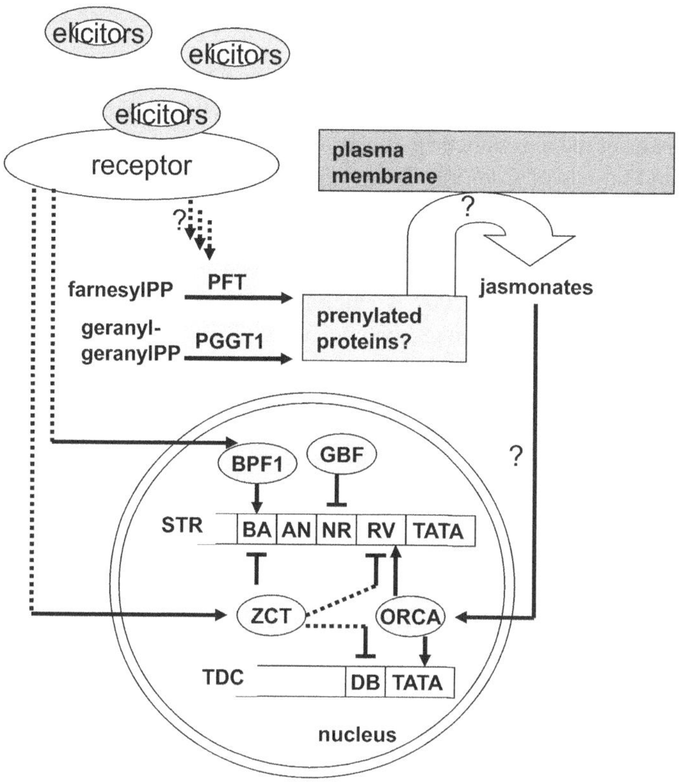

Figure 10.8 *Model of a JA signalling pathway that modulates expression of the STR and TDC components of MIA biosynthesis in* Catharanthus roseus *cell cultures (derived in part from Memelink and Gantet, 2007). The perception by a putative receptor of an elicitor will activate signalling including farnesylation and geranylgeranylation of unknown target proteins. These prenylated proteins stimulate events leading to the production and accumulation of jasmonates. The jasmonates trigger ORCA transcription factors to bind to the STR RV region containing the JERE element and to an unknown TDC element to activate their expression. The additional accumulation of mRNAs encoding ZCT repressor proteins that bind to the DB (TDC), BA (STR) and RV (STR) promoter components suggests that such proteins may play modulating roles that remain to be clearly characterised. The accumulation of CrBPF1 that may bind to the BA region of STR and GBF that binds to a G-Box in the NR region may also have a similar modulating role. Abbreviations: BPF-1,* Catharanthus roseus *box P-binding factor 1 homologue; GBF, G-box-binding factor; STR, strictosidine synthase; ZCT, zinc-finger* Catharanthus *transcription factor; ORCA, octadecanoid-responsive* Catharanthus *AP2-domain protein; TDC, tryptophan decarboxylase; STR, strictosidine synthase*

(Courdavault *et al.*, 2005). Recent studies with inhibitors of protein prenylation were shown to diminish the methyl-jasmonate-triggered expression of both the MEP and secologanin pathways and to suppress MIA accumulation (Courdavault *et al.*, 2009). Jasmonate signal transduction also appears to activate an upstream AT-hook transcription factor that may be part of the upstream components that activate ORCA3 gene expression (Vom Endt *et al.*, 2007).

Apart from the controls imposed by plant growth and development, the biosynthesis of catharanthine, tabersonine and vindoline may involve at least five subcellular compartments, and different parts of the biosynthetic pathway appear to be expressed in different cell types in above-ground tissues of *Catharanthus roseus*. In contrast, the same cell type may be involved in the biosynthesis of catharanthine and tabersonine derivatives in root tip cortical cells. While this intra- and intercellular compartmentation constitutes one level of regulation of MIA biosynthesis, little is known about the involvement of regulatory *cis* and *trans* acting elements (Memelink and Gantet, 2007; Memelink, 2009) (Figure 10.8). In the case of *STR* a jasmonate and elicitor responsive promoter element (JERE) was identified that conferred responsiveness to jasmonate. This conserved regulatory element was used to isolate and identify the *ORCA2* (octadecanoid-responsive *Catharanthus* APETALA-domain protein 2) transcription factor in a yeast one hybrid screen (Menke *et al.*, 1999) (Figure 10.8), while *ORCA3* was isolated and identified by T-DNA activation tagging (van der Fits and Memelink, 2000; van der Fits *et al.*, 2001). *Catharanthus* cell cultures transformed with *ORCA3,* which expressed increased levels of the TDC and STR, accumulated increased levels of tryptamine but not MIAs (van der Fits and Memelink, 2000). Thus, additional factors appear to be required to activate both the secologanin component of the pathway and the downstream sections of the MIA pathway. In this context, a separate jasmonate-independent STR *cis* acting element that lies upstream of the JERE regulatory element was shown to bind a CrBPF1 transcription factor (van der Fits *et al.*, 2000) (Figure 10.8). While CrBPF1 enhances elicitor-mediated *STR1* gene expression, it cannot replace ORCA3 to activate MIA biosynthesis in cell cultures (Menke *et al.*, 1999). The G-Box, a third regulatory element found between the BPF1 and ORCA binding sites in the *STR1* promotor, was used in a yeast one-hybrid screen to isolate CrGBF (G-box binding factor) (Figure 10.8) and CrMYC transcription factors (Ouwerkerk and Memelink 1999). Transient assays suggested that these transcription factors repress *STR1* gene expression, although their exact roles in MIA biosynthesis are not fully characterised. Additional yeast one-hybrid screening, using another elicitor-responsive element from the *TDC* promoter, identified three different Cys2/His2-type zinc finger proteins (ZCT) (Figure 10.8), which appear to repress both the *TDC* and *STR1* promoters (Pauw *et al.*, 2004).

10.4.6 Why has Plant Cell Culture Failed as a Commercial Production System?

The technology to convert intact plants into rapidly growing undifferentiated cells (callus; cell suspension, root or shoot cultures) has sparked intensive efforts to use them for commercial production of useful secondary metabolites. The literature on this subject is well known (Tabata, 2004; Moyano *et al.*, 2005; Zhao and Verpoorte, 2007), but in spite of a few notable successes, clearly identified industrial processes have yet to be implemented. One notable and remarkable success is the use of the alga *Crypthecodinium cohnii* by Martek Biosciences to produce high levels of docosahexanoic acid in bioreactors. This omega-3 fatty acid is incorporated into baby formula and various other products targeting adults. Another fatty acid important to infant health has been produced and purified from the fungus *Mortierella alpine* which is also cultivated in bioreactors. In spite of the fact that this alga and fungus are not cell cultures, these commercial production facilities provide examples of what might be possible within cell culture systems if valuable secondary metabolites could be produced consistently and in high yield.

In the case of *Catharanthus* cell cultures, cell lines that accumulate high levels of certain MIAs, such as serpentine, have successfully been obtained but their stability remains an issue. The major advantage of *Catharanthus* cell culture is its versatility as a source of inducible biochemical pathways for studying MIA chemistry, biochemistry and molecular biology. Several reviews have described genetic engineering experiments that have attempted to improve MIA levels in transformed *Catharanthus* cell and root cultures (Pasquali *et al.*, 2006; Zhao and Verpoorte, 2007; Zárate and Verpoorte, 2007). Early efforts to create TDC and STR overexpressing *Catharanthus* lines were technically successful, but they did not accumulate higher levels of MIAs without supplementation with appropriate precursors (Canel *et al.*, 1998; Whitmer *et al.*, 1998). While the complexity of MIA biosynthesis and the need to coordinate other pathways that supply substrates for these reactions were known, such experiments were useful in testing the feasibility of genetic engineering of cell cultures with these genes. Other experiments involving transformation of plant cell cultures with regulatory genes such as *ORCA3* were not successful in enhancing the biosynthesis and accumulation of MIAs (van der Fits and Memelink, 2000). These experiments indicate that other regulatory genes are required to accommodate expression of this complex pathway. Studies in these areas are continuing, but it seems that transforming cell cultures with individual pathway or regulatory genes will not solve the problems associated with the instability of cell cultures for MIA production and accumulation.

Much has been written about the stability of differentiated hairy root cultures and their transformation with MIA pathway genes for maintaining the production and accumulation of secondary metabolites (Shanks *et al.*, 1998; Pasquali *et al.*, 2006; Zhao and Verpoorte, 2007; Zárate and Verpoorte, 2007). *Catharanthus* hairy root cultures expressing different forms of anthranilate synthase and/or TDC accumulated higher levels of tryptamine rather than alkaloids (Hughes *et al.*, 2004). An attempt to alter the supply of isoprenoid pathway precursors by creating hairy roots expressing 3-hydroxy-3-methylglutaryl-CoA reductase produced some increases in MIA levels, but the results were not predictable (Ayora-Talavera *et al.*, 2002). In another study, RNA-mediated suppression of tryptamine biosynthesis eliminated production of MIAs in hairy root cultures that no longer had TDC activity (Runguphan *et al.*, 2009). TDC-silenced hairy root cultures appeared to accumulate unusually high levels of secologanin (0.15 g per gram dry weight) and this occurred as a result of increased expression of the secologanin pathway. The mechanism that upregulates secologanin biosynthesis and accumulation in these TDC-silenced hairy roots has not been explored, but this was a novel and unexpected finding of this experiment. Several independently transformed hairy root cultures expressing the final DAT-catalyzed step in vindoline biosynthesis have been generated (Magnotta *et al.*, 2007). While some lines accumulated up to four-fold higher levels of hörhammericine compared with controls, it is difficult to explain how this was obtained. Some evidence was provided to show that an interaction occurring between DAT and the root-specific MAT could be responsible for the lack of conversion of hörhammericine to 19-*O*-acetylhörhammericine. Since MIAs do undergo turnover during the growth of hairy root cultures (Morgan and Shanks, 1999), the accumulation of hörhammericine could be occurring if *O*-acetylation was required for turnover of tabersonine, lochnericine and hörhammericine. More studies are required to explain some of these unusual results.

The potential for enhancing the MIA content of *Catharanthus* plants has not been analysed in detail since suitable transformation systems have yet to be developed (Zárate and Verpoorte, 2007). Alternatively, little attention has been paid to the use of traditional breeding to improve the yield of MIAs. Some emphasis should be placed on the use of mutation breeding to develop high-MIA genotypes in *Catharanthus,* since such variation has already been documented in a more random manner (Dutta *et al.*, 2005; Magnotta *et al.*, 2006).

10.5 Metabolic Engineering using Enzymes with Altered Substrate Specificity

The availability of crystal structures for various enzymes involved in MIA biosynthesis, namely STR, SGD, raucaffricine glucosidase, perakine reductase and vinorine synthase (Ma *et al.*, 2005, 2006; Rosenthal *et al.*, 2006; Ruppert *et al.*, 2006; Barleben *et al.*, 2007), have set the stage for rational protein engineering and for altering the substrate specificity of these reactions. In the case of STR, crystal structures were produced with the enzyme complexed to secologanin or to tryptamine. The detailed complexes were used in the rational design of a Val208Ala mutation that accepted a larger range of substrates than tryptamine and secologanin (Loris *et al.*, 2007), and an Asp177Ala mutation that increased the turnover of the secologanin analogues (Chen *et al.*, 2006) compared with wild-type STR (McCoy and O'Connor, 2006). The rational generation of Val214Met and Phe232Leu mutants generated STRs that accepted halogenated tryptamine analogues, which yielded halogenated and variously substituted strictosidine analogues (Bernhardt *et al.*, 2007). STR-Val214Met was able to catalyze the *in vitro* biosynthesis of chloro-, methyl- or bromo-derivatives of strictosidine when supplied with appropriate substrates.

Tryptamine analogues were further investigated in transgenic hairy root cultures expressing STR-Val214Met (Runguphan *et al.*, 2009). Depending upon the tryptamine analogue being supplied, analogues of ajmalicine, tabersonine and catharanthine could be detected. In other studies, RNA-mediated suppression of tryptamine biosynthesis eliminated the production of MIAs in hairy root cultures that no longer had TDC activity (Runguphan *et al.*, 2009). TDC-silenced hairy root cultures fed with fluorotryptamine accumulated the same fluoro-MIAs of unsuppressed hairy root cultures, but these were not contaminated with their natural alkaloid partners. These examples illustrate the potential for transformed cell cultures to produce unnatural alkaloids and secondary metabolites with numerous potential uses.

10.6 Conclusion

Substantial progress has been made in our understanding of MIA biosynthesis within *Catharanthus roseus, Rauvolfia serpentina,* and to a lesser extent in *Camptotheca acuminata* and *Ophiorrhiza pumila.* In spite of this progress, gaps still exist in our knowledge of the pathways involved in the formation of terpenoid precursors as well as the many steps required for the formation of MIAs such as tabersonine, vindoline, catharanthine, camptothecin and ajmaline. Even less is known about the regulation of these pathways in terms of the combination of regulatory genes that are required to stimulate expression of entire pathways in any of the studied species. In the case of *Catharanthus*, limited knowledge of the trafficking of biosynthetic intermediates between IPAP cells, the leaf epidermis and leaf mesophyll/idioblast/laticifer cells remains a major bottleneck in elucidating the mechanisms that control MIA production. While cell and hairy root cultures provide important tools for studying the pathways/regulation of MIA biosynthesis, many of the results obtained need to be verified *in planta* in order to provide meaningful progress towards the goal of enhancing their production levels within the context of field production of medicinal crops. Much more attention needs to be paid to other MIA-producing medicinal plants, with whole plant studies involving appearance of pathways during growth and development, their compartmentation between and within cells, and the development of whole plant transformation systems for probing *in vivo* functions of candidate genes and for enhancing MIA production using genetic engineering.

The focus on regulation of MIA biosynthesis and the transport of intermediates should be enhanced. While a few promising transcription factors have been identified, little is known about how many more transcription and other regulatory factors are, in combination, necessary for activation and maintenance of entire pathways. In the case of *Catharanthus*, there are data that support suggestions about the biochemical intermediates that might be translocated between cells in order to activate the entire pathways for vindoline and catharanthine biosynthesis, but MIA transporters have yet to be identified. The subcellular organisation of MIA biosynthesis has increased in complexity, with at least five compartments being involved, but the reasons for this complexity remain elusive.

Acknowledgements

I thank the Natural Sciences and Engineering Research Council of Canada for a discovery grant that partially funds our research. VDL holds a Tier 1 Canada Research Chair in Plant Biochemistry and Biotechnology.

References

Aerts, R. and De Luca, V. (1992) Phytochrome is involved in the light-activation of vindoline biosynthesis in *Catharanthus*. *Plant Physiol.*, **100**, 1029–1032.

Ayora-Talavera, T., Chappell, J., Lozoya-Gloria, E. and Loyola-Vargas, V.M. (2002) Overexpression in *Catharanthus roseus* hairy roots of a truncated hamster 3-hydroxy-3-methylglutaryl-CoA reductase gene. *App. Biochem. Biotech.*, **97**, 135–145.

Barleben, L., Panjikar, S., Ruppert, M., Koepke, J., Stöckigt J. (2007) Molecular architecture of strictisidine glucosidase: the gateway to the biosynthesis of the monoterpenoid indole alkaloid family. *Plant Cell*, **19**, 2886–2897.

Balsevich, J., DeLuca, V. and Kurz, W.G.W. (1986) Altered alkaloid pattern in dark-grown seedlings of *Catharanthus roseus*. Isolation and characterization of 4-desacetoxyvindoline, a novel indole alkaloid and proposed precursor of vindoline. *Heterocycles*, **24**, 2415–2421.

Beninger, C.W., Cloutier, R.R., Monteiro, M.A. and Grodzinski, B. (2007) The distribution of two major iridoids in different organs of *Antirrhinum majus* L. at selected stages of development. *J. Chem. Ecol.*, **33**, 731–747.

Bernhardt, P., McCoy, E. and O'Connor, S.E. (2007) Rapid identification of enzyme variants for reengineered alkaloid biosynthesis in periwinkle. *Chem. Biol.*, **14**, 888–897.

Burlat, V., Oudin, A., Courtois, M., Rideau, M., St-Pierre, B. (2004) Co-expression of three MEP pathway genes and geraniol 10-hydroxylase in internal phloem parenchyma of *Catharanthus roseus* implicates multicellular translocation of intermediates during the biosynthesis of monoterpene indole alkaloids and isoprenoid-derived primary metabolites. *Plant J.*, **38**, 131–141.

Burse, A., Schmidt, A., Frick, S., Kuhn, J., Gershenzon, J., Boland, W. (2007) Iridoid biosynthesis in *Chrysomelina* larvae: fat body produces early terpenoid precursors. *Insect Biochem. Mol. Biol.*, **37**, 255–265.

Cacacea, S., Schröder, G., Wehingera, E., Strack, D., Schmidt, J., Schröder, J. (2003) A flavonol *O*-methyltransferase from *Catharanthus roseus*. *Phytochemistry*, **62**, 127–137.

Canel, C., Lopes-Cardoso, I., Whitmer, S., van der Fits, L., Pasquali, G., van der Heijden, R., Hoge, J.H.C., Verpoorte, R. (1998) Effects of over-expression of strictosidine synthase and tryptophan decarboxylase on alkaloid production by cell cultures of *Catharanthus roseus*. *Planta*, **205**, 414–419.

Chahed, K., Oudin, A., Guivarc'h, N., Hamdi, S., Chénieux, J.C., Rideau, M., Clastre, M. (2000) 1-deoxy-D-xylulose-5-phosphate synthase from perwinkle: cDNA identification and induced gene expression in terpenoid indole alkaloid-producing cells. *Plant. Physiol. Biochem.*, **38**, 559–566.

Chen, S., Galan, M.C., Coltharp, C. and O'Connor, S.E. (2006) Redesign of a central enzyme in alkaloid biosynthesis. *Chem. Biol.*, **13**, 1137–1141.

Collu, G., Unver, N., Peltenburg-Looman, A.M., van der Heijden, R., Verpoorte, R., Memelink, J. (2001) Geraniol-10-hydroxylase, a cytochrome P450 enzyme involved in terpenoid indole alkaloid biosynthesis. *FEBS Lett.*, **508**, 215–220.

Contin, A., van der Heijden, R., Lefeber, A.W. and Verpoorte, R. (1998) The iridoid glucoside secologanin is derived from the novel triose phosphate/pyruvate pathway in a *Catharanthus roseus* cell culture. *FEBS Lett.*, **434**, 413–416.

Courdavault, V., Thiersault, M., Courtois, M., Gantet, P., Oudin, A., Doireau, P., St-Pierre, B., Giglioli-Guivarc'h, N. (2005) CaaX-prenyltransferases are essential for expression of genes involved in the early stages of monoterpenoid biosynthetic pathway in *Catharanthus roseus* cells. *Plant Mol. Biol.*, **57**, 855–870.

Courdavault, V., Burlat, V., St-Pierre, B. and Giglioli-Guivarc'h, N. (2009) Proteins prenylated by type I protein geranylgeranyltransferase act positively on the jasmonate signalling pathway triggering the biosynthesis of monoterpene indole alkaloids in *Catharanthus roseus*. *Plant Cell Rep.*, **28**, 83–93.

De Luca, V., Balsevich, J., Tyler, R.T., Eilert, U., Panchuk, B., Kurz, W.G.W. (1986) Biosynthesis of indole alkaloids: developmental regulation of the biosynthetic pathway from tabersonine to vindoline in *Catharanthus roseus*. *J. Plant Physiol.*, **125**, 147–156.

De Luca, V. and Cutler, A.J. (1987) Subcellular localization of enzymes involved in indole alkaloid biosynthesis in *Catharanthus roseus*. *Plant Physiol.*, **85**, 1099–1102.

De Luca V., Fernandez J.A., Campbell D. and Kurz W.G.W. (1988) Developmental regulation of enzymes of indole alkaloid biosynthesis in *Catharanthus roseus. Plant Physiol.*, **86**, 447–450.

De Luca, V., Marineau, C. and Brisson, N. (1989) Molecular cloning and analysis of cDNA encoding a plant tryptophan decarboxylase: comparison with animal dopa decarboxylases. *Proc. Natl. Acad. Sci. USA*, **86**, 2582–2586.

De Luca, V. and St-Pierre, B. (2000) The cell and developmental biology of alkaloid biosynthesis. *Trends Plant Sci.*, **5**, 349–364.

Dethier, M. and De Luca, V. (1993) Partial purification of an *N*-methyltransferase involved in vindoline biosynthesis in *Catharanthus roseus. Phytochemistry*, **32**, 673–678.

Dutta, A., Batra, J., Pandey-Rai, S., Singh, D., Kumar, S., Sen, J. (2005) Expression of terpenoid indole alkaloid biosynthetic pathway genes corresponds to accumulation of related alkaloids in *Catharanthus roseus* (L.) G. Don. *Planta*, **220**, 376–383.

El-Sayed, M. and Verpoorte, R. (2007) *Catharanthus* terpenoid indole alkaloids: biosynthesis and regulation. *Phytochem. Rev.*, **6**, 277–305.

Facchini, P.J. and De Luca, V. (2008) Opium poppy and Madagascar periwinkle: model non-model systems to investigate alkaloid biosynthesis in plants. *Plant J.*, **54**, 763–784.

Fernandez, J.A., Owen, T.G., Kurz, W.G.W. and De Luca, V. (1989) Immunological detection and quantitation of tryptophan decarboxylase in developing *Catharanthus roseus* seedlings. Plant Physiol., **91**, 79–84.

Froehlich, J.E., Itoh, A. and Howe, G.A. (2001) Tomato allene oxide synthase and fatty acid hydroperoxide lyase, two cytochrome P450s involved in oxylipin metabolism, are targeted to different membranes of chloroplast envelope. *Plant Physiol.*, **125**, 306–317.

Gantet, P., Imbault, N., Thiersault, M. and Doireau, P. (1998) Necessity of a functional octadecanoic pathway for indole alkaloid synthesis by *Catharanthus roseus* cell suspensions cultured in an auxin-starved medium. *Plant Cell Physiol.*, **39**, 220–225.

Geerlings, A., Ibanez, M.M., Memelink, J., van der Heijden, R., Verpoorte, R. (2000). Molecular cloning and analysis of strictosidine β-D-glucosidase, an enzyme in terpenoid indole alkaloid biosynthesis in *Catharanthus roseus. J. Biol. Chem.*, **275**, 3051–3056.

Goossens, A. and Rischer, H. (2007) Implementation of functional genomics for gene discovery in alkaloid producing plants. *Phytochem. Rev.*, **6**, 35–49.

Guirimand. G., Burlat, V., Oudin, A., Lanoue, A., St-Pierre. B., Courdavault, V. (2009) Optimization of the transient transformation of *Catharanthus roseus* cells by particle bombardment and its application to the subcellular localization of hydroxymethylbutenyl 4-diphosphate synthase and geraniol 10-hydroxylase. *Plant Cell Rep.*, **28**, 1215–1234.

Guiso, M. and Scarpati, M.L. (1969) Iridoids. VI. Isolation of 5-*O*-β-glucosylantirrhinoside from *Antirrhinum majus* or *tortuosum. Gazz. Chim. Ital.*, **99**, 800–806.

Hedhili, S., Courdavault, V., Giglioli-Guivarc'h, N. and Gantet, P. (2007) Regulation of the terpene moiety biosynthesis of *Catharanthus roseus* terpene indole alkaloids *Phytochem. Rev.*, **6**, 341–351.

Helliwell, C.A., Sullivan, J.A., Mould R.M., Gray J.C., Peacock W.J., Dennis E.S. (2001) A plastid envelope location of *Arabidopsis ent*-kaurene oxidase links the plastid and endoplasmic reticulum steps of the gibberellin biosynthesis pathway. *Plant J.*, **28**, 201–208.

Hisiger, S. and Jolicoeur, M. (2007) Analysis of *Catharanthus roseus* alkaloids by HPLC. *Phytochem. Rev.*, **6**, 207–234.

Hsiang, Y.H. Hertzberg, R., Hecht, S. and Liu, L.F. (1985) Camptothecin induces protein-linked DNA breaks via mammalian DNA topoisomerase I. *J. Biol. Chem.*, **260**, 14873–14878.

Hughes, E.H., Hong, S.B., Gibson, S.I., Shanks, J.V., San, K.Y. (2004) Metabolic engineering of the indole pathway in *Catharanthus roseus* hairy roots and increased accumulation of tryptamine and serpentine. *Metab. Eng.*, **6**, 268–276.

Ikeda, H., Esaki, N., Nakai, S., Hashimoto, K., Uesato, S., Soda, K., Fujita, T. (1991) Acyclic monoterpene primary alcohol:NADP$^+$ oxidoreductase of *Rauwolfia serpentina* cells: the key enzyme in biosynthesis of monoterpene alcohols. *J. Biochem.*, **109**, 341–347.

Irmler, S., Schröder, G. St-Pierre, B., Crouch, N.P., Hotze, M., Schmidt, J., Strack, D., Mattern, U. Schröder, J. (2000) Indole alkaloid biosynthesis in *Catharanthus roseus*: new enzyme activities and identification of cytochrome P450 CYP72A2 as secologanin synthase. *Plant J.*, **24**, 797–804.

Katano, N., Yamamoto, H., Iio, R. and Inoue, K. (2001) 7-deoxyloganin 7-hydroxylase in Lonicera japonica cell cultures. *Phytochemistry*, **58**, 53–58.

Kutchan, T.M. (1989) Expression of enzymically active cloned strictosidine synthase from the higher plant *Rauvolfia serpentina* in *Escherichia coli*. *FEBS Lett.*, **257**, 127–130.

Laflamme, P., St-Pierre, B. and De Luca, V. (2001) Molecular and biochemical analysis of a Madagascar periwinkle root-specific minovincinine-19-hydroxy-*O*-acetyltransferase. *Plant Physiol.*, **125**, 189–198.

Lange, B.M., Wildung, M.R., Stauber, E.J., Sanchez, C., Pouchnik, D., and Croteau, R. (2000) Probing essential oil biosynthesis and secretion by functional evaluation of expressed sequence tags from mint glandular trichomes. *Proc. Natl. Acad. Sci. USA*, **97**, 2934–2939.

Levac, D., Murata, J., Kim, W.S. and De Luca, V. (2008) Application of carborundum abrasion for investigating leaf epidermis: molecular cloning of *Catharanthus roseus* 16-hydroxytabersonine-16-*O*-methyltransferase. *Plant J.*, **53**, 225–236.

Lopez-Meyer, M. and Nessler, C.L. (1997) Tryptophan decarboxylase is encoded by two autonomously regulated genes in *Camptotheca acuminata* which are differentially expressed during development and stress. *Plant J.*, **11**, 1167–1175.

Lorence, A. and Nessler, C.L. (2004) Camptothecin: over four decades of surprising findings. *Phytochemistry*, **65**, 2735–2749.

Loris, E.A., Panjikar, S., Ruppert, M., Barleben, L., Unger, M., Schuebel, H., Stöckigt, J. (2007) Structure-based engineering of strictosidine synthase: auxiliary for alkaloid libraries. *Chem. Biol.*, **14**, 979–985.

Loyola-Vargas, V.M., Galaz-Ávalos, R.M. and Kú-Cauich, R. (2007) *Catharanthus* biosynthetic enzymes: the road ahead. *Phytochem. Rev.*, **6**, 307–339.

Ma, X., Koepke, J., Bayer, A., Fritzsch, G., Michel, H., Stöckigt, J. (2005) Crystallization and preliminary X-ray analysis of native and selenomethionyl vinorine synthase from *Rauvolfia serpentina*. *Acta Crystallogr. D Biol. Crystallogr.*, **61**, 694–696.

Ma, X., Panjikar, S., Koepke, J., Loris, E., Stöckigt, J. (2006) The structure of *Rauvolfia serpentina* strictosidine synthase is a novel six-bladed beta-propeller fold in plant proteins. *Plant Cell*, **18**, 907–920.

Madyastha, K.M., Guarnaccia, R., Baxter, C. and Coscia, C.J. (1973) *S*-adenosyl-L-methionine: loganic acid methyltransferase. *J. Biol. Chem.*, **248**, 2497–2501.

Magnotta, M., Murata, J., Chen, J. and De Luca, V. (2006) Identification of a low vindoline accumulating cultivar of *Catharanthus roseus* (L.) G. Don by alkaloid and enzymatic profiling. *Phytochemistry*, **67**, 1758–1764.

Magnotta, M., Murata, J., Chen, J. and De Luca V. (2007) Expression of deacetylvindoline-4-*O*-acetyltransferase in *Catharanthus roseus* hairy roots. *Phytochemistry*, **68**, 1922–1931.

Mahroug, S., Courdavault, V., Thiersault, M., St-Pierre, B., Burlat, V. (2006) Epidermis is a pivotal site of at least four secondary metabolic pathways in *Catharanthus roseus* aerial organs. *Planta*, **223**, 1191–1200.

Mahroug, S., Burlat, V. and St-Pierre, B. (2007) Cellular and sub-cellular organisation of the monoterpenoid indole alkaloid pathway in *Catharanthus roseus*. *Phytochem. Rev.*, **6**, 363–381.

McCoy, E. and O'Connor, S.E. (2006) Directed biosynthesis of alkaloid analogs in the medicinal plant *Catharanthus roseus*. *J. Am. Chem. Soc.*, **128**, 14276–14277.

McKnight, T.D., Roessner, C.A., Devagupta, R., Scott, A.I., Nessler, C.L. (1990) Nucleotide sequence of a cDNA encoding the vacuolar protein strictosidine synthase from *Catharanthus roseus*. *Nucleic Acids Res.*, **18**, 4939.

McKnight, T.D., Bergey, D.R., Burnett, R.J. and Nessler, C.L. (1991) Expression of enzymatically active and correctly targeted strictosidine synthase in transgenic tobacco plants. *Planta*, **185**, 148–152.

Memelink, J. (2009) Regulation of gene expression by jasmonate hormones. *Phytochemistry*, **70**, 1560–1570.

Memelink, J. and Gantet, P. (2007) Transcription factors involved in terpenoid indole alkaloid biosynthesis in *Catharanthus roseus*. *Phytochem. Rev.*, **6**, 353–362.

Menke, F.L.H., Champion, A., Kijne, J.W. and Memelink, J. (1999) A novel jasmonate- and elicitor-responsive element in the periwinkle secondary metabolite biosynthetic gene Str interacts with a jasmonate- and elicitor inducible AP2-domain transcription factor, ORCA2. *EMBO J.*, **18**, 4455–4463.

Morgan, J. and Shanks, J. (1999) Inhibitor studies of tabersonine metabolism in *C. roseus* hairy roots. *Phytochemistry*, **51**, 61–68.

Moyano, E., Osuna, L., Bonfill, M., Cusido, R.M., Palazon, J., Tortoriello, J., Pinol, M.T. (2005) Bioproduction of triterpenes on plant cell cultures of *Panax ginseng* and *Galphimia glauca*. *Rec. Res. Devel. Plant Sci.*, **3**, 195–213.

Murata, J. and De Luca, V. (2005) Localization of tabersonine 16-hydroxylase and 16-OH tabersonine-16-*O*-methyltransferase to leaf epidermal cells defines them as a major site of precursor biosynthesis in the vindoline pathway in *Catharanthus roseus*. *Plant J.*, **44**, 581–594.

Murata, J., Bienzle, D., Brandle, J.E., Sensen, C.W., De Luca, V. (2006) Expressed sequence tags from Madagascar periwinkle (*Catharanthus roseus*). *FEBS Lett.*, **580**, 4501–4507.

Murata, J., Roepke J., Gordon, H. and De Luca V. (2008) 'Surfaceome' analysis of leaf epidermal cells in *Catharanthus roseus* reveals its biochemical specialization. *Plant Cell*, **38**, 131–141.

O'Connor, S.E. and Maresh, J.M. (2006) Chemistry and biology of terpene indole alkaloid biosynthesis. *Nat. Prod. Rep*,. **23**, 532–547.

Oudin, A., Courtois, M., Rideau, M. and Clastre, M. (2007a) The iridoid pathway in *Catharanthus roseus* alkaloid biosynthesis. *Phytochem. Rev.*, **6**, 259–276.

Oudin, A., Mahroug, S., Courdavault, V., Hervouet, N., Zelwer, C., Rodriguez-Concepcion, M., St-Pierre, B., Burlat, V. (2007b) Spatial distribution and hormonal regulation of gene products from methyl erythritol phosphate and monoterpene-secoiridoid pathways in *Catharanthus roseus*. *Plant Mol. Biol.*, **65**, 13–30.

Ouwerkerk, P.B.F. and Memelink, J. (1999) A G-box element from the *Catharanthus roseus* strictosidine synthase (*Str*) gene promoter confers seed-specific expression in transgenic tobacco plants. *Mol. Gen. Genet.*, **261**, 635–643.

Pasquali, G., Porto, D.D. and Fett-Neto, A.G. (2006) Metabolic engineering of cell cultures versus whole plant complexity in production of bioactive monoterpene indole alkaloids: recent progress related to an old dilemma. *J. Biosci. Bioeng.*, **101**, 287–296.

Pauw, B., Hilliou, F.A.O., Sandonis, M.V., Chatel, G., de Wolf, C.J.F., Champion, A., Pré, M., van Duijn, B., Kijne, J.W., Van Der Fits, L., Memelink, J. (2004). Zinc finger proteins act as transcriptional repressors of alkaloid biosynthesis genes in *Catharanthus roseus*. *J. Biol. Chem.*, **279**, 52940–52948.

Pereira, D.M., Ferreres, F., Oliveira, J.M., Gaspar L, Faria J, Valentão P, Sottomayor M, Andrade PB. (2010) Pharmacological effects of *Catharanthus roseus* root alkaloids in acetylcholinesterase inhibition and cholinergic neurotransmission. *Phytomedicine*, **153**, 980–987.

Ramani, S. and Jayabaskaran, C. (2008) Enhanced catharanthine and vindoline production in suspension cultures of *Catharanthus roseus* by ultraviolet-B light. *J. Mol. Signal.*, **3**, 1–9.

Rischer, H., Oreši, M., Seppänen-Laakso, T., Katajamaa, M., Lammertyn, F., Ardiles-Diaz, W., Van Montagu, M.C.E., Inzé, D., Oksman-Caldentey, K.M., Goossens, A. (2006) Gene-to-metabolite networks for terpenoid indole alkaloid biosynthesis in *Catharanthus roseus* cells. *Proc. Natl. Acad. Sci. USA*, **103**, 5614–5619.

Rodriguez, S., Compagnon, V., Crouch, N.P., St. Pierre, B., De Luca, V. (2003) Jasmonate-induced epoxidation of tabersonine by cytochrome P-450 in hairy root cultures of *Catharanthus roseus*. *Phytochemistry*, **64**, 401–409.

Rosenthal, C., Mueller, U., Panjikar, S., Sun, L., Ruppert, M., Zhao, Y., Stöckigt, J. (2006) Expression, purification, crystallization and preliminary X-ray analysis of perakine reductase, a new member of the aldo-keto reductase enzyme superfamily from higher plants. *Acta Crystallogr.*, **F62**, 1286–1289.

Roytrakul, S. and Verpoorte, R. (2007) Role of vacuolar transporter proteins in plant secondary metabolism: *Catharanthus roseus* cell culture. *Phytochem. Rev.*, **6**, 383–396.

Runguphan, W., Maresh, J.J. and O'Connor, S.E. (2009) Silencing of tryptamine biosynthesis for production of nonnatural alkaloids in plant culture. *Proc. Nat. Acad. Sci. USA*, **106**, 13673–13678.

Ruppert, M., Ma, X. and Stöckigt, J. (2005) Alkaloid biosynthesis in *Rauvolfia* – cDNA cloning of major enzymes of the ajmaline pathway. *Curr. Org. Chem.*, **9**, 1431–1444.

Ruppert, M., Panjikar, S., Barleben, L. and Stöckigt, J. (2006) Heterologous expression, purification, crystallization and preliminary X-ray analysis of raucaffricine glucosidase, a plant enzyme specifically involved in *Rauvolfia* alkaloid biosynthesis. *Acta Crystallogr. Sect. F Struct. Biol. Cryst. Commun.*, **62**, 257–260.

Sampaio-Santosa, I. and Kaplan, A.C. (2001) Biosynthetic significance of iridoids in chemosystematics. *J. Bras. Chem. Soc.*, **12**, 144–153.

Sanchez Iturbe, P., Galaz Avalos, R.M. and Loyola Vargas, V.M. (2005) Determination and partial purification of a monoterpene cyclase from *Catharanthus roseus* hairy roots. *Phyton*, 55–69.

Schilmiller, A.L., Last, R.L. and Pichersky, E. (2008) Harnessing plant trichome biochemistry for the production of useful compounds. *Plant J.*, **54**, 702–711.

Schröder, G., Unterbusch, E., Kaltenbach, M., Schmidt, J., Strack, D., De Luca, V., Schröder, J. (1999) Light-induced cytochrome P450-dependent enzyme in indole alkaloid biosynthesis: tabersonine-16-hydroxylase. *FEBS Lett.*, **458**, 97–102.

Schröder, G., Wehinger, E., Lukačin, R., Wellmann, F., Seefelder, W., Schwab, W. and Schröder, J. (2004) Flavonoid methylation: a novel 4′-*O*-methyltransferase from *Catharanthus roseus*, and evidence that partially methylated flavanones are substrates of four different flavonoid dioxygenases. Phytochem*istry*, **65**, 1085–1094.

Shanks, J.V., Bhadra, R., Morgan, J., Rijhwani, S., Vani, S. (1998) Quantification of metabolites in the indole alkaloid pathways of *Catharanthus roseus*: implications for metabolic engineering. *Biotechnol. Bioeng.*, **58**, 333–338.

Shukla, A.K., Shasany, A.K., Gupta, M.M. and Khanuja, S.P.S. (2006) Transcriptome analysis in *Catharanthus roseus* leaves and roots for comparative terpenoid indole alkaloid profiles. *J. Exp. Bot.*, **57**, 3921–3932.

Sirikantaramas, S., Asano, T., Sudo, H., Yamazaki, M., Saito, K. (2007) Camptothecin: therapeutic potential and biotechnology. *Curr. Pharm. Biotech.*, **8**, 196–202.

Sirikantaramas, S., Yamazaki, M. and Saito, K. (2008) Mutations in topoisomerase I as a self-resistance mechanism coevolved with the production of the anticancer alkaloid camptothecin in plants. *Proc. Nat Acad. Sci. USA*, **105**, 6782–6786.

Sottomayor, M., Lopez-Serrano, M., DiCosmo, F. and Ros Barcelo, A. (1998) Purification and characterization of alpha-3′,4′-anhydrovinblastine synthase (peroxidase-like) from *Catharanthus roseus* (L.) G. Don. *FEBS Lett.*, **428**, 299–303.

Stevens, L.H., Blom, T.J.M. and Verpoorte, R. (1993) Subcellular localization of tryptophan decarboxylase, strictosidine synthase and strictosidine glucosidase in suspension-cultured cells of *Catharanthus roseus* and *Tabernaemontana divaricata*. *Plant Cell Rep.*, **12**, 573–576.

Stöckigt, J., Panjikar, S., Ruppert, M., Barleben, L., Ma, X., Loris, E., Hill, M. (2007) The molecular architecture of major enzymes from ajmaline biosynthetic pathway. *Phytochem. Rev.*, **6**, 15–34.

St-Pierre, B. and De Luca, V. (1995) A cytochrome P-450 monooxygenase catalyzes the first step in the conversion of tabersonine to vindoline in *Catharanthus roseus*. *Plant Physiol.*, **109**, 131–139.

St-Pierre, B., Laflamme, P., Alarco, A.M. and De Luca, V. (1998) The terminal *O*-acetyltransferase involved in vindoline biosynthesis defines a new class of proteins responsible for coenzyme A-dependent acyl transfer. *Plant J.*, **14**, 703–713.

St-Pierre, B., Vazquez-Flota, F.A. and De Luca, V. (1999) Multicellular compartmentation of *Catharanthus roseus* alkaloid biosynthesis predicts intercellular translocation of a pathway intermediate. *Plant Cell*, **11**, 887–900.

Tabata, H. (2004) Paclitaxel production by plant-cell-culture technology. *Adv. Biochem. Engineer. Biotechnol.*, **7**, 1–23.

Tian, L., Musetti, V., Kim, J., Magallanes-Lundback, M., DellaPenna, D. (2003) The Arabidopsis *LUT1* locus encodes a member of the cytochrome P450 family that is required for carotenoid ε-ring hydroxylation activity. *Proc. Natl. Acad. Sci. USA*, **101**, 402–407.

Uesato, S., Ogawa, Y., Inouye, H., Saiki, K., Zenk, M.H. (1986) Synthesis of iridodial by cell-free extracts from *Rauwolfia serpentina* cell suspension cultures. *Tetrahedron Lett.*, **27**, 2893–2896.

Uesato, S., Ikeda, H., Fujita, T., Inouye, H., Zenk, M.H. (1987) Elucidation of iridodial formation mechanism. Partial purification and characterization of the novel monoterpene cyclase from *Rauwolfia serpentina* cell suspension cultures. *Tetrahedron Lett.*, **28**, 4431–4434.

Usia, T., Watabe, T., Kadota, S. and Tezuka, Y. (2005) Cytochrome P450 2D6 (CYP2D6) inhibitory constituents of *Catharanthus roseus*. *Biol. Pharm. Bull.*, **28**, 1021–1024.

van der Fits, L. and Memelink, J. (2000) ORCA3, a jasmonate-responsive transcriptional regulator of plant primary and secondary metabolism. *Science*, **289**, 295–297.

van der Fits, L., Zhang, H., Menke, F.L.H., Deneka, M., Memelink, J. (2000) A *Catharanthus roseus* BPF-1 homologue interacts with an elicitor-responsive region of the secondary metabolite biosynthetic gene *Str* and is induced by elicitor via a jasmonate-independent signal transduction pathway. *Plant Mol. Biol.*, **44**, 675–685.

van der Fits, L., Hilliou, F. and Memelink, J. (2001) T-DNA activation tagging as a tool to isolate regulators of a metabolic pathway from a genetically non-tractable plant species. *Transgenic Res.*, **10**, 513–521.

van der Heijden, R., Jacobs, D.I., Snoeijer, W., Hallard, D., Verpoorte R. (2004) The *Catharanthus* alkaloids: pharmacognosy and biotechnology. *Curr. Med. Chem.*, **11**, 607–628.

Vazquez-Flota, F., De Carolis, E., Alarco, A.M. and De Luca, V. (1997) Molecular cloning and characterization of desacetoxyvindoline-4-hydroxylase, a 2-oxoglutarate dependent-dioxygenase involved in the biosynthesis of vindoline in *Catharanthus roseus* (L.) G. Don. *Plant Mol. Biol.*, **34**, 935–948.

Vazquez-Flota, F. and De Luca, V. (1998) Developmental and light regulation of desacetoxyvindoline-4-hydroxylase in *Catharanthus roseus* (L.) G. Don. *Plant Physiol.*, **117**, 1351–1362.

Vazquez-Flota, F., De Luca, V., Carrillo-Pech, M., Canto-Flick, A., De Lourdes Miranda-Ham, M. (2002) Vindoline biosynthesis is transcriptionally blocked in *Catharanthus roseus* cell suspension cultures. Mol. Biotech., **22**, 1–8.

Veau, B., Courtois, M., Oudin, A., Chénieux, J.C., Rideau, M., Clastre, M. (2000) Cloning and expression of cDNAs encoding two enzymes of the MEP pathway in *Catharanthus roseus*. *Biochim. Biophys. Acta*, **1517**, 159–163.

Vom Endt, D., Soares e Silva, M., Kijne, J.W., Pasquali, G., Memelink, J. (2007) Identification of a bipartite jasmonate-responsive promoter element in the *Catharanthus roseus* ORCA3 transcription factor gene that interacts specifically with AT-Hook DNA-binding proteins. *Plant Physiol,*. **144**, 1680–1689.

Wang, W., Wang, Y., Zhang, Q., Qi, Y, Guo, D. (2009) Global characterization of Artemisia annua glandular trichome transcriptome using 454 pyrosequencing. *BMC Genomics*, **10**, 465.

Whitmer, S., Canel, C., Hallard, D., Goncalves, C., Verpoorte, R. (1998) Influence of precursor availability on alkaloid accumulation by a transgenic cell line of *Catharanthus roseus*. *Plant Physiol.*, **116**, 853–857.

Yamamoto, H., Katano, N., Ooi, A. and Inoue, K. (1998) Transformation of loganin and 7-deoxyloganin into secologanin by *Lonicera japonica* cell suspension cultures. *Phytochemistry*, **50**, 417–422.

Yamamoto, H., Katano, N., Ooi, A. and Inoue, K. (2000) Secologanin synthase, which catalyzes the oxidative cleavage of loganin into secologanin, is a cytochrome P450. *Phytochemistry*, **53**, 7–12.

Yamazaki, Y., Sudo, H., Yamazaki, M., Aimi, N., Saito, K. (2003) Camptothecin biosynthetic genes in hairy roots of *Ophiorrhiza pumila*: cloning, characterization and differential expression in tissues and by stress compounds. *Plant Cell Physiol.*, **44**, 395–403.

Yamazaki, Y., Kitajima, M., Arita, M., Takayama, H., Sudo, H., Yamazaki, M., Aimi, N., Saito, K. (2004) Biosynthesis of camptothecin. *In silico* and *in vivo* tracer study from [1-^{13}C] glucose. *Plant Physiol.*, **134**, 161–170.

Zagrobelny, M., Scheibye-Alsing, K., Bjerg Jensen, N., Lindberg Moller, B, Gorodkin, J, Bak, S. (2009) 454 pyrosequencing based transcriptome analysis of Zygaena filipendulae with focus on genes involved in biosynthesis of cyanogenic glucosides. *BMC Genomics*, **10**, 574.

Zárate, R. and Verpoorte, R. (2007) Strategies for the genetic modification of the medicinal plant *Catharanthus roseus* (L.) G. Don. *Phytochem. Rev.*, **6**, 475–491.

Zhao, J. and Verpoorte, R. (2007) Manipulating indole alkaloid production by *Catharanthus roseus* cell cultures in bioreactors: from biochemical processing to metabolic engineering. *Phytochem. Rev.*, **6**, 435–457.

Zhou, M.L., Shao, J.R. and Tang, Y.X. (2009) Production and metabolic engineering of terpenoid indole alkaloids in cell cultures of the medicinal plant *Catharanthus roseus* (L.) G. Don (Madagascar periwinkle). *Biotechnol. Appl. Biochem.*, **52**, 313–323.

11

Flavonoid Biosynthesis

Indu B. Jaganath and Alan Crozier

11.1 Introduction

Flavonoids were first discovered as components of plant pigments by Robert Boyle in 1664. In more recent times, much information has been generated on this important group of secondary metabolites in terms of their structural characterisation, chemical activities and biosynthesis, as well as on their role as dietary components with beneficial effects on health (Winkel-Shirley, 2001; Anderson and Markham, 2006).

Flavonoids constitute a relatively diverse family of aromatic molecules that are derived from the stepwise condensation of *p*-coumaroyl-coenzyme A, a product of the phenylpropanoid pathway, with three malonyl-coenzyme A residues originating from the malonic acid pathway (see Crozier *et al.*, 2006). To date, more than 7000 different flavonoid compounds have been described (Andersen and Markham, 2006). They are low molecular weight polyphenolic compounds that are widespread throughout the plant kingdom (Harborne, 1993). They are present in high concentrations in the epidermis of leaves, flowers and the skin of fruits. *In planta*, flavonoids are involved in unique survival and adaptive strategies. These include providing pigmentation for UV protection, fertility and germination of pollen, defence against pathogenic microorganisms, and acting as signal molecules in plant–microbe interactions (Dooner *et al.*, 1991; Koes *et al.*, 1994; Dixon and Paiva, 1995; Pierpoint, 2000).

The basic flavonoid skeleton comprises 15 carbons, with two aromatic rings (A and B) connected by a three-carbon bridge (C). Different structural classes of C_6-C_3-C_6 flavonoids are formed due to different degrees of oxidation of the A-ring and modifications of the central C-ring. The main classes of flavonoids are flavonols, flavones, flavan-3-ols, flavanones,

Plant Metabolism and Biotechnology, First Edition. Edited by Hiroshi Ashihara, Alan Crozier, and Atsushi Komamine.

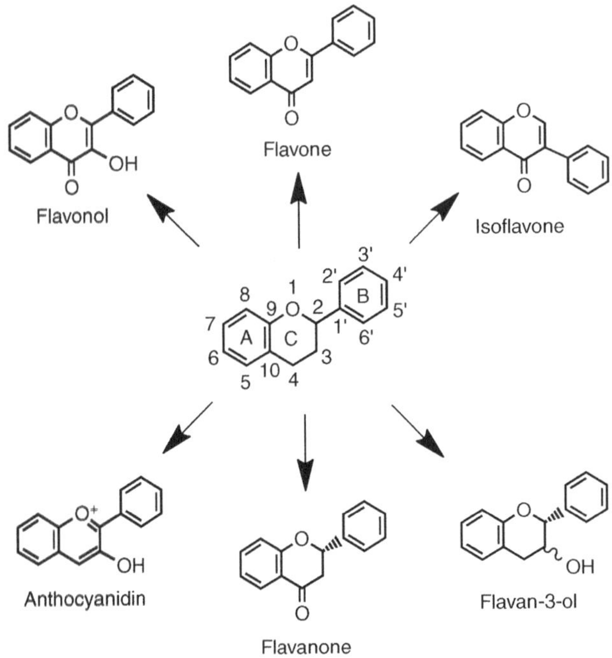

Figure 11.1 Generic structures of the major flavonoids

isoflavones and anthocyanidins (Figure 11.1). The heterocyclic ring C can occur in an isomeric open form which results in the formation of chalcones. Flavonoids normally occur as glycosides and methylated derivatives in plants. In addition, hydroxyl groups are also almost invariably present, typically at the 3, 5, 7 and/or 4′ positions. Sugars and hydroxyl groups tend to increase the water solubility of flavonoids, whereas substituents, such as methyl groups and isopentyl units, make flavonoids more lipophilic.

From the 1970s to 1990s, there was rapid and substantial progress in flavonoid research, focusing on a broad understanding of the biosynthetic pathway (Hahlbrock and Grisebach, 1975; Ebel and Hahlbrock, 1982; Heller and Forkmann, 1988). Subsequently, advances were focused on specific metabolic steps mainly to fill the remaining gaps in the pathway and characterise the enzymes involved (Heller and Forkmann, 1994). Lately, much effort has been directed at elucidating the biosynthetic pathway from a biochemical and a molecular perspective by using approaches such as transposon tagging, positional cloning, co-immunoprecipitation, affinity chromatography, and two-hybrid experiments (Winkel-Shirley, 2001). As a result of using these new methodologies, a large number of the enzymes involved in the flavonoid pathway have been identified and characterised.

11.2 Advances in Molecular Approaches for Flavonoid Biosynthetic Pathway Elucidation

Biochemistry, for most of the twentieth century, was tackled in a reductionist manner whereby the cell and its components were studied in isolation. Although a large amount of information was generated through this approach, it is nevertheless becoming increasingly evident that this is a simplistic approach compared with what actually occurs *in planta*. Biological processes are rarely controlled by a single molecular unit, but instead are regulated by processes shared across different molecular entities within cells (Carrari *et al.*, 2006). It was only through the emergence of high-throughput molecular tools in the 2000s that it became possible to achieve a more holistic approach of studying and cataloguing, in parallel, the changes in metabolites, proteins and transcripts (Mack, 2004; Fernie *et al.*, 2005; Ratcliffe and Shacher-Hill, 2005).

11.2.1 Genetic and Transgenic Approaches

In recent years, genome projects have yielded large amounts of sequence data and there is a growing need to dissect the functions of these genes. One approach to achieving this is through the application of gene-knockout technology where genes can be turned on and off, and by doing so their role and how they impact downstream networks are unravelled. The application of this technology in *Arabidopsis* has facilitated and dramatically advanced the understanding of fundamental aspects of flavonoid metabolism (Thorneycroft *et al.*, 2001). For example, Xie *et al.* (2003) revealed that the enzyme leucoanthocyanidin reductase is responsible for the brown colour of the seed coat of *Arabidopsis* and is located one step downstream from the leucoanthocyanidins. The accumulation of proanthocyanidins in the *Arabidopsis* seed coat is also dependent on the MYB transcription factor TT2, the basic-helix–loop–helix transcription factor TT8, and the WD-repeat protein TRANSPARENT TESTA GLABRA1 (TTG1). In addition, analysis of the new *tt19* mutant allowed the isolation of a glutathione-*S*-transferase-encoding gene, which is involved in the accumulation of anthocyanins and proanthocyanidins (Memelink, 2005).

To further understand and dissect the metabolic pathway for the biosynthesis of flavonoids, transgenic approaches have also been frequently used. When the phytoalexin resveratrol was cloned into alfalfa by constitutive expression of a grapevine *stilbene synthase* gene, there was a reduction in symptoms following infection by the leaf spot pathogen *Phoma medicaginis* (Hipskind and Paiva, 2000). In another study, the constitutive overexpression of *isoflavone-O-methyltransferase* in transgenic alfalfa resulted in more rapid and increased production of the pterocarpan phytoalexin medicarpin after infection by *P. medicaginis*, resulting in amelioration of symptoms (He and Dixon, 2000).

In addition to elucidating the flavonoid pathway, there have been numerous attempts to modify flavonoid biosynthesis for a variety of reasons. The flavonoid contents and compositions of plants vary according to species. For example, isoflavonoids, which are known for their antimicrobial phytoalexins properties, are found mainly in legumes, and attempts have been made to introduce this branch pathway into non-leguminous species (Hain *et al.*, 1993). Research has also focused on the production of genetically-modified ornamentals as a means of developing new flower colours (Davies *et al.*, 1998; Mol *et al.*, 1998; Aida *et al.*, 2000; Suzuki *et al.*, 2002; Fukui *et al.*, 2003). Flavonoid-3′,5′-hydroxylase (F3′5′H), one of the key enzymes necessary for the expression of blue or purple flower colour, was investigated by Shimada *et al.* (2001). Upon introducing F3′5′H under the control of the CaMV 35S promoter into pink-flowered varieties of petunia that are deficient in the enzyme, the flower colour of the transgenic plants changes from pink to magenta. On the other hand, when the same enzyme was introduced into blue petunia varieties, the flowers of the transgenic plants changed from deep blue to a pale blue or pink colour, indicating the functionality of this transgene (Shimada *et al.*, 2001).

Due to their potential health benefits, there is increasing interest in the development of agronomically important food crops with optimised levels and composition of flavonoids. Attempts have been made with plants such as tomato (*Lycospersicon esculentum*) where a *Petunia hybrida CHI* gene was transformed and this resulted in increased levels of quercetin glycosides and smaller, but still substantial, increases in kaempferol glycosides in the skin of the fruit. Overall this resulted up to a 78-fold elevation in flavonol concentration (Muir *et al.*, 2001). In another study, Liu *et al.* (2007) introduced the *isoflavone synthase* gene into the non-legume lettuce (*Latuca sativa*) in order to convert naringenin, which is ubiquitous in higher plants, to the isoflavone genistein. The overexpression of regulatory genes or transcriptional factors such as the LC and C1 brought about a 20-fold increase in flavonols in the flesh of tomato fruit (Bovy *et al.*, 2002; Le Gall *et al.*, 2003). Expression LC and C1 in potatoes (*Solanum tuberosum*) resulted in an enhanced accumulation of kaempferol and anthocyanins in the tubers (de Vos *et al.*, 2000).

Besides transcription factors which increase the activity of the flavonoid pathway, negative regulators of flavonoid biosynthesis have also been used to modify the pathway to attain desired results. Mutation of the DE-ETIOLATED1 gene (DET1) produced highly pigmented (hp-2) tomatoes where the darker colour of the fruit was a consequence of elevated levels of both flavonoids and carotenoids (Bino *et al.*, 2005). Flavonoid levels were found to increase up to 3.5-fold, whereas lycopene content was doubled and β-carotene levels increased ten-fold compared with wild-type fruits (Davuluri *et al.*, 2005).

Although there has been some success in engineering enhanced levels of certain flavonoids in fruits and vegetables, because of the formation of complex networks, feedback loops for enzymatic activity and modulation of protein complexes for metabolic channelling, it nonetheless remains a challenging task not only to generate the desired phytochemical profile but to produce it consistently in subsequent generations. In several cases, overexpression of introduced genes has resulted in the production of unexpected products, demonstrating the complexity of the metabolic networks and our lack of knowledge of these networks and how they are regulated. This was the case in a study by Bovy *et al.* (2002) in which the C1 and R transcriptional factors were introduced into tomato. This led to the induction of several flavonoid genes, but was not sufficient to induce F3′5′H activity, which appeared to be essential for the production of anthocyanins in tomato fruit. Also, the host plant or tissue may be 'incapable' of producing certain compounds due to the substrate specificity of endogenous enzymes, as was reported for the tomato dihydroflavonol-4-reductase that was restricted in its substrate specificity to dihydromyricetin and, thus, can only give rise to the production of delphinidin-type anthocyanins (Bovy *et al.*, 2002). These results indicate that a full understanding and complete dissection of the complex network of flavonoid metabolic pathways is far from complete. For example, recent reports underline that important questions still remain to be answered in the field of proanthocyanidins (Xie and Dixon, 2005). In addition, pathways for sequential modification, such as glycosylation, acylation and methylation, which are essential for the stable accumulation of flavonoids *in planta*, remain relatively unexplored.

However, with the recent advances in molecular technology, aided by the enormous power of large-scale genomics, metabolomics and proteomics, voluminous amounts of data have been generated. The information has been analysed in detail using bioinformatics tools which have revolutionised biological investigations, particularly in mapping out a more compete picture of metabolic pathways (Fukushima *et al.*, 2009).

11.2.2 Metabolomics

Recent advances in mass spectrometry (MS) and nuclear magnetic resonance (NMR) have permitted the development and establishment of methods that offer both high accuracy and sensitivity for the non-targeted measurement and identification of metabolites in complex mixtures (Fernie *et al.*, 2004). This has given rise to a new area of analytical science referred to as metabolomics. Metabolomics can be defined as a non-targeted and holistic approach with the ultimate aim of an unbiased and comprehensive monitoring of all metabolites in an organism (Dunn and Ellis, 2005; Morgenthal *et al.*, 2006). Since this approach deals with all cellular metabolites, it has been recently recognised as an important sector of post-genomic science and is now widely used as a key tool to monitor gene expression products. Metabolomics offers a 'snapshot' or fingerprint of a cell in a given physiological state (Morgenthal *et al.*, 2006). Through exhaustive metabolic profiling, metabolomics allows phenotyping even in the absence of any visible change in a cell or an individual plant and it has been suggested that it is comparable to whole-genome sequencing (Oksman-Caldentey and Saito, 2005).

Since metabolites have widely differing chemical properties, such as polarity, molecular weight, volatility and chemical reactivity, no single existing technique is able to profile all the metabolites in a biological system. A combination of different analytical methods of high sensitivity, such as NMR, gas chromatography and MS, are generally used for comprehensive non-targeted chemical profiling. Use of these techniques can simultaneously profile the effect of, for instance, stress, nutritional status and any genetic or environmental perturbation, on hundreds of metabolites to create massive and complex data-sets (Dixon *et al.*, 2006). Analysis of these data-sets for the interpretation of metabolic profiles in terms of the underlying biochemical network of reactions and regulations represents an equally important component of metabolomics. Chemometrics and multivariate analysis, such as a principal component analysis, hierarchical cluster analysis, and self-organisation mapping, are often used for data mining (Kose *et al.*, 2001; Duran *et al.*, 2003; Jonsson *et al.*, 2004). In addition to these classical methods, a novel concept for the analysis of metabolite dynamics based on the comprehensive analysis of differential correlations between metabolites across different genotypes or environmental conditions has been proposed (Morgenthal *et al.*, 2006; Ursem *et al.*, 2008). Metabolic data can be used to construct and interpret simple correlation networks and elucidate biosynthetic pathways in plants and, thus, help to unravel the biological basis of certain traits or regulation in plants.

11.2.3 Systems Biology Approach

Systems biology as defined by Sweetlove *et al.* (2003) is the comprehensive multidimensional representation of all major biosynthetic pathways of the cell. The ultimate goal is to elucidate overall plant metabolism as an integrated system, and this can be achieved by identifying and characterising the genes, proteins and metabolites involved in the plant system. Although still in its infancy, there has been an increasing amount of research utilising this integrative approach when trying to acquire a more complete understanding of the diversity and complexity of plant biosynthetic pathways. The development of systems biology is undoubtedly based on the technological advances and improvements in the coverage of the metabolome. This approach is becoming increasingly appealing to researchers since the conventional technique of analysis of single compounds, or groups of compounds, is reaching its limit.

11.2.3.1 Integration of Transcriptomics and Metabolomics Data

Integrative genomics approaches, especially between metabolomics and transcriptomics, have generated rich descriptive data networks especially for the flavonoid biosynthetic pathway. In addition to the characterisation of the systematic interaction of biological processes, integrative study is also used to assist in prediction of gene functions. Through this integration process, not only are the gaps in our knowledge of enzymatic function and the regulation of pathways filled, but the extensive coordination and communication between pathways can be elucidated (Saito *et al.*, 2007; Fukushima *et al.*, 2009).

An example of this approach is seen with the research carried out by Tohge *et al.* (2005). Integration of metabolomics and transcriptomics data was used to reveal gene-to-metabolite networks for identifying the function of unknown genes involved in flavonoid biosynthesis. The *PAP1* gene encoding a MYB transcription factor was overexpressed in *Arabidopsis thaliana,* and through metabolomics, an array of 1800 putative metabolites, including eight novel anthocyanins, was identified. Transcriptome analysis revealed that in addition to the induction of well-known genes involved in anthocyanin production, several genes with unidentified functions or annotated with putative functions, encoding putative glycosyltransferases, acyltransferases, glutathione *S*-transferases, sugar transporters and transcription factors, were also upregulated. Of these, two putative glycosyltransferase genes (*At5g17050* and *At4g14090*) induced by *PAP1* expression were shown to encode flavonoid 3-*O*-glucosyltransferase and anthocyanin 5-*O*-glucosyltransferase (Tohge *et al.*, 2005).

11.2.3.2 Phytochemical Genomics

Another approach used for the functional identification of flavonoid biosynthesis genes is through phytochemical genomics. In this approach, *in silico* analysis of genes and metabolites using publicly available databases are utilised to further construct gene-to-metabolite networks in order to accelerate studies required to fill in the missing areas in our knowledge of how metabolic pathways are regulated. This technology can be utilised if large transcriptome data-sets such as ATTED-II (Obayashi *et al.*, 2007) are available and used instead of limited in-house data-sets that are specific to particular conditions. In this context, *Arabidopsis thaliana* is a good model plant as its entire genome has been sequenced. This method is an extremely powerful technique for the prediction and identification of gene function when combined with metabolic profiling (Saito *et al.*, 2007).

A phytochemical genomics approach was recently utilised by Tohge *et al.* (2007). Flavonoid profiling was carried out on wild-type *Arabidopsis* plants and T-DNA insertion mutants. *In silico* co-expression analysis of genes and metabolites obtained from this study was carried out using a publicly-available transcriptome database of DNA microarrays. With this approach, functions of specific genes have been revealed. Co-expression analysis showed that a methyltransferase gene, *AtOMT1* (At5g54160) was not only involved in the production of lignins and sinapoyl esters, but also in the methylation of flavonols for the formation of isorhamnetin. In another study by Yonekura-Sakakibara *et al.* (2008), transcriptome coexpression analysis using 139 genes known to be involved in flavonoid metabolism, and integration with 15 newly identified and eight known flavonols, a gene encoding flavonol 3-*O*-arabinosyltransferase was identified and the physiological role of a gene coding for UDP-rhamnose synthase in flavonoid biosynthesis was revealed.

11.2.3.3 Bioinformatics Approaches for Integration of 'Omics' Data

In order to extract maximal benefit from the integration of 'omics' data (transcriptomics, proteomics and metabolomics), several statistical and computational methods supported by bioinformatics tools are required. These tools will aid data storage, visualisation and integration and finally analysis of data. With the advancement of high throughput 'omics' technologies, there is an exponential growth of data generated and in these situations, the utilisation of bioinformatics becomes even more critical. Bioinformatics will soon be responsible for the evolution of biology from a descriptive to a predictive science (Shinozaki and Sakakibara, 2009).

Public research databases for the various omics are continually increasing. There have been several attempts by different groups of researchers to standardise their databases in order to facilitate data sharing. An attempt was made by an international organisation, the Microarray Gene Expression Data (MGED) Society, to facilitate the sharing of functional genomics and proteomics array data. The Min. Information About a Microarray Expt. (MIAME) and Microarray Gene Expression Object Model (MAGE) standard have been proposed for transcriptomics data (Mehrotra and Mendes, 2006; Ball and Brazma, 2006). Similar attempts were made for the standardisation of metabolomics data, by ArMet (Jenkins *et al.*, 2004) and MIAMET (Bino *et al.*, 2004). Sharing mass spectra data and retention-time index libraries of plant metabolite profiles generated from gas chromatography-electron impact MS has been proposed, and the Golm Metabolome Database (GMD) has been established (Kopka *et al.*, 2005).

For data integration of the various 'omics' research to estimate gene functions and relationships among cellular elements, such as transcripts and metabolites, it is important that all omics data be managed in a single database. A successful attempt with regard to the data integration of metabolomics and transcriptomics is seen with the establishment of the Platform for RIKEN Metabolomics, which is a web-based service for metabolomics and transcriptomics in discovering novel metabolic networks and their transcriptional regulation systems (http://prime.psc.riken.jp/). In addition, there are a number of free and commercially available web-resourced databases, such as KEGG (Kanehisa and Goto, 2000; Kanehisa *et al.*, 2006), MetaCyc (Zhang *et al.*, 2005; Caspi *et al.*, 2008) and KaPPA-View3 (Tokimatsu *et al.*, 2005), built on information in the current literature that illustrate numerous biosynthetic routes, including the flavonoid biosynthetic pathways

11.3 The Flavonoid Biosynthetic Pathway as it is Today

11.3.1 Gateway into the Flavonoid Pathway

The flavonoid biosynthetic pathway starts with the condensation of one molecule of 4-coumaroyl-CoA and three molecules of malonyl-CoA, yielding a tetraketide. The resulting

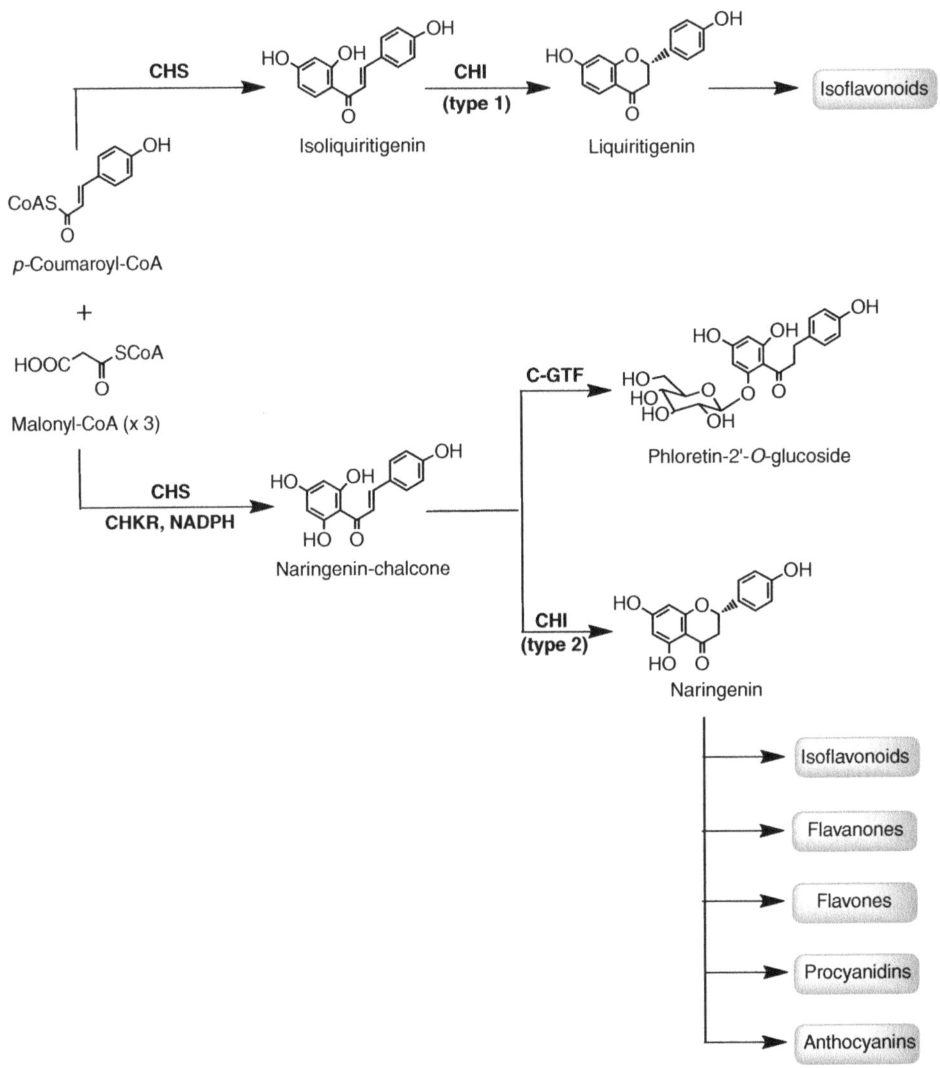

Figure 11.2 *Gateway into the flavonoid biosynthetic pathway. Enzyme abbreviations: CHS, chalcone synthase; CHI, chalcone isomerase; CHKR, chalcone ketide reductase; NADPH, nicotinamide adenine dinucleotide phosphate; C-GTF, chalcone 2′-O-glucosyltransferase*

tetraketide intermediate undergoes intramolecular cyclisation to form naringenin-chalcone (Figure 11.2). This reaction is catalyzed by chalcone synthase (CHS). During the condensation reaction, decarboxylation of malonyl-CoA to an acetyl-CoA carbanion occurs (Kreuzaler *et al.*, 1978). However, for the production of isoflavonoids, there is a slight modification in the reaction catalyzed by CHS as isoflavonoids are derived from 2′,4,4′-trihydroxychalcone (isoliquiritigenin), which lacks the 2′-hydroxyl group. The reduction of the hydroxyl group probably occurs during the polyketide stage prior to cyclisation (Dewick *et al.*, 1982), and this reaction is catalyzed by chalcone ketide reductase, an NADPH-dependent enzyme that presumably interacts with CHS (Welle and Grisebach, 1988).

CHS, being the entry point into the flavonoid pathway, is the most thoroughly investigated enzyme of the phenylpropanoid pathway and it is the first flavonoid enzyme whose crystal structure was successfully resolved (Ferrer *et al.*, 1999). Through genetic analyses, various CHS mutants have been identified in many plant species. Typically, mutants that completely lack CHS activity have a white flower colour, white pollen, and altered seed coat colour. The importance of CHS as the first enzyme in the flavonoid pathway was further established through molecular techniques. Gene silencing of CHS, as expected, led to completely white flowers. In addition, through molecular cloning, novel CHS mutants with altered activities were obtained. *In vitro* mutagenesis of CHS has resulted in the identification of important residues leading to altered substrate specificity (Jez *et al.*, 2002), thereby producing novel coloured flavonoids in flowers (Yu *et al.*, 2006).

The next step in the flavonoid biosynthesis is the stereospecific conversion of naringenin-chalcone to (2*S*)-5,7,4′-trihydroxyflavanone (naringenin) by chalcone isomerase (CHI). However, in legumes, CHI also catalyses metabolism of isoliquiritigenin to (2*S*)-4′,7-dihydroxyflavanone (liquiritigenin) (Forkmann and Heller, 1999). The conversion of naringenin-chalcone to naringenin is very rapid compared with the rather slow isomerisation of isoliquiritigenin to liquiritigenin, due to the intramolecular hydrogen bond in the substrate molecule (Shimada *et al.*, 2003). CHI enzymes isolated from non-legume plants are unable to catalyze the conversion of isoliquiritigenin to liquiritigenin. As a consequence, CHI enzymes have been subdivided into two groups. Type 1 CHI, found in both legumes and non-legumes, isomerise only 2′-hydroxychalcones, while type II CHIs which are exclusive to legumes accept both 2′-deoxy- and 2′-hydroxychalcone as a substrate (Shimada *et al.*, 2003) (Figure 11.2).

In some plants, such as carnations (*Dianthus caryophyllus)*, the yellow flower colour is due to the accumulation of chalcone-2′-*O*-glucoside which is formed from naringenin-chalcone in a reaction catalyzed by chalcone 2′-*O*-glucosyltransferase (Forkmann and Dangelmayr, 1980; Ogata *et al.*, 2004). Glucosylated chalcones occur in other species including false waterwillow (*Andrographis echioides*) (Jayaprakasam *et al.*, 2001), Canadian wildginger (*Asarum canadense)* (Iwashina and Kitajima, 2000), Glory Bower (aka Arni) (*Clerodendron phlomidis*) (Roy and Pandey, 1994) and bubinga (*Guibourtia tessmanii*) (Fuendjiep *et al.*, 2002)

Naringenin is a key central intermediate, as from this point onwards the flavonoid biosynthetic pathway diverges into several side-branches, each resulting in the production of different class of flavonoids, including isoflavones, flavanones, flavones, flavonols, proanthocyanidins and anthocyanins (Figure 11.2).

11.3.2 Isoflavonoid Branch Pathway

The isoflavonoid pathway operates almost exclusively in leguminous plants, and most of the enzymes involved in this pathway have been identified and their encoding genes cloned (Jung *et al.*, 2003). The microsomal cytochrome P450 enzyme, 2-hydroxyisoflavanone synthase, commonly known as isoflavone synthase, mediates the first step in this branch pathway, converting naringenin and isoliquiritigenin to their respective isoflavones genistein and daidzein.

Naringenin/liquiritigenin undergo abstraction of a hydrogen radical at C-3 followed by B-ring migration from C-2 to C-3 and the subsequent hydroxylation of the resulting C-2 radical to yield 2-hydroxy-2,3-dihydrogenistein and 2′-hydroxyisoflavanone, respectively (Dixon and Ferreira, 2002). These molecules are unstable and undergo dehydration to yield genistein or daidzein (Figure 11.3). Daidzein and genistein are important precursors in the biosynthesis of antimicrobial phytoalexins. In addition, they also play a major role in establishing the symbiotic relationship between the plant and rhizobial bacteria and inducers of *nod* gene expression.

4′-*O*-Methylation of daidzein by 2,7,4′-trihydroxyisoflavanone-4′-*O*-methyltransferase yields formononetin (Akashi *et al.*, 2003). Formononetin is a crucial intermediate in the biosynthesis of phytoalexins, and as such is the starting point for the formation of various pterocarpans. Formononetin undergoes a series of reactions including hydroxylation, reduction and dehydration to form medicarpin and vestitol (Figure 11.3). The enzymes involved are: isoflavone-2′-hydroxylase, 2′-hydroxyisoflavone reductase, pterocarpan synthase and pterocarpan reductase (Guo *et al.*, 1994; Akashi *et al.*, 2006). Since the pterocarpan phytoalexins are typically involved in the defence against pathogens of leguminous plants, they are not usually expressed constitutively but induced by biotic and abiotic stresses (Dewick, 1994). Daidzein is also a crucial intermediate in the biosynthesis of various complex pterocarpans, such as phaseollin and glyceollin. Phaseollins are the major phytoalexins found in beans (*Phaseolus vulgaris*), while glyceollins are occur mainly in soybean (*Glycine max*) and, like their other counterparts, they are only induced in response to plant pathogens and stress.

Isoflavone-*O*-methyltransferase is also responsible of the 4′-*O*-methylation of genistein yielding biochanin A and 7′-*O*-methylation to yield prunetin (Figure 11.3). For storage purposes, typically, glucosides and malonylglucosides of isoflavones such as genistein, daidzein and biochanin are produced (Kim *et al.*, 2005). Hydroxylation of genistein to 2′-hydroxygenistein is carried out by isoflavone-2′-hydroxylase in legumes such as soybean (Kochs and Grisebach, 1986) and licorice (*Glycyrrhiza glabra*) (Akashi *et al.*, 1998). 2′-Hydroxygenistein represents a branch point leading to the biosynthesis of prenylated isoflavanones such as dalbergioidin and kievitone. These prenylated isoflavanones together with prenylated isoflavones such as wighteone and luteone, have been observed to be more active against pathogens than their non-prenylated analogues (O'Neill *et al.*, 1983; Gagnon and Ibrahim, 1997). A recent review by Yazaki *et al.* (2009) stressed the importance of prenylation of flavonoids, especially in terms of enhanced beneficial effects on human health such as anticancer, anti-androgen, anti-leishmania, and anti-nitric oxide production. Prenyltransferases that catalyze the key step of flavonoid prenylation are membrane-bound enzymes, and the first prenyltransferase gene *SfN8DT* has only recently been identified. (Yazaki *et al.*, 2009).

11.3.3 Flavanone Branch Pathway

Flavanones are the dominant flavonoids in citrus fruits. The flavanone structure, with the absence of the $\Delta^{2,3}$ double bond and presence of a chiral centre at C2, is highly reactive and is readily subjected to hydroxylation, glycosylation and *O*-methylation reactions. The 4′-*O*-methyl derivative of naringenin, ponciretin, is thought to be involved in defence against

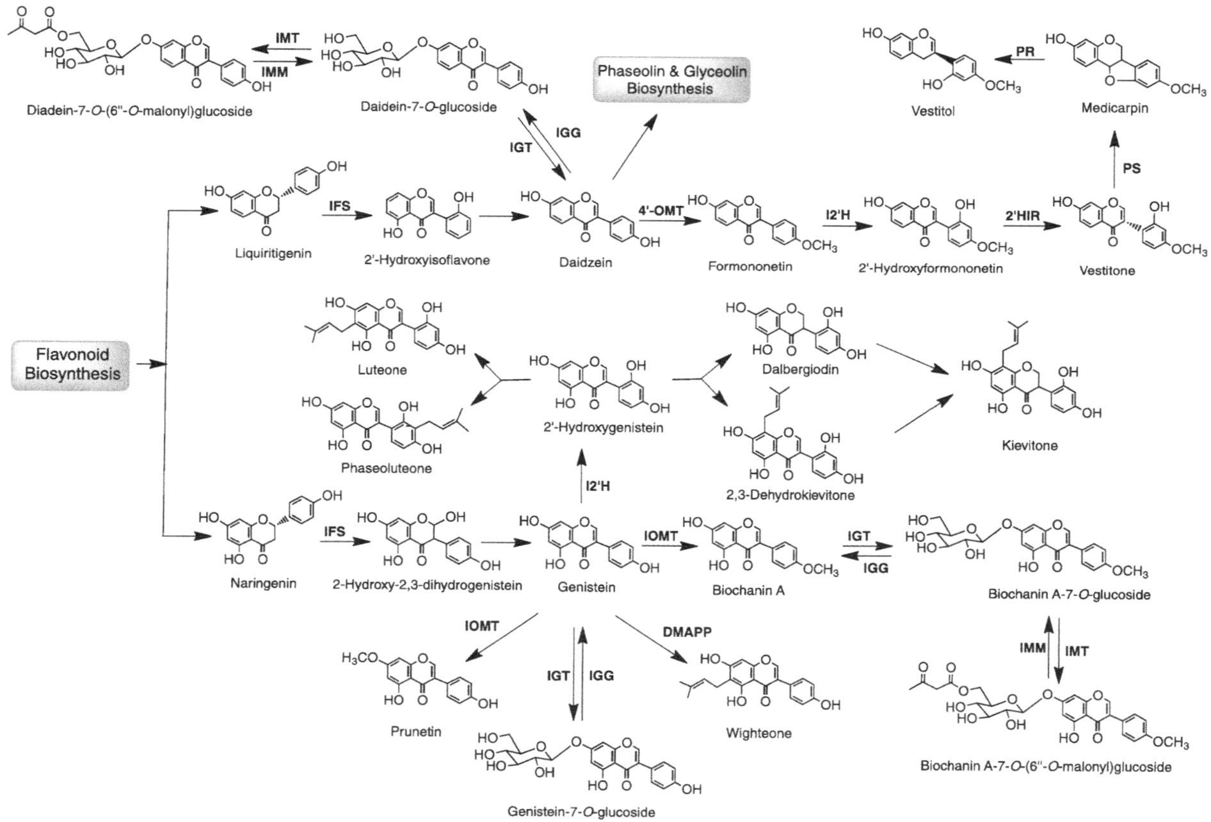

Figure 11.3 *Isoflavonoid branch pathway. Enzyme abbreviations: IFS, isoflavone synthase; 4′-OMT, 2,7,4′-trihydroxyisoflavanone-4′-O-methyltransferase; IGG, isoflavone 7-O-glucoside β-glucosidase; IGT, isoflavone 7-O-glucosyltransferase; IMT, isoflavone 7-O-glucoside-6″-O-malonyltransferase; IMM, isoflavone 7-O-glucoside-6″-O-malonate malonylesterase; I2′H, isoflavone-2′-hydroxylase; 2′HIR, 2′-hydroxyisoflavone reductase; PS, pterocarpan synthase; PR, pterocarpan reductase; IOMT, isoflavone-O-methyltransferase; DMAPP, isoflavone dimethylallyl transferase*

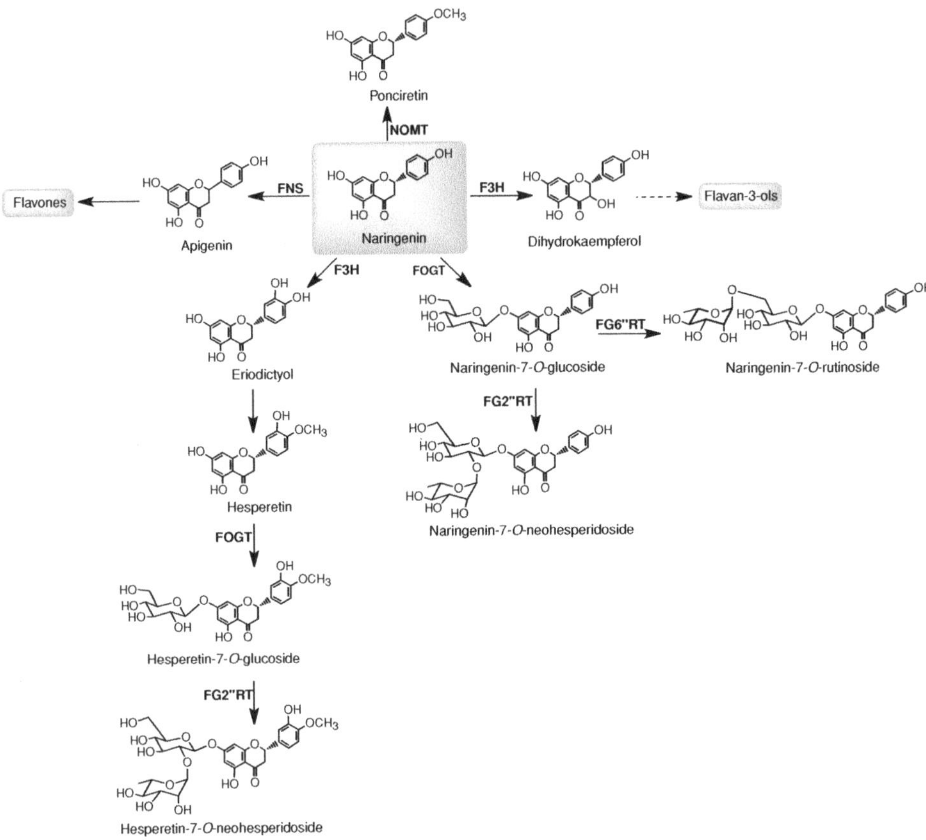

Figure 11.4 *Flavanone branch pathway. Enzyme abbreviations: FNS, flavone synthase; N-OMT, naringenin 4′-O-methyltransferase; F3H, flavanone 3-hydroxylase; FOGT, flavanone 7-O-glucosyltransferase; FG2″RT, flavanone 7-O-glucoside-2″-O-rhamnosyltransferase; FG6″RT, flavanone 7-O-glucoside-6″-O-rhamnosyltransferase*

pathogens. It exhibits antifungal activity and is synthesized principally during pathogen attacks (Middleton and Kandaswami, 1994).

The taste of the citrus fruit is predominantly determined by the types of sugar substitution of the flavanone glycoside. Naringenin-7-*O*-rutinoside (narirutin) is tasteless while naringenin-7-*O*-neohesperidoside (naringin) is responsible for the bitter taste of citrus fruits (Horowitz and Gentili, 1969; Frydman *et al.*, 2004). Catalysed by flavanone 7-*O*-glucosyltransferase, naringenin is converted to prunin (naringenin-7-*O*-glucoside), which being unstable is converted rapidly to the disaccharide conjugates naringenin-7-*O*-neohesperidoside (naringin) and naringenin-7-*O*-rutinoside (narirutin) (Lewinsohn *et al.*, 1986) (Figure 11.4). Hesperitin, another flavanone, is synthesized from eriodictyol, which is produced through 3′-hydroxylation of naringenin. Hesperitin is glycosylated to form hesperitin-7-*O*-glucoside which is further metabolised with the attachment of L-rhamnose to the glucosyl moiety to form hesperetin-7-*O*-rutinoside (hesperidin) and hesperetin-7-*O*-neohesperidoside (neohesperidin) (Figure 11.4) (Lewinsohn *et al.*, 1989; Bar-Peled *et al.*, 1993).

11.3.4 Flavone Branch Pathway

In a reaction requiring NADPH and oxygen, flavone synthase (FNS) is the enzyme responsible for the introduction of a $\Delta^{2,3}$ double bond that converts flavanones to flavones (Figure 11.5) (Heller and Forkmann, 1994). Two types of FNS have so far been identified, FNS-I and FNS-II. FNS-II is the enzyme responsible for flavone formation in many plants, while FNS-I appears to be confined to species of the Apiacea (Martens *et al.*, 2001). Flavones have varied functions *in planta,* acting as co-pigments in flowers, antioxidants to protect plants from UV damage, and phytoalexins with antimicrobial activity (Dixon, 1986; Schmelzer *et al.*, 1988; Yu *et al.*, 2006). They also play a role as rhizobial signal molecules during the establishment of symbiosis between legumes and nitrogen-fixing rhizobia (Dakora *et al.*, 1993).

The various flavone structures are formed due to substitutions such as methylation, hydroxylation, isoprenylation and glucosylation (Martens and Mithofer, 2005). An example is acacetin, the 4′-*O*-methylated derivative of apigenin (Wollenweber, 1994). Acacetin has been found in a number of species across various plant families, most prominently in the Asteracea and Leguminosae. The 6-*C*-glucosylated form of apigenin, isovitexin, has also been reported. Isovitexin can be further glycosylated at the 7-hydroxyl position with either glucose, rhamnose or xylose, yielding the conjugates shown in Figure 11.5. FNS converts the flavanone eriodictyol to luteolin which is also synthesized from apigenin in a reaction catalyzed by flavonoid 3′-hydroxylase (F3′H) (Figure 11.5) (Kitada *et al.*, 2001; Ueyama *et al.*, 2002). Lutein accumulates in plants as sugar conjugates, with glucose and rhamnose being the most commonly occurring *O*-glycoside conjugates. The presence of glucuronide and methyl derivatives of luteolin have also been reported (Schulz and Weissenb, 1988).

11.3.5 Flavonol Branch Pathway

The flavanone naringenin is converted to dihydrokaempferol by a 3-*O*-hydroxylation catalyzed by flavanone 3-hydroxylase (F3H) (Heller and Forkmann, 1994). F3′H and F3′5′H then catalyze the introduction of B-ring hydroxyl groups at the appropriate positions to respectively yield dihydroquercetin and dihydromyricetin (Figure 11.6). Dihydroflavonols are biosynthetic intermediates in the formation of flavonols, flavan-3-ol monomers, proanthocyanidins and anthocyanidins (Springob *et al.*, 2003). The enzymes F3′H and F3′5′H are, therefore, important regulators in the flavonoid biosynthetic pathway, especially in the production of anthocyanins influencing flower colour by controlling the ratio of pelargonidin (red to orange), cyanidin (red to violet), and delphinidin (violet to blue) (Holton and Cornish, 1995; Winkel-Shirley, 2001). Flavonols are then formed by the introduction of a double bond between C-2 and C-3 positions by the enzyme flavonol synthase to produce myricetin, quercetin and kaempferol (Figure 11.6). Later in the pathway, dihydroflavonol 4-reductase is the key enzyme controlling flux into biosynthetic branches leading to anthocyanins and proanthocyanidins (see sections 11.3.6 and 11.3.7).

Quercetin-3-*O*-rutinoside (rutin) is synthesized from its immediate precursor quercetin-3-*O*-glucoside (isoquercetin) in a reaction catalyzed by flavonol 3-*O*-glucosyltransferase

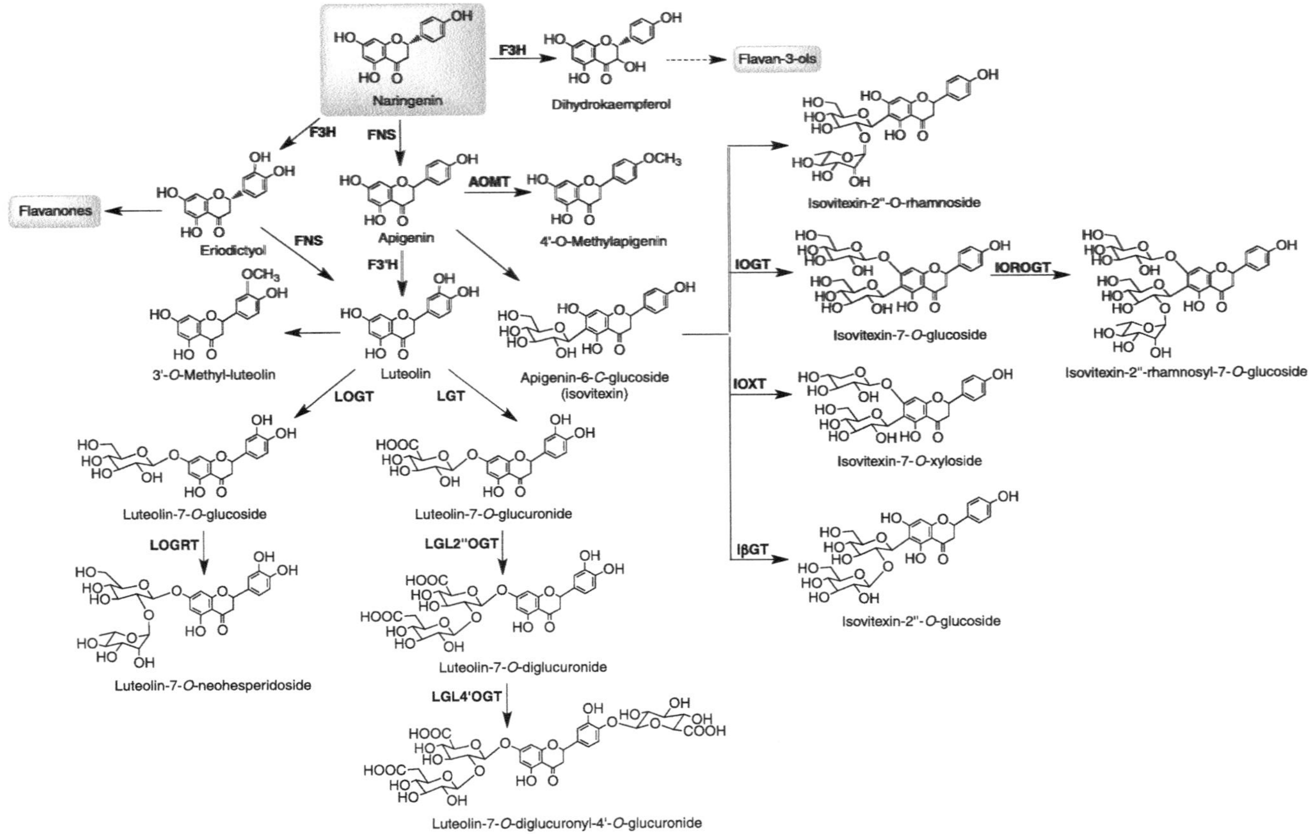

Figure 11.5 *Flavone branch pathway. Enzyme abbreviations: F3H, flavanone 3-hydroxylase; FNS, flavone synthase; AOMT, apigenin 4′-O-methyltransferase; F3′H, flavonoid 3′-hydroxylase; LGT, luteolin 7-O-glucuronosyltransferase; LOGRT, luteolin 7-O-glucoside 2″-O-rhamnosyltransferase; LGL2″OGT, luteolin 7-O-glucuronide 2″-O-glucuronosyltransferase; LGL4′OGT, luteolin 7-O-glucuronide-4′-O-glucuronosyltransferase; IOGT, isovitexin 7-O-glucosyltransferase; IOROGT, isovitexin 2′-O-rhamnoside 7-O-glucosyltransferase; IOXT, isovitexin 7-O-xylosyltransferase; IβGT, isovitexin b-glucosyltransferase*

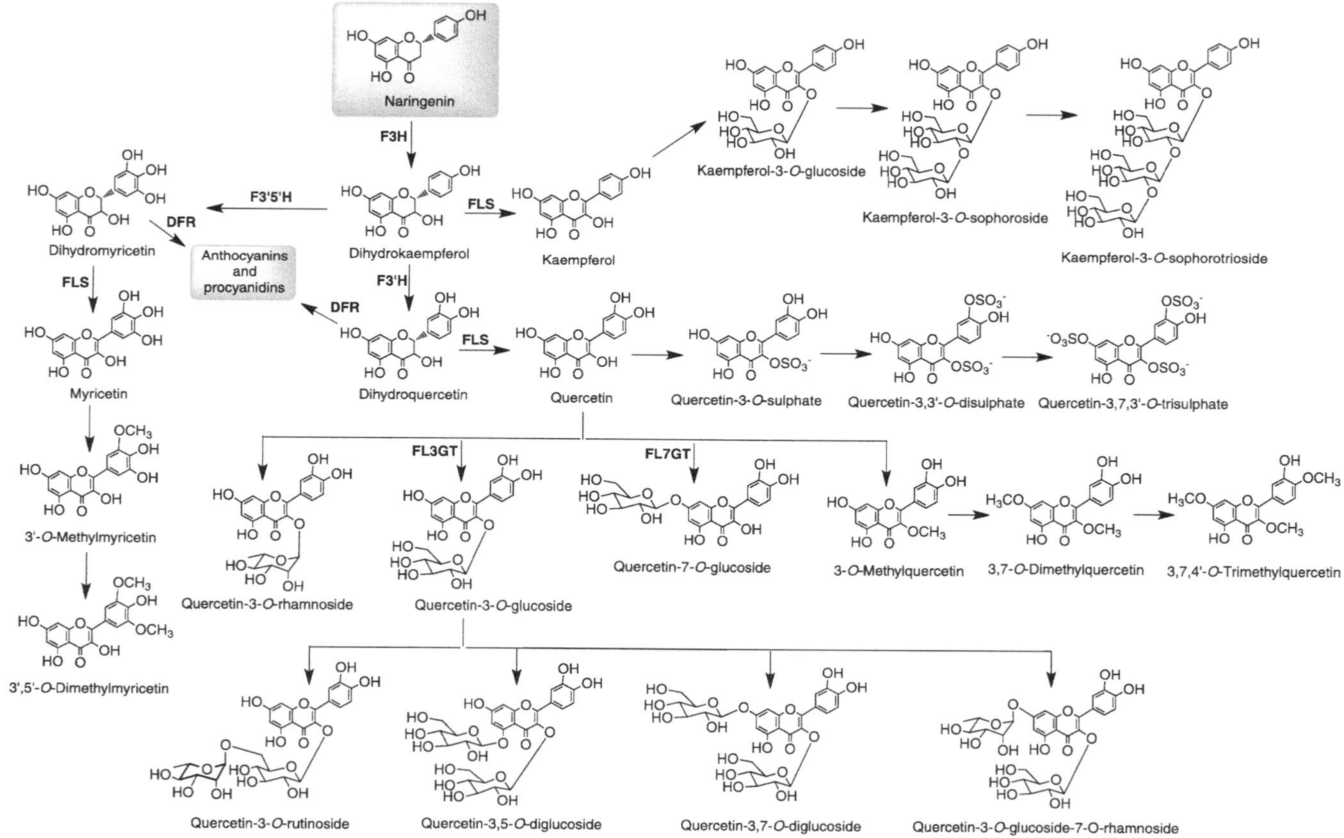

Figure 11.6 *Flavonol branch pathway. Enzyme abbreviations: F3H, flavanone 3-hydroxylase; F3′H, flavonoid 3′-hydroxylase; F3′5′H, flavonoid 3′,5′-hydroxylase; DFR, dihydroflavonol 4-reductase; FLS, flavonol synthase; FL3GT, flavonol 3-O-glucosyltransferase; FL7GT, flavonol 7-O-glucosyltransferase*

(Barber and Behrman, 1991). Other glucosides which have been found in *Arabidopsis thaliana* are quercetin 7-*O*-glucoside, quercetin-3-*O*-rhamnoside, quercetin-3-*O*-glucoside-7-*O*-rhamnoside, quercetin 3,7-*O*-diglucoside and quercetin 3,5-*O*-diglucoside (Figure 11.6) (Routaboul *et al.*, 2006; Kerhoas *et al.*, 2006). Glucosyltransferases in *Arabidopsis thaliana* preferentially glucosylate the 3- and 7-hydroxyl groups of the C- and A-rings of the flavonol skeleton. The enzymes catalyzing these steps have been isolated from *Arabidopsis* (Kim *et al.*, 2006a,b).

11.3.6 Proanthocyanidin Branch Pathway

Flavan-3-ols represent a class of flavonoids and are the building blocks of condensed tannins or proanthocyanidins. Flavan-3-ols exist in two stereo-isoforms, namely 2,3-*trans*-flavan-3-ols and 2,3-*cis*-flavan-3-ols. The first committed step in the synthesis of proanthocyanidin that diverges from the pathway common with anthocyanins is believed to be the synthesis of flavan-3-ols, such as (+)-catechin or (–)-epicatechin, either directly or indirectly from leucocyanidin and leucodelpdinidin which are derived from dihydrokaempferol (Figure 11.7). Three enzymes, anthocyanidin synthase (ANS), leucoanthocyanidin reductase (LAR) and anthocyanidin reductase (ANR), are believed to act at the entrance into the two major stereo-specific pathways of proanthocyanin biosynthesis, catalyzing the formation of (+)-catechin, (+)-gallocatechin, (–)-epicatechin and (–)-epigallocatechin (Tanner *et al.*, 2003). (–)-Epicatechin and (–)-epigallocatechin can be converted to (–)-epicatechin-3-*O*-gallate and (–)-epigallocatechin-3-*O*-gallate by esterification with gallic acid (Figure 11.7).

11.3.7 Anthocyanidin Branch Pathway

Anthocyanidin synthase catalyzes the oxidation of the colourless leucoanthocyanidin to a precursor of the coloured anthocyanidin, either through a hydroxylation at the C-3 position of leucoanthocyanidin or through the direct formation of a C-3 ketone (Springob *et al.*, 2003). Subsequently, it was suggested that dehydration or enolisation to 2-flaven-3,4-diol leads to anthocyanidin, while formation of 3-flavan-3,4-diol can lead either to (2*R*,3*S*)-*cis*-dihydroflavonol or (2*R*,3*R*)-*trans*-dihydroflavonol (Welford *et al.*, 2001).

Anthocyanin biosynthesis can originate from cyanidin, delphinidin or pelargonidin. Anthocyanins arising from cyanidin progress via glycosylation, forming cyanidin-3-*O*-glucoside and cyanidin-5-*O*-glucoside, which then leads into the biosynthesis of various compounds such as cyanidin-3,5-*O*-diglucoside, cyanidin-3-O-sophoroside and cyanidin-3-*O*-(6-*O*-*p*-coumaroyl)glucoside-5-*O*-glucoside (shisonin). The formation of cyanidin-*O*-3,5-diglucoside (rose anthocyanin) is catalyzed by anthocyanidin 3,5-*O*-glycosyltransferase, a distinctive enzyme that glucosylates both the flavonoid A-ring at the 5-*O*-position and the 3-*O*-position of the C-ring (Figure 11.8). Initially the intermediate cyanidin-5-*O*-glucoside is formed, and it is further glucosylated to form cyanidin-3,5-*O*-diglucoside (Ogata *et al.*, 2005). Alternatively, cyanidin-3-*O*-sophoroside can be produced when another glucose moiety is added to cyanidin-3-*O*-glucoside at the 2″-*O*-position of the glucose moiety by the enzyme anthocyanidin-3-*O*-glucoside-2″-*O*-glucosyltransferase. This feature confers the bright blue colour as seen in *Ipomoea*. In plant families such as Lamiacea, Convolvulaceae and Ranunculacea, the sophorosides are further acylated and methylated (Terahara *et al.*, 2001, 2004; Saito *et al.*, 2002, 2005).

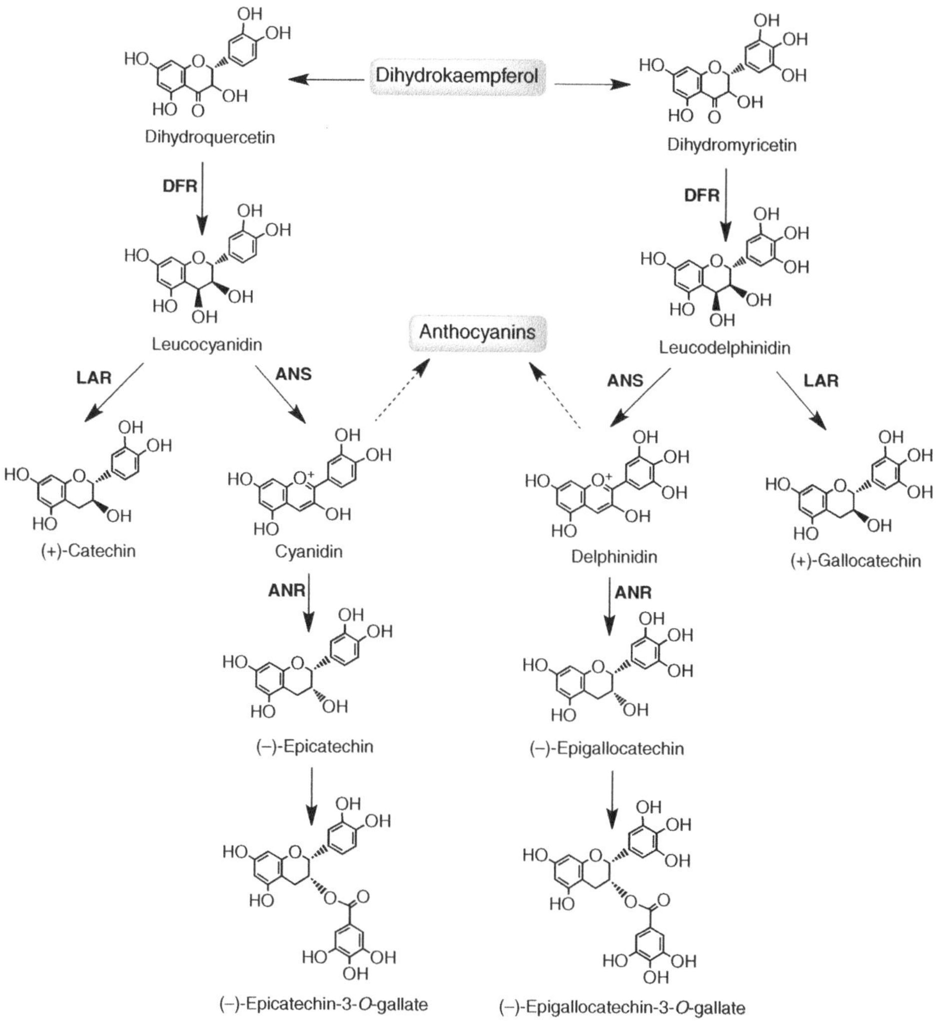

Figure 11.7 Proanthocyanidin branch pathway. Enzyme abbreviations: DFR, dihydroflavonol 4-reductase; LAR, leucoanthocyanidin reductase; ANS, anthocyanidin synthase; ANR, anthocyanidin reductase

The blue colouration of flowers of vascular plants is due mainly to the presence of delphinidin-3-*O*-glucoside. The enzyme anthocyanidin 3-*O*-glucosyltransferase catalyzes the glucosylation of delphinidin (Kitamura, 2006). Delphinidin-3-*O*-glucoside can give rise to stable anthocyanins such as delphinidin-3-*O*-(6″-*O*-malonyl)glucosyl-3′,5′-*O*-diglucoside (ternatin C5) and delphinidin-3-*O*-glucosyl-3′,5′-*O*-di(6″-*O*-caffeoyl)glucoside (gentiodelphin) (Figure 11.8), which contribute to a wide variety of flower colours. The addition of another glucose moiety at the 2″-*O*-position of the 3-*O*-glucose moiety produces delphinidin-3-*O*-sophoroside.

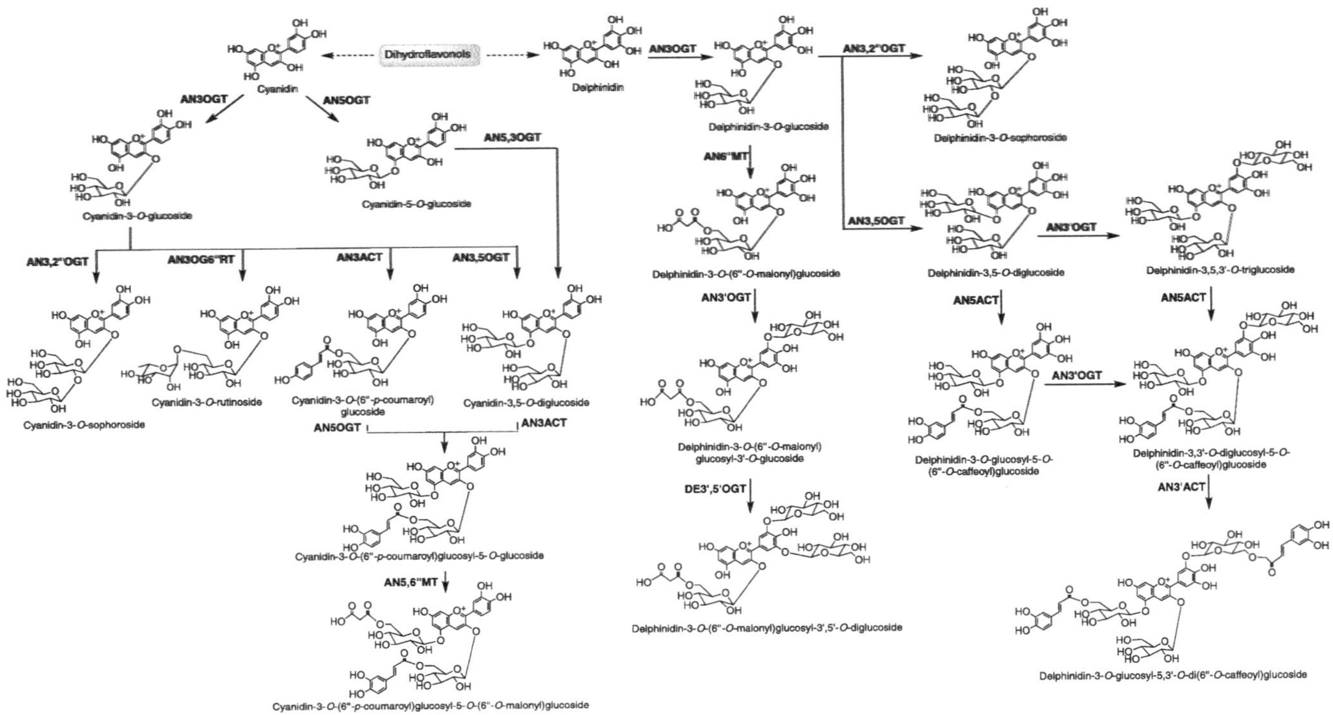

Figure 11.8 *Anthocyanidin branch pathway I. Enzyme abbreviations: AN3OGT, anthocyanin 3-O-glycosyltransferase; AN5OGT, anthocyanin 5-O-glycosyltransferase; AN5,3OGT, anthocyanidin 5,3-O-glycosyltransferase; AN3OG6″RT, anthocyanidin-3-O-glucosyl-6″-O-rhamnosyltransferase; AN3,2″OGT, anthocyanidin 3-O-glucosyl-2″-O-glucosyltransferase; AN3ACT, anthocyanin 3-aromatic acyltransferase; AN5′OGT, anthocyanin 5′-O-glycosyltransferase; AN5, 6″MT, anthocyanin 5-O-glucosyl-6″-O-malonyltransferase; AN3OGT, anthocyanidin 3-O-glucosyltransferase; AN3,6″MT, anthocyanin 3-O-glucosyl-6″-O-malonyltransferase; AN5ACT, anthocyanin 5-aromatic acyltransferase; DE3′,5′OGT, delphinidin 3′,5′-O-glucosyltransferase*

Ternatins are a family of blue anthocyanins found in the petals of *Clitoria ternata* (butterfly pea). Among them, ternatin C5, the simplest form of ternatin, is produced by the stepwise transfer of two glucose residues to the 3′- and 5′-positions of delphinidin-3-*O*-(6″-*O*-malonyl)glucoside by the enzymes anthocyanin 3′-*O*-glycosyltransferase and delphinidin 3′,5′-*O*-glucosyltransferase (Kazuma *et al.*, 2004). The first committed step in the biosynthesis of gentiodelphin is the 5-*O*-glucosylation of delphinidin-3-*O*-glucoside, catalysed by anthocyanin-5′-*O*-glycosyltransferase which has been characterised in *Perilla frutescens* by Yamazaki *et al.* (1999). The subsequent steps towards the formation of gentiodelphin involve 3-*O*-, 5-*O*- and 3′-*O*-glucosylations of the delphinidin backbone (Fukuchi-Mizutani *et al.*, 2003) and aromatic acylation with two caffeoyl residues (Fujiwara *et al.*, 1998) (Figure 11.8).

Pelargonidin-3-*O*-glucoside, synthesized from pelargonidin, undergoes further modification that enhances the stability of the molecule (Figure 11.9). This is seemingly the case with the 5-*O*-glucosylation of pelargonidin-3-*O*-glucoside, catalyzed by anthocyanin-5-*O*-glucosyltransferase to form pelargonidin-3,5-*O*-diglucoside (Yamazaki *et al.*, 2002). Sometimes the modification of the anthocyanin moiety contributes to the refinement or intensification of colour as achieved by aromatic acylation with hydroxycinnamic acids such as caffeic acid and *p*-coumaric acid to form pelargonidin-3-*O*-(6″-O-caffeoyl)glucoside and pelargonidin-3-*O*-(6″-O-*p*-coumaroyl)glucoside (Figure 11.9). The enzyme catalyzing this step, anthocyanin-3-*O*-glucoside-6″-*O*-acyltransferase (anthocyanin 3-aromatic acyltransferase), has been characterised in *Perilla frutescens* (Yonekura-Sakakibara *et al.*, 2000). Other modifications to the anthocyanin structure include malonylation to form pelargonidin-3-*O*-(6″-*O*-malonyl)glucoside. Salvianin (pelargonidin-3-*O*-(6″-*O*-caffeoyl)glucosyl-5-*O*-(4″,6″-*O*-dimalonyl)glucoside, a polyacylated anthocyanin, is synthesized through a number of steps that involve glucosylation, aromatic acylation and malonylation, which increase stability, solubility and colour intensity. Salvianin is the predominant anthocyanin in scarlet sage (*Salvia splendens*) and is responsible for the intense scarlet colour of the flowers (Yonekura-Sakakibara *et al.*, 2000). Pelargonidin-3-*O*-glucoside can also undergo rhamnosylation catalyzed by anthocyanidin-3-*O*-glucoside-6″-O-rhamnosyltransferase, and the resultant pelargonidin-3-O-rutinoside is further converted by anthocyanidin-3-*O*-rutinoside-5-*O*-glucosyltransferase to pelargonidin-3-*O*-rutinosyl-5-*O*-glucoside (Figure 11.9).

11.4 Conclusions

Although knowledge of the flavonoid pathway is extensive, it is nonetheless still incomplete. One of the next main targets for metabolic systems is filling the gaps in our knowledge of enzymatic function and the regulation of pathways. Part of this includes the elucidation of the extensive networking and coordination between pathways. In this context, metabolomics and its integration into other 'omics' (genomics, transcriptomics and proteomics) is predicted to play an important future role in identifying the key regulatory steps and characterising pathway networking. This approach will also allow scientists to draw links from the genome to the activity of metabolites. In contrast to the availability of public transcriptome data, public datasets for the accumulation of metabolome data are quite limited. The combination of data on co-expression of genes and co-accumulation

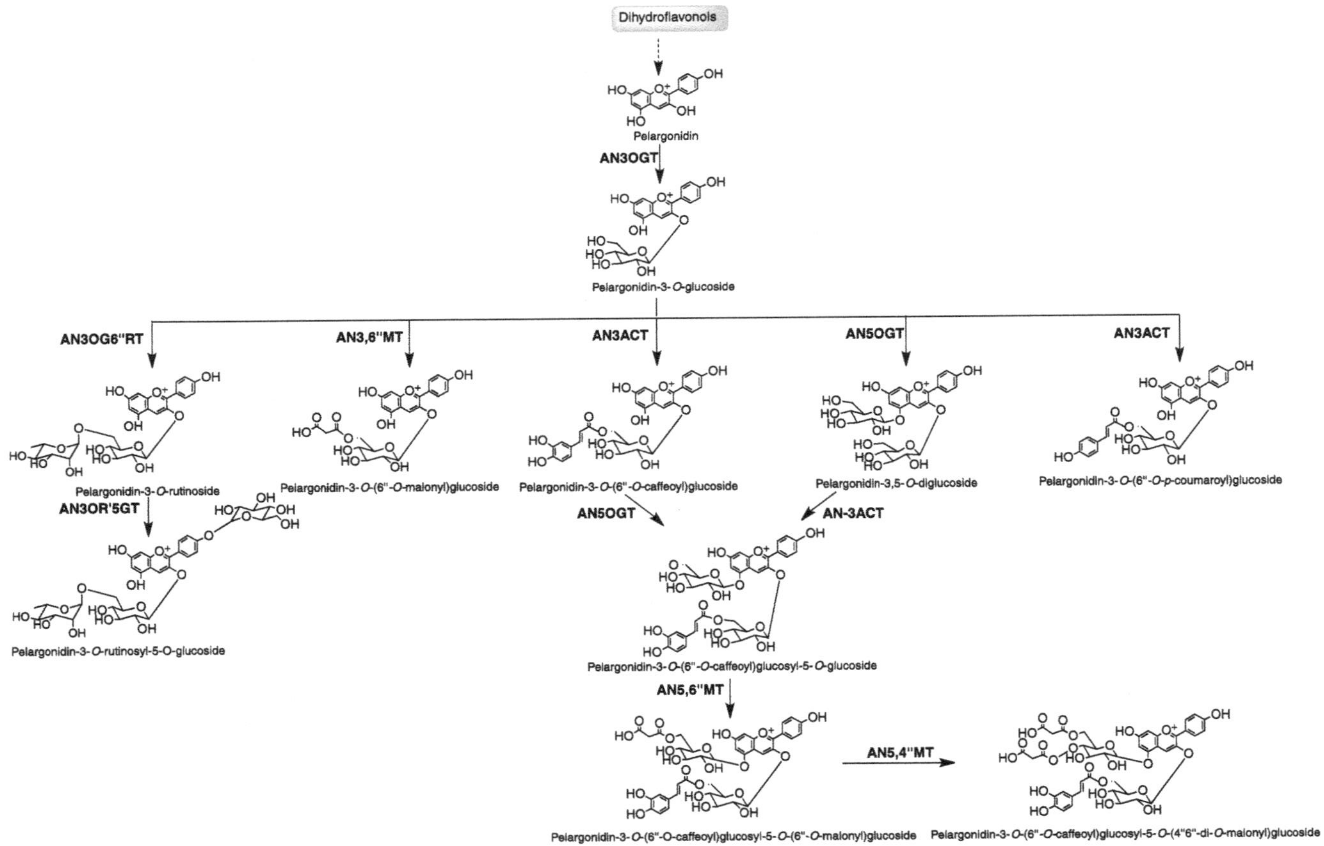

Figure 11.9 *Anthocyanidin branch pathway II. Enzyme abbreviations: AN3OGT, anthocyanidin 3-O-glucosyltransferase; AN3OG6″RT, anthocyanidin-3-O-glucosyl-6″-O-rhamnosyltransferase; AN3ACT, anthocyanin 3-aromatic acyltransferase; AN3OR5GT, anthocyanidin-3-O-rutinosyl-5-O-glucosyltransferase; AN5′OGT, anthocyanin 5′-O-glycosyltransferase; AN5,6″MT, anthocyanin 5-O-glucosyl-6″-O-malonyltransferase; AN5,4″MT, anthocyanin 5-O-glucosyl-4″-O-malonyltransferase; AN3,6″MT, anthocyanin 3-O-glucosyl-6″-O-malonyltransferase*

of metabolites from large public databases should reinforce research efforts for decoding gene functions, which is necessary for the high-throughput discovery of plant-based pharmaceuticals and for the development of functional foods and stress-resistant plants.

References

Aida, R., Kishimoto, S., Tanaka, Y., and Shibata, M. (2000) Modification of flower color in torenia (*Torenia fournieri Lind.*) by genetic transformation. *Plant Sci.*, **153**, 33–42.

Akashi, T., Aoki, T. and Ayabe, S. (1998) CYP81E1, a cytochrome P450 cDNA of licorice (*Glycyrrhiza echinata* L.), encodes isoflavone 2'-hydroxylase. *Biochem. Biophys. Res. Commun.*, **251**, 67–70.

Akashi, T., Sawada, Y., Shimada, N., and Ayabe, S. (2003) cDNA cloning and biochemical characterization of S-adenosyl-L-methionine: 2,7,4'-trihydroxyisoflavanone 4'-*O*-methyltransferase, a critical enzyme of the legume isoflavonoid phytoalexin pathway. *Plant Cell Physiol.*, **44**, 103–112.

Akashi, T., Koshimizu, S., Aoki, T., Sakurai, N., Aoki, T. and Ayabe, S. (2006) Identification of cDNAs encoding pterocarpan reductase involved in isoflavan phytoalexin biosynthesis in *Lotus japonicus* by EST minin. *FEBS Lett.*, **580**, 5666–5670.

Andersen, Ø.M. and Markham, K.R. (ed.) (2006) *Flavonoids: Chemistry, Biochemistry, and Applications*, CRC Taylor & Francis, Boca Raton, FL.

Ball, C.A. and Brazma, A. (2006) MGED Standards: Work in Progress; Omics. *J. Integr. Biol.*, **10**, 138–144.

Barber, G.A. and Behrman, E.J. (1991) The synthesis and characterization of uridine 5′-(β-L-rhamnopyranosyl diphosphate) and its role in the enzymic synthesis of rutin. *Arch. Biochem. Biophys.*, **288**, 239–242.

Bar-Peled, M., Fluhr, R. and Gressel, J. (1993) Juvenile-specific localization and accumulation of a rhamnosyltransferase and its bitter flavonoid in foliage, flowers, and young citrus fruits; *Plant Physiol.*, **103**, 1377–1384.

Bino, R.J., Hall, R.D., Fiehn, O., Hall, R.D., Bovy, A., Jonker, H.H., Tikunov, Y., Lommen, A., Moco, S. and Ilevin, I. (2004) Potential of metabolomics as a functional genomics tool. *Trends Plant Sci.*, **9**, 418–425.

Bino, R.J., de Vos, C.H.R., Lieberman, M., Kopka, J., Saito, K., Draper, J., Nikolau, B. J., Mendes, P., Roessner-Tunali, U., Beale, M. H., Trethewey, R. N., Lange, B. M., Wurtele, Lloyd, E.S. and Sumner, W. (2005) The light hyper-responsive high-pigment 2dg mutation of tomato: alterations in the fruit metabolome. *New Phytol.*, **166**, 427–438.

Bovy, A., de Vos, R., Kemper, M., Schijlen, E., Pertejo, M. A., Muir, S., Collins, G., Robinson, S., Verhoeyen, M., Hughes, S., Santos-Buelga, C. and van Tunen, A. (2002) High-flavonol tomatoes resulting from the heterologous expression of the maize transcription factor genes *LC* and *C1*. *Plant Cell*, **14**, 2509–2526.

Carrari, F., Schauer, N., Willmitzer, L., and Fernie, A.R. (2006) Systems biology: a renaissance of the top-down approach for plant analysis, in *Plant Metabolomics* (*Biotechnology in Agriculture and Forestry Vol. 57*) (eds K. Saito, R.A. Dixon and L. Willmitzer), Springer, Berlin, pp. 185–198.

Caspi, R., Foerster, H., Fulcher, C.A., Kaipa, P., Krummenacker, M., Latendresse, M., Paley, S., Rhee, S.Y., Shearer, A.G., Tissier, C., Walk, T.C., Zhang, P. and Karp, P.D. (2008) The MetaCyc database of metabolic pathways and enzymes and the BioCyc collection of pathway/genome databases. *Nucleic Acids Res.*, **36**, D623–D631.

Crozier, A., Jaganath, I.B. and Clifford, M.N. (2006) Phenols, polyphenols and tannins: an overview, in *Plant Secondary Metabolites and the Human Diet* (eds A. Crozier, H. Ashihara and M.N. Clifford), Blackwell Publishing, Oxford, pp. 1–24.

Dakora, F.D., Joseph, C.M. and Phillips, D.A. (1993) Alfalfa (*Medicago sativa* L.) root exudates contain isoflavonoids in the presence of *Rhizobium meliloti*. *Plant Physiol.*, **101**, 819–824.

Davies, K.M., Bloor, S.J., Spiller, G.B., and Deroles, S.C. (1998) Production of yellow colour in flowers: redirection of flavonoid biosynthesis in *Petunia*. *Plant J.*, **13**, 259–266.

Davuluri, G.R., van Tuinen, A., Fraser, P.D., Manfredonia, A., Newman, R., Burgess, D., Brummell, D.A., King, S.R., Palys, J., Uhlig, J., Bramley, P.M., Pennings, H. and Bowler, C. (2005) Fruit-specific RNAi-mediated suppression of DET1 enhances carotenoid and flavonoid content in tomatoes. *Nature Biotechnol.*, **23**, 890–895.

de Vos, R., Bovy, A., Busink, H., Muir, S., Collins, G. and Verhoeyen, M. (2000) Improving health potential of crop plants by means of flavonoid pathway engineering. *Polyphenols Commun.*, **1**, 25–26.

Dewick, P.M. (1994) Isoflavonoids, in *Flavonoid: Advances in Research Since 1986* (ed. J.B. Harborne), Chapman and Hall, London, pp. 117–238.

Dewick, P.M., Steele, M.J., Dixon, R.A. and Whitehead, I.M. (1982) Biosynthesis of isoflavonoid phytoalexins: incorporation of sodium [1,2-$^{13}C_2$]-acetate into phaseollin and kievitone. *Z. Naturforsch.*, **37C**, 363–368.

Dixon, R.A. (1986) The phytoalexin response: eliciting, signalling and control of host gene expression. *Biol. Rev.*, **61**, 239–291.

Dixon, R.A. and Paiva, N.L. (1995) Stress-induced phenylpropanoid metabolism. *Plant Cell*, **7**, 1085–1097.

Dixon, R.A. and Steele, C.L. (1999) Flavonoids and isoflavonoids – a gold mine for metabolic engineering. *Trends Plant Sci.*, **4**, 394–400.

Dixon, R.A. and Ferreira, D. (2002) Genistein. *Phytochemistry*, **60**, 205–211.

Dixon, R.A., Gang, D.R. and Adrian, J. (2006) Applications of metabolomics in agriculture. *J. Agric. Food Chem.*, **54**, 8984–8994.

Dooner, H.K., Robbins, T P. and Jorgensen, R.A. (1991) Genetic and developmental control of anthocyanin biosynthesis. *Ann. Rev. Gen.*, **25**, 173–199.

Dunn, W.B. and Ellis, D.I. (2005) Metabolomics: current analytical platforms and methodologies. *Trends Anal. Chem.*, **24**, 285–294.

Duran, A.L., Yang, J., Wang, L., and Sumner, L.W. (2003) Metabolomics spectral formatting, alignment and conversion tools (MSFACTs). *Bioinformatics*, **19**, 2283–2293.

Ebel, J. and Hahlbrock, K. (1982) Biosynthesis, in *The Flavonoids: Advances in Research* (eds J.B. Harborne and T.J. Malory), Chapman and Hall, London, pp. 641–679.

Fernie, A.R., Trethewey, R.N., Krotzky, A.J., and Willmitzer, L. (2004) Metabolite profiling: from diagnostics to systems biology. *Nat. Rev. Mol. Cell Biol.*, **5**, 763–769.

Fernie, A.R., Geigenberger, P. and Stitt, M. (2005) Flux an important, but neglected, component of functional genomics. *Curr. Opin. Plant Biol.*, **8**, 174–182.

Ferrer, J.L., Jez, J.M., Bowman, M.E., Bowman, M.E., Dixon, R.A., and Noel, J.P. (1999) Structure of chalcone synthase and the molecular basis of plant polyketide biosynthesis. *Nat. Struct. Biol.*, **6**, 775–784.

Forkmann, G. and Dangelmayr, B. (1980) Genetic control of chalcone isomerase activity in flowers of *Dianthus caryophyllus*. *Biochem. Genet.*, **18**, 519–527.

Forkmann, G. and Heller, W. (1999) Biosynthesis of flavonoids, in *Polyketides and Other Secondary Metabolites including Fatty Acids and their Derivatives; Comprehensive Natural Products Chemistry Vol. 1* (eds D. Barton, *et al.*), Elsevier, Amsterdam, pp. 713–748.

Frydman, A., Weisshaus, O., Bar-Peled, M., Huhman, D.V., Sumner, L.W., Marin, F.R., Lewinsohn, E., Fluhr, R., Gressel, J. and Eyal, Y. (2004) Citrus fruit bitter flavors: isolation and functional characterization of the gene Cm1,2RhaT encoding a 1,2 rhamnosyltransferase, a key enzyme in the biosynthesis of the bitter flavonoids of citrus. *Plant J.*, **40**, 88–100.

Fuendjiep, V., Wandji, J., Tillequin, F., Mulholland, D.A., Budzikiewicz, Z., Fomum, T., Nyemba, A.M. and Koch, N. (2002) Chalconoid and stilbenoid glycosides from *Guibourtia tessmanii* . *Phytochemistry*, **60**, 803–806.

Fujiwara, H., Tanaka, Y., Yonekura-Sakakibara, K., Fukuchi-Mizutani, M., Nakao, M., Fukui, Y., Yamaguchi, M., Ashikari, T. and Kusumi, T. (1998) cDNA cloning, gene expression and subcellular localization of anthocyanin 5-aromatic acyltransferase from Gentiana triflora. *Plant J.*, **16**, 421–431.

Fukuchi-Mizutani, M., Okuhara, H., Fukui, Y., Nakao, M., Katsumoto, Y., Yonekura-Sakakibara, K., Kusumi, T., Hase, T. and Tanaka, Y. (2003) Biochemical and molecular characterization of a novel UDP-glucose:anthocyanin 3′-O-glucosyltransferase, a key enzyme for blue anthocyanin biosynthesis, from gentian. *Plant Physiol.*, **132**, 1652–1663.

Fukui, Y., Tanaka, Y., Kusumi, T., Iwashita, T. and Nomoto, K. (2003) A rationale for the shift in colour towards blue in transgenic carnation flowers expressing the flavonoid 3′,5′-hydroxylase gene. *Phytochemistry*, **63**, 15–23.

Fukushima, A., Kusano, M., Redestig, H., Arita, M. and Saito, K. (2009) Integrated omics approaches in plant systems biology. *Curr. Opin. Chem. Biol.*, **13**, 532–538.

Gagnon, H. and Ibrahim, R.K. (1997) Effects of various elicitors on the accumulation and secretion of isoflavonoids in white lupin. *Phytochemistry*, **44**, 1463–1467.

Guo, L., Dixon, R.A. and Paiva, N.L. (1994) The 'pterocarpan synthase' of alfalfa: association and co-induction of vestitone reductase and 7,2′-dihydroxy-4′-methoxy-isoflavanol (DMI) dehydratase, the two final enzymes in medicarpin biosynthesis. *FEBS Lett.*, **356**, 221–225.

Hahlbrock, H. and Grisebach, H. (1975) Biosynthesis of flavonoids, in *The Flavonoids* (eds J.B. Harborne, T.J. Mabry and H. Mabry), Academic Press, San Diego, pp. 866–915.

Hain, R., Reif, H.J., Krause, E., Langebartels, R., Kindl, H., Vornam, B., Wiese, W., Schmelzer, E., Schreier, P.H., Stöcker, R.H. and Stenzel, K. (1993) Disease resistance results from foreign phytoalexin expression in a novel plant. *Nature*, **361**, 153–156.

Harborne, J.B. (1993) *The Flavonoids: Advances in Research Since 1986*, Chapman and Hall, London.

He, X.-Z. and Dixon, R.A. (2000) Genetic manipulation of isoflavone 7-*O*-methyltransferase enhances the biosynthesis of 48 *O*-methylated isoflavonoid phytoalexins and disease resistance in alfalfa. *Plant Cell*, **12**, 1689–1702.

Heller, W. and Forkmann, G. (1988) Biosynthesis, in *The Flavonoids* (ed. J.B. Harborne), Chapman and Hall, London, pp. 399–425.

Heller, W. and Forkmann, G. (1994) Biosynthesis of flavonoids, in *The Flavonoids* (ed. J.B. Harborne), Chapman and Hall, London, pp. 499–536.

Hipskind, J.D. and Paiva, N.L. (2000) Constitutive accumulation of a resveratrol-glucoside in transgenic alfalfa increases resistance to *Phoma medicaginis. Mol. Plant-Microbe Interact.*, **13**, 551–562.

Holton, T.A. and Cornish, E. (1995) Genetics and biochemistry of anthocyanin biosynthesis. *Plant Cell*, **7**, 1071–1083.

Horowitz, R.M. and Gentili, B. (1969) Taste and structure in phenolic glycosides. *J. Agric. Food Chem.*, **17**, 696–770.

Iwashina, T. and Kitajima, J. (2000) Chalcone and flavonol glycosides from *Asarum canadense* (Aristolochiaceae). *Phytochemistry*, **55**, 971–974.

Jayaprakasam, B., Gunasekar, D., Rao, K.V., Blond, A. and Bodo, B. (2001) Androechin, a new chalcone glucoside from *Andrographis echioides. J. Asian Nat. Prod. Res.*, **3**, 43–48.

Jenkins, H., Hardy, N., Beckmann, M., Draper, J., Smith, A.R., Taylor, J., Fiehn, O., Goodacre, R., Bino, R.J., Hall, R., Kopka, J., Lane, G.A., Lange, B.M., Liu, J.R., Mendes, P., Nikolau, B.J., Oliver, S.G., Paton, N.W., Rhee, S., Roessner-Tunali, U., Saito, K., Smedsgaard, J., Sumner, L.W., Wang, T., Walsh, S., Wurtele, E.S. and Kell, D.B. (2004) A proposed framework for the description of plant metabolomics experiments and their results. *Nat. Biotechnol.*, **2**, 1601–1606.

Jez, J.M., Bowman, M.E. and Noel, J.P. (2002) Expanding the biosynthetic repertoire of plant type III polyketide synthases by altering starter molecule specificity. *Proc. Nat. Acad. Sci. USA*, **99**, 5319–5324.

Jonsson, P., Gullberg, J., Nordström, A., Kusano, M., Kowalczyk, M., Sjöström, M. and Moritz, T. (2004) A strategy for identifying differences in large series of metabolomics samples analyzed GC/MS. *Anal. Chem.*, **76**, 1738–1745.

Jung, W.S., Chung, Ill-M. and Heo, H-Y. (2003) Manipulating isoflavone levels in plants. *J. Plant Biotechnol.*, **5**, 149–155.

Kanehisa, M. and Goto, S. (2000) KEGG: Kyoto Encyclopedia of Genes and Genomes. *Nucleic Acids Res.*, **28**, 27–30.

Kanehisa, M., Goto, S., Hattori, M., Aoki-Kinoshita, K.F., Itoh, M., Kawashima, S., Katayama, T., Araki, M., and Hirakawa, M. (2006) From genomics to chemical genomics: new developments in KEGG. *Nucleic Acids Res.*, **34**, D354–D357.

Kazuma, K., Kogawa, K., Noda, N., Kato, N. and Suzuki, M. (2004) Identification of delphinidin 3-*O*-(6″-*O*-malonyl)-β-glucoside-3′-*O*-β-glucoside, a postulated intermediate in the biosynthesis of ternatin C5 in the blue petals of Clitoria ternatea (butterfly pea). *Chem. Biodiversity*, **1**, 1762–1770.

Kerhoas, L., Aouak, D., Cingoz, A., Routaboul, J.M., Lepiniec, L., Einhorn, J. and Birlirakis, N. (2006) Structural characterization of the major flavonoid glycosides from *Arabidopsis thaliana* seeds. *J. Agric. Food Chem.*, **54**, 6603–6612.

Kim, J.J., Kim, S.H., Hahn, S.J. and Chung, M. (2005) Changing soybean isoflavone composition and concentrations under two different storage conditions over three years. *Food Res. Internat.*, **38**, 435–444.

Kim, J.H., Kim, B.G., Park, Y. Ko, J.H., Lim, C.E., Lim, J., Lim, Y. and Ahn, J. (2006a) Characterization of flavonoid-7-*O*-glucosyltransferase from *Arabidopsis thaliana. Biosci. Biotechnol. Biochem.*, **70**, 1471–1477.

Kim, J.H., Kim, B.G., Ko, J.H., Lee, Y., Hur, H., Lim, Y., and Ahn, J. (2006b) Molecular cloning, expression, and characterization of a flavonoid glycosyltransferase from *Arabidopsis thaliana*. *Plant Sci.*, **170**, 897–903.

Kitada, C., Gong, Z., Tanaka, Y., Yamazaki, M. and Saito, K. (2001) Differential expression of two cytochrome P450s involved in the biosynthesis of flavones and anthocyanins in chemo-varietal forms of *Perilla frutescens*. *Plant Cell Physiol.*, **42**, 1338–1344.

Kitamura, S. (2006) Transport of flavonoids, in *The Science of Flavonoids* (ed. E. Grotewold), Springer, Berlin, pp. 123–146.

Kochs, G. and Grisebach, H. (1986) Enzymic synthesis of isoflavones. *Eur. J. Biochem.*, **155**, 311–318.

Koes, R.E., Quattrocchio, F. and Mol, J.N.M. (1994) The flavonoid biosynthetic pathway in plants: function and evolution. *BioEssays*, **16**, 123–132.

Kopka, J., Schauer, N., Krueger, S., Birkemeyer, C., Usadel, B., Bergmüller, E., Dörmann, P., Weckwerth, W., Gibon, Y., Stitt, M., Willmitzer, Fernie, A.R. and Steinhauser D. (2005) GMD@CSB.DB: the Golm metabolome database. *Bioinformatics*, **21**, 1635–1638.

Kose, F., Weckwerth, W., Linke, T. and Fiehn, O. (2001) Visualizing plant metabolomics correlation networks using clique-metabolite matrices. *Bioinformatics*, **17**, 1198–1208.

Kreuzaler, F., Light, R.J. and Hahlbrock, K. (1978) Flavanone synthase catalyzes carbon dioxide exchange and decarboxylation of malonyl-CoA. *FEBS Lett.*, **94**, 175–178.

Le Gall, G., Colquhoun, I.J., Davis, A.L., Collins, G.J. and Verhoeyen, M.E. (2003) Metabolite profiling of tomato (*Lycopersicon esculentum*) using H-1 NMR spectroscopy as a tool to detect potential unintended effects following a genetic modification. *J. Agric. Food Chem.*, **51**, 2447–2456.

Lewinsohn, E., Berman, E., Mazur, Y. and Gressel, J. (1986) Glucosylation of exogenous flavanones by grapefruit (*Citrus paradisi*) cell cultures. *Phytochemistry*, **25**, 2531–2535.

Lewinsohn, E., Britsch, L., Mazur, Y. and Gressel, J. (1989) Flavanone glycoside biosynthesis in citrus chalcone synthase, UDP-glucose:flavanone-7-*O*-glucosyl-transferase and -rhamnosyl-transferase activities in cell-free extracts. *Plant Physiol.* **91**, 1323–1328.

Liu, R., Hu, Y., Li, J., *et al.* (2007) Production of soybean isoflavone genistein in non-legume plants via genetically modified secondary metabolism pathway. *Metabol. Engineering*, **9**, 1–7.

Mack, G.S. (2004) Can complexity be commercialized? *Nature Biotech.*, **22**, 1223–1229.

Martens, S., Forkmann, G., Matern, U. and Lukacin, R. (2001) Cloning of parsley *flavone synthase I*. *Phytochemistry*, **58**, 43–46.

Martens, S. and Mithofer, A. (2005) Flavones and flavone synthases. *Phytochemistry*, **66**, 2399–2407.

Mehrotra, B. and Mendes, P. (2006) Bioinformatics approaches to integrate metabolomics and other systems biology data, in *Biotechnology in Agriculture and Forestry, Vol. 57: Plant Metabolomics* (eds K. Saito, R.A. Dixon and L. Willmitzer), Springer, Berlin, pp. 105–115.

Memelink, J. (2005) The use of genetics to dissect plant secondary pathways. *Curr. Opin. Plant Biol.*, **8**, 230–235.

Middleton, Jr E. and Kandaswami, C. (1994) The impact of plant flavonoids on mammalian biology: implications for immunity, inflammation and cancer, in *The Flavonoids: Advances in Research Since 1986* (ed. J.B. Harborne), Chapman and Hall, London, pp. 619–652.

Mol, J., Grotewald, E. and Koes, R. (1998) How genes paint flowers and seeds. *Trends Plant Sci.*, **3**, 212–217.

Morgenthal, K., Weckwerth, W. and Steuer, R. (2006) Metabolomic networks in plants: transitions from pattern recognition to biological interpretation. *Biosystems*, **83**, 108–117.

Muir, S R., Collins, G.J., Robinson, S., Hughes, S., Bovy, A., De Vos, C.H.R., van Tunen, A.J. and Verhoeyen, M.E. (2001) Overexpression of petunia *chalcone isomerase* in tomato results in fruit containing increased levels of flavonols. *Nature Biotechnol.*, **19**, 470–474.

Obayashi, T., Kinoshita, K., Nakai, K., Shibaoka, M., Hayashi, S., Saeki, M., Shibata, D., Saito, K. and Ohta, H. (2007) ATTED-II: a database of co-expressed genes and cis elements for identifying co-regulated gene groups in *Arabidopsis*. *Nucleic Acids Res.*, **35**, D863–D869.

Ogata, J., Itoh, Y., Ishida, M., Yoshida, H. and Ozeki, Y. (2004) Cloning and heterologous expression of cDNAs encoding flavonoid glucosyltransferases from *Dianthus caryophyllus*. *Plant Biotech.*, **21**, 367–375.

Ogata, J., Kanno, Y., Itoh, Y., Tsugawa, H. and Suzuki, M. (2005) Plant biochemistry: anthocyanin biosynthesis in roses. *Nature*, **435**, 757–758.

Oksman-Caldentey, K-M. and Saito, K. (2005) Integrating genomics and metabolomics for engineering plant metabolic pathways. *Curr. Opin. Biotech.*, **16**, 174–179.

O'Neill, M.J., Adesanya, S.A. and Roberts, M.F. (1983) Antifungal phytoalexins in *Phaseolus aureus* Roxb. *Z. Naturforsch.*, **38c**, 693–697.

Pierpoint, W.S. (2000) Why do plants make medicines. *Biochemist*, **22**, 37–40.

Ratcliffe, R.G. and Shacher-Hill, Y. (2005) Revealing metabolic phenotypes in plants: inputs from NMR analysis. *Biol. Rev. Camb. Philos.*, **80**, 27–43.

Routaboul, J.M., Kerhoas, L., Debeaujon, I., Pourcel, L., Caboche, M., Einhorn, J. and Lepiniec, L. (2006) Flavonoid diversity and biosynthesis in seed of *Arabidopsis thaliana*. *Planta*, **224**, 96–107.

Roy, R. and Pandey, V.B. (1994) A chalcone glycoside from *Clerodendron phlomidis*. *Phytochemistry*, **37**, 1775–1776.

Saito, N., Toki, K., Moriyama, H., Shigihara, A. and Honda, T. (2002) Acylated anthocyanins from the blue-violet flowers of *Anemone coronaria*. *Phytochemistry*, **60**, 365–373.

Saito, N., Toki, K., Morita, Y., Hoshino, A., Iida, S., Shigihara, A. and Honda, T. (2005) Acylated peonidin glycosides from duskish mutant flowers of *Ipomoea nil*. *Phytochemistry*, **66**, 1852–1860.

Saito, K., Hirai, Y.M. and Yonekura-Sakakibara, K. (2007) Decoding genes with coexpression networks and metabolomics – 'majority report by precogs'. *Trends Plant Sci.*, **13**, 36–43.

Schmelzer, E., Jahnen, W. and Hahlbrock, K. (1988) *In situ* localization of light-induced chalcone synthase mRNA, chalcone synthase, and flavonoid end products in epidermal cells of parsley leaves. *Proc. Natl. Acad. Sci. USA*, **85**, 2989–2993.

Schulz, M. and Weissenb, G. (1988) Three specific UDP-glucuronate-flavone-glucuronosyl-transferases from primary leaves of *Secale cereale*. *Phytochemistry*, **27**, 1261–1267.

Shimada, M., Ohbayashi, R., Nakano-Shimada, Y., *et al.* (2001) Genetic engineering of the anthocyanin biosynthetic pathway with flavonoid-3′,5′-hydroxylase: specific switching of the pathway in petunia. *Plant Cell Rep.*, **20**, 456–462.

Shimada, N., Aoki, T., Sato, S., Hoshino, A., Iida, S., Shigihara, A. and Honda, T. (2003) A cluster of genes encodes the two types of chalcone isomerase involved in the biosynthesis of general flavonoids and legume-specific 5-deoxy(iso)flavonoids in *Lotus japonicus*. Plant Physiol., **131**, 941–951.

Shinozaki, K. and Sakakibara, H. (2009) Omics and bioinformatics: an essential toolbox for systems analyses of plant functions beyond 2010. *Plant Cell Physiol.*, **50**, 1177–1180.

Springob, K., Nakajima, J.-I., Yamazaki, M., and Saito, K. (2003) Recent advances in the biosynthesis and accumulation of anthocyanins. *Nat. Prod. Rep.*, **20**, 288–303.

Suzuki, H., Nakayama, T., Yonekura-Sakakibara, K., Fukui, Y., Nakamura, N., Yamaguchi, M., Tanaka, Y., Kusumi, T. and Nishino, T. (2002) cDNA cloning, heterologous expressions, and functional characterization of malonyl-coenzyme A:anthocyanidin 3-*O*-glucoside-6″-*O*-malonyltransferase from dahlia flowers. *Plant Physiol.*, **130**, 2142–2151.

Sweetlove, L.J., Last, R.L. and Fernie, A.R. (2003) Predictive metabolic engineering: a goal for systems biology. *Plant Physiol.*, **132**, 420–425.

Tanner, G.J., Francki, K.T., Abrahams, S., Watson, J.M., Larkin, J.P. and Ashton, A.R. (2003) Proanthocyanidin biosynthesis in plants: purification of legume leucoanthocyanidin reductase and molecular cloning of its cDNA. *J. Biol. Chem.*, **278**, 31647–31656.

Terahara, N., Callebaut, A., Ohba, R., Nagata, T., Ohnishi-Kameyama, M. and Suzuki, M. (2001) Acylated anthocyanidin 3-sophoroside-5-glucosides from *Ajuga reptans* flowers and the corresponding cell cultures. *Phytochemistry*, **58**, 493–500.

Terahara, N., Konczak, I., Ono, H., Yoshimoto, M. and Yamakawa, O. (2004) Characterization of acylated anthocyanins in callus induced from storage root of purple-fleshed sweet potato, *Ipomoea batatas* L. *J. Biomed. Biotechnol.*, **5**, 279–286.

Thorneycroft, D., Sherson, S.M. and Smith, S.M. (2001) Using gene knockouts to investigate plant metabolism. *J. Expt. Bot.*, **52**, 1593–1601.

Tohge, T., Nishiyama, Y., Hirai, M.Y., Yano, M., Nakajima, J., Awazuhara, M., Inoue, E., Yamazaki, M. and Saito, K. (2005) Functional genomics by integrated analysis of metabolome and transcriptome of Arabidopsis plants over-expressing an MYB transcription factor. *Plant J.*, **42**, 218–235.

Tohge, T., Yonekura-Sakakibara, K., Niida, R., Watanabe-Takahashi, A. and Saito, K. (2007) Phytochemical genomics in *Arabidopsis thaliana*: a case study for functional identification of flavonoid biosynthesis genes. *Pure Appl. Chem.*, **78**, 811–823.

Tokimatsu, N., Sakurai, H., Suzuki, H., Ohta, H., Nishitani, K., Koyama, T., Umezawa, T., Misawa, N., Saito, K. and Shibata, D. (2005) KaPPA-view: a web-based analysis tool for integration of transcript and metabolite data on plant metabolic pathway maps. *Plant Physiol.*, **138**, 289–1300.

Ueyama, Y., Fukuchi-Mizutani, M., Fukui, Y., Miyazaki, K., Ohkawa, H., Kusumi, T. and Tanaka, Y. (2002) Molecular and biochemical characterization of torenia flavonoid 3′-hydroxylase and flavone synthase II and modification of flower color by modulating the expression of these genes. *Plant Sci.*, **163**, 253–263.

Ursem, R., Tikunov, Y., Bovy, A., van Berloo, R. and van Eeuwijk, F. (2008) A correlation network approach to metabolic data analysis for tomato fruits. *Euphytica*, **161**, 181–193.

Welford, R.W.D., Turnbull, J.J., Claridge, T.D.W., Prescott, A.G. and Schofield, C.J. (2001) Evidence for oxidation at C-3 of the flavonoid C-ring during anthocyanin biosynthesis. *Chem. Commun.*, 1828–1829.

Welle, R. and Grisebach, H. (1988) Isolation of a novel NADPH-dependent reductase which coacts with chalcone synthase in the biosynthesis of 6′-deoxychalcone. *FEBS Lett.*, **236**, 221–225.

Winkel-Shirley, B. (2001) Flavonoid biosynthesis. A colorful model for genetics, biochemistry, cell biology, and biotechnology. *Plant Physiol.*, **126**, 485–493.

Wollenweber, E. (1994) Flavones and flavonols, in *The Flavonoids: Advances in Research Since 1986* (ed. J.B. Harborne), Chapman and Hall, London, pp. 259–336.

Xie, D.Y., Sharma, S.B., Paiva, N.L., Ferreira, D. and Dixon, R.A. (2003) Role of anthocyanidin reductase, encoded by *BANYULS* in plant flavonoid biosynthesis. *Science*, **299**, 396–399.

Xie, D.-Y. and Dixon, R.A. (2005) Proanthocyanidin biosynthesis – still more questions than answers? *Phytochemistry*, **66**, 2127–2144.

Yamazaki, M., Gong, Z., Fukuchi-Mizutani, M., Fukui, Y., Tanaka, Y., Kusumi, T. and Saito, K. (1999) Molecular cloning and biochemical characterization of a novel anthocyanin 5-*O*-glucosyltransferase by mRNA differential display for plant forms regarding anthocyanins. *J. Biol. Chem.*, **274**, 7405–7411.

Yamazaki, M., Yamagishi, E., Gong, Z., Fukuchi-Mizutani, M., Fukui, Y., Tanaka, Y., Kusumi, T., Yamaguchi, M., and Saito, K. (2002) Two flavonoid glucosyltransferases from *Petunia hybrida*: molecular cloning, biochemical properties and developmentally regulated expression. *Plant Mol. Biol.*, **48**, 401–411.

Yazaki, K., Sasaki, K. and Tsurumaru, Y. (2009) Prenylation of aromatic compounds, a key diversification of plant secondary metabolites. *Phytochemistry*, **70**, 1739–1745.

Yonekura-Sakakibara, K., Tanaka, Y., Fukuchi-Mizutani, M., Fujiwara, H., Fukui, Y., Ashikari, T., Murakami, Y., Yamaguchi, M. and Kusumi, M. (2000) Molecular and biochemical characterization of a novel hydroxycinnamoyl-CoA: anthocyanin 3-*O*-glucoside-6″-*O*-acyltransferase from *Perilla frutescens*. *Plant Cell Physiol.*, **41**, 495–502.

Yonekura-Sakakibara, K., Tohge, T., Matsuda, F., Nakabayashi, R., Takayama, H., Niida, R., Watanabe-Takahashi, A., Inoue, E. and Saito, K. (2008) Comprehensive flavonol profiling and transcriptome coexpression analysis leading to decoding gene-metabolite correlations in *Arabidopsis*. *Plant Cell*, **20**, 2160–2176.

Yu, O., Matsuno, M. and Subramanian, S. (2006) Flavonoid compounds in flowers: genetics and biochemistry, in *Floriculture, Ornamental and Plant Biotechnology Advances and Topical Issues* (ed. J.A. Teixeira da Silva), Global Science Books, London, pp. 282–292.

Zhang, P., Foerster, H., Tissier, C.P., Mueller, L., Paley, S., Karp, P.D. and. Rhee, S.Y. (2005) MetaCyc and AraCyc: metabolic pathway databases for plant research. *Plant Physiol.*, **138**, 27–37.

12

Pigment Biosynthesis I. Anthocyanins

Yoshihiro Ozeki, Yuki Matsuba, Yutaka Abe, Naoyuki Umemoto and Nobuhiro Sasaki

12.1 Introduction

Plants display a variety of colours, especially their flower petals, fruit skins and autumn leaves. Petals are the most colourful parts, and can be orange, red, purple, blue, green or yellow. These colours result from three major plant pigments, anthocyanins/flavonoids, carotenoids/chlorophylls and betalains (Grotewold, 2006; Tanaka *et al.*, 2008). Chlorophylls are necessary for photosynthesis and are found mainly in the leaves and stems. They are degraded in most flowers and fruits during development, whereas anthocyanins/flavonoids, as well as carotenoids and betalains (which are discussed in Chapter 13), are synthesized so that they accumulate, giving a wide diversity of colours.

Anthocyanins and flavonoids are widely distributed in higher plants. Both have a C_6-C_3-C_6 backbone structure. Some flavonoids are important in protecting against environmental stresses such as ultraviolet light and pathogen attack (Treutter, 2006). While some flavonoids are colourless, others, such as those found in yellow carnation flower petals, can be pale or vivid yellow. There are six main anthocyanidin aglycones, namely pelargonidin, cyanidin, delphinidin, peonidin, petunidin and malvidin, which form conjugates with sugars and organic acids to generate thousands of anthocyanins of differing colours, ranging from orange, red and purple to blue. Synthesis of anthocyanidins is one of the main routes in the flavonoid biosynthesis pathway. The genes encoding most of the enzymes involved in flavonoid and anthocyanidin biosynthesis have now been cloned. Some of these genes have been targeted biotechnologically in order to modify flower colour and generate blue carnations (*Dianthus caryophyllus*) (Fukui *et al.*, 2003) and blue roses (*Rosa* spp.) (Katsumoto *et al.*, 2007). The *in vivo* colours of anthocyanins are influenced by the situation in which

Plant Metabolism and Biotechnology, First Edition. Edited by Hiroshi Ashihara, Alan Crozier, and Atsushi Komamine.

they accumulate, and vary with factors such as pH and metal ions, as well as the presence of other flavonoids that act as copigments in vacuoles (Yoshida *et al.*, 2009). For instance, vacuolar pH affects the colour of the 'Heavenly Blue' variety of morning glory (*Pharbitis nil*). Before the flowers open, the vacuolar pH is 6.6, giving rise to purple-red coloured buds, but following opening the pH increases to 7.7 and the petals become blue (Yoshida *et al.*, 2009). The rise in vacuolar pH in the 'Heavenly Blue' morning glory is caused by the Na^+/H^+ antiporter (Fukada-Tanaka *et al.*, 2000). In the petunia (*Petunia x hybrida*), H^+P-ATPase at the tonoplast determines the vacuolar pH and flower petal colour (Verweij *et al.*, 2008). Metals also affect flower colour. Expression of the iron ion transporter TgVit1 increases the Fe^{+++} content of vacuoles, turning the petals sky blue (Shoji *et al.*, 2007; Momonoi *et al.*, 2009). Downregulation of expression of the gene that encodes the protein ferritin, which complexes with Fe^{+++}, leads to an increase in the free Fe^{+++} content of petals, and this in turn induces a change in colour from sky blue to marine blue (Shoji *et al.*, 2010). Co-pigmentation of flavonoids is important for the blue colour of petals of *Commelina communis* flowers (Kondo *et al.*, 1992), and varieties with an elevated flavonol content were chosen for the development of transgenic, blue flowering roses (Katsumoto *et al.*, 2007). Thus, in addition to the genes involved in anthocyanin biosynthesis, genes that affect pH, metal ions and the accumulation of co-pigmenting compounds in vacuoles which accumulate anthocyanins are also important in modifying flower colour.

The colours of flowers and fruits (kernels) are recognised visually, so that it is easy to detect pigment synthesis mutants. As a result, colours have been used as genetic markers for more than 100 years. Genetic loci have been identified for anthocyanin synthesis in maize (*Zea mays*), petunia, snapdragon (*Antirrhinum majus*), morning glory and carnation. Some studies indicate that mutable and variegated phenotypes are caused by movement of transposable elements (Grotewold, 2006). Modern molecular biology has revealed the molecular characteristics of transposable elements which are now commonly used as natural tag sequences or as molecular markers for genes involved in anthocyanin synthesis.

This chapter details those anthocyanin biosynthetic pathways which are related principally to flower colour. The anthocyanin biosynthetic pathway up to flavanone 3-hydroxylase (F3H) has been discussed in detail in Chapter 11. The present chapter will focus on enzymes in the flavonoid biosynthesis pathway post-F3H, namely flavonoid 3′-hydroxylase (F3′H) and flavonoid 3′,5′-hydroxylase (F3′5′H).

12.2 The Anthocyanin Biosynthetic Pathway

The flavonoid synthetic pathway begins with 4-coumaroyl-CoA, as the compound to which three malonyl-CoAs are sequentially condensed and decarboxylated by chalcone synthase (CHS) so as to form the C_6-C_3-C_6 compound, naringenin-chalcone. This chalcone is isomerised by chalcone isomerase (CHI), and the C-ring is reduced by F3H to form dihydrokaempferol (see Figure 12.1). The B-ring is then hydroxylated by F3′H to form dihydroquercetin or by F3′5′H to form dihydromyricetin. These are the starting materials of anthocyanin biosynthesis, and the subsequent steps are catalyzed by dihydroflavonol reductase (DFR) and anthocyanidin synthase (ANS) (also known as leucoanthocyanidin dioxygenase), yielding anthocyanin aglycones. Catechin is also synthesized from

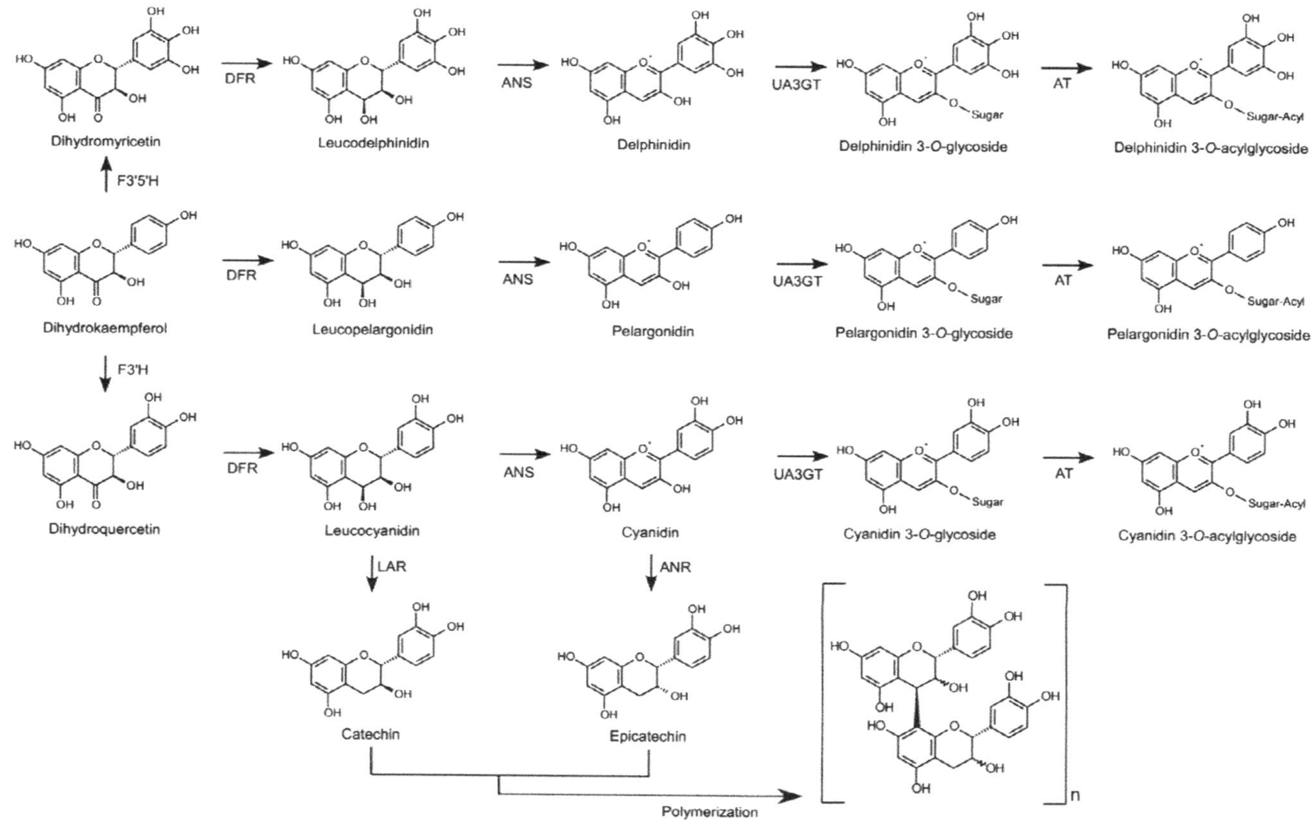

Figure 12.1 *Anthocyanin and proanthocyanidin biosynthetic pathways. Abbreviations: F3′H, flavonoid 3′-hydroxylase; F3′5′H, flavonoid 3′,5′-hydroxylase; DFR, dihydroflavonol reductase; ANS, anthocyanidin synthase (also known as leucoanthocyanidin dioxygenase); LAR, leucoanthocyanidin reductase; ANR, anthocyanidin reductase; UA3GT, UDP-glycose-dependent anthocyanidin 3-O-glycosyltransferases; AT, acyltransferase. Anthocyanins are further modified by glycosylation at the 5, 7 and 3′ positions in some plant species; these glycosylation steps are not shown in this schematic pathway*

leucocyanidin by leucoanthocyanidin reductase (LAR). Epicatechin is synthesized from cyanidin by anthocyanidin reductase (ANR). Catechin and epicatechin molecules are polymerised to form proanthocyanidins. DFR, ANS and ANR were first identified as a result of studies by geneticists and molecular biologists, and not from biochemical evidence.

DFR activity was first detected in crude extracts prepared from suspension-cultured cells of Douglas fir (*Pseudotsuga menziesii*), a species which synthesizes leucocyanidin from dihydroquercetin using NADPH as a reductant (Stafford and Lester, 1982). The gene encoding *DFR* was identified as the responsible nucleotide sequence of the mutable allele, *pallida*^{recurrens}, in snapdragon, by *Tam3* transposon tagging (Martin *et al.*, 1985). By using this nucleotide sequence as a probe, *DFR* homologues have been isolated from many plant species, and their enzyme properties have been studied using recombinant DFR proteins.

Differing isoforms of DFR determine the content and ratios of pelargonidin, cyanidin and delphinidin. Blue roses have been produced in two ways: by introducing the *F3′5′H* gene and by a change in DFR. Rose DFR prefers dihydroquercetin to dihydromyricetin; it follows that dihydrokaempferol is converted to leucocyanidin more rapidly than leucodelphinidin (see Figure 12.1). Cyanidin is therefore produced from dihydroquercetin more readily than delphinidin from dihydromyricetin. This is evident even in transgenic roses containing the *F3′5′H* gene. To produce blue roses it was necessary to eliminate rose DFR, and then to introduce a *DFR* gene derived from a plant with a high preference for dihydromyricetin, so as to promote the metabolic flux from dihydromyricetin to delphinidin. The RNAi construct to downregulate the expression of the rose *DFR* gene, and the construct to overexpress the *Iris* × *hollandica DFR* gene, which has a high preference for dihydromyricetin, were both introduced together with the *Viola F3′5′H* gene (Katsumoto *et al.*, 2007). The anthocyanin profile of these transgenic roses comprised 98% delphinidin-*O*-glycosides.

Unlike DFR, ANS does not exhibit any strong substrate preference. Since no ANS enzyme activity was detected, a biochemical approach was not feasible; the *ANS* gene was first identified in the maize *A2* gene by transposon tagging (Messen *et al.*, 1990). Almost ten years were required for biochemical characterisation of the activity of the ANS enzyme activity, using a recombinant protein, which was shown to be a 2-oxoglutarate-dependent oxygenase (Saito *et al.*, 1999).

Although the flavonoid/anthocyanin biosynthetic pathway was almost completely understood by the end of the twentieth century, progress in elucidating the proanthocyanidin biosynthetic pathway was slower (Dixon *et al.*, 2005; Xie and Dixon, 2005). Proanthocyanidins are highly polymerised, water-insoluble, and are responsible for the brown colour of seed coats. The less polymerised proanthocyanidins have a less intense colour, are more water-soluble, and are responsible for the astringent taste of wines and some fruit juices. Characterisation of the chemical structure of proanthocyanidins using nuclear magnetic resonance is difficult and, as a consequence, there are few examples, one of which is the structural elucidation of a pentamer from apple juice (Abe *et al.*, 2008a). Mutant genes encoding enzymes involved in proanthocyanidin biosynthesis, and in transcriptional regulatory factors and transporters, give rise to colourless seed coats, a readily visible phenotype that has led to the discovery of *Transparent Testa* (*TT*), *Transparent Testa Glabra* (*TTG*), *tannin deficient seed* (*tds*) and *BANYULS* mutants of *Arabidopsis*. Monomeric building blocks of proanthocyanidins are the flavan-3-ol catechin and/or epicatechin monomers, although epiafzelichin and afzelichin subunits do occur less commonly. The enzyme catalyzing synthesis of the catechin monomer from leucocyanidin, LAR, was purified from

Desmodium uncinatum (Spanish tick-clover), and the encoding cDNA was isolated by Tanner *et al.* (2003). This enzyme belongs to the reductase–epimerase–dehydrogenase (RED) protein family, because of its NADPH-dependent reduction. The gene involved in epicatechin synthesis from cyanidin, *ANR*, was identified as the *BAN* gene of *Arabidopsis* (*AtANR*) (Devic *et al.*, 1999), and was subsequently isolated from *Medicago truncatula* (Barrel clover) (*MtANR*) (Xie *et al.*, 2003). Unlike the recombinant AtANR protein prepared from *Escherichia coli*, which specifically uses NADPH as a reductant, recombinant MtANR can use both NADPH and NADH, although NADPH is preferred (Xie *et al.*, 2004). After synthesis of the monomer, the detailed reaction mechanisms for polymerisation remain unclear. Recent studies of transporters (see Section 12.5) suggest that monomers are first glucosylated (Pang *et al.*, 2008) and are then transported into the vacuoles, where the polymerisation reaction takes place (Zhao and Dixon, 2009). The enzyme(s) involved in the condensation reactions have not yet been identified, however.

12.3 Glycosylation of Anthocyanidins

Anthocyanidins are hydrophobic and lipophilic, and can therefore permeate cellular and vacuolar membranes. Conjugation of the aglycones with sugars and organic acids may cause the molecules to become more hydrophilic, so that they cannot permeate membranes and become sequestered and compartmented in vacuoles. Modifications to anthocyanins can affect colours by changing absorbance, stability and solubility. Almost all anthocyanins in plants have a C3 sugar, and the conjugation is catalyzed by UDP-sugar-dependent anthocyanidin-3-*O*-glycosyltransferases (UA3GT). Defects in UA3GT activity lead, in some cases, to inhibition of anthocyanin synthesis. For example, anthocyanin-deficient grape varieties, such as Muscat, have a reduced expression of UA3GT (Kobayashi *et al.*, 2004). Roses are an exception as they do not have UA3GT, but do have a UDP-glucose-dependent glycosyltransferase (UGT) which transfers glucose to the C5 position and then to the C3 position (Ogata *et al.*, 2005).

The UA3GT gene was first identified in maize (Furtek *et al.*, 1988). It has since been found that plants have over 100 UGT genes. UGT genes comprise a super family in the plant genome. In the small genome of *Arabidopsis* there are 107 UGT genes (Yonekura-Sakakibara, 2009). The typical character of UGTs is a consensus amino-acid sequence, which defines the Plant Secondary Product Glycosyltransferases (PSPG)-box (Figure 12.2). X-ray crystallography analysis of UGTs has confirmed that the PSPG-box region is responsible for the sugar donor-binding domain (Shao *et al.*, 2005; Offen *et al.*, 2006). For UGT of *Aralia cordata* (the Japanese spikenard), the point mutation of one amino-acid sequence in the PSPG-box leads to an altered substrate specificity from UDP-galactose to UDP-glucose (Kubo *et al.*, 2004).

Conservation of the PSPG-box sequence in UGT facilitates the design of PCR primers for amplifying UGT cDNAs in plants across all species. cDNA fragments derived from these primers give nucleotide sequences followed by 5′- and 3′- rapid amplification of cDNA ends (RACE) so as to yield full-length UGT cDNAs. The PCR strategy for isolating glucuronic acid transferase for anthocyanins/flavonoids using degenerated primers corresponding to the PSPG-box was not successful, however. The first identification of an UDP-glucuronic acid:anthocyanin glucuronosyltransferase was made from the red daisy (*Bellis perennis*).

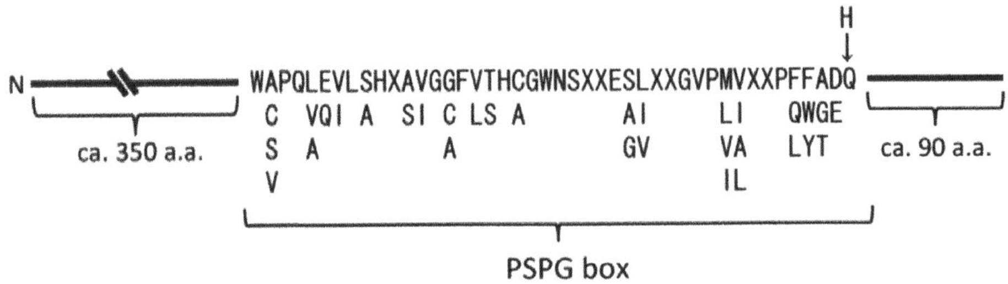

Figure 12.2 *Typical amino-acid sequences for the Plant Secondary Product Glycosyltransferases (PSPG)-box. An alteration from the glutamine (Q) residue to the histidine (H) residue resulted in the change of UDP-glucose to UDP-galactose as a donor molecule (Kubo* et al., *2004)*

The enzyme was purified, the partial amino-acid sequences were determined, and the primers corresponding to these sequences used to isolate the cDNA (Sawada *et al.*, 2005). Subsequent experiments with UDP-glucuronic acid:flavonoid glucuronosyltransferases in Lamiales found that changes to the amino-acid sequence in the PSPG-box (Noguchi *et al.*, 2009) led to failure of the PCR strategy based on the previously identified consensus sequence of the PSPG-box.

Recombinant proteins produced by the *E. coli* expression system can be produced using cDNAs, enabling their *in vitro* enzyme activity to be investigated and the substrate preference determined. Recent enzyme characterisation of recombinant UGT proteins found that the substrate specificity of donors (UDP-glucose, UDP-galactose, UDP-rhamnose, etc.) is strict, whereas that of the acceptor molecules is broad. In some cases, the *in vitro* enzyme reaction of recombinant UGT can transfer sugar to a wide range of molecular species. For example, carnation F3GT *in vitro* can transfer a glucose moiety to other flavonoids as well as anthocyanins (Ogata *et al.*, 2004). This implies that *in vitro* substrate specificity does not directly reflect reactions *in vivo*.

The entire genome sequence of several plant species has recently been determined. For example, in *Arabidopsis*, rice (*Oryza sativa*) and the grape (*Vitis vinifera*), all *UGT* candidate genes have been identified (Yonekura-Sakakibara, 2009). The expression profile of these genes can also be identified by transcriptome analysis, using the nucleotide sequence information whereby DNA arrays can be developed. Recombinant technology can be utilised *in vitro* to determine the enzyme substrate specificities encoding the *UGT* genes. This may not be the case *in vivo*, however, as the substrate specificity of the enzyme encoding *UGT* genes may be different, since they now take part in the catalytic steps of the metabolic pathways described above. Transcriptomic and metabolomic data and transcriptome coexpression analysis are powerful technologies that are currently being used to identify particular UGTs in specific metabolic pathways. UGT73C6 and UGT78D1 in *Arabidopsis* were both predicted to be related to the catalysis of flavonoids. Through the use of T-DNA insertional mutant lines of *UGT* genes, gene substrate specificity can be predicted from the accumulated flavonoid products. As an example, UGT73C6 and UGT78D1 were found to be flavonol UGTs (Jones *et al.*, 2003). T-DNA insertional mutant lines of *UGT* genes are available for *Arabidopsis* and rice, and are readily utilised to characterise the specific UGT that is under study. Where secondary metabolites other than those found in *Arabidopsis* and rice are involved, an alternative strategy is necessary, such as reverse genetics in transgenic plants harbouring RNAi constructs that downregulate the expression of candidate UGT genes. Analysis of the products that accumulate in transgenic plants then permits the characterisation of the types of UGTs that are involved in each biosynthetic pathway.

12.4 Acylation of Anthocyanin Glycosides

Acylation by aliphatic and aromatic moieties is another way to modify anthocyanin structure. Aliphatic moieties are found in China aster (*Callistephus chinensis*) (malonyl) and carnation petals (malyl). Aromatic moieties occur in gentian (*p*-coumaroyl) and delphinium (*p*-hydroxybenzoyl) petals, and in the suspension-cultured cells of carrot (*Daucus carota*) (sinapoyl) and beach silvertop (*Glehnia littoralis*) (feruloyl) (Matsuba *et al.*, 2008). Acylation is important in promoting the stability and solubility of anthocyanins in vacuoles (Abe *et al.*, 2008b). For aromatic moieties, intramolecular stacking, such as sandwiching of the anthocyanidin between two aromatic molecules (Yoshida *et al.*, 1992), improves stability even in weak alkaline solutions, and strengthens the blue colouration (Dangles *et al.*, 1993). The acylation reaction which transfers the acyl unit to the acceptor glycosyl moiety of the anthocyanin is catalyzed by anthocyanin acyltransferase (AAT). Flavonoid acyltransferase (AT) activity was first detected about 40 years ago using an aliphatic-CoA, malonyl-CoA:flavonoid-7-*O*-glycoside malonyltransferase, as a donor (Hahlbrock, 1972). The first identification of aliphatic AAT was in crude extracts prepared from petals of *Callistephus chinensis* (China aster) (Teusch and Forkmann, 1987). Aliphatic AAT was first purified from the petals of the scarlet sage (*Salvia splendens*), and its encoding cDNA isolated by Suzuki *et al.* (2001). Aromatic AAT was first demonstrated in a crude extract prepared from petals of *Silene*, using hydroxycinnamoyl-CoA as the donor molecule (Kamsteeg *et al.*, 1980). Aromatic AAT protein from gentian was the first to be purified to homogeneity (Fujiwara *et al.* 1997), and the encoding cDNA was isolated by Fujiwara *et al.*, (1998). Using acyl-CoA as a donor, cDNAs for ATs have been identified not only for anthocyanins but also for other secondary metabolites. Their amino-acid sequences are under study, and indicate that acyl-CoA-dependent ATs, including AAT, belong to the BAHD family. This family is named after benzylalcohol-*O*-acetyltransferase (BEAT), anthocyanin-hydroxycinnamoyltransferase (AHCT), anthranilate *N*-*O*-hydroxycinnamoyl/benzoyltransferase (HCBT) and deacetylvindoline-4-*O*-acetyltransferase (DAT) (Pierre and De Luca, 2000; Nakayama *et al.*, 2003).

In plants another type of AT, which uses acyl-glucoses as donor substrates, has recently been found. To this end, acyl-glucose-dependent anthocyanin acyltransferase (AGAAT) was purified, and the encoding cDNA was isolated from butterfly pea (*Clitoria ternatea*) (Noda *et al.*, 2006). The amino-acid sequence showed that this AGAAT contains serine carboxypeptidase-like (SCPL) proteins. SCPL proteins are located in vacuoles, suggesting that AGAAT occurs in vacuoles rather than the cytosol. The key question regarding the AGAAT reaction *in vivo* is how to synthesize the acyl-glucose and transport it to the vacuoles. Acyl-glucoses, such as hydroxycinnamic acid-glucoses (HCAGs), are synthesized by UDP-glucose dependent UGT in the cytosol. An example is sinapate glucosyltransferase from *Gomphrena globosa* (Chinese aster) (Matsuba *et al.*, 2008). This UGT also has a PSPG-box, for which the cDNA can be isolated by PCR using degenerated primers by means of the same strategy used for the isolation of flavonoid and anthocyanin UGT cDNAs. This implies that donor molecules of the AGAAT reaction are synthesized in the cytosol and are then transported into the vacuoles. It is known that vacuolar HCAG species differ between plant species and varieties. A supplement of HCAG molecules as a donor plays an important role in the AGAAT reaction *in vivo* by arranging the acyl moiety of anthocyanin. For the anthocyanin-synthesizing cultured cells of carrot and beach silvertop, which are closely related Umbelliferous species, the major anthocyanin molecules are respectively cyanidin-3-*O*-(2″-*O*-xylosyl-6″-*O*-sinapoylglucosyl) galactoside) (Harborne *et al.*, 1983) and cyanidin-3-(2″-*O*-xylosyl-6″-*O*-feruloyl-glucosyl)galactoside (Miura *et al.*, 1998). Consequently the only difference between the anthocyanins in the two plants is their synapoyl and feruloyl acyl groups; the aglycone and sugar moieties are the same. The substrate preference of AGAAT enzymes in crude extracts prepared from carrot and beach silvertop cultured cells was for feruloyl-glucose over other HCAGs. Analysis of HCAGs revealed that carrot cells contain large amounts of sinapoyl-glucose and silvertop cells contain high levels of feruloyl-glucose, coincident with the acyl moieties of the accumulated anthocyanins in the two species. These results suggest that the availability of donor molecules, rather than the substrate specificity of AGAAT, determines the type of acylated anthocyanins that accumulate (Matsuba *et al.*, 2008) (Figure 12.3).

Red and pink carnation petals synthesize and accumulate the aliphatic anthocyanin, pelargonidin-3,5-di-(6″-*O*-4, 6‴-*O*-1-cyclic malonyl)glucoside, and its cyanidin equivalent (Nakayama *et al.*, 2000). These unusual anthocyanins are water-soluble and are found in the vacuolar sap of cells of carnation petals (Figure 12.4, left). Some carnation varieties have petals with metallic colours as a result of the presence of insoluble anthocyanins which are condensed so as to make up anthocyanic vacuolar inclusions (AVIs) (Figure 12.4, right). Visible light entering the vacuoles undergoes diffuse reflection from the AVIs, producing the metallic colours (Figure 12.4, lower right). AVIs have been observed in many plant species, including carnation (*Dianthus caryophyllus*), rose, Texas bluebell (*Eustoma exaltatum* ssp. *russellianum*), grape (*Vitis vinifera*), sweet potato (*Ipomoea batatas*) and *Arabidopsis* (Nozue *et al.*, 1993; Markham *et al.*, 2000; Conn *et al.*, 2003; Poustka *et al.*, 2007; Pourcel *et al.*, 2010). Some AVIs are smooth and round, but others, as in carnation, have a rough surface. The mechanism of formation of AVIs is unknown, but among plant species different mechanisms might be responsible for the diverse shapes. For anthocyanin-synthesizing suspension-cultured cells of sweet potato (*Ipomoea batatas*), it was found that a protein, VP24, is likely to be central to the formation of AVIs (Nozue *et al.*, 1997).

Most carnation varieties that form AVIs in petal vacuoles accumulate anthocyanins lacking a malyl moiety (that is, pelargonidin- or cyanidin-3,5-*O*-diglucosides), suggesting that a malyl moiety is required to increase solubility in the vacuolar sap. Pelargonidin and cyanidin-3,5-*O*-diglucosides are soluble in the vacuolar sap of rose petals. They are only weakly soluble, however, in carnation vacuolar sap, in which they form AVIs (Figure 12.4).

Malyl-CoA is not available commercially, so that, although it is believed that aliphatic moieties are transferred from aliphatic acyl-CoA by AAT, it has not been possible to detect acyl-CoA dependent AAT activity in carnation petals. Isolation and purification of acyl donor molecules from carnation petals has, however, revealed that the acyl donor is not a CoA ester but the glucose ester of malate. Chemically synthesized malyl-1-α- and

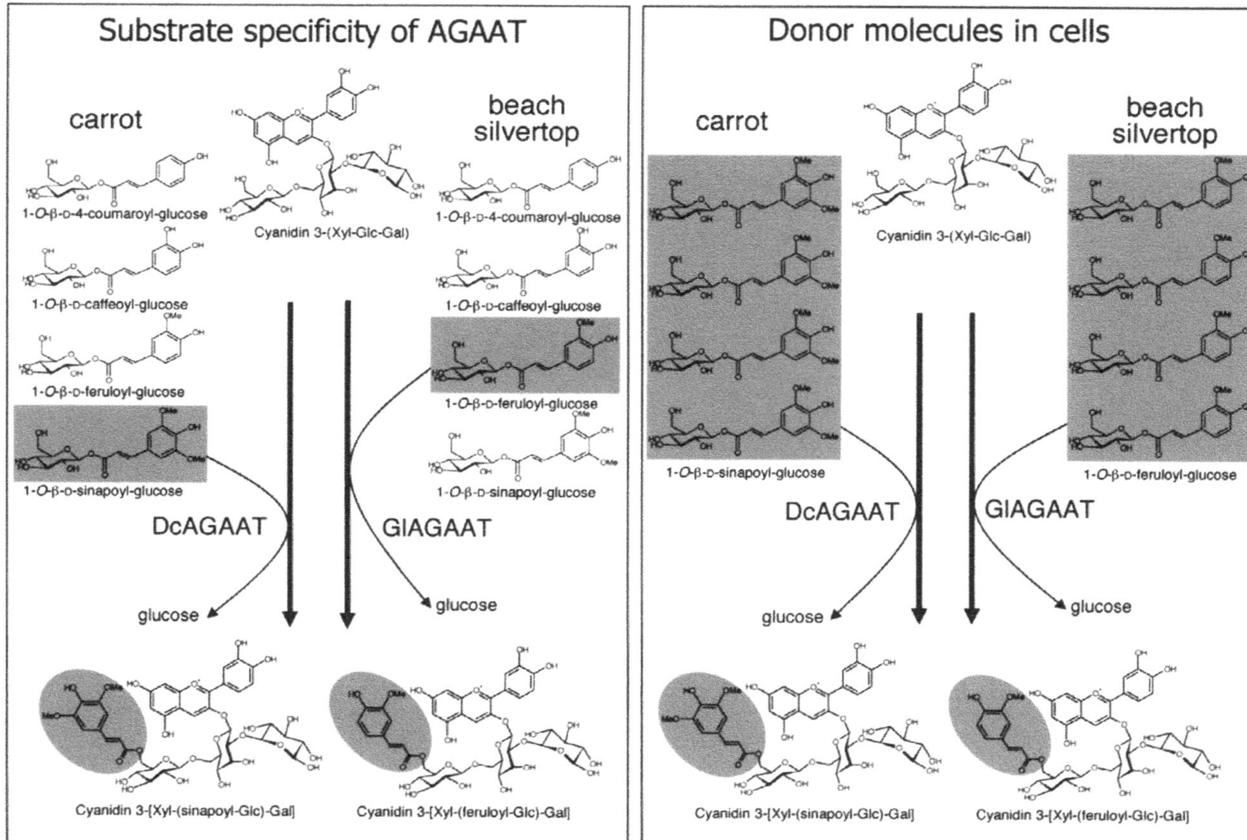

Figure 12.3 *Proposed mechanism for the determination of acyl moieties by acyl-glucose-dependent acyltransferase in the carrot and beach silvertop. The supplement of donor molecules might predominantly determine the acylated product species (right) rather than the substrate specificity of AGAAT (left) (Matsuba* et al.*, 2008)*

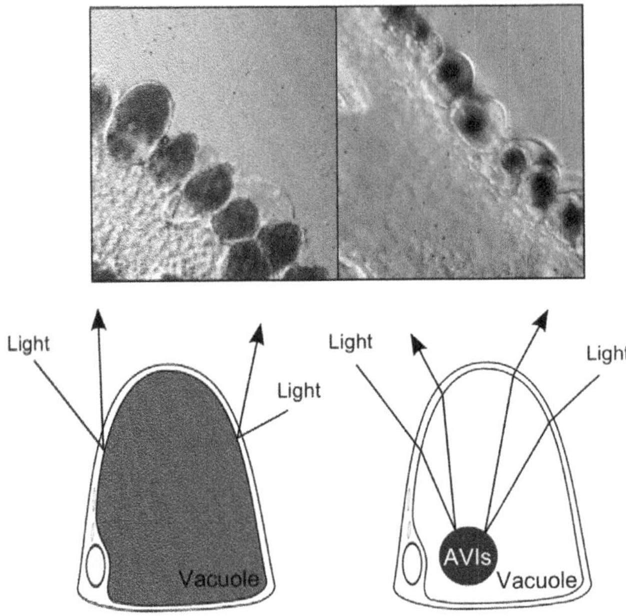

Figure 12.4 *Anthocyanic vacuolar inclusions (AVIs) in carnation petals. Pelargonidin- and cyanidin-3,5-di-(6''-O-4, 6''-O-1-cyclic malonyl)glucosides are soluble in vacuolar sap (left) but pelargonidin- and cyanidin-3,5-O-diglucosides, without a malyl moiety, are not. Instead they form AVIs in vacuoles (right)*

1-β-glucoses were reacted with pelargonidin-3-*O*-glucoside and cyanidin-3-*O*-glucoside in crude protein extracts prepared from the petals of wild-type carnation. Malyl-1-β-glucose, but not malyl-1-α-glucose, yielded anthocyanin 3-*O*-malylglucosides, suggesting that the enzyme belongs to the SCPL family with a strictly refined β stoichiometry (Abe *et al.*, 2008b). Carnation varieties that form AVIs in vacuoles were used to detect acyl transfer activity in crude protein extracts prepared from petals, although no acyl transfer activity was detected in most varieties that formed AVIs (Abe *et al.*, 2008b). These results suggest that malyl transfer in carnation petals takes place by SCPL-type AGAAT. To test this hypothesis, degenerated primers corresponding to the conserved amino-acid sequence for SCPL proteins were designed and used for PCR of genomic DNA prepared from carnation. This process yielded the nucleotide sequences of the *AMalT* gene. Transposable elements were found inserted in the genomes of some carnation varieties that form AVIs and lack AMalT activity, resulting in disruption of the *AMalT* gene. In the genomes of varieties with a variegated phenotype, footprint sequences left at the inserted position of the moving event were found. These results suggest that the identified gene was responsible for AMalT. According to the amino-acid sequence of the *AMalT* gene, this enzyme belongs to the SCPL family (Umemoto *et al.*, 2009).

Why can AMalT malylate anthocyanin-3-*O*-glucosides but not anthocyanin-3,5-*O*-diglucosides? (see Abe *et al.*, 2008b). It is possible that anthocyanin-3-*O*-glucosides synthesized in the cytosol are transported into a non-cytosolic compartment (the endoplasmic reticulum, prevacuolar compartments or vacuoles, as described in Section 12.5), where AMalT catalyzes conversion to anthocyanin-3-*O*-malylglucoside, which in turn is glycosylated at the C5 position. The glucosyltransferase involved remains unidentified, however.

A major problem with the isolated *AMalT* gene is the failure to detect enzymatic activity in recombinant proteins. Many recombinant enzymic proteins involved in the anthocyanin biosynthetic pathway have been produced by *E. coli* and yeast, and their *in vitro* enzyme properties investigated. AMalT cDNA that was introduced into expression vectors and transformed into *E. coli* and yeast did not exhibit enzyme activity. In another experiment, the construct harbouring AMalT cDNA driven by the 35S promoter was introduced into petunia and tobacco (*Nicotianum tabacum*) plants. Expression of the AMalT transcripts was detected, but no AMalT enzyme activity was found in the crude protein extracts (Ozeki, unpublished data). There is not yet any biochemical proof of AMalT activity in recombinant proteins.

A recent *Arabidopsis* mutant analysis for AVI formation has provided new insights into the sequestration of anthocyanins. Addition of naringenin to *Arabidopsis* seedlings induced anthocyanin synthesis, so as to form AVIs (Poustka *et al.*, 2007). Seedlings of the *5gt Arabidopsis* mutant, in which the gene encoding anthocyanidin-5-*O*-glucosyltransferase was disrupted by T-DNA insertion, had a greater accumulation of anthocyanins as AVIs than in the wild-type. However, the number of AVIs in mutant seedlings related to the autophagic process (*atg* mutants) following addition of naringenin was less than in wild-type, while vanadate increased the number of AVIs in the mutant. These results suggest that autophagy in prevacuolar compartments is involved in the transport of anthocyanins and the formation of AVIs (Pourcel *et al.*, 2010).

12.5 Transport of Anthocyanins from Cytosol to Vacuoles

Anthocyanins are sequestrated in vacuoles, but the mechanism by which they are transported into vacuoles is a matter of debate (Figure 12.5). Since transporter proteins are localised in the tonoplast membrane, they are difficult to purify using buffers containing detergents. As a result there is limited availability of columns and gel matrixes for purification. It has recently been shown that two major transporter systems localised in the tonoplast membrane, multidrug resistance-associated protein (MRP/ABCC [ATP binding cassette subfamily C]) (Martinoia *et al.*, 2002; Klein *et al.*, 2006; Rea, 2007; Verrier *et al.*, 2008) and multidrug and toxic compound extrusion (MATE) (Yazaki, 2005; Yazaki *et al.*, 2008), are important in transporting plant secondary metabolites into the vacuoles. The first of these directly uses energy derived from ATP hydrolysis, whereas the latter uses a [H^+] electrochemical gradient driving force across the membrane created by proton pumps, such as vacuolar H^+-ATPase (V-ATPase) and vacuolar H^+-pyrophosphatase (V-PPase) (Gaxiola *et al.*, 2007; Martinoia *et al.*, 2007). Plants have many more genes encoding transporter proteins than bacteria and animals: 15 MRP/ABCC genes in 129 ABC and 56 MATE transporter genes have been identified in the *Arabidopsis* genome (Sánchez-Fernández *et al.*, 2001; Yazaki *et al.*, 2008). Although there is increasing genetic and molecular biological evidence for the genes encoding these transporter proteins, clear-cut biochemical phenotypic evidence is hard to obtain. This is because the measurement of *in vitro* transporter activity is critical, and in some cases transporter activity has been found for a larger than expected number of substrates.

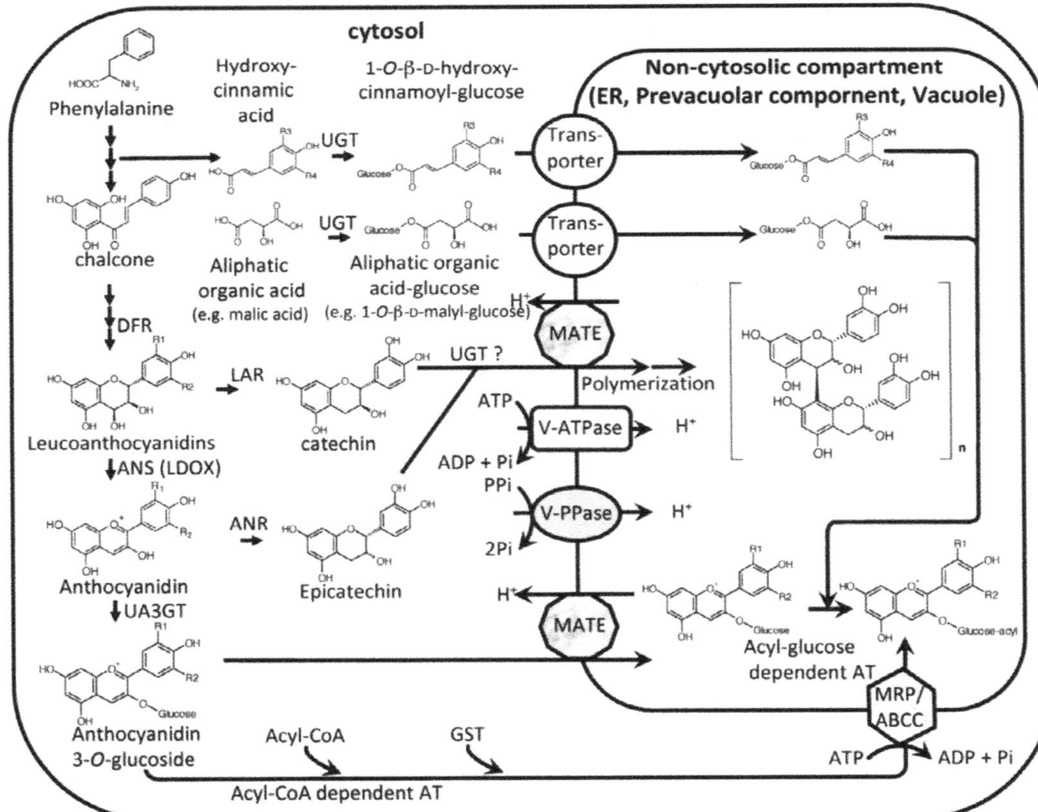

Figure 12.5 *Diagrammatic representation of anthocyanin and procyanidin transport. Abbreviations: MRP/ABCC, multidrug resistance-associated protein (ATP binding cassette subfamily C); MATE, multidrug and toxic compound extrusion; V-ATPase, vacuolar H^+-ATPase; V-PPase, vacuolar H^+-pyrophosphatase; GST, glutathione S-transferase; UGT, UDP-glucose-dependent glucosyltransferase; AT, acyltransferase; ER, endoplasmic reticulum. Other abbreviations are as shown in Figure 12.1*

The first experiment to detect anthocyanin transport activity used isolated and purified vacuoles prepared from anthocyanin-synthesizing carrot suspension-cultured cells and a radio-labelled anthocyanin (Hopp and Seitz, 1987). Uptake of [^{3}H]cyanidin-3-*O*-(2″-*O*-xylosyl-6″-*O*-sinapoylglucosyl)galactoside into isolated carrot vacuoles, via the tonoplast membrane, was observed in a time-dependent manner against a concentration gradient. There was no marked stimulation of transport activity by addition of ATP, but transport activity was reduced upon adding the protonophore carbonylcyanide-*m*-chlorophenylhydrazone. This change indicates that uptake of anthocyanins into vacuoles is a [H^+] energy-dependent process, which might involve MATE but not MRP/ABCC. However, deacylated [^{3}H]cyanidin-3-*O*-(2″-*O*-xylosyl-6″-*O*-sinapoylglucosyl)galactoside, and the -glucosyl-galactoside and the -galactoside, were not taken up by isolated vacuoles. This result implies that cytosolic acylation of anthocyanins with sinapic acid is essential for their transport into vacuoles. These *in vitro* observations are not consistent with the location of the acylation reaction stated in the previous section; since AGAAT might be an SCPL protein localised in non-cytosolic compartments, the glycosylated anthocyanin could be transported into the vacuoles (or other compartments, such as the prevacuolar compartment), where it could react with AGAAT. The *in vitro* biochemical evidence of the transport activity using isolated vacuoles and the catalytic activity of AGAAT using the crude protein extract do not agree, such that the *in vivo* mechanism is yet to be settled.

The first gene known to be involved in the transport of anthocyanins was the maize mutant *Bz2*. The nucleotide sequence of *Bz2* was identified by transposon tagging, which revealed a mutable colour phenotype in the kernel skin anthocyanins. This gene was found to encode a glutathione *S*-transferase (GST) (Marrs *et al.*, 1995). GSTs are not transporters *per se*, but act as non-enzymatic carrier proteins called 'ligandins', which escort compounds to the transporters in the tonoplast. Since *Bz2* was first identified, many GST homologues have been detected. *Arabidopsis* contains 54 GST genes, and those responsible for flavonoid and anthocyanin transport belong to class phi GSTs (Dixon *et al.*, 2010). *Bz2*-deficient maize kernels, having yellow skin colour, exhibited red spots due to the accumulation of anthocyanin following transient expression by microprojectile bombardment of *Bz2* cDNA driven by the 35S promoter. Similar red spots on *Bz2*-deficient maize kernels were observed when petunia *A9*, which was identified by *dTph1* element insertion, was used in transient expression instead of *Bz2*. The GST encoded in *A9* is able to complement the *Bz2* deficiency (Alfenito *et al.*, 1998). The transient expression of *Bz2* and *A9* appear to complement the carnation *fl3* mutant phenotype (which has pale pink petals), since deep red spots were observed against the pale pink background where the cells received the construct (Larsen *et al.*, 2003). *Arabidopsis TT19* encoding GST is responsible for the transport of both anthocyanins and proanthocyanidins in the seed coat (Kitamura *et al.*, 2004). Recent advances in mass spectrometry (MS)/Edman sequencing used in the genome project have generated a new strategy for GST investigation. The amino-acid sequences of five GST proteins in anthocyanin-synthesizing suspension-cultured cells of *Vitis vinifera* were determined by MS/Edman sequencing, and their full-length nucleotide sequences were identified in the Institute of Genomic Research (TIGR) grape database. Two of the five GST proteins were able to complement *Bz2*-deficient maize kernels so as to produce red spots in the transient expression system (Conn *et al.*, 2008).

The transporters for anthocyanins escorted by GSTs have been identified and characterised in *Arabidopsis* and maize. *Arabidopsis AtMRP1* was isolated by PCR, using degenerate primers corresponding to the amino-acid sequence of the second ATP-binding cassette of yeast and human MRPs. When *AtMRP1* was expressed in yeast it resulted in the transport of [^{3}H]glutathione-conjugated cyanidin-3-*O*-glucoside into membrane-enriched vesicles (Lu *et al.*, 1997). In a similar way, *AtMRP2* was isolated and shown to have transport activity for [^{3}H]glutathione-conjugated cyanidin-3-*O*-glucoside (Lu *et al.*, 1998). A total of ten MRP homologous sequences were found in maize expression sequence tag (EST) sequences. Of these, ZmMRP3 was responsible for the transportation of anthocyanin; this was demonstrated by expression analysis of several anthocyanin-deficient mutants and reverse genetics (Goodman *et al.*, 2004).

MATE-type flavonoid/anthocyanin (proanthocyanidin) transporters were first identified as *TT12* in *Arabidopsis* by T-DNA tagging (Debeaujon *et al.*, 2001). Two pieces of evidence suggest that the TT12-encoded MATE protein homologue is responsible for the transport of proanthocyanidins: the characterisation of expression profiles of *TT12* in wild-type plants, and a reverse genetic strategy to change the colourless phenotype of the *tt12* mutant to a pigmented form by introduction of the wild-type *TT12* gene into the mutant plants (Debeaujon *et al.*, 2001). No biochemical analysis has been undertaken to identify which compounds were transported. The transport activity and preferences of the TT12 transporter were analysed *in vitro* using membrane vesicles prepared from yeast expressing TT12 (Marinova *et al.*, 2007). TT12 expressed on the yeast membrane did not transport proanthocyanidins, proanthocyanidin-*O*-glucosides or flavonol-*O*-glycosides, but could transport cyanidin-3-*O*-glucoside into yeast vesicles in the presence of ATP. This property has been confirmed by other investigators (Gomez *et al.*, 2009; Zhao and Dixon, 2009). Inhibitor analysis showed that transport involved the driving force of a [H^+] gradient created by V-ATPase or V-PPase, which has the typical properties of MATE transporters. In the yeast vesicle system, no direct evidence for the transportation of catechin and catechin-3-*O*-glucoside has been found, although catechin-3-*O*-glucoside did inhibit transport of cyanidin-3-*O*-glucoside. This suggests that *in vivo*, but not *in vitro*, TT12 plays an important role in catechin-3-*O*-glucoside transport (Marinova *et al.*, 2007). In the entire genome sequence of *Vitis vinifera*, 65 MATE genes have been identified (Gomez *et al.*, 2009). According to phylogenetic analysis, two MATE genes were possible anthocyanin MATE transporters. Both were expressed in yeast membrane vesicles and, in contrast to TT12, transported anthocyanin *p*-coumaroylglucosides, but not unacylated anthocyanin glucosides, into the vesicles in a [H^+] gradient-dependent manner (Gomez *et al.*, 2009). These observations on the acylation of anthocyanins are vital for *in vitro* transport activity in *Vitis*, and correspond to data showing a preference for the transportation of anthocyanins in carrot vacuoles (Hopp and Seitz, 1987). In membrane vesicles prepared from hairy roots of *Medicago truncatula*, cyanidin-3-*O*-glucoside and daidzein-7-*O*-glucoside, but not epicatechin-3′-*O*-glucoside, were taken up in an ATP-dependent manner. Epicatechin-3′-*O*-glucoside was, however, taken up by membrane vesicles prepared from TT2 (Myb)-overexpressing *Medicago truncatula* hairy roots (Zhao and Dixon, 2009). The *Medicago MATE1* gene was identified as a *TT12* homologue which could complement the *Arabidopsis tt12* mutation in accumulating proanthocyanidins. TT12 expressed in the yeast membrane vesicles could also transport epicatechin-3′-*O*-glucoside, suggesting that *Medicago* MATE1 is responsible for proanthocyanidin transporters *in vivo*, rather than anthocyanin transporters *in vitro* (Zhao and Dixon, 2009).

Recent advances in transcription regulatory factors for upregulating anthocyanin synthesis provide an opportunity to identify anthocyanin transporter genes upregulated concomitantly with the expression of the structural genes for anthocyanin biosynthesis. The candidate grape MATEs for anthocyanin transport were estimated in the ectopic expression of the hairy roots of *VlmybA1* (Cutanda-Perez *et al.*, 2009). Overexpression of *Ant1*, tomato Myb, in the enhancer trapping system revealed upregulation of the *MATE* gene (Mathews *et al.*, 2003). These results suggest that the genes for anthocyanin transporters are under the same (or related) regulatory mechanisms as the structural genes for anthocyanin biosynthesis.

Two questions arise regarding the transport of anthocyanins. The first is whether both MRP/ABCC and MATE operate in anthocyanin transport, or only one of the two? (and if so, which?). One possibility prevails in our system in which anthocyanin biosynthesis was induced in regenerated torenia (*Torenia*) shoots. When regenerated torenia shoots grown in darkness were transferred to a medium having high sucrose content, and were then cultured in the light, anthocyanin synthesis and chlorophyll degradation occurred simultaneously (Nagira and Ozeki, 2004). Suppression subtractive hybridisation of mRNAs for anthocyanin-synthesizing shoots, minus green shoots, was used to isolate a set of cDNAs. Microarray analysis using these cDNAs showed that the *GST* and *MRP/ABCC* genes were upregulated in anthocyanin-synthesizing shoots (Nagira *et al.*, 2006). RNAi experiments with transgenic torenia revealed that downregulation of either *GST* genes or *MRP/ABCC* genes suppressed anthocyanin synthesis in shoots, but the flower colours of RNAi transformants for either *GST* or *MRP/ABCC* plants were not altered. A small decrease in anthocyanins was observed in the petals of both types of plant (Nagira, personal communication). These results suggest that GST and MRP/ABCC operate during anthocyanin synthesis in shoots growing in a high sucrose medium and illumination, but has little effect on anthocyanin synthesis in petals. Different transporters may operate in different organs and tissues of a plant, as well as in different plant species.

The second question concerns the evidence that anthocyanin transport occurs in the ER and prevacuolar compartments, not in the vacuoles (Grotewold, 2004). The data considered above regarding MRP/ABCC and MATE indicate that anthocyanin transport is located in the vacuoles. Two routes have been proposed, a ligandin transporter model and a vesicular transport model (Grotewold and Davies, 2008). Grotewold's group presented data indicating that the location of anthocyanin transport was ER, not vacuoles, and that anthocyanins in the ER are transported using trafficking pathways of an ER-to-vacuole protein-sorting route (Poustka *et al.*, 2007). AVI formation in *Arabidopsis* mutants suggests that the prevacuolar compartments play an important role in anthocyanin transport (Pourcel *et al.*, 2010). Transportation not only to the vacuoles directly, but also to other cellular compartments, is a candidate for the pathway of anthocyanin transport. It is possible that MRP/ABCC and/or MATE act as transporters in the other cellular compartments. Future experiments will settle these two questions.

12.6 Concluding Remarks

In the original identification of C_6-C_3-C_6 molecules it was believed that anthocyanins, proanthocyanidins and isoflavonoids were flavonoids. All textbooks indicate that the anthocyanin biosynthetic pathway is derived from the flavonoid biosynthetic pathway, as are the isoflavonoid and proanthocyanidin biosynthetic routes. The last pathway shown in

Figure 12.5 is presented as the branching point from the anthocyanin biosynthetic pathway. In fact, recent research shows that each is an independent route regulated by different regulatory mechanisms *in vivo*. Some enzymes prepared from spatially or temporally distinct tissues, or from plants experiencing different environmental stresses (such as pathogen attack), catalyze the same reaction *in vitro*, but different genes are responsible for the enzymes showing the same *in vitro* reaction. In other cases, the same genes encoding the enzymes are differentially regulated by different transcriptional regulatory factors; some of the factors are common, but others contribute in different combinations. The flavonoid, isoflavonoid, anthocyanin and proanthocyanidin biosynthetic pathways are all independent of each other. Legume plants have all four pathways, but other plant species lack the isoflavonoid biosynthetic pathway. Caryophyllales, with the exception of Caryophyllaceae and Molluginaceae, have the flavonoid and the proanthocyanidin biosynthetic pathways, but do not synthesize anthocyanins. During their evolution the betalain biosynthetic pathway has developed, however. It has been proposed that the enzymes for these four pathways *in vivo* are located and operate work in different cellular compartments, called 'metabolons' (Winkel-Shirley, 1999; Saslowsky and Winkel-Shirley, 2001; Ralston and Yu, 2006). According to this hypothesis, soluble cellular enzyme proteins associate with membrane enzyme proteins, such as cinnamate 4-hydroxylase (C4H) and F3′(5′)H, as multienzyme complexes on metabolons. It is possible that the enzymes of each of the four pathways are located on independent metabolons *in vivo*, followed by transportation and sequestration via different transporters.

References

Abe, Y., Shoji, T., Kawahara, N., Kamakura, H., Kanda, T., Goda, Y. and Ozeki, Y. (2008a) Structural characterization of a procyanidin tetramer and pentamer from the apple by low-temperature NMR analysis. *Tetrahedron Lett.*, **49**, 6413–6418.

Abe, Y., Tera, M., Sasaki, N., Okamura, M., Umemoto, N., Momose, M., Kawahara, N., Kamakura, H., Goda, Y., Nagasawa, K. and Ozeki, Y. (2008b) Detection of 1-*O*-malylglucose: pelargonidin 3-*O*-glucose-6″-*O*-malyltransferase activity in carnation (*Dianthus caryophyllus*). *Biochem. Biophys. Res. Comm.*, **373**, 473–477.

Alfenito, M.R., Souer, E., Goodman, C.D., Buell, R., Mol, J., Koes, R. and Walbot V. (1998) Functional complementation of anthocyanin sequestration in the vacuole by widely divergent glutathione *S*-transferases. *Plant Cell*, **10**, 1135–1150.

Conn, S., Zhang, W. and Franco, C. (2003) Anthocyanin vacuolar inclusions (AVIs) selectively bind acylated anthocyanins in *Vitis vinifera* L. (grapevine) suspension culture. *Biotech. Lett.*, **25**, 835–839.

Conn, S., Curtin, C., Bézier, A., Franco, C. and Zhang, W. (2008) Purification, molecular cloning, and characterization of glutathione *S*-transferases (GSTs) from pigmented *Vitis vinifera* L. cell suspension cultures as putative anthocyanin transport proteins *J. Exp. Bot.*, **59**, 3621–3634.

Cutanda-Perez, M.C., Ageorges, A., Gomez, C., Vialet, S., Terrier, N., Romieu, C. and Torregrosa, L. (2009) Ectopic expression of *VlmybA1* in grapevine activates a narrow set of genes involved in anthocyanin synthesis and transport. *Plant Mol. Biol.*, **69**, 633–648.

Dangles, O., Saito, N. and Brouillard, R. (1993) Anthocyanin intramolecular copigment effect. *Phytochemistry*, **34**, 119–124.

Debeaujon, I., Peeters, A.J.M., Léon-Kloosterziel, K.M. and Koornneef, M. (2001) The *TRANSPARENT TESTA12* gene of Arabidopsis encodes a multidrug secondary transporter-like protein required for flavonoid sequestration in vacuoles of the seed coat endothelium. *Plant Cell*, **13**, 852–871.

Devic, M., Guilleminot, J., Debeaujon, I., Bechtold, N., Bensaude, E., Koornneef, M., Pelletier, G., Pelletier, G. and Delseny, M. (1999) The *BANYULS* gene encodes a DFR-like protein and is a marker of early seed coat development. *Plant J.*, **19**, 387–398.

Dixon, R.A., Xie, D.Y. and Sharma, S.B. (2005) Proanthocyanidins – a final frontier in flavonoid research? *New Phytol.*, **165**, 9–28.

Dixon, D.P., Skipsey, M. and Edwards, R. (2010). Roles for glutathione transferases in plant secondary metabolism. *Phytochemistry*, **71**, 338–350.

Fujiwara, H., Tanaka, Y., Fukui, Y., Nakao, M., Ashikari, T. and Kusumi, T. (1997) Anthocyanin 5-aromatic acyltransferase from *Gentiana triflora*. Purification, characterization and its role in anthocyanin biosynthesis. *Eur. J. Biochem.*, **249**, 45–51.

Fujiwara, H., Tanaka, Y., Yonekura-Sakakibara, K., Fukuchi-Mizutani, M., Nakao, M., Fukui, Y., Yamaguchi, M., Ashikari, T. and Kusumi, T. (1998) cDNA cloning, gene expression and subcellular localization of anthocyanin 5-aromatic acyltransferase from *Gentiana triflora*. *Plant J.*, **16**, 421–431.

Fukada-Tanaka, S., Inagaki, Y., Yamaguchi, T., Saito, N. and Iida, S. (2000) Colour-enhancing protein in blue petal. *Nature*, **407**, 581.

Fukui, Y., Tanaka, Y., Kusumi, T., Iwashita, T. and Nomoto, K. (2003) A rationale for the shift in colour towards blue in transgenic carnation flowers expressing the flavonoid 3′,5′-hydroxylase gene. *Phytochemistry*, **63**, 15–23.

Furtek, D., Schiefelbein, J.W., Johnston, F. and Nelson, O.E. Jr (1988) Sequence comparisons of 3 wild-type *bronze-1* alleles from *Zea mays*. *Plant Mol. Biol.*, **11**, 473–481.

Gaxiola, R.A., Palmgren, M.G. and Schumacher, K. (2007) Plant proton pumps. *FEBS Lett.*, **581**, 2204–2214.

Gomez, C., Terrier, N., Torregrosa, L., Vialet, S., Fournier-Level, A., Verriès, C., Souquet, J.-M., Mazauric, J.-P., Klein, M., Cheynier, V. and Ageorges, A. (2009) Grapevine MATE-type proteins act as vacuolar H^+-dependent acylated anthocyanin transporters. *Plant Physiol.*, **150**, 402–415.

Goodman, C.D., Casati, D. and Walbot, V. (2004) A multidrug resistance-associated protein involved in anthocyanin transport in *Zea mays*. *Plant Cell*, **16**, 1812–1826.

Grotewold, E. (2004) The challenges of moving chemicals within and out of cells: insights into the transport of plant natural products. *Planta*, **219**, 906–909.

Grotewold, E. (2006) The genetics and biochemistry of floral pigments. *Ann. Rev. Plant Biol.*, **57**, 761–780.

Grotewold, E. and Davies, K. (2008) Trafficking and sequestration of anthocyanins. *Nat. Prod. Commun.* **3**, 1251–1258.

Hahlbrock, K. (1972) Malonyl coenzyme A: flavanone glucoside malonyl-transferase from illuminated cell suspension cultures of parsley. *FEBS Lett.*, **28**, 65–68.

Harborne, J.B., Mayer, A.M. and Bar-Nun, N. (1983) Identification of the major anthocyanin of carrot cells in tissue culture as cyanidin 3-(sinapoylxylosylglucosylgalactoside). *Z. Naturforsch.*, **38C**, 1055–1056.

Hopp, W. and Seitz, H.U. (1987) The uptake of acylated anthocyanin into isolated vacuoles from a cell suspension culture of *Daucus carota*. *Planta*, **170**, 74–85.

Jones, P., Messner, B., Nakajima, J., Schaffner, A.R. and Saito, K. (2003) UGT73C6 and UGT78D1, glycosyltransferases involved in flavonol glycoside biosynthesis in *Arabidopsis thaliana*. *J. Biol. Chem.*, **278**, 43910–43918.

Kamsteeg, J., van Brederode, J., Hommels, C.H. and van Nigtevecht, G. (1980) Identification, properties and genetic control of hydroxycinnamoyl-coenzyme A: anthocyanindin 3-rhamnosyl (1 $\rightarrow$ 6) glucoside, 4‴-hydroxycinnamoyl transferase isolated from petals of *Silene dioica*. *Biochem. Physiol. Pflanz.*, **175**, 403–411.

Katsumoto, Y., Fukuchi-Mizutani, M., Fukui, Y., Brugliera, F., Holton, T.A., Karan, M., Nakamura, N., Yonekura-Sakakibara, K., Togami, J., Pigeaire, A., Tao, G.-Q., Nehra, N.S., Lu, C.-Y., Dyson, B.K., Tsuda, S., Ashikari, T., Kusumi, T., Mason, J.G. and Tanaka, Y. (2007) Engineering of the rose flavonoid biosynthetic pathway successfully generated blue-hued flowers accumulating delphinidin. *Plant Cell Physiol.*, **48**, 1589–1600.

Kitamura, S., Shikazono, N. and Tanaka, A. (2004) *TRANSPARENT TESTA 19* is involved in the accumulation of both anthocyanins and proanthocyanidins in *Arabidopsis*. *Plant J.*, **37**, 104–114.

Klein, M., Burla, B. and Martinoia, E. (2006) The multidrug resistance-associated protein (MRP/ABCC) subfamily of ATP-binding cassette transporters in plants. *FEBS Lett.*, **580**, 1112–1122.

Kobayashi, S., Goto-Yamamoto, N. and Hirochika, H. (2004) Retrotransposon-induced mutations in grape skin colour. *Science*, **304**, 982.

Kondo, T., Yoshida, K., Nakagawa, A., Kawai, T., Tamura, H. and Goto, T. (1992) Structural basis of blue-colour development in flower petals from *Commelina communis*. *Nature*, **358**, 515–518.

Kubo, A., Arai, Y., Nagashima, S. and Yoshikawa, T. (2004) Alteration of sugar donor specificities of plant glycosyltransferases by a single point mutation. *Arch. Biochem. Biophys.*, **429**, 198–203.

Larsen, E.S., Alfenito, M.R., Briggs, W.R. and Walbot, V. (2003) A carnation anthocyanin mutant is complemented by the glutathione *S*-transferases encoded by maize *Bz2* and petunia *An9*. *Plant Cell Rep.*, **21**, 900–904.

Lu, Y.-P., Li, Z.-S. and Rea, P.A. (1997) *AtMRP1* gene of *Arabidopsis* encodes a glutathione *S*-conjugate pump: Isolation and functional definition of a plant ATP-binding cassette transporter gene. *Proc. Natl. Acad. Sci. USA*, **94**, 8243–8248.

Lu, Y.-P., Li, Z.-S., Drozdowicz, Y.M., Hörtensteiner, S., Martinoia, E. and Rea, P.A. (1998) AtMRP2, an *Arabidopsis* ATP binding cassette transporter able to transport glutathione *S*-conjugates and chlorophyll catabolites: functional comparisons with AtMRP1. *Plant Cell*, **10**, 267–282.

Marinova, K., Pourcel, L., Weder, B., Schwarz, M., Barron, D., Routaboul, J.-M., Debeaujon, I. and Klein, M. (2007) The *Arabidopsis* MATE transporter TT12 acts as a vacuolar flavonoid/H^+-antiporter active in proanthocyanidin-accumulating cells of the seed coat. *Plant Cell*, **19**, 2023–2038.

Makham, K.R., Gould, K.S., Winefield, C.S., Mitchell, K.A., Bloor, S.J. and Boase, M.R. (2000) Anthocyanic vacuolar inclusions – their nature and significance in flower colouration. *Phytochemistry*, **55**, 327–336.

Marrs, K.A., Alfenito, M.R., Lloyd, A.M. and Walbot V. (1995) A glutathione *S*-transferase involved in vacuolar transfer encoded by the maize gene *Bronze-2*. *Nature*, **375**, 397–400.

Martin, C., Carpenter, R., Sommer, H., Saedlerl, H.and Coen, E.S. (1985) Molecular analysis of instability in flower pigmentation of *Antirrhinum majus*, following isolation of the *pallida* locus by transposon tagging. *EMBO J.*, **4**, 1625–1630.

Martinoia, E., Klein, M., Geisler, M., Bovet, L., Forestier, C., Kolukisaoglu, Ü., Müller-Röberand, B. and Schulz, B. (2002) Multifunctionality of plant ABC transporters – more than just detoxifiers. *Planta*, **214**, 345–355.

Martinoia, E., Maeshima, M. and Neuhaus, H.E. (2007) Vacuolar transporters and their essential role in plant metabolism. *J. Exp. Bot.*, **58**, 83–102.

Mathews, H., Clendennen, S.K., Caldwell, C.G., Liu, X.L., Connors, K., Matheis, N., Schuster, D.K., Menasco, D.J., Wagoner, W., Lightner, J. and Wagner, D.R. (2003) Activation tagging in tomato identifies a transcriptional regulator of anthocyanin biosynthesis, modification, and transport. *Plant Cell*, **15**, 1689–1703.

Matsuba, Y., Okuda, Y., Abe, Y., Kitamura, Y., Terasaka, K., Mizukami, H., Kamakura, H., Kawahara, N., Goda, Y., Sasaki, N. and Ozeki, Y. (2008) Enzymatic preparation of 1-O-hydroxycinnamoyl-β-D-glucoses and their application to the study of 1-O-hydroxycinnamoyl-β-D-glucose-dependent acyltransferase in anthocyanin-producing cultured cells of *Daucus carota* and *Glehnia littoralis*. *Plant Biotechnol.*, **25**, 369–375.

Messen, A., Höhmann, S., Martin, W., Schnable, P.S., Peterson, P.A., Saedler, H. and Gierl, A. (1990) The *En/Spm* transposable element of *Zea mays* contains splice sites at the termini generating a novel intron from a *dSpm* element in the *A2* gene. *EMBO J.*, **9**, 3051–3057.

Miura, H., Kitamura, Y., Ikenaga, T., Mizobe, K., Shimizu, T., Nakamura, M., Kato, Y., Yamada, T., Maitani, T. and Goda, Y. (1998) Anthocyanin production of *Glehnia littoralis* callus cultures. *Phytochemistry*, **48**, 279–283.

Momonoi, K., Yoshida, K., Mano, S., Takahashi, H., Nakamori, C., Shoji, K., Nitta, A. and Nishimura, M. (2009) A vacuolar iron transporter in tulip, TgVit1, is responsible for blue colouration in petal cells through iron accumulation. *Plant J.*, **59**, 437–447.

Nagira, Y. and Ozeki, Y. (2004) A system in which anthocyanin synthesis is induced in regenerated torenia shoots. *J. Plant Res.*, **117**, 377–383.

Nagira, Y., Shimamura, K., Hirai, S., Shimanuki, M., Kodama, H. and Ozeki, Y. (2006) Identification and characterization of genes induced for anthocyanin synthesis and chlorophyll degradation in regenerated torenia shoots using suppression subtractive hybridization, cDNA microarrays, and RNAi techniques. *J. Plant Res.*, **119**, 217–230.

Nakayama, M., Koshioka, M., Yoshida, H., Kan, Y., Fukui, Y., Koike, A. and Yamaguchi, M. (2000) Cyclic malyl anthocyanins in *Dianthus caryophyllus*. *Phytochemistry*, **55**, 937–939.

Nakayama, T., Suzuki, H. and Nishino, T. (2003) Anthocyanin acyltransferases: specificities, mechanism, phylogenetics, and applications. *J. Mol. Catalysis B: Enzymatic*, **23**, 117–132.

Noda, N., Kazuma, K., Sasaki, T., Tsugawa, H and Suzuki, M. (2006) Molecular cloning of 1-*O*-acetylglucose dependent anthocyanin aromatic acyltransferase in ternatin biosynthesis of butterfly pea (*Clitoria ternatea*). *Plant Cell Physiol.*, **47s**, s109.

Noguchi, A., Horikawa, M., Fukui, Y., Fukuchi-Mizutani, M., Iuchi-Okada, A., Ishiguro, M., Kiso, Y., Nakayama, T. and Ono, E. (2009) Local differentiation of sugar donor specificity of flavonoid glycosyltransferase in Lamiales. *Plant Cell*, **21**, 1556–1572.

Nozue, M., Kubo, H., Nishimura, M., Katou, A., Hattori, C., Usuda, N., Nagata, T. and Yasuda, H. (1993) Characterization of intravacuolar pigmented structures in anthocyanin-containing cells of sweet potato suspension cultures. *Plant Cell Physiol.*, **34**, 803–808.

Nozue, M., Yamada, K., Nakamura, T., Kubo, H., Kondo, M. and Nishimura, M. (1997) Expression of a vacuolar protein (VP24) in anthocyanin-producing cells of sweet potato in suspension culture. *Plant Physiol.*, **115**, 1065–1072.

Offen, W., Martinez-Fleites, C., Yang, M., Kiat-Lim, E., Davis, B.G., Tarlin, C.A., Ford, C.M., Bowles, D.J. and Davies, G.J. (2006) Structure of a flavonoid glucosyltransferase reveals the basis for plant natural product modification. *EMBO J.* **25**, 1396–1405.

Ogata, J., Itoh, Y., Ishida, M., Yoshida, H. and Ozeki, Y. (2004) Cloning and heterologous expression of a cDNA encoding flavonoid glucosyltransferases from *Dianthus caryophyllus*. *Plant Biothechnol.*, **21**, 367–375.

Ogata, J., Kanno, Y., Itoh, Y., Tsugawa, H. and Suzuki, M. (2005) Plant biochemistry: Anthocyanin biosynthesis in roses. *Nature*, **435**, 757–758.

Pang, Y., Peel, G.J., Sharma, S.B., Tang, Y. and Dixon, R.A. (2008). A transcript profiling approach reveals an epicatechin-specific glucosyltransferase expressed in the seed coat of *Medicago truncatula*. *Proc. Natl. Acad. Sci. USA*, **105**, 14210–14215.

Pierre, B. St. and De Luca, V. (2000) Evolution of acyltransferase genes: origin and diversification of the BAHD superfamily of acyltransferases involved in secondary metabolism. *Rec. Adv. Phytochem.*, **34**, 285–315.

Pourcel, L., Irani, N.G., Lu, Y., Riedl, K., Schwartz, S. and Grotewold, E. (2010) The formation of anthocyanic vacuolar inclusions in *Arabidopsis thaliana* and implications for the sequestration of anthocyanin pigments. *Mol. Plant*, **3**, 78–90.

Poustka, F., Irani, N.G., Feller, A., Lu, Y., Pourcel, L., Frame, K. and Grotewold, E. (2007) A trafficking pathway for anthocyanins overlaps with the endoplasmic reticulum-to-vacuole protein sorting route in Arabidopsis and contributes to the formation of vacuolar inclusions. *Plant Physiol.*, **145**, 1323–1335.

Ralston, L. and Yu, O. (2006) Metabolons involving plant cytochrome P450s. *Phytochem. Rev.*, **5**, 459–472.

Rea, P.A. (2007) Plant ATP-binding cassette transporters. *Ann. Rev. Plant Biol.* **58**, 347–375.

Saito, K., Kobayashi, M., Gong, Z., Tanaka, Y. and Tamazaki, M. (1999) Direct evidence for anthocyanidin synthase as a 2-oxoglutarate-dependent oxygenase: molecular cloning and functional expression of cDNA from a red form of *Perilla frutescens*. *Plant J.*, **17**, 181–189.

Sánchez-Fernández, R., Davies, T.G.E., Coleman, J.O.D. and Rea, P.A. (2001) The *Arabidopsis thaliana* ABC protein superfamily, a complete inventory *J. Biol. Chem.*, **276**, 30231–30244.

Saslowsky, D. and Winkel-Shirley, B. (2001) Localization of flavonoid enzymes in Arabidopsis roots. *Plant J.*, **27**, 37–48.

Sawada, S., Suzuki, H., Ichimaida, F., Yamaguchi, M., Iwashita, T., Fukui, Y., Hemmi, H., Nishino, T. and Nakayama, T. (2005) UDP-glucuronic acid:anthocyanin glucuronosyltransferase from red daisy (*Bellis perennis*) flowers. Enzymology and phylogenetics of a novel glucuronosyltransferase involved in flower pigment biosynthesis. *J. Biol. Chem.*, **14**, 899–906.

Shao, H., He, X., Achnine, L., Blount, J.W., Dixon, RA. and Wang, X. (2005) Crystal structures of a multifunctional triterpene/flavonoid glycosyltransferase from *Medicago truncatula*. *Plant Cell*, **17**, 3141–3154.

Shoji, K., Miki, N., Nakajima, N., Momonoi, K., Kato, C. and Yoshida, K. (2007) Perianth bottom-specific blue colour development in tulip cv. Murasakizuisho requires ferric ions. *Plant Cell Physiol.*, **48**, 243–251.

Shoji, K., Momonoi, K. and Tsuji, T. (2010) Alternative expression of vacuolar iron transporter and ferritin genes leads to blue/purple colouration of flowers in tulip cv. 'Murasakizuisho'. *Plant Cell Physiol.*, **51**, 215–224.

Stafford, H.A. and Lester, H.H. (1982) Enzymic and nonenzymic reduction of (+)-dihydroquercetin to its 3,4,-diol. *Plant Physiol.*, **70**, 695–698.

Suzuki, H., Nakayama, T., Fukui, N., Nakamura, N., Nakao, M., Tanaka, Y., Yamaguchi, M., Kusumi, T. and Nishino, T. (2001) Malonyl-CoA:anthocyanin 5-*O*-glucoside-6-O-malonyltransferase from Scarlet Sage (*Salvia splendens*) flowers. Enzyme purification, gene cloning, expression, and characterization. *J. Biol. Chem.*, **276**, 49013–49019.

Tanaka, Y., Sasaki, N. and Ohmiya, A. (2008) Biosynthesis of plant pigments: anthocyanins, betalains and carotenoids. *Plant J.*, **54**, 733–749.

Tanner, G.J., Francki, K.T., Abrahams, S., Watson, J.M., Larkin, P.J. and Ashton, A.R. (2003) Proanthocyanidin biosynthesis in plants. Purification of legume leucoanthocyanidin reductase and molecular cloning of its cDNA. *J. Biol. Chem.*, **278**, 31647–31656.

Teusch, M. and Forkmann, G. (1987) Malonyl-coenzyme A: anthocyanidin 3-glucoside malonyltransferase from flowers of *Callistephus chinensis*. *Phytochemistry*, **26**, 2181–2183.

Treutter, D. (2006) Significance of flavonoids in plant resistance: a review. *Environ. Chem. Lett.*, **4**, 147–157.

Umemoto, N., Abe, Y., Cano, E.A., Okamura, M., Sasaki, N., Yoshida, S. and Ozeki, Y. (2009) *Carnation serine carboxypeptidase-like acyltransferase is important for anthocyanin malyltransferase activity and formation of anthocyanic vacuolar inclusions.* 5th International Workshop on Anthocyanins, Japan, 2009, pp. 115.

Verrier, P.J., Bird, D., Burla, B., *et al.* (2008) Plant ABC proteins – a unified nomenclature and updated inventory. *Trends Plant Sci.*, **13**, 151–159.

Verweij, W., Spelt, C., Di Sansebastiano, G.P., Vermeer, J., Reale, L., Ferranti, F., Koes, R. and Quattrocchio, F. (2008) An H^+ P-ATPase on the tonoplast determines vacuolar pH and flower colour. *Nature Cell Biol.*, **10**, 1456–1462.

Winkel-Shirley, B. (1999) Evidence for enzyme complexes in the phenylpropanoid and flavonoid pathways. *Physiol. Plantarum*, **107**, 142–149.

Xie, D.-Y., Sharma, S.B., Paiva, N.L., Ferreira, D. and Dixon, R.A. (2003) Role of anthocyanidin reductase, encoded by *BANYULS* in plant flavonoid biosynthesis. *Science*, **299**, 396–399.

Xie, D.-Y., Sharma, S.B. and Dixon, R.A. (2004) Anthocyanidin reductases from *Medicago truncatula* and *Arabidopsis thaliana*. *Arch. Biochem. Biophys.*, **422**, 91–102.

Xie, D.-Y. and Dixon, R.A. (2005) Proanthocyanidin biosynthesis – still more questions than answers? *Phytochemistry*, **66**, 2127–2144.

Yazaki, K. (2005) Transporters of *secondary* metabolites. *Curr. Opin. Plant Biol.*, **8**, 301–307.

Yazaki, K., Sugiyama, A., Morita, M. and Shitan, N. (2008) Secondary transport as an efficient membrane transport mechanism for plant secondary metabolites. *Phytochem. Rev.*, **7**, 513–524.

Yonekura-Sakakibara, K. (2009) Functional genomics of natural product glycosyltransferases in Arabidopsis. *Plant Biotechnol.*, **26**, 267–274.

Yoshida, K., Kondo, T. and Goto, T. (1992) Intramolecular stacking conformation of gentiodelphin, a diacylated anthocyanin from *Gentiana makinoi*. *Tetrahedron*, **48**, 4313–4326.

Yoshida, K., Mori, M. and Kondo, T. (2009) Blue flower colour development by anthocyanins: from chemical structure to cell physiology. *Nat. Prod. Rep.*, **26**, 884–915.

Zhao, J. and Dixon, R.A. (2009) MATE transporters facilitate vacuolar uptake of epicatechin 3′-*O*-glucoside for proanthocyanidin biosynthesis in *Medicago truncatula* and *Arabidopsis*. *Plant Cell*, **21**, 2323–2340.

13

Pigment Biosynthesis II: Betacyanins and Carotenoids

Masaaki Sakuta and Akemi Ohmiya

13.1 Betacyanins

Red colours in flowers are mainly produced by two types of pigments: anthocyanins and betacyanins. These two pigments are stored in vacuoles, and serve important functions in plant reproduction through recruiting pollinators and seed dispersers. Betacyanins, red-violet pigments, and betaxanthins, yellow pigments, are members of the betalains, in which the basic structure is betalamic acid. The betacyanin molecule contains two nitrogen atoms, whereas anthocyanins contain no nitrogen. Betacyanins were originally known as nitrogenous anthocyanins (Piattelli, 1976; Mabry, 1980). Betacyanins are ammonium conjugates of dihydroindole (*cyclo*-DOPA) and dihydropyridine (betalamic acid) moieties, the latter also occurring in betaxanthins. In the betaxanthin molecule *cyclo*-DOPA moieties are replaced by amino acids or amines. All betacyanins are based on two isomeric aglycones, betabidin and isobetanidin, which differ only in their stereochemistry at C15. Glycosylation and subsequent acylation of the C5 or C6 hydroxyl group of aglycones generates a diversity of betacyanin structures (Figure 13.1).

Although anthocyanins are widely found in higher plants as red, purple and blue pigments of flowers, fruits, seeds, hypocotyls of seedlings, autumn leaves, and so on, betacyanins have largely replaced anthocyanins in the Caryophyllales, excluding the families Caryophyllaceae and Molluginaceae (Strack *et al.*, 2003). The occurrence of anthocyanins in the betacyanin-producing Caryophyllales has not been reported (Harborne, 1996). Thus, these two red pigments, anthocyanins and betacyanins, have never been demonstrated to

Plant Metabolism and Biotechnology, First Edition. Edited by Hiroshi Ashihara, Alan Crozier, and Atsushi Komamine.

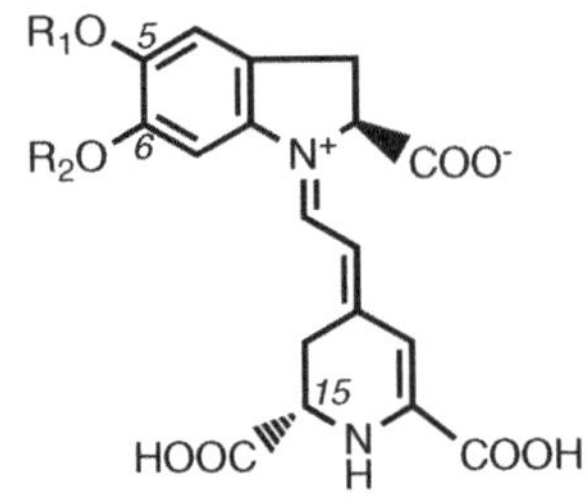

R_1 = H, R_2 = H Betanidin
R_1 = glucose, R_2 = H Betanin
R_1 = glucuronic acid, glucose, R_2 = H Amaranthin
R_1 = malonlylglucose, R_2 = H Phyllocactin
R_1 = feruloylglucose, R_2 = H Lampranthin II
R_1 = H, R_2 = glucose Gomphenrenin I

Figure 13.1 *Structures of betacyanins*

co-exist in one plant, and the evolutionary mechanism of the mutual exclusion of biosynthetic pathways remains a mystery. Betalains are also found in some fungi, including the poisonous mushroom fly agaric (*Amanitia muscaria*). The specific occurrence of betalains in phylogenetically distinct plants is another taxonomic mystery.

13.1.1 Biosynthesis

The biochemistry and genetics of the betalain biosynthetic pathway are relatively uncharacterised, while those of the anthocyanin biosynthetic pathway are probably one of the best-studied examples of secondary metabolism in higher plants (see Chapters 11 and 12). Radioactive feeding experiments have revealed that betalains are synthesized from tyrosine *via* DOPA (Garay and Towers, 1966; Nassif-Makki and Constabel, 1972), and DOPA is converted to the intermediates of betalains, betalamic acid and *cyclo*-DOPA (Miller *et al.*, 1968). Feeding experiments using doubly-labelled tyrosine have shown that extradiol cleavage, a step common to the synthesis of stezolobic acid (Saito and Komamine, 1976, 1978), and subsequent closure by the bounding of nitrogen to carbon 3, gives rise to betalamic acid (Fischer and Dreiding, 1972; Impellizzeri and Piattelli, 1972). The conjugation of betalamic acid and amino acids or amines leads to the formation of yellow betaxanthins, while the conjugation of betalamic acid and *cyclo*-DOPA results in the formation of betanidin, the aglycone of red betacyanins (Figure 13.2). These condensation reactions occur spontaneously (Schliemann *et al.*, 1999; Strack *et al.*, 2003). Betacyanins are normally stored as glycosides in vacuoles, after two different forms of glucosylation reactions at the betanidin step (Sciuto *et al.*, 1972; Heuer and Strack, 1992) or at the *cyclo*-DOPA step (Wyler *et al.*, 1984; Sasaki *et al.*, 2004).

Extensive studies on the biosynthetic reactions of betalains led to the discovery of several enzymes that function in early betalain biosynthesis: the polyphenol oxidase (PPO)-type tyrosinase (Steiner *et al.*, 1996, 1999) and the extradiol DOPA dioxygenase (DOD).

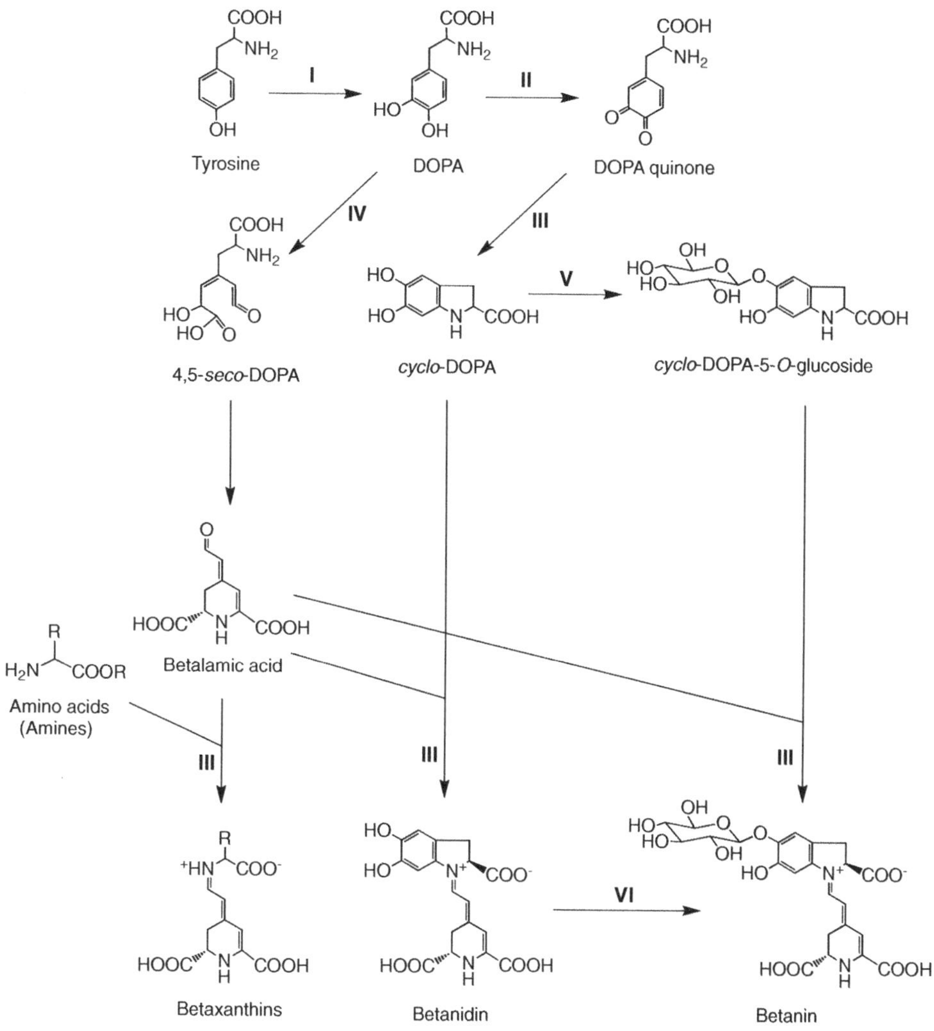

Figure 13.2 *Biosynthetic pathway of betalains. (I) tyrosine hydroxylase; (II) polyphenol oxidase; (III) spontaneous reaction; (IV) DOPA 4,5-dioxygenase; (V)* cyclo-*DOPA 5-O-glucosyltransferase; (VI) betanidin 5-O-glucosyltransferase*

Tyrosinase is a copper-type bifunctional enzyme that catalyzes both hydroxylation of tyrosine to DOPA and the subsequent oxidation reaction that converts DOPA to *o*-quinone (Strack and Schliemann, 2001). The cyclisation of *o*-quinone to *cyclo*-DOPA is assumed to occur spontaneously (Strack *et al.*, 2003). Tyrosine hydroxylation activity, converting tyrosine to DOPA, is separated from the oxidation of DOPA to DOPA *o*-quinone in *Portulaca grandiflora* (Yamamoto *et al.*, 2001), but the corresponding gene is as yet unknown.

The extradiolic 4,5-cleavage of DOPA is required for the formation of betalamic acid (Figure 13.2) and is catalyzed by the extradiol DOPA-4,5-dioxygenase. DOD was first detected in the betalain-producing fungus *Amanita muscaria* (Girod and Zrÿd 1991), and later characterised as extradiol DOPA-2,3- and -4,5-dioxygenase (Hinz *et al.*, 1997; Mueller *et al.*, 1997b). Genetic complementation of white petals of *Portulaca grandiflora* by particle bombardment transformation indicated that *Amanita* DOD catalyzed both 4,5- and 2,3-aromatic ring cleavage. This result showed that the formation of not only betalains, but also muscaflavin, a pigment that is not found in higher plants, is catalyzed by *Amanita* DOD (Mueller *et al.*, 1997a). A novel plant *DOD*, distinct from *Amanita DOD*, was originally isolated as a gene that encodes the DOPA- 4,5-dioxygenase from *Portulaca grandiflora* (Christinet *et al.*, 2004). The role of *Portulaca DOD* in the betalain biosynthetic pathway was determined by genetic complementation in white petals of *Portulaca grandiflora*, in which the set of genes for colour formation are missing (Christinet *et al.*, 2004). Recently, the enzymatic activity of recombinant *Mirabilis jalapa* DOD was shown to be able to catalyze the conversion of DOPA to betalamic acid *in vitro* (Sasaki *et al.*, 2009). In addition, some DODs from non-betalain-producing plants possess DOD activity *in vitro* (Tanaka *et al.*, 2008), although the functions of these homologues *in vivo* remain elusive. Therefore, further biochemical and physiological studies are required to assess the importance of *DOD* in betalain biosynthesis. The analyses of *DOD* homologues in species of both betalain-producing and non-producing plants should provide a better understanding of the evolutionary mechanism of *DOD* genes in betalain biosynthesis of the Caryophyllales.

Betacyanins are normally stored as glycosides in vacuoles. In their biosynthetic pathway, two different forms of glucosylation reactions occur. The first is at the *cyclo*-DOPA step and the second is at the betanidin step, which matches the biosynthetic processes of other plant secondary products such as anthocyanins. Two regio-specific betanidin glucosylating enzymes, UDP-glucose:betanidin-5-*O*-glucosyltransferase and UDP-glucose:betanidin-6-*O*-glucosyltransferase (5-GT and 6-GT), were partially purified from Livingstone daisy (*Dorotheanthus bellidiformis*) cell cultures (Heuer and Strack, 1992). Although both 5-GT and 6-GT were shown to discriminate between individual hydroxyl groups of the respective substrates, these enzymes catalyze the indiscriminate transfer of glucose from UDP-glucose to hydroxyl groups of betanidin, flavonols and anthocyanidins (Vogt *et al.*, 1997; Vogt, 2002).

On the other hand, it has been reported that *cyclo*-DOPA-5-*O*-glucoside rather than betanidin or betanin was an efficient precursor of amaranthin in *Celosia cristata* (Sciuto *et al.*, 1974). In addition, the accumulation of *cyclo*-DOPA-5-*O*-glucoside has been shown in young beet plants (Wyler *et al.*, 1984) and root peels of red beet (Kujala *et al.*, 2001). Activity of *cyclo*-DOPA GT has been detected in crude extracts prepared from red petals of *Mirabilis jalapa* and several betacyanin-producing plants (Sasaki *et al.*, 2004). Furthermore, cDNAs encoding *cyclo*-DOPA5GT were isolated from the petals of *Mirabilis jalapa* and the inflorescences of *Celosia cristata* (Sasaki *et al.*, 2005b). The expression profile of the *cyclo-DOPA5GT* gene agreed with the *cyclo*-DOPA5GT activity during the development of petals of *Mirabilis jalapa*. UDP-glucuronic acid:*cyclo*-DOPA-5-glucoside glucuronosyltransferase activity was detected in *Celosia cristata* (Sasaki *et al.*, 2005a), indicating that modification with the glucuronic acid moiety occurs at *cyclo*-DOPA. These facts suggest the dual pathway of glycosylation in betacyanin biosynthesis. Further studies on glycosylation and subsequent acylation in betalain biosynthesis may elucidate which pathway is the main route for batalain production, or whether the step at which glucosylation and acylation occur depends on the species.

13.1.2 Factors Controlling Betacyanin Biosynthesis

Betacyanin accumulation in the Caryophyllales is affected markedly by environmental factors. In the last quarter of the twentieth century, the effects of various environmental factors on betacyanin accumulation were investigated. Light-stimulated betacyanin synthesis mediated by the red/far red reversible action of phytochrome has been reported (Nicola *et al.*, 1973a,b, 1974; Elliot, 1979c; Spasic *et al.*, 1985). Plant growth regulators also affect betacyanin accumulation. In particular, it is well known that cytokinins markedly promote betacyanin accumulation (Piattelli, 1976). The action of light and kinetin on betacyanin synthesis has been analysed (Koehler, 1972a,b; Koehler *et al.*, 1981). In etiolated *Amaranthus* seedlings, exogenously supplied cytokinins strongly promote betacyanin accumulation in the dark (Bauberger and Mayer, 1960). This occurs mainly in two specific tissues, the lower epidermal cells of cotyledons, excluding guard cells, and hypocotyl endodermis (Elliott, 1983). A bioassay for cytokinins using betacyanin accumulation as a marker in *Amaranthus* seedlings was reported (Biddington and Thomas, 1973) and the variability of this method was discussed (Elliott, 1979a–d). It has also been reported that betacyanin accumulation is affected by various growth regulators other than cytokinins. In suspension cultures of *Phytolacca americana*, betacyanin accumulation was markedly stimulated by 2,4-D at a concentration of 5 μM (Sakuta *et al.*, 1991), whereas betacyanin accumulation in callus cultures of *Beta vulgaris* was suppressed by 2,4-D (Constabel and Nassif-Makki, 1971). It has also been shown that betacyanin accumulation in callus cultures of *Portulacca grandiflora* is increased by NAA, another synthetic auxin, at a concentration of 1 ppm (Endress, 1976). Inhibitory effects of gibberellic acid and abscisic acid on betacyanin accumulation have been shown in seedlings of *Amaranthus* (Biddington and Thomas, 1977; Stobart and Kinsman, 1977; Guruprasad and Laloraya, 1980) and suspension cultures of *Phytolacca americana* (Hirano *et al.*, 1996). However, details of the mode of action of growth regulators on betacyanin accumulation are poorly understood.

13.1.3 Molecular Mechanism of the Mutual Exclusion of Anthocyanins and Betacyanins

The two types of red pigment, anthocyanins and betacyanins, never occur together in the same plant. Although anthocyanins are widely distributed in higher plants, betacyanins have replaced anthocyanins in the Caryophyllales, except in the families Caryophyllaceae and Molluginaceae (Strack *et al.*, 2003). The occurrence of anthocyanins in the betacyanin-producing Caryophyllales has not been reported (Harborne, 1996). This curious mutual exclusion has been examined from genetic and evolutionary perspectives (Koes *et al.*, 1994; Stafford, 1990, 1994); nevertheless, little is known about it at the molecular level.

Although the molecular mechanism of betacyanin biosynthesis is still relatively poorly understood, the biosynthetic pathway of flavonoids is probably one of the best-studied examples of secondary metabolism in higher plants (see Chapters 11 and 12). With few exceptions, flavonoid biosynthetic genes have been cloned and analysed, and factors that control transcription of the genes have also been isolated by genetic means. The regulatory mechanism of flavonoid biosynthesis has been revealed in several species. Whereas anthocyanins are absent in almost all members of the Caryophyllales, species of this family do contain other flavonoids, especially flavonols (Iwashina, 2001). For example, the yellow tepals of *Astrophytum* species contain the flavonol glycosides quercetin-3-*O*-galactoside and quercetin-3-*O*-rhamnosylglucoside, together with the aglycones quercetin, kaempferol and isorhamnetin, in the form of spherical crystals (Iwashina *et al.*, 1988). Dihydroflavonols are at the branching point of flavonols and anthocyanins in the flavonoid biosynthetic pathway (Figure 13.3). This suggests that anthocyanin biosynthesis from dihydroflavonols to anthocyanins may be blocked in the Caryophyllales. Some insights can be gained from dihydroflavonol-4-reductase (DFR) and anthocyanidin synthase (ANS), which are involved in the conversion of dihydroflavonols to anthocyanins. The isolation and functional identification of *DFR* and *ANS* genes from spinach (*Spinacia oleracea*) and pokeweed (*Phytolacca americana*), which are non-anthocyanin-producing plants of the Caryophyllales, have been reported (Shimada *et al.*, 2004, 2005). The expression profile revealed that in *Spinacia oleracea, DFR* and *ANS* were not expressed in most tissues and organs, except seeds.

One possible explanation for the lack of anthocyanin synthesis in the Caryophyllales may therefore be the suppression or limited expression of the *DFR* and *ANS* (Shimada *et al.*, 2005). The evolution of *cis*-regulatory elements has been proposed to be a major source of morphological diversification, as mutations in *cis*-regulatory elements often lead to dramatic tissue-specific pattern changes while preserving the essential roles of these genes in other processes (Shapiro *et al.*, 2004; Gompel *et al.*, 2005). The modification of *DFR* and *ANS cis*-regulatory elements may lead to the limited expression of *DFR* and *ANS*, resulting in defective anthocyanin synthesis in the Caryophyllales (Shimada *et al.*, 2007). A more detailed analysis of the promoters and the characterisation of regulators for flavonoid biosynthesis will provide further understanding of the regulatory mechanism of flavonoid biosynthesis in the Caryophyllales and may elucidate why these plants do not produce anthocyanins.

13.1.4 Betacyanins as Food Colourants

In recent years, naturally occurring pigments have been used increasingly as quality food colourant, since consumers are wary of synthetic dyes and have a clear preference for natural pigments. When natural pigments are used as food colourants, colour stability is a major concern. There are several factors that have been recognised to affect the stability of natural pigments (Delgado-Vargas *et al.*, 2000). The colour of anthocyanins is directly influenced by pH; red in acid solutions, violet or purple in neutral solutions, and blue in alkaline pH. This is the reason that most colourants containing anthocyanins can only be used at pH values of <4. In contrast, the colour hue of betalains is unaffected at pH 3.5–7.0, which is within the range of most foods. Betalains have been used as natural colourants in food products such as ice cream, candies, yogurt, and so on. Besides imparting attractive colour to food products, betalains have been shown to confer free-radical scavenging and allied antioxidant activities (Cai *et al.*, 2003; Stintzing *et al.*, 2005; Tesoriere *et al.*, 2005, 2008), and it has been suggested that betalains as food colourants may provide protection against certain oxidative stress-related disorders in human (Kanner *et al.*, 2001; Lee *et al.*, 2005; Allegra *et al.*, 2007) .

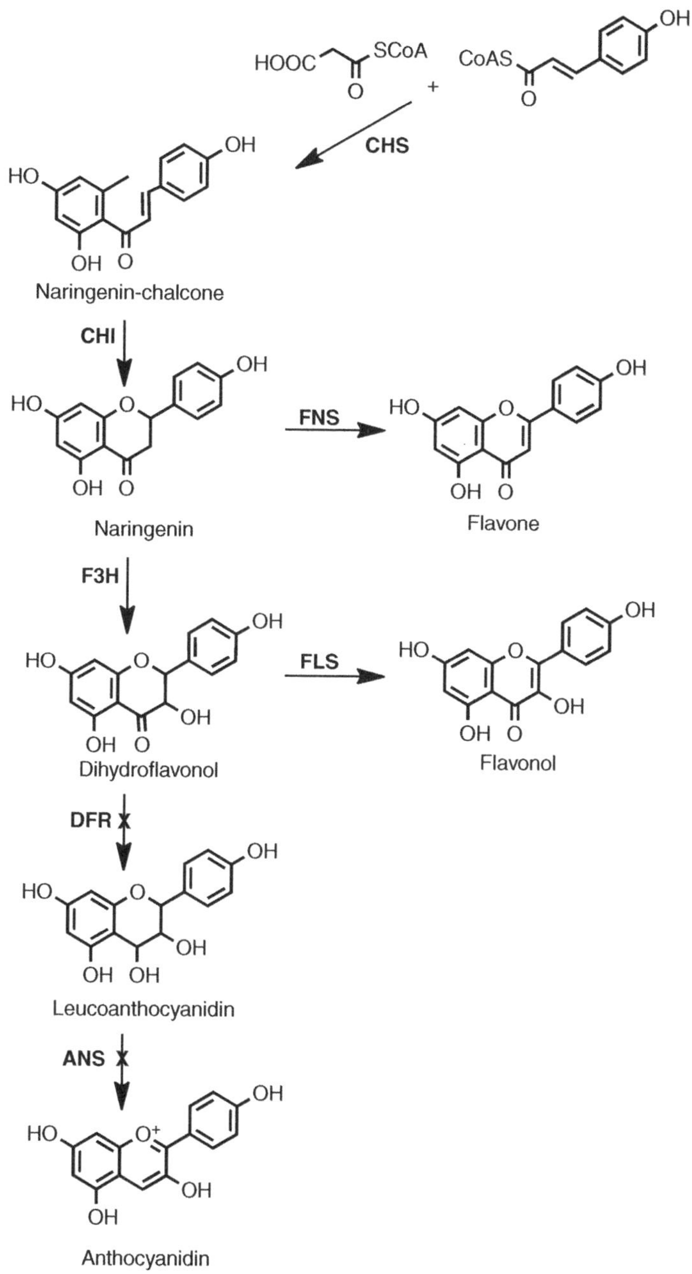

Figure 13.3 *Flavonoid biosynthetic pathway in the Caryophyllales. CHS, chalcone synthase; CHI, chalcone isomerase; F3H, flavanone 3-hydroxylase; FNS, flavone synthase; DFR, dihydroflavonol 4-reductase; ANS, anthocyanidin synthase*

13.2 Carotenoids

Carotenoids are C_{40} isoprenoid pigments with or without epoxy, hydroxy and keto groups, and provide the bright yellow, orange or red colours. More than 700 naturally occurring carotenoids have been identified, widely distributed in plants, animals and microorganisms (Britton *et al.*, 2004). They furnish flowers and fruits with distinct colours which, along with anthocyanins, attract pollinators and seed dispersers. In photosynthetic tissues, carotenoids are essential structural components of the photosynthetic antenna and reaction centre complexes and protect against potentially harmful photo-oxidative processes (Green and Durnford, 1996; Niyogi, 2000). Carotenoids are made up with 5-carbone isoprene units. They are synthesized as part of the terpenoid biosynthesis pathway, sharing a precursor, geranylgeranyl diphosphate (GGPP), with other essential pathways that lead to the synthesis of gibberellins, brassinosteroids, chlorophylls and vitamin E. Some carotenoids provide substrates for the biosynthesis of the plant hormones abscisic acid (ABA) and strigolactone (Nambara and Marion-Poll, 2005; Dun *et al.*, 2009). Carotenoids also play an important role in human nutrition and health, serving as provitamin A and anticancer activities (Mayne, 1996). Some carotenoids are used as food colourants, cosmetics and pharmaceuticals.

13.2.1 Carotenoid Diversity

Carotenoids accumulate in photosynthetic tissues of all the green plants, although their yellow/red colours are masked by the chlorophylls. Flowers, fruits and roots of some plants accumulate specific carotenoids depending on the plant species and/or cultivars. They furnish unique colours ranging from yellow and orange to red, and are at their brilliant best in the autumn colours of senescing leaves which lose chlorophyll prior to abscission.

13.2.1.1 Leaves

The green tissues of most plants have similar carotenoid profiles: carotenoids that are essential for photosynthesis, such as lutein, β-carotene, violaxanthin and neoxanthin, are invariably found. In leaves, carotenoids are located in chloroplasts, form chlorophyll–carotenoid–protein complexes, and are involved in light harvesting and chlorophyll photoprotection (Vishnevetsky *et al.*, 1999).

13.2.1.2 Roots

Most plants do not accumulate carotenoids in their roots. The storage roots of carrot (*Daucus carota*) (Baranska *et al.*, 2006) and sweet potato (*Ipomoea batatas*) (Hagenimana *et al.*, 1999) are exceptions. They accumulate a high concentration of β-carotene, and serve as an important source of vitamin A in the human diet. Because of an extremely high concentration, β-carotene in carrot root is stored in a large crystal in a chromoplast (Straus, 1961).

13.2.1.3 Fruits

Fruit maturation is accompanied by colour change: the accumulation of carotenoid and/or anthocyanin pigments takes place concomitantly with a decrease in chlorophyll content.

Such colour changes in the ripening fruits contrasts with the vegetative colour of plants, and signals to seed dispersers that the fruits are ready to eat. During the ripening process, transition from chloroplasts to chromoplasts occurs, and carotenoids are sequestered in the plastogobule or crystalline structure (Tevini and Steinmuller, 1985).

Carotenoid compositions contained in fruits vary widely among plant species and cultivars. Fruits of tomato (*Solanum lycopersicum*) accumulate a large amount of lycopene (Fraser *et al.*, 1994). Capsanthin and capsorbin, ketocarotenoids which contain one and two acyl-cyclo-pentanol rings, respectively, are typical carotenoids in the fruit of red pepper (*Capsicum annuum*) (Hornero-Méndez *et al.*, 2000). Citrus fruits display a wide range of colourations due to the accumulation of specific carotenoids. Juice sacs of mandarin orange fruits predominantly accumulate β-cryptoxanthin, while those of naval orange fruits predominantly accumulate violaxanthin (Matsumoto *et al.*, 2007). The pink or red colour of juice sacs in some citrus fruit cultivars is due to accumulation of lycopene (Xu *et al.*, 2006).

Starchy seeds generally possess little or no carotenoids. Maize (*Zea mays*) is a unique plant accumulating substantial amount of carotenoids, mainly lutein and zeaxanthin, in the amyloplasts of seeds (Kurilich and Juvik, 1999). Oilseeds of pumpkin (*Cucurbita pepo*), sunflower (*Helianthus annuus*) and canola (*Brasica napus*) predominantly accumulate lutein in the elioplasts, lipid-storing plastids (Matus *et al.*, 1993; Shewmaker *et al.*, 1999; McGraw *et al.*, 2001). *Bixa orellana* is the only plant that accumulates bixin in its seeds (Bouvier *et al.*, 2003a). Bixin is a dicarboxyl monomethyl ester apocarotenoid, produced by a cleavage of lycopene. It also known as annatto, and is used in food and cosmetics as a red colour additive.

13.2.1.4 Flowers

Major carotenoids contained in the flower petals are xanthophylls, pale- to deep-yellow in colour. Some well-known examples are yellow flowers of chrysanthemum (*Chrysanthemum morifolium*), rose (*Rosa hybrida*), viola (*Viola tricolor*), and narcissus (*Narcissus pseudonarcissus*) (Eugster and Märki-Fisher, 1991; Kishimoto *et al.*, 2004; Molnar and Szabolcs, 1980; Valadon and Mummery, 1968). In most flowers, xanthophylls are present in an esterified form. They are associated with carotenoid-lipoprotein structures and are sequestered within the chromoplasts (Vishnevetsky *et al.*, 1999).

Petals of some plants have a modified ability of carotenoid biosynthesis and accumulate unique carotenoids which are rarely found in other plant species. The petals of *Adonis aestivalis* and *A. annua* accumulate a large amount of astaxanthin, a red ketocarotenoid (Cunningham and Gantt, 2005). They also contain smaller quantities of adonirubin and adonixanthin, signature pigments of these species, and 3-hydroxyechinenon. In petals of Asiatic hybrid lily (*Lillium* spp.), the major carotenoids of a yellow-flowered cultivar are yellow xanthophylls, while capsanthin accumulates in a red-flowered cultivar (Yamagishi *et al.*, 2010). Capsanthin occurs in the fruit of red pepper, but it is rarely found in flowers. The orange petals of calendula (*Calendula officinalis*) contain reddish carotenoids that are absent in yellow petals. Some have a *cis*-structure at C5 or C5′, which is very rare in plants (Kishimoto *et al.*, 2005).

Some flowers accumulate apocarotenoids, carotenoid cleavage products, with an orange to red colour. The red style branches of crocus (*Crocus sativus*), from which the spice saffron is produced, accumulate unique apocarotenoids, crocetin glycosides, picrocrocin and safranal. They are responsible for the colour, taste and aroma of saffron (Bouvier *et al.*, 2003b). Flowers of *Boronia megastigma* are a complex source of apocarotenoids including β-ionone (Cooper *et al.*, 2009).

13.2.2 Carotenoid Biosynthesis

Biosynthesis of carotenoids takes place in plastids catalyzed by nuclear-encoded enzymes (see reviews by Cunningham and Gantt, 1998; Fraser and Bramley, 2004; Howitt and Pogson, 2006; Tanaka *et al.*, 2008; Cazzonelli and Pogson, 2010). The initial steps in the pathway leading to the formation of 5-carbon compound, isopentenyl diphosphate (IPP), are described in Chapter 9.

Figure 13.4 summarises the carotenoid biosynthesis pathway starting from IPP. Four IPPs are condensed to C_{20}-geranylgeranyl-diphosphate (GGPP). Condensation of two GGPP molecules by phytoene synthase (PSY) yields the first C_{40} carotenoid, phytoene. Two or more PSY homologues are found in some plant species, each showing organ-specific and stress-inducible expressions. Tomato has two different types of PSYs (Psy-1 and Psy-2) (Fraser *et al.*, 1999). Psy-1 encodes a fruit- and flower-specific isoform and is responsible for carotenogenesis in chromoplasts. In green tissues, Psy-2, which is homologous to Psy-1 but highly divergent from it, is expressed predominantly, and it makes the major contribution to carotenogenesis in chloroplasts. Chromoplast-specific PSYs have also been identified in pepper (Römer *et al.*, 1993), maize (Buckner *et al.*, 1996) and daffodil (*Narcissus pseudonarcissus*) (Schledz *et al.*, 1996).

Conjugated double bonds are subsequently added by two structurally similar enzymes, phytoene desaturase (PDS) and ζ-carotene desaturase (ZDS). These desaturation reactions yield the intermediates phytofluene, ζ-carotene, neurosporene, and lycopene; these carotenoids contain 5, 7, 9 and 11 conjugated double bonds, respectively. Increasing the number of conjugated double bonds shifts the absorption toward longer wavelengths, resulting in colourless phytoene and phytofluene, pale-yellow ζ-carotene, orange-yellow neurosporene, and red lycopene. During the desaturation steps, several reaction intermediates with a *cis*-configuration are produced (Isaacson *et al.*, 2004). Because all-*trans*-lycopene configuration is required for the following lycopene cyclisation steps, conversion of *cis*- to *trans*-configuration is carried out by two types of isomerases, CRTISO and Z-ISO, as well as light-mediated photoisomerisation (Isaacson *et al.*, 2002; Park *et al.*, 2002). Z-ISO was recently identified in maize; its activity occurs upstream of CRTISO, catalyzing the *cis* to *trans* conversion of 15-*cis*-bond in 9,15,9′-tri-*cis*-ζ-carotene, the product of PDS, to form 9,9′-di-*cis*-ζ-carotene, the substrate of ZDS (Li *et al.*,, 2007). CRTISO isomerises the *cis* double bonds at 7,9 and 7′,9′ of 7,9,9′-tri-*cis*-neurosporene and 7′,9′-di-*cis*-lycopene, the products of ZDS, to produce all-*trans*-lycopene (Isaacson *et al.*, 2004).

The cyclisation of lycopene is a branch point in the pathway, catalyzed by two lycopene cyclases, lycopene β-cyclase (LCYB) and lycopene ε-cyclase (LCYE). Because LCYE of most plants adds only one ε-ring to lycopene (Cunningham *et al.*, 1996; Cunningham and Gantt, 2001), the pathway in plants typically proceeds only along branches leading to carotenoids with one β- and one ε-ring (β,ε-carotenoid) or two β-rings (β,β-carotenoid). Lycopene-ε-cyclase in romaine lettuce (*Lactuca sativa*) has the unique ability to add two ε-rings to lycopene and yields bicyclic ε-carotene, lactucaxanthin (Cunningham and Gantt, 2001). A single amino-acid mutation is shown to be sufficient to form bicyclic ε-carotene.

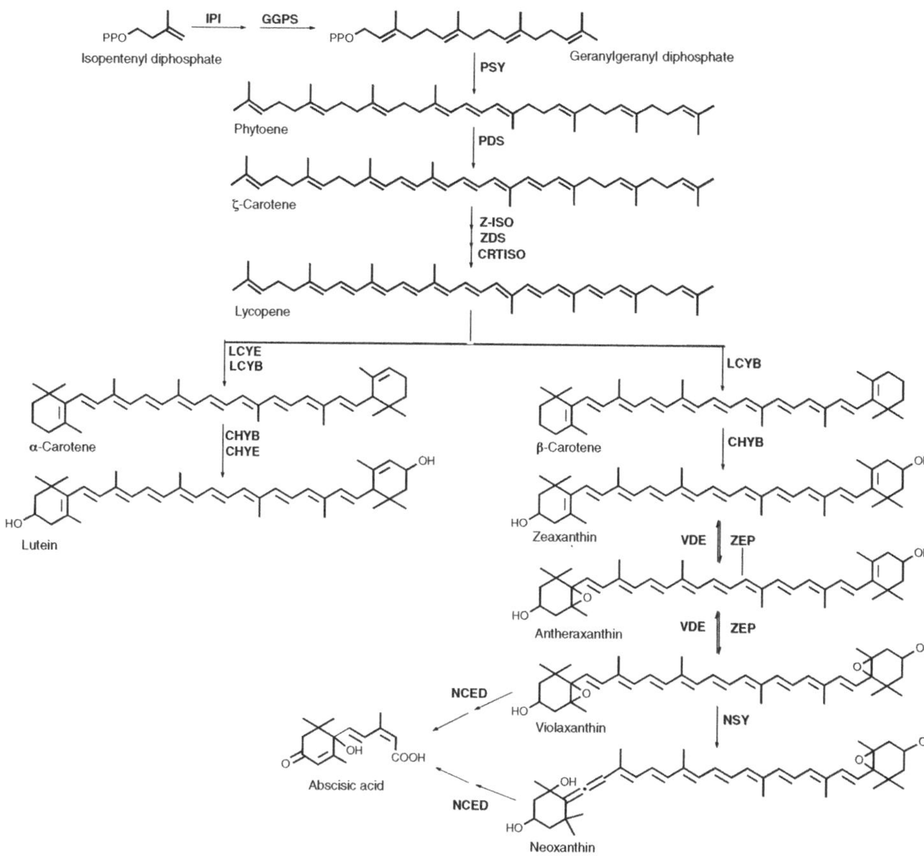

Figure 13.4 *Schematic of the carotenoid biosynthesis pathway in plants. IPI, isopentenyl pyrophosphate isomerase; GGPS, geranylgeranyl pyrophosphate synthase; PSY, phytoene synthase; PDS, phytoene desaturase; Z-ISO, 15-*cis*-ζ-CRTISO; ZDS, ζ-carotene desaturase; CRTISO, carotenoid isomerase; LCYE, lycopene ε-cyclase; LCYB, lycopene β-cyclase; CHYE, ε-ring hydroxylase; CHYB, β-ring hydroxylase, ZEP, zeaxanthin epoxidase; VDE, violaxanthin deepoxidase; NSY, neoxanthin synthase; NCED, 9-cis epoxycarotenoid dioxygenase*

β- and α-carotenes are further modified by hydroxylation or epoxidation to produce a variety of structural features. Those oxygenated derivatives of carotene are called xanthophylls. Two types of hydroxylases are involved in hydroxylation at 3 and 3' positions of α- and β-rings. β-hydroxylase (CHYB), a nonheme diiron monooxygenase, acts on β-rings, while ε-ring hydroxylase (CHYE), a P450-type monooxygenase, CYP97C1, acts on ε-rings (Tian *et al.*, 2004). The β-ring of α-carotene is hydroxylated by CHYB to produce zeinoxanthin, and then the ε-ring is hydroxylated by CHYE to yields lutein. Two β-rings of β-carotene are hydroxylated in two steps by CHYB to yield β-cryptoxanthin and then zeaxanthin. Epoxidation at the 5,6 and 5′,6′ of the β-ring of zeaxanthin, catalyzed by zeaxanthin epoxidase (ZEP), yields violaxanthin. Violaxanthin is converted to neoxanthin by the action of neoxanthin synthase (NSY). Both violaxanthin and neoxanthin are cleaved by NCED to form xanthoxin, a precursor of plant hormone ABA (Nambara and Marion-Poll, 2005).

13.2.3 Carotenoid Degradation

Specific enzymatic cleavage of carotenoids produces various types of apocarotenoids, some of which play important biological functions in the growth and development of both plants and animals. The most well-known apocarotenoids are vitamin A and ABA. Vitamin A is an important visual pigment and signalling molecule in animals (Della Penna and Pogson, 2006). The plant hormone ABA plays a key role in seed development and in responses to environmental stresses related to loss of water (Nambara and Marion-Poll, 2005). Some carotenoids provide substrates for the biosynthesis of strygolactone, a signalling molecule that inhibits auxillary branch outgrowth (Dun *et al.*, 2009). In addition to these bioactive compounds, they provide flowers and fruits with unique aromas and colours for attracting pollinators and seed dispersers.

The first protein found to specifically cleave carotenoids, viviparous14 (VP14), was identified by an analysis of a viviparous ABA-deficient mutant of maize (Schwartz *et al.*, 1997; Tan *et al.*, 1997). VP14 cleaves the 11,12(11′,12′) double bonds of the 9-*cis* isomers of neoxanthin and violaxanthin to yield xanthoxin, the precursor of ABA. The pioneering work on VP14 facilitated the discovery of related enzymes in different plant species and other organisms. In *Arabidopsis*, enzymes specifically cleave carotenoids fall into nine clades (Tan *et al.*, 2003). Five of these, the 9-*cis* epoxycarotenoid dioxygenases (NCED2, NCED3, NCED5, NCED6, and NCED9) are closely related to, and have the same activity as VP14. The remaining four carotenoid cleavage dioxygenases (CCD1, CCD4, CCD7, and CCD8) have low sequence homologies to the NCEDs, and their enzyme activities and substrate specificities also differ from those of the NCEDs. Orthologues belonging to the NCED and CCD subfamilies have been identified in many plant species, and they are named on the basis of their homology to those of *Arabidopsis*.

Recent studies concerning CCD enzymes have made significant progress in relation to their enzymatic activities and biological significance of their apocarotenoid products (Ohmiya, 2009). CCD1 contributes to the formation of apocarotenoid volatiles in the fruits and flowers. Orthologues of AtCCD1 have been found in a variety of plant species. PhCCD1 from petunia (*Petunia hybrida*), and LeCCD1 from tomato, cleave carotenoids at 9,10 (9′,10′) double bonds and contribute to the formation of β-ionone and geranylacetone, important flavour constituents (Simkin *et al.*, 2004a,b).

Some CCD4s are involved in the pigmentation in flowers and seeds. In crocus stigma, red-coloured apocarotenoids are produced by cleavage of zeaxanthin catalyzed by zeaxanthin cleavage enzyme (CsZCD) (Bouvier *et al.*, 2003b). The seed of *Bixia orellana* accumulate a red apocarotenoid called bixin, a colour additive used in foods and cosmetics. Bixin is produced by the cleavage of lycopene catalyzed by lycopene dioxygenase (BoLCD) (Bouvier *et al.*, 2003a). Both CsZCD and BoLCD fall into the same clade as AtCCD4. Chrysanthemum CCD4 (CmCCD4) contributes white petal colour formation by cleaving carotenoid into colourless compounds, as described in Section 13.2.5 (Ohmiya *et al.*, 2006).

Heterologous expression studies in *Escherichia coli* showed that CCD7 cleaves carotenoids asymmetrically at the 9,10 double bond to yields β-ionone and C_{27} aldehyde (Schwartz *et al.*, 2004). The C_{27} aldehyde is then cleaved by CCD8 at the 13,14 double bond to produce a C_{18} ketone and a C_9 dialdehyde. Sequential activity of CCD7 and CCD8 have been assumed to produce strigolactone because mutations of these genes result in reduced levels of strigolactone and increased branching phenotypes in every plant species studied to date (Dun *et al.*, 2009).

13.2.4 Regulation of Carotenoid Biosynthesis

13.2.4.1 Vegetative Tissues

In photosynthetic tissues, carotenoid biosynthesis is enhanced by light and suppressed by prolonged darkness (Simkin *et al.*, 2003). In these cases, carotenogenesis is predominantly regulated at the transcriptional level. For example, expression levels of *PSY* and *PDS* are extremely low in dark-adapted leaves of pepper. In the seedlings of mustard (*Sinapis alba*) and *Arabidopsis*, an enhanced level of *PSY* expression was observed when the seedlings were exposed to light (von Lintig *et al.*, 1997). Upregulation of PSY expression in the light has also been reported with tomato leaves (Giuliano *et al.*, 1993). These results suggest strongly that PSY is a key enzyme that controls carotenogenesis in leaves.

In the promoter of *Arabidopsis PSY*, *cis*-regulatory element ATCTA was identified and RAP2.2 was isolated as a transcription factor that binds to the element. The ATCTA element is also present in the promoter of *PDS*, which suggests that a common regulatory mechanism exists for both genes (Welsch *et al.*, 2007). Using a yeast two-hybrid assay, the RING finger protein SEVEN IN ABSENTIA OF ARABIDOPSIS2 (SINAT2) was identified as a RAP2.2 interaction partner. Suppression of either *RAP2.2* or *SINAT2* expression levels by T-DNA insertion resulted in only small alterations of carotenoid content in root calli, suggesting that these factors are components of a more complex regulatory network controlling carotenoid biosynthesis.

Analysis of a *carotenoid chloroplast regulatory1* (*ccr1*) mutant of *Arabidopsis* has shown that carotenoid biosynthesis is epigenetically regulated (Cazzonelli *et al.*, 2009). The *CCR1* gene encodes a chromatin-modifying histone methyltransferase enzyme (SET DOMAIN GROUP 8, SDG8). Mutation of *SDG8* causes altered methylation of the *CRTISO* gene surrounding the translation start site, which results in reduced levels of *CRTISO* expression. The altered level of *CRTISO* mRNA not only causes a substantial decrease of lutein in leaves but also results in increased shoot branching, possibly by changing the level of strigolactone, a carotenoid-derived branch-inhibiting hormone.

13.2.4.2 Fruits and Flowers

Flowers and fruits have a wide variety of carotenoid contents, ranging from little or none to large amounts. In marigold (*Tagetes erecta*), the difference in petal colour from pale-yellow to orange-red is caused by the degree of accumulation of the xanthophyll lutein. Moehs *et al.* (2001) showed that an increase in *PSY* and 1-deoxy-D-xylulose-5-phosphate synthase (*DXS*) expressions might be responsible for the difference in the amount of lutein in the petals. PSY has also been identified as a rate-limiting enzyme of carotenoid biosynthesis in canola seeds (Shewmaker *et al.*, 1999) and tomato fruits (Fraser *et al.*, 1994).

Each flower has a distinct balance of β,ε- and β,β-carotenoids. For example, in petals of marigold and chrysanthemum, the main carotenoids are lutein and its derivatives (β,ε-carotenoids) (Moehs *et al.*, 2001; Kishimoto and Ohmiya, 2006), On the other hand, yellow tepals of Asiatic hybrid lily contain predominantly β,β-carotenoids, such as antheraxanthin and violxanthin (Yamagishi *et al.*, 2010). It is generally thought that the ratio of β,ε- and β,β-carotenoids can be controlled simply by the relative amounts or activities of LCYB and LCYE. Petals of both marigold and chrysanthemum show higher expression levels of *LCYE* compared with that of *LCYB*. In contrast, expression of *LCYB* is higher than that of *LCYE* in yellow tepals of Asiatic hybrid lily. There is a shift from β,ε-carotenoids to β,β-carotenoids during development of *Gentiana lutea* flowers and citrus fruits (Zhu *et al.*, 2003; Kato *et al.*, 2004). This compositional change is accompanied by an increase in *LCYB* transcripts and a decrease in *LCYE* transcripts.

There is increasing evidence that CHYB is a key regulatory enzyme for carotenoid sequestration into the chromoplasts. The *wf* mutant in tomato, which produces a white flower phenotype, is caused by a mutation in *CrtR-b2*, a *CHYB* homologue that is expressed specifically in petals (Galpaz *et al.*, 2006). The *wf* mutant shows a significant decrease in the levels of β-carotene derivatives in petals, whereas the transcript levels of genes encoding the rate-limiting enzymes involved in isoprenoid and carotenoid biosyntheses show the same transcript levels as in the petals of wild-type plants. In *Arabidopsis*, the homologue of *CHYB* is not expressed in the floral organ, while other carotenogenic genes are substantially expressed (Kim *et al.*, 2009; AtGenExpress database: http://www.arabidopsis.org/info/expression/ATGenExpress.jsp). In white flowers of *Ipomoea nil*, most of the carotenogenic gene expression is especially low; in particular, expression of *CHYB* was suppressed very strongly (Yamamizo *et al.*, 2010). In tepals of Asiatic hybrid lily (*Lilium* spp.), the expression level of the *CHYB* homologue in various cultivars correlates well with the carotenoid content and more closely than any other carotenogenic genes (Yamagishi *et al.*, 2010). Furthermore, in chrysanthemum, lily and *G. lutea*, *CHYB* expression is shown to be upregulated in parallel with an increase in the carotenoid level during petal development (Zhu *et al.*, 2003; Kishimoto and Ohmiya, 2006; Yamagishi *et al.*, 2010). Hydroxylation is an important process for the esterification of carotenoids, which in turn is necessary for carotenoid sequestration and stabilisation in the lipid-rich plastoglobule of the chromoplast. The upregulation of *CHYB*, which encodes for an enzyme that catalyzes the addition of hydroxyl residues required for esterification, may be a key event in carotenoid sequestration into the chromoplasts. These findings strongly suggest that the extent of carotenoid accumulation may not rely solely on the rate of biosynthesis, but on the extent to which the carotenoid structure is suitable for the sequestration in the chromoplasts.

Although knowledge of the regulatory mechanism of the carotenogenesis is limited, several components that affect carotenoid levels in fruit chromoplasts have been identified. In tomato fruits, the genes responsible for *hp1* and *hp2* (mutations conferring a high level of carotenoids) have been shown to encode the proteins UV-DAMAGED DNA-BINDING PROTEIN 1 (DDB1) and DEETIOLATED 1 (DET1), respectively (Liu *et al.*, 2004). These are the components that are involved in the light-signal transduction pathway. In addition, other light-signalling components, such as HY5 and COP1, have been shown to antagonistically regulate carotenoid levels (Liu *et al.*, 2004; Davuluri *et al.*, 2005). In *Arabidopsis*, DDB1 and DET1 form a complex with Cullin4 (CUL4), a ubiquitin-conjugating E3 ligase, to repress photomorphogenesis (Bernhardt *et al.*, 2006; Chen *et al.*, 2006). CUL4 homologue has also been identified in tomato fruits and is suggested to be in association with DET1 and DDB1 (Wang *et al.*, 2008). Downregulation of these components by RNA interference (RNAi) increased the carotenoid content of tomato fruits (Liu *et al.*, 2004; Davuluri *et al.*, 2005; Wang *et al.*, 2008). Although precise information on mechanism(s) by which these components affect carotenoid levels remains to be elucidated, these genes may become useful genetic tools for the manipulation of carotenoid content.

13.2.5 Regulation of Carotenoid Accumulation Other than via Biosynthesis

There are some white flowers which, despite having the ability to synthesize carotenoids, contain only negligible amounts. Factors affecting carotenoid content other than biosynthesis may exist in such flowers. Two different regulatory factors have been reported in chrysanthemum and cauliflower (*Brassica oleracea*).

In the case of chrysanthemums, a gene encoding carotenoid cleavage dioxygenase 4 (CmCCD4a) is specifically expressed in white petals (Ohmiya *et al.*, 2006). Suppression of *CmCCD4a* expression by RNAi results in a change in the petal colour from white to yellow. In addition, carotenogenic gene expression normally occurs in white petals (Kishimoto and Ohmiya 2006). These findings indicate that white petals synthesize carotenoids but immediately cleave them enzymatically into colourless compounds. Little is known about the CCD expressed in petals of other plants. In petals of rose (*Rosa damascena*), 15 apocarotenoids have been identified, and the existence of a CCD that cleaves diverse carotenoid substrates has been postulated (Eugster and Märki-Fisher, 1991). Recently, orthologues of CCD1 and CCD4 were identified in rose, and the possible involvement in carotenoid degradation in flowers was demonstrated (Huang *et al.*, 2009a,b).

Analysis of cauliflower Orange (Or) mutant demonstrated the importance of sink capacity for carotenoid accumulation. *Or* is a gain-of-function mutation, and single-locus *Or* mutation confers a high level of β-carotene accumulation in curd tissues (floral meristems) where carotenoids are normally absent (Li *et al.*, 2001). The *Or* gene encodes a novel protein containing a cysteine-rich zinc finger domain, which is found in DnaJ-like molecular chaperones (Lu *et al.*, 2006). This protein plays an important role in triggering differentiation of proplastids, which in turn act as a metabolic sink for carotenoids. Transformation of the *Or* gene into wild-type cauliflower (or) converts the white curd tissue into an orange colour with increased levels of β-carotene.

There are many plant species that lack carotenoids in their flowers and fruits. Few data are available, however, as to the key event responsible for the absence of carotenoid in such tissues. The two examples described above raise the possibility that there are diverse mechanisms that control carotenoid levels in the chromoplasts.

13.3 Metabolic Engineering of Carotenoids

13.3.1 Genetic Manipulation for Elevated β-Carotene

Deficiency in provitamin A can cause growth retardation in children and frequently impairs vision, leading to blindness (Fraser and Bramley, 2004). Most staple foods such as rice, wheat and potato accumulate negligible amounts of β-carotene, a precursor of vitamin A. Thus, hundreds of millions of people in developing countries who rely on such staple foods are suffering from vitamin A deficiency. Increasing the level of β-carotene in major staple crops is, therefore, expected to have a broad and significant impact on human nutrition and health.

13.3.1.1 Rice Grain

Numerous attempts have been made to manipulate carotenoid content in food crops in order to improve their nutritional value for the human diet (Sandmann *et al.*, 2006). The most well-known example is 'Golden Rice' (*Oryza sativa*), a genetically modified rice engineered to produce β-carotene in the endosperm. The first attempt was made to introduce daffodil *PSY* gene under the control of the endosperm-specific glutelin promoter and *crtI* with a plastid-targeting sequence under the control of CaMV35S promoter (Ye *et al.*, 2000). The carotenoid level in the transgenic endosperm has been estimated at 1.6 μg/g dry weight. Although the enzymatic activity of *crtI* converts phytoene to lycopene, the synthesized carotenoids were mostly β-carotene. Schaub *et al.* (2005) have shown that the most of the carotenogenic genes, such as *PDS*, *ZDS*, *CRTISO*, and *LCYB* are expressed in wild-type rice endosperm, whereas *PSY* transcripts are virtually absent. Subsequently, Paine *et al.* (2005) have found that maize PSY has higher activity than daffodil PSY in rice grain. Overexpression of maize *PSY* gene together with *crtI* produced transgenic rice with a high carotenoid content in the endosperm (<37 μg/g dry weight), which has the potential to alleviate vitamin A deficiency. The result indicated that the phytoene synthesis is a rate-limiting step in β-carotene synthesis in the rice grain.

13.3.1.2 Tomato Fruits

Early attempts to manipulate carotenoid biosynthesis in tomato were conducted using promoters constitutively expressing transgenes. However, altering carotenoid content and/or composition in vegetative tissues will have detrimental effects on plant growth because the carotenoid biosynthetic pathway shares GGPP with other essential metabolic pathways that lead to synthesis of gibberellins, chlorophylls, and vitamin E. The vegetative tissues of transgenic tomato constitutively overexpressing the phytoene synthase (*PSY*) gene have an increased carotenoid content. The transformant, however, has a dwarf phenotype due to the depletion of the endogenous GGPP pool, resulting in a shortage in gibberellins (Fray *et al.*, 1995). In later experiments, bacterial phytoene synthase gene (*crtB*) was introduced under the control of the ripening-specific promoter of tomato polygalacturonase (Fraser *et al.*, 2002). Spatial and temporal expression of *PSY* successfully increased carotenoid content in ripening fruit without dwarfism. These results suggest strongly that the manipulation of carotenoid biosynthesis should, therefore, be carried out in a tissue-specific manner in order to circumvent undesired effects.

In order to increase the level of β-carotene, the lycopene β-cyclase gene was introduced to tomato using an rRNA operon promoter combined with the strong ribosome binding site, the Shine-Dalgarno sequence (Apel and Bock, 2009). Introduction of the bacterial lycopene β-cyclase gene (*crtY*) did not strongly alter the carotenoid composition in transformed tomato fruits. On the other hand, most of the lycopene was converted into β-carotene when daffodil *LCYB* was introduced. In addition, total carotenoid content was increased 1.5-fold, indicating that *LCYB* expression enhanced the flux through the carotenoid biosynthesis pathway in chromoplasts.

13.3.1.3 Canola Seeds

The most successful attempts at increasing carotenoid levels by manipulating biosynthetic genes have been achieved with canola, one of the major oil crops worldwide. Wild-type canola seeds contain low level of carotenoids (33 μg/g fresh weight), mainly as lutein. Seed-specific overexpression of *Erwinia uredovora* phytoene synthase (*crtB*) gene with napin promoter produces transformants containing up to a 50-fold increase in total carotenoids in their seeds (Shewmaker *et al.*, 1999). Transgenic seeds were a bright orange colour, accumulating mainly α- and β-carotenes. The amount of lutein and other xanthophylls remained low in the transformed seeds, indicating that not only PSY, but also CHYB, is a rate-limiting step in canola seeds.

13.3.1.4 Potato Tubers

In tubers of *Solanum tuberosum* and *S. phureja*, violaxanthin is the most abundant carotenoids with virtually no carotenes. When the *crtB* gene with a plastid-targeting sequence was introduced under the control of the tuber-specific patatin promoter, carotenoid levels in transgenic tubers of both *S. tuberosum* and *S. phureja* increased (Ducreux *et al.*, 2005). The carotenoid composition also changed dramatically in both transformants, with an increased level of β-carotene.

Orange-yellow potato tubers were obtained by overexpression of the cauliflower *Or* gene using a tuber-specific promoter of the granule-bound starch synthase gene (Lopez *et al.*, 2008). Transgenic potato tubers contain six-fold higher carotenoids with enhanced levels of β-carotene, lutein and violaxanthin compared with wild-type tubers.

The results show that *Or* functions across plant species and can be a useful tool for the manipulation of the carotenoid content of food crops.

13.3.2 Genetic Manipulation for Ketocarotenoid Production

Astaxanthin and canthaxanthin are red ketocarotenoids produced in many microorganisms. They have strong antioxidant activity, and when consumed can have beneficial effects on human health. Ketocarotenoids are derived from β-carotene by 3-hydroxylation and 4-ketolation at both ionone end groups, catalyzed by β-carotene hydroxylase and β-carotene ketolase, respectively (Figure 13.5). They are rarely found in higher plants, because the genes encoding β-carotene ketolase are restricted to some bacteria, fungi and algae. Therefore many attempts have been made to produce ketocarotenoids *in planta* by introducing β-carotene ketolase genes from bacteria (*CrtW* or *CrtO*) (Misawa, 2009; Zhu *et al.*, 2009).

13.3.2.1 Canola Seeds

In order to produce ketocarotenoid-accumulating canola seed, seven genes from bacteria were tandemly connected in an expression vector and introduced (Fujisawa *et al.*, 2009). In addition to *crtW* and β-carotene hydroxylase (*crtZ*) genes to divert flux towards ketocarotenoids, the genes for isopentenyl pyrophosphate isomerase (*idi*), GGPP synthase (*crtE*), *crtB*, *crtI* and *crtY* was introduced to increase metabolic flux to β-carotene. Transgenic seeds contain up to a 19- to 30-fold increase in total carotenoids, including ketocarotenoids such as echinenone, canthaxanthin, astaxanthin and adonixanthin.

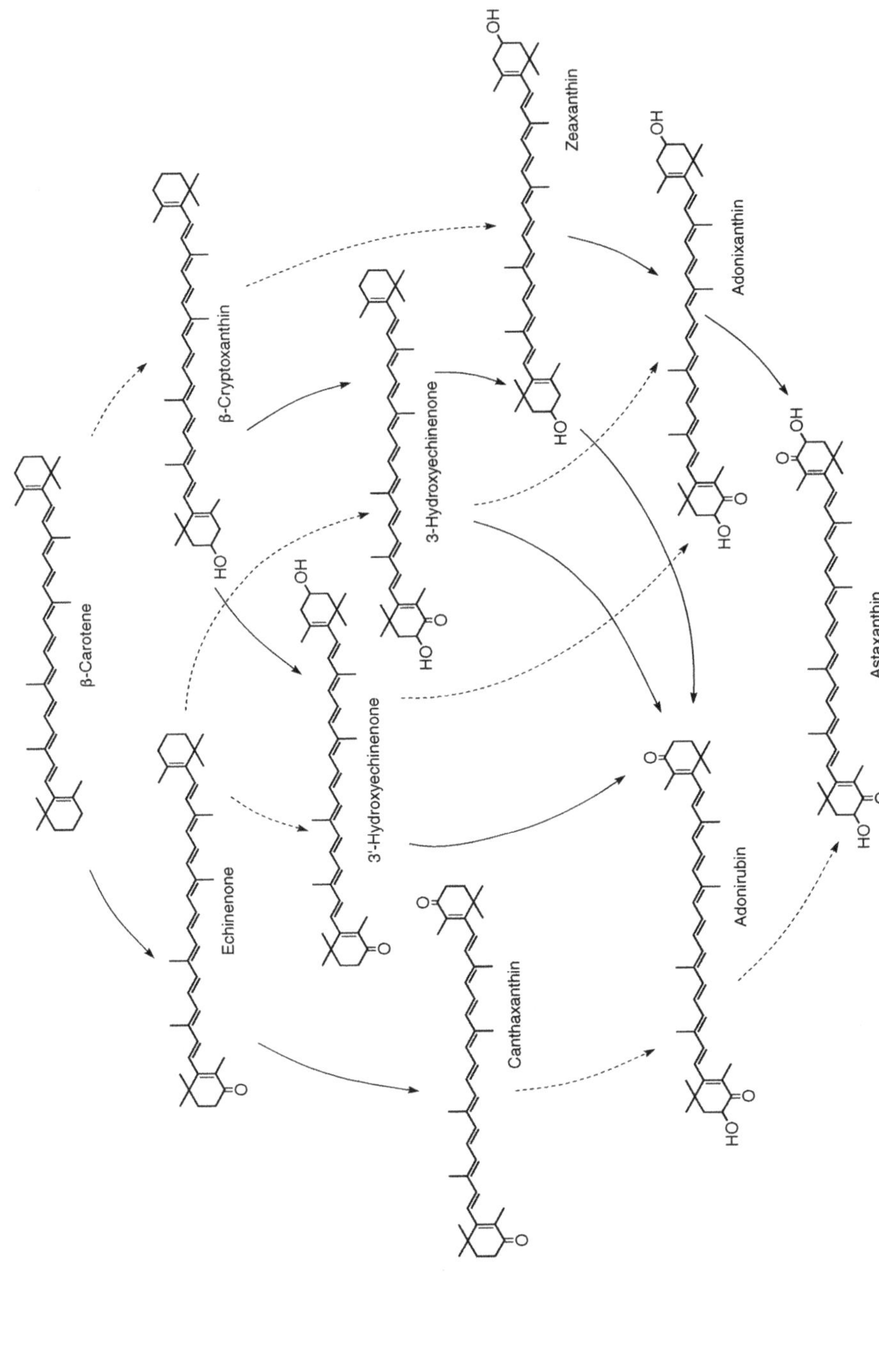

Figure 13.5 *Ketocarotenoid biosynthesis pathway. Bold and dotted lines indicate reactions catalyzed by β-carotene ketolase and β-carotene hydroxylase, respectively*

13.3.2.2 Maize Kernel

Multi-gene transfer was also conducted using a white maize variety deficient for carotenoid synthesis in the endosperm (Zhu *et al.*, 2008). When *G. lutea LCYB* was introduced together with *crtI*, *crtW* and maize *PSY1*, transgenic endosperm accumulated various ketocarotenoids such as adonixanthin, echinenone, 3-hydroxyechinenone, and astaxanthin. On the other hand, when *G. lutea CHYB* was introduced instead of *G. lutea LCYB*, the ketocarotenoid accumulated in the transgenic endosperm was adonixanthin alone. Transgenic plants with all five genes introduced produced adonixanthin, echinenone and 3-hydroxyechinenone, but lacked astaxanthin. Conversion of adonixanthin to astaxanthin did not proceed when *CHYB* was introduced together with *crtW*. The result suggests that β-carotene hydroxylase and bacterial β-carotene ketolase compete for the unsubstituted β-ionone rings as substrates in the pathway, and the second ketolation leading to astaxanthin may be inhibited when the hydroxylase activity is high.

13.3.2.3 Potato Tubers

Because wild-type potato tubers do not contain zeaxanthin, a substrate of β-ketolase to produce ketocarotenoids, *crtO* was re-introduced into a transgenic potato accumulating zeaxanthin by inactivation of zeaxanthin epoxidase (Römer *et al.*, 2002; Gerjets and Sandmann, 2006). Transgenic plants expressing *crtO* constitutively accumulated several types of ketocarotenoids including astaxanthin, echinenone, 3′-hydroxyechinenone, and 4-ketozeaxanthin in tubers. The newly-formed ketocarotenoids comprised approximately 10–12% of total carotenoids in leaves and tubers.

13.3.2.4 Tobacco and Lotus Flowers

In contrast to food crops, genetic engineering aimed at alteration of carotenoid composition in the flower has been paid little attention. Only a few attempts have been made to introduce bacterial ketolases to produce novel flower-colour accumulating ketocarotenoids. Bacterial *crtW* was introduced to *Lotus japonicus* (Suzuki *et al.*, 2007) and *Nicotiana glauca* (Gerjets *et al.*, 2007), model plants that have carotenogenic flowers. Transgenic flowers of *L. japonicus* accumulated astaxanthin and the petal colour changed from yellow to orange, while those of *N. glauca* accumulated only a small amount of adonixanthin. Then Zhu *et al.* (2007) introduced *crtO* from cyanobacteria to *N. glauca* as an alternative, and obtained transgenic plants with improved ketocarotenoid production. Transgenic flowers accumulated higher levels of ketocarotenoids than those with introduced *crtW*, with a wider spectrum of yellow-coloured monoketocarotenoids comprising 4-ketolutein, echinenone, 3′-hydroxyechinenone, and 4-ketozeaxanthin. However, they lack red-coloured diketocarotenoids such as canthaxanthin and astaxanthin, suggesting limited activity of β-ketolase.

References

Allegra, M., Tesoriere, L. and Livrea, M.A. (2007) Betanin inhibits the myrloperoxidase/nitrite-induced oxidation of human low density lipoproteins. *Free Rad. Res.*, **41**, 335–341.

Apel, W. and Bock, R. (2009) Enhancement of carotenoid biosynthesis in transplastomic tomatoes by induced lycopene-to-provitamin A conversion. *Plant Physiol.*, **151**, 59–66.

Baranska, M., Baranski, R., Shultz, H. and Nothnagel, T. (2006) Tissue-specific accumulation of carotenoids in carrot roots. *Planta*, **224**, 1028–1037.

Bauberger, E. and Mayer, A.M. (1960) Effect of kinetin on formation of red pigment in seedlings of *Amaranthus retroflexus*. *Science*, **141**, 1094–1095.

Bernhardt, A., Lechner, E., Hano, P., Schade, V., Dieterle, M., Anders, M., Dubin, M. J., Benvenuto, G., Bowler, C., Genschik, P. and Hellmann, H. (2006) CUL4 associates with DDB1 and DET1 and its downregulation affects diverse aspects of development in *Arabidopsis thaliana*. *Plant J.*, **47**, 591–603.

Biddington, N.K. and Thomas, T.H. (1973) A modified *Amaranthus* betacyanin bioassay for the rapid determination of cytokinins in plant extracts. *Planta*, **111**, 183–186.

Biddington, N.K. and Thomas, T.H. (1977) Interaction of abscisic acid, cytokinin and gibberellins in the control of betacyanin synthesis in seedlings of *Amaranthus caudatus*. *Physiol. Plant.*, **40**, 312–314.

Bouvier, F., Dogbo, O. and Camara, B. (2003a) Biosynthesis of the food and cosmetic plant pigment bixin. *Science*, **300**, 2089–2091.

Bouvier, F., Suire, C., Mutterer, J. and Camara, B. (2003b) Oxidative remodeling of chromoplast carotenoids: identification of the carotenoid dioxygenase *CsCCD* and CsZCD genes involved in crocus secondary metabolite biogenesis. *Plant Cell*, **15**, 47–62.

Britton, G., Liaaen-Jensen, S. and Pfander, H. (2004) *Carotenoids Handbook*. Birkhauser, Basel, Switzerland.

Buckner, B., Miguel, P.S., Janick-Buckner, D. and Bennetzen, J.L. (1996) The *y1* gene of maize codes for phytoene synthase. *Genetics*, **143**, 479–488.

Cai, Y., Sun, M. and Corke, H. (2003) Antioxidant activity of betalains from plants of the Amaranthaceae. *J. Agric. Food Chem.*, **51**, 2288–2294.

Cazzonelli, C.I., Cuttriss, A. J., Cossetto, S. B., Pye, W., Crisp, P., Whelan, J., Finnegan, E. J., Turnbull, C. and Pogsona, B. J. (2009) Regulation of carotenoid composition and shoot branching in *Arabidopsis* by a chromatin modifying histone methyltransferase, SDG8. *Plant Cell*, **21**, 39–53.

Cazzonelli, C.I. and Pogson, B.J. (2010) Source to sink: regulation of carotenoid biosynthesis in plants. *Trends Plant Sci.*, **15**, 266–274.

Chen, H., Shen, Y., Tan, X., Yu, L., Wang, J., Guo, L., Zhang, Y., Zhang, H., Feng, S., Strickland, E., Zheng, N. and Deng, X. W. (2006) *Arabidopsis* CULLIN4 forms an E3 ubiquitin ligase with RBX1 and the CDD complex in mediating light control of development. *Plant Cell*, **18**, 1991–2004.

Christinet, L., Burdet, F.X., Zaiko, M., Hinz, U. and Zrÿd, J.P. (2004) Characterization and functional identification of a novel plant 4,5-extradiol dioxygenase involved in betalain pigment biosynthesis in *Portulaca grandiflora*. *Plant Physiol.*, **134**, 265–274.

Constabel, F. and Nassif-Makki, K. (1971) Betalainbildung in *Beta*-Calluskulturen. *Ber. Deutsch Bot. Ges.*, **84**, 629–636.

Cooper, C.M., Davies, N.W. and Menary, R.C. (2009) Changes in some carotenoids and apocarotenoids during flower development in *Boronia megastigma* (Nees). *J. Agric. Food Chem.*, **57**, 1513–1520.

Cunningham, F.X.J., Pogson, B., Sun, Z.R., *et al.* (1996) Functional analysis of the β and ε lycopene cyclase enzymes of Arabidopsis reveals a mechanism for control of cyclic carotenoid formation. *Plant Cell*, **8**, 1613–1626.

Cunningham, F.X.J. and Gantt, E. (1998) Genes and enzymes of carotenoid biosynthesis in plants. Ann. Rev. Plant Physiol. Plant Mol. Biol., **49**, 557–583.

Cunningham, F.X.J. and Gantt, E. (2001) One ring or two? Determination of ring number in carotenoids by lycopene ε-cyclases. *Proc. Natl. Acad. Sci. USA*, **98**, 2905–2910.

Cunningham, F.X.J. and Gantt, E. (2005) A study in scarlet: enzymes of ketocarotenoid biosynthesis in the flowers of *Adonis aestivalis*. *Plant J.*, **41**, 478–492.

Davuluri, G.R., van Tuinen, A., Fraser, P. D., Manfredonia, A., Newman, R., Burgess, D., Brummell, D.A., King, S. R., Palys, J., Uhlig, J., Bramley, P. M., Pennings, H. M. J. and Bowler, C. (2005) Fruit-specific RNAi-mediated suppression of DET1 enhances carotenoid and flavonoid content in tomatoes. *Nat. Biotechnol.*, **23**, 890–895.

Delgado-Vargas, F., Jimenez, A.R. and Paredes-Lopez, O. (2000) Natural pigments: carotenoids, anthocyanins, and betalains – characteristics, biosynthesis, processing, and stability. *Crit. Rev. Food Sci. Nutr.*, **40**, 173–289.

Della Penna, D. and Pogson, B.J. (2006) Vitamin synthesis in plants: tocopherols and carotenoids. *Ann. Rev. Plant Biol.*, **57**, 711–738.

Ducreux, L.J.M., Morris, W.L., Hedley, P.E., Shepherd, T., Davies, H.V., Millam, S. and Taylor, M.A. (2005) Metabolic engineering of high carotenoid potato tubers containing enhanced levels of ß-carotene and lutein. *J. Exp. Bot.*, **56**, 81–89.

Dun, E.A., Philip, B., Brewer, P.B. and Beveridge, C.A. (2009) Strigolactones: discovery of the elusive shoot branching hormone. *Trends Plant Sci.*, **14**, 364–372.

Elliott, D.C. (1979a) Analysis of variability in the *Amaranthus* bioassay for cytokinins: effects of water stress on benzyladenin- and fusicoccin-dependent responses. *Plant Physiol.*, **63**, 269–273.

Elliott, D.C. (1979b) Analysis of variability in the *Amaranthus* bioassay for cytokinins: effects of "aging" excised cotyledons. *Plant Physiol.* **63**, 274–276.

Elliott, D.C. (1979c) Temperature-dependent expression of betacyanin synthesis in *Amaranthus* seedlings. *Plant Physiol.*, **63**, 277–279.

Elliott, D.C. (1979d) Temperature-sensitive responses of red light-dependent betacyanin synthesis. *Plant Physiol.*, **64**, 521–524.

Elliott, D.C. (1983) Accumulation of cytokinin-induced betacyanin in specific cells of *Amaranthus tricolor* seedlings. J. Exp. Bot., **34**, 67–73.

Endress, R. (1976) Betacyan-Akkumulation in Kallus von Portulaca grandiflora var. JR unter dem Einfluss von Phytohormonen und Cu^{2+}-Ionen auf unterschiedlichen Grundmedium. *Biochem. Physiol. Pflanzen*, **169**, 87–98.

Eugster, C.H. and Märki-Fisher, E. (1991) The chemistry of rose pigments. *Angew. Chem. Int. Ed. Engl.*, **30**, 654–672.

Fischer, N. and Dreiding, A.S. (1972) Biosynthesis of betaleins. On the cleavage of the aromatic ring during enxymic transformation of dopa into betalamic acid. *Helv. Chem. Acta*, **55**, 649–658.

Fraser, P.D., Truesdale, M.R., Bird, C.R., Schuch, W. and Bramley, P.M. (1994) Carotenoid biosynthesis during tomato fruit development. *Plant Physiol.*, **105**, 405–413.

Fraser, P.D., Kiano, J.W., Truesdale, M.R., Schuch, W. and Bramley, P.M. (1999) Phytoene synthase-2 enzyme activity in tomato does not contribute to carotenoid synthesis in ripening fruit. *Plant Mol. Biol.*, **40**, 687–698.

Fraser, P. D., Romer, S., Shipton, C. A., Mills, P. B., Kiano, J.W., Misawa, N., Drake, R.G., Schuch, W. and Bramley, P.M. (2002) Evaluation of transgenic tomato plants expressing an additional phytoene synthase in a fruit-specific manner. *Proc. Natl. Acad. Sci. USA*, **99**, 1092–1097.

Fraser, P.D. and Bramley, P.M. (2004) The biosynthesis and nutritional uses of carotenoids. *Prog. Lipid Res.*, **43**, 228–265.

Fray, R. G., Wallace, A., Fraser, P.D., Valero, D., Hedden, P., Bramley, P. M. and Grierson, D. (1995) Constitutive expression of a fruit phytoene synthase gene in transgenic tomatoes causes dwarfism by redirecting metabolites from the gibberellin pathway. *Plant J.*, **8**, 693–701.

Fujisawa, M., Takita, E., Harada, H., Sakurai, N., Suzuki, H., Ohyama, K., Shibata, D. and Misawa, N. (2009) Pathway engineering of *Brassica napus* seeds using multiple key enzyme genes involved in ketocarotenoid formation. *J. Exp. Bot.*, **60**, 1319–1332.

Galpaz, N., Ronen, G., Khalfa, Z., Zamir, D. and Hirschberg, J. (2006) A chromoplast-specific carotenoid biosynthesis pathway is revealed by cloning of the tomato white-flower locus. *Plant Cell*, **18**, 1947–1960.

Garay, A.S. and Towers, G.H.N. (1966) Studies on the biosynthesis of amaranthin. *Can. J. Bot.*, **44**, 231–236.

Gerjets, T. and Sandmann, G. (2006) Ketocarotenoid formation in transgenic potato. *J. Exp. Bot.*, **57**, 3639–3645.

Gerjets, T., Sandmann, M., Zhu, C. and Sandmann, G. (2007) Metabolic engineering of ketocarotenoid biosynthesis in leaves and flowers of tobacco species *Biotech. J.*, **2**, 1263–1269.

Girod, P.A. and Zrÿd, J.P. (1991) Biogenesis of betalains: purification and partial characterization of Dopa 4,5-dioxygenase from *Amanita muscaria*. *Phytochemistry*, **30**, 169–174.

Giuliano, G., Bartley, G.E. and Scolnik, P.A. (1993) Regulation of carotenoid biosynthesis during tomato development. *Plant Cell*, **5**, 379–387.

Gompel, N., Prud'homme, B., Wittkopp, P.J., Kassner, V.A. and Carroll, S.B. (2005) Chance caught on the wing: *cis*-regulatory evolution and the origin of pigment patterns in *Drosophila*. *Nature*, **433**, 481–487.

Green, B.R. and Durnford, D.G. (1996) The chlorophyll-carotenoid proteins of oxygenic photosynthesis. *Ann. Rev. Plant Physiol. Plant Mol. Biol.*, **47**, 685–714.

Guruprasad, K.N. and Laloraya, M.M. (1980) Dissmilarity in the inhibition of betacyanin synthesis caused by gibberellic acid and abscisic acid. *Biochem. Physiol. Pflanzen*, **175**, 582–586.

Hagenimana, V., Carey, E.E., Gichuki, S.T., Oyunga, M.A. and Inumgi, J.K. (1999) Carotenoid contents in fresh, dried and processed sweetpotato products. *Ecol. Food Nutr.*, **37**, 455–473.

Harborne, J.B. (1996) The evolution of flavonoid pigments in plants, in *Comparative Phytochemistry* (ed. T. Swain), Academic Press, London, pp. 271–295.

Heuer, S. and Strack, D. (1992) Synthesis of betanin from betanidin and UDP-glucose by a protein preparation from cell suspension cultures of *Dorotheanthus bellidiformis* (Burm.f.) N. E. Br. *Planta*, **186**, 626–628.

Hinz, U.G., Fivaz, J., Girod, P.A. and Zrÿd, J.P. (1997) The gene coding for the DOPA dioxygenase involved in betalain biosynthesis in *Amanita muscaria* and its regulation. *Mol. Gen. Genet.*, **256**, 1–6.

Hirano, H., Sakuta, M. and Komamine, A. (1996) Inhibition of betacyanin accummulation by abscisic acid in suspension cultures of *Phytolacca americana. Z. Naturforsch.*, **51c**, 818–822.

Hornero-Méndez, D., Gómez-Ladrón de Guevara, R. and Minguez-Mosquera, M.I. (2000) Carotenoid biosynthesis changes in five red pepper (*Capsicum annuum* L.) cultivars during ripening. Cultivar selection for breeding. *J. Agric. Food Chem.*, **48**, 3857–3864.

Howitt, C.A. and Pogson, B. (2006) Carotenoid accumulation and function in seeds and non-green tissues. *Plant Cell Environ.*, **29**, 435–445.

Huang, F-C., Horvath, G., Molnar, P., Turcsi, E., Deli, J., Schrader, J., Sandmann, G., Schmidt, H. and Schwab, W. (2009a) Substrate promiscuity of RdCCD1, a carotenoid cleavage oxygenase from *Rosa damascena. Phytochemistry*, **70**, 457–464.

Huang, F.-C., Molnar, P. and Schwab, W. (2009b) Cloning and functional characterization of carotenoid cleavage dioxygenase 4 genes. *J. Exp. Bot.*, **60**, 3011–3022.

Impellizzeri, G. and Piattelli, M. (1972) Biosynthesis of indicaxanthin in *Opuntia ficus-indica* fruit. Phytochemistry, **11**, 2499–2502.

Isaacson, T., Ronen, G., Zamir, D. and Hirschberg, J. (2002) Cloning of tangerine from tomato reveals a carotenoid isomerase essential for the production of β-carotene and xanthophylls in plants. *Plant Cell*, **14**, 333–342.

Isaacson, T., Ohad, I., Beyer, P. and Hirschberg, J. (2004) Analysis in vitro of the enzyme CRTISO establishes a poly-cis-carotenoid biosynthesis pathway in plants. *Plant Physiol.*, **136**, 4246–4255.

Iwashina, T. (2001) Flavonoids and their distribution in plant families containing the betalain pigments (Review). *Ann. Tsukuba Bot. Gard.*, **20**, 11–74.

Iwashina, T., Ootani, S. and Hayashi, K. (1988) On the pigmentation spherical bodies and crystals in tepals of Cactaceous species in reference to the nature of betalains or flavonols. *Bot. Mag. Tokyo*, **101**, 175–184.

Kanner, J., Harel, S. and Granit, R. (2001) Betalains – a new class of dietary cationized antioxidants. *J. Agric. Food Chem.*, **49**, 5178–5185.

Kato, M., Ikoma, Y., Matsumoto, H., Sugiura, M., Hyodo, H. and Yano, M. (2004) Accumulation of carotenoids and expression of carotenoid biosynthetic genes during maturation in citrus fruit. *Plant Physiol.*, **134**, 824–837.

Kim, J., Smith, J.J., Tian, L. and Dellapenna, D. (2009) The evolution and function of carotenoid hydroxylases in *Arabidopsis. Plant Cell Physiol.*, **50**, 463–479.

Kishimoto, S., Maoka, T., Nakayama, M. and Ohmiya, A. (2004) Carotenoid composition in petals of chrysanthemum (*Dendranthema grandiflorum* (Ramat.) Kitamura). *Phytochemistry*, **65**, 2781–2787.

Kishimoto, S., Maoka, T., Sumitomo, K. and Ohmiya, A. (2005) Analysis of carotenoid composition in petals of calendula (*Calendula officinalis L.*). *Biosci. Biotechnol. Biochem.*, **69**, 2122–2128.

Kishimoto, S. and Ohmiya, A. (2006) Regulation of carotenoid biosynthesis in petals and leaves of chrysanthemum (*Chrysanthemum morifolium* Ramat.). *Physiol. Plant.*, **128**, 437–447.

Koelher, H.K. (1972a) Action of inhibitors of protein and nucleic acid synthesis on light-dependent and kinetin-stimulated betcyanin synthesis. *Phytochemistry*, **11**, 127–132.

Koelher, H.K. (1972b) Phytocontrol of betacyanin synthesis in *Amaranthus caudatus* seedlings in the presence of kinetin. *Phytochemistry*, **11**, 133–137.

Koelher, H.K., Kochhar, S. and Mohr, H. (1981) Action of light and kinetin on betalain synthesis in seedlings of *Amaranthus caudatus*: A two-factor analysis. *Ber. Deutsch Bot. Ges.*, **94**, 27–34.

Koes, R., Quattrocchio, F. and Mol, J. (1994) The flavonoid biosynthetic pathway in plants: function and evolution. *BioEssays*, **16**, 123–132.

Kujala, T., Loponen, J. and Pihlaja, K. (2001) Betalains and phenolics in red beetroot (*Beta vulgaris*) peel extracts: extraction and characterisation. *Z. Naturforsch.*, **56c**, 343–348.

Kurilich, A.C. and Juvik, J.A. (1999) Quantification of carotenoid and tocopherol antioxidants in *Zea mays*. *J. Agric. Food Chem.*, **47**, 1948–1955.

Lee, C.H., Wettasinghe, M., Bolling, B.W., Ji, L.L. and Parkin, K.L. (2005) Betalains, phase II enzyme-inducing components from red beetroot (*Beta vulgaris* L.) extracts. *Nutr. Cancer*, **53**, 91–103.

Li, F., Murillo, C. and Wurtzel, E.T. (2007) Maize *Y9* encodes a product essential for 15-cis-ζ-carotene isomerization. *Plant Physiol.*, **144**, 1181–1189.

Li, L., Paolillo, D.J., Parthasarathy, M.V., DiMuzio, E.M. and Garvin, D.F. (2001) A novel gene mutation that confers abnormal patterns of β-carotene accumulation in cauliflower (*Brassica oleracea* var. botrytis). *Plant J.*, **26**, 59–67.

Liu, Y., Roof, S., Ye, Z., Barry, C., Tuinen, A., Vrebalov, J., Bowler, C. and Giovannoni, J. (2004) Manipulation of light signal transduction as a means of modifying fruit nutritional quality in tomato. *Proc. Natl. Acad. Sci. USA*, **101**, 9897–9902.

Lopez, A.B., Van Eck, J., Conlin, B.J., Paolillo, D.J., O'Neill, J. and Li, L. (2008) Effect of the cauliflower *Or* transgene on carotenoid accumulation and chromoplast formation in transgenic potato tubers. *J. Exp. Bot.*, **59**, 213–223.

Lu, S., Van Eck, J., Zhou, X., Lopez, A.B., O'Halloran, D.M., Cosman, K.M, Conlin, B.J., Paolillo, D. J., Garvin, D.F., Vrebalov, J., Kochian, L.V., Küpper, H., Earle, E.D., Cao, J. and Li, L. (2006) The cauliflower *Or* gene encodes a DnaJ cysteine-rich domain-containing protein that mediates high levels of β-carotene accumulation. *Plant Cell*, **18**, 3594–3605.

Mabry, T.J. (1980) Betalains, in *Secondary Plant Products* (eds E.A. Bell and B.V. Charlwood), *Encyclopedia of Plant Physiology Vol. 8*, Springer-Verlag, Berlin, pp. 513–533.

Matsumoto, H., Ikoma, Y., Kato, M., Kuniga, T., Nakajima, N. and Yoshida, T. (2007) Quantification of carotenoids in *Citrus* fruit by LC–MS and comparison of patterns of seasonal changes for carotenoids among *Citrus* varieties. *J. Agric. Food Chem.*, **55**, 2356–2368.

Matus, Z., Molnar, P. and Szabo, L. (1993) Main carotenoids in pressed seeds (*Cucurbitae semen*) of pumpkin (*Cucurbita pepo convar.pepo* var. styriaca). *Acta Pharm. Hung.*, **63**, 247–256.

Mayne, S.T. (1996) β-Carotene, carotenoids and disease prevention in humans. *FASEB J.*, **10**, 690–701.

McGraw, K.J., Hill, G.E., Stradi, R. and Parker, R.S. (2001) The influence of carotenoid acquisition and utilization on the maintenance of species-typical plumage pigmentation in male American goldfinches (*Carduelis tristis*) and northern cardinals (*Cardinalis cardinalis*). *Physiol. Biochem. Zool.*, **74**, 843–852.

Miller, H.E., Rosler, H., Wohlpart, A., Wyler, H., Wilcox, M.E., Frohofer, H., Mabry, T.J. and Dreiding, A.S. (1968) Biogenese der betalaine. Biotransformation von Dopa und tyrosin in den betalaminsaureteil des betanins. Vorläufige Mitteilung. *Helv. Chim. Acta*, **51**, 1470–1474.

Misawa, N. (2009) Pathway engineering of plants toward astaxanthin production. *Plant Biotech.*, **26**, 93–99.

Moehs, C.P., Tian, L., Osteryoung, K.W. and Della Penna, D. (2001) Analysis of carotenoid biosynthetic gene expression during marigold petal development. *Plant Mol. Biol.*, **45**, 281–293.

Molnar, P. and Szabolcs, J. (1980) Occurrence of 15-*cis*-violaxanthin in *Viola tricolor*. *Phytochemistry*, **19**, 623–627.

Mueller, L.A., Hinz, U., Uze, M., Sautter, C. and Zrÿd, J.P. (1997a) Biochemical complementation of the betalain biosynthetic pathway in *Portulaca grandiflora* by a fungal 3,4-dihydroxyphenylalanine dioxygenase. *Planta*, **203**, 260–263.

Mueller, L.A., Hinz, U. and Zrÿd, J.P. (1997b) The formation of betalamic acid and muscaflavin by recombinant dopa-dioxygenase from *Amanita*. *Phytochemistry*, **44**, 567–569.

Nambara, E. and Marion-Poll, A. (2005) Abscisic acid biosynthesis and catabolism. *Ann. Rev. Plant Biol.*, **56**, 165–185.

Nassif-Makki, H. and Constabel, F. (1972) Zur Bedeutung von tyrosin als vorstufe der betalainsynthese. *Z. Pflanzenphysiol.* **67**, 201–206.

Nicola, M.G., Piattelli, M. and Amico, V. (1973a) Phytocontrol of betaxanthin synthesis in *Celosia plumose* seedlings. *Phytochemistry*, **12**, 353–357.

Nicola, M.G., Piattelli, M. and Amico, V. (1973b) Effect of continuous far red on betaxanthin and betacyanin synthesis. *Phytochemistry*, **12**, 2163–2166.

Nicola, M.G., Amico, V. and Piattelli, M. (1974) Effect of white and far-red light on betalain formation. *Phytochemistry*, **13**, 439–442.

Niyogi, K. (2000) Safety valves for photosynthesis. *Curr. Opin. Plant Biol.*, **3**, 455–460.

Ohmiya, A. (2009) Carotenoid cleavage dioxygenases and their apocarotenoid products in plants. *Plant Biotech.*, **26**, 351–358.

Ohmiya, A., Kishimoto, S., Aida, R., Yoshioka, S. and Sumitomo, K. (2006) Carotenoid cleavage dioxygenase (CmCCD4a) contributes to white color formation in chrysanthemum petals. *Plant Physiol.*, **142**, 1193–1201.

Paine, J.A., Shipton, C.A., Chaggar, S., Howells, R.M, Kennedy, M.J., Vernon, G., Wright, S.Y., Hinchuliffe, E., Adams, J.L., Silverstone, A.L. and Drake, R. (2005) Improving the nutritional value of golden rice through increased pro-vitamin A content. *Nat. Biotechnol.*, **23**, 482–487.

Park, H., Kreunen, S.S., Cuttriss, A.J., DellaPenna, D. and Pogson, B.J. (2002) Identification of the carotenoid isomerase provides insight into carotenoid biosynthesis, prolamellar formation, and photomorphogenesis. *Plant Cell*, **14**, 321–332.

Piattelli, M. (1976) Betalains, in *Chemistry and Biochemistry of Plant Pigments, Vol.* **1** (2nd edn) (ed. T.W. Goodwin), Academic Press, New York, pp. 560–596.

Römer, S., Hugueney, P., Bouvier, F., Camara, B. and Kuntz, M. (1993) Expression of the genes encoding the early carotenoid biosynthetic-enzymes in *Capsicum annuum*. *Biochem. Biophys. Res. Commun.*, **196**, 1414–1421.

Römer, S., Lübeck, J., Kauder, F., Steiger, S., Adomat, C. and Sandmann, G. (2002) Genetic engineering of a zeaxanthin-rich potato by antisence inactivation and co-supression of carotenoid epoxidation. *Metab. Eng.*, **4**, 263–272.

Saito, K. and Komamine, A. (1976) Biosynthesis of stizolobinic acid and stizolobic acid in higher plants: A enzyme system(s) catalyzing the conversion of dihydroxyphenylalanine into stizolobinic acid and stizolobic acid from etiolated seedlings of *Stizolobium hassjoo. Eur. J. Biochem.*, **68**, 237–243.

Saito, K. and Komamine, A. (1978) Biosynthesis of stizolobinic acid and stizolobic acid in higher plants: Stizolobinic acid synthase and stizolobic acid synthase, new enzymes which catalyze the reaction sequences leading to the formation of stizolobinic acid and stizolobic acid from 3,4-dihydroxyphenylalanine in *Stizolobium hassjoo. Eur. J. Biochem.*, **82**, 385–392.

Sakuta, M., Hirano, H. and Komamine, A. (1991) Stimulation by 2,4-dichlorophenoxyacetic acid of betacyanin accumulation in suspension cultures of *Phytolacca americana. Physiol. Plant.*, **83**, 154–158.

Sandmann, G., Romer, S. and Fraser, P.D. (2006) Understanding carotenoid metabolism as a necessity for genetic engineering of crop plants. *Metabolic Eng.*, **8**, 291–302.

Sasaki, N., Adachi, T., Koda, T. and Ozeki, Y. (2004) Detection of UDP-glucose:*cyclo*-DOPA 5-O-glucosyltransferase activity in four o'clocks (*Mirabilis jalapa* L.). *FEBS Lett.*, **568**, 159–162.

Sasaki, N., Abe, Y., Wada, K., Koda, T., Goda, Y., Adachi, T. and Ozeki, Y. (2005a) Amaranthin in feather cockscombs is synthesized via glucuronylation at the cyclo-DOPA glucoside step in the betacyanin biosynthetic pathway. *J. Plant Res.*, **118**, 439–442.

Sasaki, N., Wada, K., Koda, T., Kasahara, K., Adachi, T. and Ozeki, Y. (2005b) Isolation and characterization of cDNAs encoding an enzyme with glucosyltransferase activity for *cyclo*-DOPA from four o'clocks and feather cockscombs. *Plant Cell Physiol.*, **46**, 666–670.

Sasaki, N., Abe, Y., Goda, Y., Adachi, T., Kasahara, K. and Ozeki, Y. (2009) Detection of DOPA 4,5-dioxygenase(DOD) activity using recombinant protein prepared from *Escherichia coli* cells harboring cDNA encoding DOD from *Mirabilis jalapa. Plant Cell Physiol.*, **50**, 1012–1016.

Schaub, P., Al-Babili, S., Drake, R. and Beyer, P. (2005) Why is Golden Rice (yellow) instead of red? *Plant Physiol.*, **138**, 441–450.

Schledz, M., Al-Babili, S., von Lintig, J, Haubruck, H., Rabbani, S., Kleinig, H. and Beyer, P. (1996) Phytoene synthase from *Narcissus pseudonarcissus*: functional expression, galactolipid requirement, topological distribution in chromoplasts and induction during flowering. *Plant J.*, **10**, 781–792

Schliemann, W., Kobayashi, N. and Strack, D. (1999) The decisive step in betaxanthin biosynthesis is a spontaneous reaction. *Plant Physiol.*, **119**, 1217–1232.

Schwartz, S.H., Tan, B.C., Gage, D.A., Zeevaart, J.A.D. and McCarty, D.R. (1997) Specific oxidative cleavage of carotenoids by VP14 of maize. *Science*, **276**, 1872–1874.

Schwartz, S.H., Qin, X. and Loewen, M. (2004) The biochemical characterization of two carotenoid cleavage enzymes from *Arabidopsis* indicates that a carotenoid-derived compound inhibits lateral branching. *J. Biol. Chem.*, **279**, 46940–46945.

Sciuto, S., Oriente, G. and Piatelli, M. (1972) Betanidin glycosylation in *Opuntia dillenii. Phytochemistry*, **11**, 2259–2262.

Sciuto, S., Oriente, G., Piattelli, M., Impelizzeri, G. and Amico, V. (1974) Biosynthesis of amaranthin in *Celosia plumosa*. *Phytochemistry*, **13**, 947–951.

Shapiro, M.D., Marks, M.E., Peichel, C.L., Blackman, B.K., Nereng, K.S., Jonsson, B., Schluter, D. and Kingsley, D.M. (2004) Genetic and developmental basis of evolutionary pelvic reduction in threespine sticklebacks. *Nature*, **428**, 717–723.

Shewmaker, C.K., Sheehy, J.A., Daley, M., Colburn, S. and Ke, D.Y. (1999) Seed-specific overexpression of phytoene synthase: increase in carotenoids and other metabolic effects. *Plant J.*, **20**, 401–412.

Shimada, S., Takahashi, K., Sato, Y. and Sakuta, M. (2004) Dihydroflavonol 4-reductase cDNA from non-anthocyanin-producing species in the Caryophyllales. *Plant Cell Physiol.*, **45**, 1290–1298.

Shimada, S., Inoue, T.Y. and Sakuta, M. (2005) Anthocyanidin synthase in non-anthocyanin-producing Caryophyllales species. *Plant J.*, **44**, 950–959.

Shimada, S., Otsuki, H. and Sakuta, M. (2007) Transcriptional control of anthocyanin biosynthetic genes in the Caryophyllales. *J. Exp. Bot.*, **58**, 957–967.

Simkin, A.J., Zhu, C., Kuntz, M. and Sandmann, G. (2003) Light-dark regulation of carotenoid biosynthesis in pepper (*Capsicum annuum*) leaves. *J. Plant Physiol.*, **160**, 439–443.

Simkin, A.J., Schwartz, S. H., Auldridge, M., Taylor, M. G. and Klee, H. J. (2004a) The tomato carotenoid cleavage dioxygenase 1 genes contribute to the formation of the flavor volatiles β-ionone, pseudoionone, and geranylacetone. *Plant J.*, **40**, 882–892.

Simkin, A. J., Underwoo, B. A., Auldridge, M., Loucas, H.M., Shibuya, K., Schmelz, E., Clark, D.G. and Klee, H. J. (2004b) Circadian regulation of the PhCCD1 carotenoid cleavage dioxygenase controls emission of β-ionone, a fragrance volatile of petunia flowers. *Plant Physiol.*, **136**, 3504–3514.

Spasic, M., Milic, B. and Obrenivic, S. (1985) Superoxide dismutase activity versus betacyanin induction under continuous red illumination in *Amaranthus* seedlings. *Biochem. Physiol. Pflanzen*, **180**, 319–322.

Stafford, H.A. (1990) Flavonoid evolution: an enzymic approach. *Plant Physiol.*, **96**, 680–685.

Stafford, H.A. (1994) Anthocyanins and betalains: evolution of the mutually exclusive pathways. *Plant Sci.*, **101**, 91–98.

Steiner, U., Schliemann, W. and Strack, D. (1996) Assay for tyrosine hydroxylation activity of tyrosinase from betalain-forming plants and cell cultures. *Anal. Biochem.*, **238**, 72–75.

Steiner, U., Schliemann, W., Böhm, H. and Strack, D. (1999) Tyrosinase involved in betalain biosynthesis of higher plants. *Planta*, **208**, 114–124.

Stintzing, F.C., Herbach, K.M., Mosshammer, M.R., Carle, R., Yi, W., Sellappan, S., Akoh, C.C., Bunch, R. and Felker, P. (2005) Color, betalain pattern, and antioxidant properties of cactus pear (*Opuntia* spp.) clones. *J. Agric. Food Chem.*, **53**, 442–451.

Stobart, A.K. and Kinsman, L.T. (1977) The hormonal control of betacyanin synthesis in *Amaranthus caudatus*. *Phytochemistry*, **16**, 1139–1142.

Strack, D. and Schliemann, W. (2001) Bifunctional polyphenol oxidases: novel functions in plant pigment biosynthesis. *Angew. Chem.* Int. Ed., **40**, 3791–3794.

Strack, D., Vogt, T. and Schliemann, W. (2003) Recent advance in betalain research. *Phytochemistry*, **62**, 247–269.

Straus, W. (1961) Studies on the chromoplasts of carrots. *Protoplasma*, **53**, 405–421.

Suzuki, S., Nishihara, M., Nakatsuka, T., Misawa, N., Ogiwara, I., and Yamamura, S. (2007) Flower color alteration in *Lotus Japonicus* by modification of the carotenoid biosynthetic pathway. *Plant Cell Rep.*, **26**, 951–959.

Tan, B.C., Schwartz, S.H., Zeevaart, J.A.D. and McCarty, D.R. (1997) Genetic control of abscisic acid biosynthesis in maize. *Proc. Natl. Acad. Sci. USA*, **94**, 12235–12240.

Tan, B.C., Joseph, L. M., Deng, W.T., Liu, L., Li, Q.B., Cline, K. and McCarty, D.R. (2003) Molecular characterization of the *Arabidopsis* 9-*cis* epoxycarotenoid dioxygenase gene family. *Plant J.*, **35**, 44–56.

Tanaka, Y., Sasaki, N. and Ohmiya, A. (2008) Biosynthesis of plant pigments: anthocyanins, betalains and carotenoids. *Plant J.*, **54**, 733–749.

Tesoriere, L., Butera, D., Allegra, M., Fazzari, M. and Livrea, M.A. (2005) Distribution of betalain pigments in red blood cells after consumption of cactus pear fruits and increased resistance of the cells to ex vivo induced oxidative hemolysis in humans. *J. Agric. Food Chem.*, **53**, 1266–1270.

Tesoriere, L., Fazzari, M., Angileri, F., Gentile, C. and Livrea, M.A. (2008) In vitro digestion of betalainic foods. Stability and bioaccessibility of betaxanthins and betacyanins and antioxidative potential of food digesta. *J. Agric. Food Chem.*, **56**, 10487–10492.

Tevini, M. and Steinmuller, D. (1985) Composition and function of plastoglobuli. II. Lipid composition of leaves and plastogulobuli during senescence. *Planta*, **163**, 91–96.

Tian, L., Musetti, V., Kim, J., Magallanes-Lundback, M. and DellaPenna, D. (2004) The *Arabidopsis LUT1* locus encodes a member of the cytochrome P450 family that is required for carotenoid ε-ring hydroxylation activity. *Proc. Natl. Acad. Sci. USA*, **101**, 402–407.

Valadon, L.R.G. and Mummery, R.S. (1968) Carotenoids in floral parts of a narcissus, a daffodil and a tulip. *Biochem. J.*, **106**, 479–484.

Vishnevetsky, M., Marianna Ovadis, M. and Vainstein, A. (1999) Carotenoid sequestration in plants: the role of carotenoid associated proteins. *Trends Plant Sci.*, **4**, 232–235.

Vogt, T. (2002) Substrate specificity and sequence analysis define a polyphyletic origin of betanidin 5- and 6-*O*-glucosyltransferase from *Dorotheanthus bellidiformis*. *Planta*, **214**, 492–495.

Vogt, T., Zimmermann, E., Grimm, R., Meyer, M. and Strack, D. (1997) Are the characteristics of betanidin glucosyltransferases from cell-suspension cultures of *Dorotheanthus bellidiformis* indicative of their phylogenetic relationship with flavonoid glucosyltransferases? *Planta*, **203**, 349–361.

von Lintig, J., Welsch, R., Bonk, M., Giuliano, G., Batschauer, A. and Kleinig, H. (1997) Light-dependent regulation of carotenoid biosynthesis occurs at the level of phytoene synthase expression and is mediated by phytochrome in *Sinapis alba* and *Arabidopsis thaliana* seedlings. *Plant J.*, **12**, 625–634.

Wang, S., Liu, J., Feng, Y., Niu, X., Giovannoni, J. and Liu, Y. (2008) Altered plastid levels and potential for improved fruit nutrient content by downregulation of the tomato DDB1-interacting protein CUL4. *Plant J.*, **55**, 89–103.

Welsch, R., Maass, D., Voegel, T., DellaPenna, D. and Beyer (2007) Transcription factor RAP2.2 and its interacting partner SINAT2: stable elements in the carotenogenesis of *Arabidopsis* leaves. *Plant Physiol.*, **145**, 1073–1085.

Wyler, H., Meuer, U., Bauer, J. and Stravas-Mombelli, L. (1984) *Cyclo*dopa glucoside (=(2S)-5-(-β-D-glucopyranosyloxy)-6-hydroxyindoline-2-carboxylic acid) and its occurrence in red beet (*Beta vulgaris* var. *rubra* L.). *Helv. Chim. Acta*, **67**, 1348–1355.

Xu, J., Tao, N., Liu, Q. and Deng, X. (2006) Presence of diverse ratios of lycopene/β-carotene in five pink or red-fleshed citrus cultivars. *Sci. Hort.*, **108**, 181–184.

Yamagishi, M., Kishimoto, S. and Nakayama, M. (2010) Carotenoid composition and changes in expression of carotenoid biosynthetic genes in tepals of Asiatic hybrid lily. *Plant Breed.*, **129**, 100–107.

Yamamizo, C., Kishimoto, S. and Ohmiya, A. (2010) Carotenoid composition and carotenogenic gene expression during *Ipomoea* petal development. *J. Exp. Bot.*, **61**, 709–719.

Yamamoto, K., Kobayashi, N., Yoshitama, K., Teramoto, S. and Komamine, A. (2001) Isolation and purification of tyrosine hydroxylase from callus cultures of *Portulaca grandiflora*. *Plant Cell Physiol.*, **42**, 969–975.

Ye, X., Al-Babili, S., Kloti, A., Zhang, J., Lucca, P., Beyer, P. and Potrykus, I. (2000) Engineering the provitamin A (β-carotene) biosynthetic pathway into (carotenoid-free) rice endosperm. *Science*, **287**, 303–305.

Zhu, C., Yamamura, S., Nishihara, M., Koiwa, H. and Sandmann, G. (2003) cDNA for the synthesis of cyclic carotenoids in petals of *Gentiana lutea* and their regulation during flower development. *Biochim. Biophys. Acta*, **1625**, 305–308.

Zhu, C., Gerjets, T. and Sandmann, G. (2007) *Nicotiana glauca* engineered for the production of ketocarotenoids in flowers and leaves by expressing the cyanobacterial *crtO* ketolase gene. *Transgenic Res.*, **16**, 813–821.

Zhu, C., Naqvia, S., Breitenbach, J., Sandmannb, G., Christoua, P. and Capella, T. (2008) Combinatorial genetic transformation generates a library of metabolic phenotypes for the carotenoid pathway in maize. *Proc. Natl. Acad. Sci. USA*, **105**, 18232–18237.

Zhu, C., Naqvi, S., Capell, T. and Christou, P. (2009) Metabolic engineering of keto-carotenoid biosynthesis in higher plants. *Arch. Biochem. Biophys.*, **483**, 182–190.

14

Metabolomics in Plant Biotechnology

Yozo Okazaki, Akira Oikawa, Miyako Kusano, Fumio Matsuda and Kazuki Saito

14.1 Introduction

The metabolome represents the ultimate cell phenotype which is controlled through the transmission of genetic information encoded in DNA. Thus, one has to analyse a snapshot of the metabolome to know the exact status of cells. In this context, metabolomics plays a key role in the field of plant biotechnology, where plant cells are modified by the expression of engineered foreign gene(s) (Oksman-Caldentey and Saito, 2005; Saito and Matsuda, 2010). The effects of transgene(s) should be evaluated by the changes of metabolome, whether the intended alteration is achieved and whether unintended changes take place – often referred to as substantial equivalence of genetically modified organisms. However, no single analytical technique can deal with all metabolites found in plant cells because of the extreme variety of phytochemicals in higher plants (Yonekura-Sakakibara and Saito, 2009). Therefore, combinations of multiple platforms for metabolome analysis are necessary for better coverage of a wide range of compounds (Saito and Matsuda, 2010). In this chapter, we describe analytical techniques of metabolomics, bioinformatics of metabolomics, and then applications to plant biotechnology.

14.2 Analytical Technologies

14.2.1 Gas Chromatography-Mass Spectrometry

Combinations of chromatographic and mass spectrometry (MS) techniques are utilised for metabolite profiling. Capillary gas chromatography (GC) has extremely high resolution

Plant Metabolism and Biotechnology, First Edition. Edited by Hiroshi Ashihara, Alan Crozier, and Atsushi Komamine.

which enables thousands of compounds to be detected with a high sample throughput. GC-MS-based metabolomics detects mainly primary metabolites and volatile compounds. Analysis of most primary metabolites requires an appropriate derivation step to ensure that all the metabolites are suitably volatile for GC analysis. MS is the detection system of choice for GC, and various MS analysers can be employed such as quadruple, time-of-flight (TOF), sector and ion-trap MS. TOF/MS is frequently used for metabolomics because it is fast and able to generate mass-to-charge ratio (*m/z*) information and corresponding GC retention time data. Plants produce numerous and varied primary metabolites, including sugars, amino acids, organic acids, fatty acids and steroids, which have different physicochemical properties. It is important to detect as many of these compounds as possible using the same chromatographic separation, without any analytical bias. In this context, GC-TOF/MS has become a powerful tool in plant metabolomics (Fiehn *et al.*, 2000; Roessner *et al.*, 2001; Fernie *et al.*, 2004; Keurentjes *et al.*, 2006; Schauer *et al.*, 2006, 2008).

GC-MS technology has been continually developed for more than 30 years, and recently new techniques such as two-dimensional gas chromatography (GC x GC) have come to the fore in metabolomics research. GC x GC offers high-resolution separations coupled with a high peak capacity, and with TOF/MS has been applied to metabolomic analysis of mouse, rice and other species (Shellie *et al.*, 2005; Kusano *et al.*, 2007; Ralston-Hooper *et al.*, 2008).

14.2.2 Liquid Chromatography-Mass Spectrometry

Liquid chromatography (LC)-MS systems offer an alternative method of analysis, with the chromatographic system linked to the MS by atmospheric pressure ionisation (API) interfaces such as electrospray-ionisation (ESI) and atmospheric pressure chemical ionisation (APCI) (Codrea *et al.*, 2007; Dunn, 2008). In the case of plant metabolomics studies with secondary metabolites, LC-MS is the analytical method of choice and, as described by De Vos *et al.* (2007), Naoumkina *et al.* (2007), Bottcher *et al.* (2008) and Matsuda *et al.* (2009), is typically performed in the following manner. Metabolites are extracted from plant tissues with water-methanol and, after simple pretreatments such as filtration/solid-phase extraction, components in the crude extract are separated by LC employing an octadecylsilyl (ODS) reversed-phase column eluted with water-acetonitrile-formic acid mobile phase gradient. Metabolites eluting from the column are ionised by an ESI interface using mild conditions that produce either a negatively or positively charged molecular-related ion ($[M+H]^+$ or $[M-H]^-$). The mass analyser is operated at 'scan' mode (e.g. scanning from *m/z* 100 to 1000) to detect a wide range of metabolites. The obtained raw chromatographic data are processed to produce data matrices. Several types of peak-picking software have been developed to analyse LC-MS data (Broeckling *et al.*, 2006; Smith *et al.*, 2006; Codrea *et al.*, 2007). Metabolite signals are assigned by additional information such as LC retention data of standard compounds and tandem mass spectral data (MS^2). The current bottleneck for LC-MS metabolomics is metabolite identification because of a shortage of authentic standards of phytochemicals and a subsequent absence of reference retention times and MS^2 data. Thus, although approximately 1500 metabolite signals were detected by the metabolome analyses of *Arabidopsis*, only a few hundred have so far been identified (Bottcher *et al.*, 2008; Matsuda *et al.*, 2009). Much effort is required for more comprehensive identification of metabolites by obtaining additional reference compounds and generating much more comprehensive databases (Kind and Fiehn, 2006; Moco *et al.*, 2006; Shinbo *et al.*, 2006; Bocker and Rasche, 2008).

14.2.3 Capillary Electrophoresis-Mass Spectrometry

Capillary electrophoresis (CE)-MS is utilised to analyse a wide spectrum of ionic metabolites. In capillary zone electrophoresis, constituent ions migrate on the basis of the electrostatic force resulting from the charge and size of ions, in addition to electro-osmotic flow derived from the capillary and the type of electrolyte used. Ionic compounds are separated with high resolution in a narrow capillary, but those with hydrophilic functional groups, such as hydroxyl and carboxyl groups, have to be derivatised prior to analysis. In metabolomic analyses using CE-MS, samples are often divided for cation and anion analyses. For cation analysis, Soga's method using formic acid as an electrolyte is convenient experimentally, and provides excellent chromatographic resolution of metabolites with good reproducibility in replicate analyses (Soga *et al.*, 2006). In contrast, routine CE-MS methods for high-resolution analysis of anions have not yet been established, although various approaches, including the use of coated capillaries, have been assessed (Soga *et al.*, 2002a,b; Harada *et al.*, 2006).

Target ionic metabolites in CE-MS analyses include amino acids, organic acids, nucleotides and sugar phosphates. Because these metabolites are physiologically important and common to all organisms, CE-MS has been applied in a variety of metabolomic studies (Monton and Soga, 2007; Oikawa *et al.*, 2008; Ramautar *et al.*, 2009), including identification of biomarkers for the progression of prostate cancer (Sreekumar *et al.*, 2009), oxidative stress (Soga *et al.*, 2006), measurement of internal body time (Minami *et al.*, 2009), and identification of unknown gene functions in *Arabidopsis* (Watanabe *et al.*, 2008).

14.2.4 Fourier Transform Ion Cyclotron Resonance Mass Spectrometry (FT-ICR MS)

Accurate *m/z* values obtained with the ultra-high resolving power of Fourier transform ion cyclotron resonance mass spectrometry (FT-ICR MS) are useful not only for identification of chemical structures of detected compounds, but are also of value in metabolomic studies for the following two reasons. Firstly, separate detection of different compounds which have very similar molecular masses can be acquired by direct infusion without any chromatographic steps, and this can result in the detection of a number of metabolites with rapid analysis. Secondly, accurate estimation of the chemical formulae of detected peaks can be acquired, which can lead to identification of unknown metabolites. MS^2 analysis coupled with FT-ICR MS is helpful for estimation of chemical formulae, because fragment ions are also detected with high resolution and high accuracy. In fact, several metabolites which accumulated in *Arabidopsis* following treatment with herbicides were identified by MS^2 analyses of FT-ICR MS (Oikawa *et al.*, 2006). However, because of difficulties in hardware handling and the processing of the vast amounts of data acquired, the number of reports on metabolomic studies using FT-ICR MS is limited. Non-targeted metabolite analysis of strawberry fruits was the first report in metabolomic study using FT-ICR MS (Aharoni *et al.*, 2002). The technique has also been applied to *Arabidopsis* functional genomics (Hirai *et al.*, 2004, 2005; Tohge *et al.*, 2005), transgenic tobacco (Mungur *et al.*, 2005), herbicide-treated *Arabidopsis* including development of software for data processing (Oikawa *et al.*, 2006), and identification of metabolic biomarkers of Crohn's disease (Jansson *et al.*, 2009). Recently, FT-ICR MS connected with LC has been used for metabolomic studies (Iijima *et al.*, 2008; Suzuki *et al.*, 2008) which may result in the identification of large numbers of metabolites including isomers. The ultra-high resolution provided by FT-ICR MS has the potential to develop a new field of metabolomics.

14.2.5 Nuclear Magnetic Resonance Spectroscopy

Metabolites almost without exception are composed of hydrogen, carbon, nitrogen, oxygen and phosphorus. These elements have isotopes that yield nuclear magnetic resonance (NMR) signals when placed in a strong magnetic field and pulsed with radio-frequency electromagnetic radiation. NMR spectroscopy is also one of the key analytical tools for metabolomics because it can yield detailed information about the quantities and identities of the metabolites present in extracts or *in vivo* (Kikuchi *et al.*, 2004; Lindon *et al.*, 2004; Wang *et al.*, 2004; Krishnan *et al.*, 2005; Clayton *et al.*, 2006; Ratcliffe and Shachar-Hill, 2006; Sekiyama and Kikuchi, 2007; Tian *et al.*, 2007; Hagel *et al.*, 2008; Sekiyama *et al.*, 2010). Among the various pulse sequence programs, ^{1}H NMR is used most routinely for high-throughput metabolomic studies due to its relatively short acquisition time per analysis. The advantages of NMR over MS-based methods include the fact that it is non-destructive, non-biased (any compounds with isotopes detectable by NMR can be analysed), easily quantifiable and permits the identification of novel compounds. Since NMR is non-destructive and usually does not require any derivatisation process, it allows the sample to be analysed subsequently by other spectroscopic methods. It is also possible to record a NMR spectrum from living tissue (Kikuchi *et al.*, 2004; Mesnard and Ratcliffe, 2005) and solid samples (Blaise *et al.*, 2009; Xu *et al.*, 2009; Sekiyama *et al.*, 2010). The major disadvantage of NMR, relative to MS, is its low sensitivity. Overlapping of the signals derived from many similar molecules from biological samples is another major problem which inhibits the accurate assignment of NMR signals. Disadvantages of lack of sensitivity and resolution are gradually being overcome by the development of cryogenic probes (Kovacs *et al.*, 2005) and multidimensional NMR techniques (Kikuchi *et al.*, 2004; Sekiyama and Kikuchi, 2007; Chikayama *et al.*, 2008). Stable isotope labelling of the samples with ^{13}C and ^{15}N, which is frequently used for the structural analysis of proteins by NMR, is also a useful technique which enhances the sensitivity of NMR (Kikuchi and Hirayama, 2007).

14.3 Informatics Techniques

Extremely large amounts of data are generated by instrumental analysis, particularly in the case of high-performance instruments frequently used for metabolome analysis which can detect tiny signals with high resolution. To handle the large data-sets and comprehend the metabolome data, automated software is needed which is capable of picking up peaks from mass or NMR spectra, aligning the peaks among the samples, and identifying and quantifying each metabolite. Therefore, informatics is an essential tool for processing large metabolomic data-sets (Fukushima *et al.*, 2009; Tohge and Fernie, 2009). For GC-MS data, automated deconvolution and identification systems using NIST mass spectral search programs or AMDIS are publicly available (Ausloos *et al.*, 1999). For processing of raw MS data acquired by HPLC-MS or CE-MS, several kinds of software are available including MetAlign (Vorst *et al.*, 2005), MZmine2 (Katajamaa *et al.*, 2006) and XCMS (Smith *et al.*, 2006). To analyse the data-sets from FT-ICR MS, DrDmassPlus (Oikawa *et al.*, 2006) can be used for peak picking, peak alignment and some statistical analyses. This software can facilitate comprehensive data analyses for non-targeted metabolomics approaches.

To identify peaks processed by the software, several mass spectral databases can be used including MassBank (Horai *et al.*, 2008), METLIN (Smith *et al.*, 2005), MS2T (Matsuda *et al.*, 2009), MSRI@CSB.DB (Kopka *et al.*, 2005), NIST Chem WebBook (Linstrom and Mallard, 2001), Lipid Search (Taguchi *et al.*, 2007), FiehnLib (Kind *et al.*, 2009) and MASSFinder library (http://www.massfinder.com). Some of these databases have a data management system designed to assist metabolite identification by providing public access to its depository. In addition to these mass spectral databases, several compound databases that are readily available include KNApSAcK (Shinbo *et al.*, 2006), KEGG (Kanehisa *et al.*, 2008), PubChem (Wheeler *et al.*, 2008), LipidBank (Yasugi and Watanabe, 2002) and LIPIDMAPS (Fahy *et al.*, 2007). Information on compound name, chemical formula and molecular weight deposited in these databases also can be used for peak identification if authentic compounds are not available. Use can also be made of several databases that focus mainly on the mass spectra or information about metabolites from several particular plant species. KOMICS (Iijima *et al.*, 2008) and MotoDB (Moco *et al.*, 2006) provide information on detected ions from tomato, and ARMeC (http://www.armec.org/MetaboliteLibrary/) has databases of metabolites from *Arabidopsis* and potato.

Although the collection of the metabolite abundance information is a challenging task, the interpretation of these data is both time-consuming and daunting. Procedures for interpretation of metabolome data can be facilitated by using metabolic pathway databases such as AraCyc (Mueller *et al.*, 2003), MapMan (Thimm *et al.*, 2004), KaPPA-View (Tokimatsu *et al.*, 2005) and KEGG (Kanehisa *et al.*, 2008). Some of these databases can import the abundance information of metabolites (and transcriptome information), and display an integrated overview of metabolic state by reflecting the abundance of each metabolite on the possible metabolic pathway in plants. These databases also include information on enzymatic reactions underlying these metabolic pathways, and information on proteins or genes involved in each reaction step. Although these pathway databases are helpful, it is necessary to bear in mind that many metabolic pathways in plants are divided by subcellular and intercellular compartmentalisation when attempting to understand the state of metabolism in plants from metabolome data.

In metabolomic studies, statistical analysis is often employed to evaluate the comprehensive differences in the detected metabolites in different samples (Fiehn, 2002; Fukusaki and Kobayashi, 2005; Hall, 2006). Various statistical methods used in conventional genetic studies are applicable to metabolomic data by considering the amount of each metabolite as a trait value. Principal component analysis (PCA), one method of multivariate analysis, is commonly used in metabolomic studies. There have been many reports on the application of PCA to metabolomic data (Catchpole *et al.*, 2005; Takahashi *et al.*, 2005; Tarpley *et al.*, 2005; Tohge *et al.*, 2005; Baker *et al.*, 2006; Dixon *et al.*, 2006; Oikawa *et al.*, 2006; Kim *et al.*, 2007; Kusano *et al.*, 2007; Moco *et al.*, 2007). In addition, several statistical analytical methods have been used for the analysis of metabolomic data-sets, for example: hierarchical cluster analysis (HCA) (Grata *et al.*, 2007; Parveen *et al.*, 2007), partial least squares discriminant analysis (PLS-DA) (Jonsson *et al.*, 2004; Kusano *et al.*, 2007), and batch-learning self-organising map (BL-SOM) (Hirai *et al.*, 2004, 2005; Kim *et al.*, 2007). Depending on the objective of each study, the most appropriate statistical analytical method should be exploited to evaluate the metabolomic data.

14.4 Biotechnological Application

14.4.1 Application for Functional Genomics

Functional characterisation of genes on a genome scale is one of the most important and challenging tasks in the post-genomic era. In modern *Arabidopsis* research, the loss-of-function or gain-of-function mutant lines play a very important role in the study of functional genomics. A combination of the metabolomics approach and these bioresources has been demonstrated to be an effective strategy in uncovering the role of the genes of unknown function. Using GC-MS and CE-MS, Watanabe *et al.* (2008) obtained a metabolome data-set from wild-type *Arabidopsis* and T-DNA insertion mutants having immature *bsas* (β-substituted alanine synthase) genes. Statistical analyses revealed that one unknown metabolite, which was usually found in wild type, did not accumulate in one of the *bsas* mutants, *bsas3;1*. The compound was eventually identified as a unique dipeptide, and the *bsas3;1* gene was shown to be involved in the biosynthesis of the dipeptide. Functional characterisation of unknown genes using the metabolomics approach can also be highly accelerated by other '-omics' approach such as transcriptome analysis. Several groups used transcriptome coexpression analysis with metabolome data to find new metabolic genes using a limited number of known genes, based on an assumption that the genes involved in the same metabolic pathway are coexpressed by a shared regulatory mechanism (Saito *et al.*, 2008). This strategy, based on the correlation of 'gene and metabolite', has successfully revealed the function of genes involved in secondary metabolism (Hirai *et al.*, 2005, 2007; Tohge *et al.*, 2005; Yonekura-Sakakibara *et al.*, 2007, 2008; Sawada *et al.*, 2009) and primary metabolism of plants (Persson *et al.*, 2005; Okazaki *et al.*, 2009).

14.4.2 Application for Metabolome QTL Analysis

Since metabolite levels in plant tissues (m-trait) is also a quantitative trait, quantitative trait loci (QTL) analysis of m-traits, such as the level of seed vitamin E, revealed the QTLs responsible for the control of the metabolite level and its genetic principles (Gilliland *et al.*, 2006). Recently, metabolome QTL (mQTL) analyses have made possible a comprehensive understanding of the genetic background of m-traits (Keurentjes *et al.*, 2006; Schauer *et al.*, 2006, 2008; Wentzell *et al.*, 2007; Lisec *et al.*, 2008; Rowe *et al.*, 2008). The mQTL analysis of *Arabidopsis* revealed that the QTLs are unevenly distributed in the genome, and there are several QTL hot-spot regions (Keurentjes *et al.*, 2006; Lisec *et al.*, 2008; Rowe *et al.*, 2008). The existence of the QTL hotspots suggests that the overall composition of the plant metabolome can be controlled by the manipulation of small genomic regions. In addition, it has been reported that several candidate genes with plausible functional annotation could be deduced from gene ontology (GO) information (Keurentjes *et al.*, 2006; Lisec *et al.*, 2008). Further advances in gene sequencing techniques will allow the determination of variations in genome sequences between two parents of experimental lines prior to QTL analysis (Clark *et al.*, 2007; Lizardi, 2008). A relationship between m-traits and other important traits, such as yield, taste and biomass, has been paid much attention, since these traits are likely to interact closely with plant metabolism (Schauer *et al.*, 2006; Meyer *et al.*, 2007; Lisec *et al.*, 2008). The analysis of tomato fruit traits with metabolome data indicated that there are weak correlations among these traits (Schauer *et al.*, 2006, 2008). The regression analysis of metabolome data to *Arabidopsis* biomass traits demonstrated that the growth rate of *Arabidopsis* seedlings is predictable from metabolome signature to some extent (Meyer *et al.*, 2007). These pioneering works suggest that interactions between metabolite compositions and other traits may be predictable through future advanced metabolome analysis.

14.4.3 Application for Evaluation of Genetically Modified Organisms

For improvement in crop properties, such as an increase in yield or accumulation of nutritional metabolites, two different strategies can be employed: traditional breeding and genetic modification (GM). Furthermore, both strategies have the possibility to generate 'unintended effects' in phenotypes, including changes of the metabolome of crops. Unintended effects can represent statistically significant differences in phenotypes between GM material and the parental and/or sibling lines. Genetic crosses between closely related species were estimated to produce a wider range of variance than transformation of genes, because selection from homogeneous populations with transformation of genes is likely to produce the smallest number of unintended effects. Furthermore, genetic modification toward closely related species can be considered to give similar outcomes to those observed by genetic crosses between existing germplasm pools. Since metabolomics is a technology that aims to identify and quantify the metabolome in an organism, metabolomic analysis techniques can be applied to evaluate the substantial compositional similarity between genetically-modified and conventional crops.

MS-based and NMR-based metabolite profiling has been performed for the evaluation of substantial equivalence of genetically modified crops, such as tomato, potato, maize, potato and wheat (Le Gall *et al.*, 2003; Catchpole *et al.*, 2005; Baker *et al.*, 2006; Levandi *et al.*, 2008). However, a single analytical method enabling complete metabolome analysis does not exist. Hence, data collection from many analytical platforms is important to give wide coverage of metabolite profiles. For this purpose, we have developed a multiple MS-based metabolomics pipeline that consists of different chromatographic techniques connected to TOF/MS. Figure 14.1 presents an example for evaluation of substantial equivalence of genetically modified tomatoes, which are overexpressing a certain protein. By using data obtained from the multiple MS-based metabolomics pipeline, PCA was performed for visualisation of substantial compositional similarity in metabolite profiles between transgenic and control (genetic background of the transgenic line) tomato fruits together with two different cultivars. When the transgenic and control samples form a tight cluster in the first and second component in PCA, it indicates that these metabolite compositions can be considered to be similar. As shown in Figure 14.1, transgenic and control samples were well clustered in the score scatter plot, while each group of other different cultivars shows clear separations. Thus, metabolite profiling by the combination of unbiased analytical technology and appropriate statistical data analysis has become a powerful tool for the assessment of safety of GM crops.

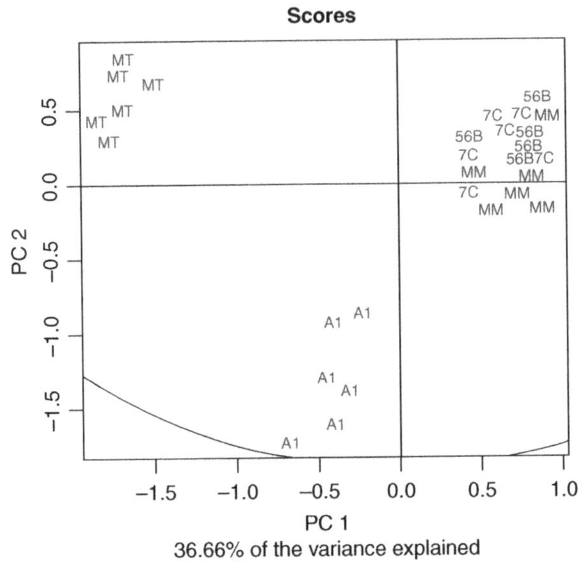

Figure 14.1 *Score scatter plot for principal components PC1 and PC2 generated from the data (30 tomato fruit samples × 3697 peaks) obtained by principal component analysis (PCA). A1 and MT are two different tomato cultivars, whereas others are genetically modified lines (56B and 7C) and its genetic background (MM)*

14.4.4 Application for Identification of Biomarkers

The discovery of biomarkers in the life sciences helps to determine the condition of disease at the level of metabolites, leading to the development of new drugs. Comparative metabolomics, for example, between diseased and healthy organs, after and before drug treatment, has often found new biomarkers (Koulman *et al.*, 2009). Sreekumar and coworkers applied metabolomic techniques based on GC-MS and LC-MS to find the role of sarcosine as a biomarker in prostate cancer progression (Sreekumar *et al.*, 2009). They analysed 262 samples and detected more than 1126 metabolites. Profiling analysis of these detected metabolites among benign, localised and metastatic prostate cancer revealed that the amounts of sarcosine increased substantially during prostate cancer progression. Finally, coupled with the results of the molecular biological and physiological experiments, they identified sarcosine as a potentially important metabolic intermediary of cancer cell invasion and aggressivity. As in this case and the finding of ophthalmic acid as an oxidative stress biomarker (Soga *et al.*, 2006), only one biomarker was sometimes identified in metabolomic analyses. However, in most metabolomic studies, the levels of several metabolites, including unknowns, have been identified as potential biomarkers (Koulman *et al.*, 2009). For example, in a comparison of healthy and Crohn's diseased samples, 21 discriminating metabolites were selected as biomarkers from 18,706 measured masses detected by the FT-ICR MS analysis and statistical data processing (Jansson *et al.*, 2009). To refine genuine biomarker(s) from these candidates, traditional genetic or physiological experiments in addition to other omics studies such as proteome and transcriptome analyses may be needed. Metabolomic analysis and statistical data processing is one of the most promising methods for the discovery of biomarkers.

Acknowledgement

We thank Dr Henning Redestig at RIKEN Plant Science Center for sharing his results prior to publication.

References

Aharoni, A., Ric de Vos, C.H., Verhoeven, H.A., Maliepaard, C.A., Kruppa, G., Bino, R. and Goodenowe, D.B. (2002) Nontargeted metabolome analysis by use of Fourier Transform Ion Cyclotron Mass Spectrometry. *Omics*, **6**, 217–234.

Ausloos, P., Clifton, C.L., Lias, S.G., Mikaya, A.I., Stein, S.E., Tchekhovskoi, D.V., Sparkman, O.D., Zaikin, V. and Zhu, D. (1999) The critical evaluation of a comprehensive mass spectral library. *J. Am. Soc. Mass Spectrom.*, **10**, 287–299.

Baker, J.M., Hawkins, N.D., Ward, J.L., Lovegrove, A., Napier, J.A., Shewry, P.R. and Beale, M.H. (2006) A metabolomic study of substantial equivalence of field-grown genetically modified wheat. *Plant Biotechnol. J.*, **4**, 381–392.

Blaise, B.J., Giacomotto, J., Triba, M.N., Toulhoat, P., Piotto, M., Emsley, L., Segalat, L., Dumas, M.E. and Elena, B. (2009) Metabolic profiling strategy of *Caenorhabditis elegans* by whole-organism nuclear magnetic resonance. *J. Proteome Res.*, **8**, 2542–2550.

Bocker, S. and Rasche, F. (2008) Towards de novo identification of metabolites by analyzing tandem mass spectra. *Bioinformatics*, **24**, i49–i55.

Bottcher, C., von Roepenack-Lahaye, E., Schmidt, J., Schmotz, C., Neumann, S., Scheel, D. and Clemens, S. (2008) Metabolome analysis of biosynthetic mutants reveals a diversity of metabolic changes and allows identification of a large number of new compounds in Arabidopsis. *Plant Physiol.*, **147**, 2107–2120.

Broeckling, C.D., Reddy, I.R., Duran, A.L., Zhao, X. and Sumner, L.W. (2006) MET-IDEA: data extraction tool for mass spectrometry-based metabolomics. *Anal. Chem.*, **78**, 4334–4341.

Catchpole, G.S., Beckmann, M., Enot, D.P., Mondhe, M., Zywicki, B., Taylor, J., Hardy, N., Smith, A., King, R.D., Kell, D.B., Fiehn, O. and Draper, J. (2005) Hierarchical metabolomics demonstrates substantial compositional similarity between genetically modified and conventional potato crops. *Proc. Natl. Acad. Sci. USA*, **102**, 14458–14462.

Chikayama, E., Suto, M., Nishihara, T., Shinozaki, K. and Kikuchi, J. (2008) Systematic NMR analysis of stable isotope labeled metabolite mixtures in plant and animal systems: coarse grained views of metabolic pathways. *PLoS One*, **3**, e3805.

Clark, R.M., Schweikert, G., Toomajian, C., Ossowski, S., Zeller, G., Shinn, P., Warthmann, N., Hu, T.T., Fu, G., Hinds, D.A., Chen, H., Frazer, K.A., Huson, D.H., Scholkopf, B., Nordborg, M., Ratsch, G., Ecker, J.R. and Weigel, D. (2007) Common sequence polymorphisms shaping genetic diversity in *Arabidopsis thaliana*. *Science*, **317**, 338–342.

Clayton, T.A., Lindon, J.C., Cloarec, O., Antti, H., Charuel, C., Hanton, G., Provost, J.P., Le Net, J.L., Baker, D., Walley, R.J., Everett, J.R. and Nicholson, J.K. (2006) Pharmaco-metabonomic phenotyping and personalized drug treatment. *Nature*, **440**, 1073–1077.

Codrea, M.C., Jimenez, C.R., Heringa, J. and Marchiori, E. (2007) Tools for computational processing of LC-MS datasets: a user's perspective. *Comput. Methods Programs Biomed.*, **86**, 281–290.

De Vos, R.C., Moco, S., Lommen, A., Keurentjes, J.J., Bino, R.J. and Hall, R.D. (2007) Untargeted large-scale plant metabolomics using liquid chromatography coupled to mass spectrometry. *Nat. Protoc.*, **2**, 778–791.

Dixon, R.A., Gang, D.R., Charlton, A.J., Fiehn, O., Kuiper, H.A., Reynolds, T.L., Tjeerdema, R.S., Jeffery, E.H., German, J.B., Ridley, W.P. and Seiber, J.N. (2006) Applications of metabolomics in agriculture. *J. Agric. Food Chem.*, **54**, 8984–8994.

Dunn, W.B. (2008) Current trends and future requirements for the mass spectrometric investigation of microbial, mammalian and plant metabolomes. *Phys. Biol.*, **5**, 11001.

Fahy, E., Sud, M., Cotter, D. and Subramaniam, S. (2007) LIPID MAPS online tools for lipid research. *Nucleic Acids Res.*, **35**, W606–612.

Fernie, A.R., Trethewey, R.N., Krotzky, A.J. and Willmitzer, L. (2004) Metabolite profiling: from diagnostics to systems biology. *Nat. Rev. Mol. Cell Biol.*, **5**, 763–769.

Fiehn, O. (2002) Metabolomics – the link between genotypes and phenotypes. *Plant Mol. Biol.*, **48**, 155–171.

Fiehn, O., Kopka, J., Dörmann, P., Altmann, T., Trethewey, R.N. and Willmitzer, L. (2000) Metabolite profiling for plant functional genomics. *Nat. Biotechnol.*, **18**, 1157–1161.

Fukusaki, E. and Kobayashi, A. (2005) Plant metabolomics: potential for practical operation. *J. Biosci. Bioeng.*, **100**, 347–354.

Fukushima, A., Kusano, M., Redestig, H., Arita, M. and Saito, K. (2009) Integrated omics approaches in plant systems biology. *Curr. Opin. Chem. Biol.*, **13**, 532–538.

Gilliland, L.U., Magallanes-Lundback, M., Hemming, C., Supplee, A., Koornneef, M., Bentsink, L. and Dellapenna, D. (2006) Genetic basis for natural variation in seed vitamin E levels in *Arabidopsis thaliana*. *Proc. Natl. Acad. Sci. USA*, **103**, 18834–18841.

Grata, E., Boccard, J., Glauser, G., Carrupt, P.A., Farmer, E.E., Wolfender, J.L. and Rudaz, S. (2007) Development of a two-step screening ESI-TOF-MS method for rapid determination of significant stress-induced metabolome modifications in plant leaf extracts: the wound response in *Arabidopsis thaliana* as a case study. *J. Sep. Sci.*, **30**, 2268–2278.

Hagel, J.M., Weljie, A.M., Vogel, H.J. and Facchini, P.J. (2008) Quantitative ^{1}H nuclear magnetic resonance metabolite profiling as a functional genomics platform to investigate alkaloid biosynthesis in opium poppy. *Plant Physiol.*, **147**, 1805–1821.

Hall, R.D. (2006) Plant metabolomics: from holistic hope, to hype, to hot topic. *New Phytol.*, **169**, 453–468.

Harada, K., Fukusaki, E. and Kobayashi, A. (2006) Pressure-assisted capillary electrophoresis mass spectrometry using combination of polarity reversion and electroosmotic flow for metabolomics anion analysis. *J. Biosci. Bioeng.*, **101**, 403–409.

Hirai, M.Y., Yano, M., Goodenowe, D.B., Kanaya, S., Kimura, T., Awazuhara, M., Arita, M., Fujiwara, T. and Saito, K. (2004) Integration of transcriptomics and metabolomics for understanding of global responses to nutritional stresses in *Arabidopsis thaliana*. *Proc. Natl. Acad. Sci. USA*, **101**, 10205–10210.

Hirai, M.Y., Klein, M., Fujikawa, Y., Yano, M., Goodenowe, D.B., Yamazaki, Y., Kanaya, S., Nakamura, Y., Kitayama, M., Suzuki, H., Sakurai, N., Shibata, D., Tokuhisa, J., Reichelt, M., Gershenzon, J., Papenbrock, J. and Saito, K. (2005) Elucidation of gene-to-gene and metabolite-to-gene networks in *Arabidopsis* by integration of metabolomics and transcriptomics. *J. Biol. Chem.*, **280**, 25590–25595.

Hirai, M.Y., Sugiyama, K., Sawada, Y., Tohge, T., Obayashi, T., Suzuki, A., Araki, R., Sakurai, N., Suzuki, H., Aoki, K., Goda, H., Nishizawa, O.I., Shibata, D. and Saito, K. (2007) Omics-based identification of *Arabidopsis* Myb transcription factors regulating aliphatic glucosinolate biosynthesis. *Proc. Natl. Acad. Sci. USA*, **104**, 6478–6483.

Horai, H., Arita, M. and Nishioka, T. (2008) Comparison of ESI-MS spectra in MassBank Database. *Proc. Int. Conf. Biomed. Eng. Inform.*, **2**, 853–857.

Iijima, Y., Nakamura, Y., Ogata, Y., Tanaka, K., Sakurai, N., Suda, K., Suzuki, T., Suzuki, H., Okazaki, K., Kitayama, M., Kanaya, S., Aoki, K. and Shibata, D. (2008) Metabolite annotations based on the integration of mass spectral information. *Plant J.*, **54**, 949–962.

Jansson, J., Willing, B., Lucio, M., Fekete, A., Dicksved, J., Halfvarson, J., Tysk, C. and Schmitt-Kopplin, P. (2009) Metabolomics reveals metabolic biomarkers of Crohn's disease. *PLoS One*, **4**, e6386.

Jonsson, P., Gullberg, J., Nordstrom, A., Kusano, M., Kowalczyk, M., Sjostrom, M. and Moritz, T. (2004) A strategy for identifying differences in large series of metabolomic samples analyzed by GC/MS. *Anal. Chem.*, **76**, 1738–1745.

Kanehisa, M., Araki, M., Goto, S., Hattori, M., Hirakawa, M., Itoh, M., Katayama, T., Kawashima, S., Okuda, S., Tokimatsu, T. and Yamanishi, Y. (2008) KEGG for linking genomes to life and the environment. *Nucleic Acids Res.*, **36**, D480–484.

Katajamaa, M., Miettinen, J. and Oresic, M. (2006) MZmine: toolbox for processing and visualization of mass spectrometry based molecular profile data. *Bioinformatics*, **22**, 634–636.

Keurentjes, J.J., Fu, J., de Vos, C.H., Lommen, A., Hall, R.D., Bino, R.J., Van Der Plas, L.H., Jansen, R.C., Vreugdenhil, D. and Koornneef, M. (2006) The genetics of plant metabolism. *Nat. Genet.*, **38**, 842–849.

Kikuchi, J., Shinozaki, K. and Hirayama, T. (2004) Stable isotope labeling of *Arabidopsis thaliana* for an NMR-based metabolomics approach. *Plant Cell Physiol.*, **45**, 1099–1104.

Kikuchi, J. and Hirayama, T. (2007) Practical aspects of uniform stable isotope labeling of higher plants for heteronuclear NMR-based metabolomics. *Methods Mol. Biol.*, **358**, 273–286.

Kim, J.K., Bamba, T., Harada, K., Fukusaki, E. and Kobayashi, A. (2007) Time-course metabolic profiling in *Arabidopsis thaliana* cell cultures after salt stress treatment. *J. Exp. Bot.*, **58**, 415–424.

Kind, T. and Fiehn, O. (2006) Metabolomic database annotations via query of elemental compositions: mass accuracy is insufficient even at less than 1 ppm. *BMC Bioinformatics*, **7**, 234.

Kind, T., Wohlgemuth, G., Lee do, Y., Lu, Y., Palazoglu, M., Shahbaz, S. and Fiehn, O. (2009) FiehnLib: mass spectral and retention index libraries for metabolomics based on quadrupole and time-of-flight gas chromatography/mass spectrometry. *Anal. Chem.*, **81**, 10038–10048.

Kopka, J., Schauer, N., Krueger, S., Birkemeyer, C., Usadel, B., Bergmuller, E., Dörmann, P., Weckwerth, W., Gibon, Y., Stitt, M., Willmitzer, L., Fernie, A.R. and Steinhauser, D. (2005) GMD@CSB.DB: the Golm Metabolome Database. *Bioinformatics*, **21**, 1635–1638.

Koulman, A., Lane, G.A., Harrison, S.J. and Volmer, D.A. (2009) From differentiating metabolites to biomarkers. *Anal. Bioanal. Chem.*, **394**, 663–670.

Kovacs, H., Moskau, D. and Spraul, M. (2005) Cryogenically cooled probes – a leap in NMR technology. *Prog. Nucl. Magn. Reson. Spectrosc.*, **46**, 131–155.

Krishnan, P., Kruger, N.J. and Ratcliffe, R.G. (2005) Metabolite fingerprinting and profiling in plants using NMR. *J. Exp. Bot.*, **56**, 255–265.

Kusano, M., Fukushima, A., Kobayashi, M., Hayashi, N., Jonsson, P., Moritz, T., Ebana, K. and Saito, K. (2007) Application of a metabolomic method combining one-dimensional and two-dimensional gas chromatography-time-of-flight/mass spectrometry to metabolic phenotyping of natural variants in rice. *J. Chromatogr. B Analyt. Technol. Biomed. Life Sci.*, **855**, 71–79.

Le Gall, G., Colquhoun, I.J., Davis, A.L., Collins, G.J. and Verhoeyen, M.E. (2003) Metabolite profiling of tomato (*Lycopersicon esculentum*) using ^{1}H NMR spectroscopy as a tool to detect potential unintended effects following a genetic modification. *J. Agric. Food Chem.*, **51**, 2447–2456.

Levandi, T., Leon, C., Kaljurand, M., Garcia-Canas, V. and Cifuentes, A. (2008) Capillary electrophoresis time-of-flight mass spectrometry for comparative metabolomics of transgenic versus conventional maize. *Anal. Chem.*, **80**, 6329–6335.

Lindon, J.C., Holmes, E. and Nicholson, J.K. (2004) Metabonomics: systems biology in pharmaceutical research and development. *Curr. Opin. Mol. Ther.*, **6**, 265–272.

Linstrom, P.J. and Mallard, W.G. (2001) The NIST Chemistry WebBook: a chemical data resource on the Internet. *J. Chem. Eng. Data*, **46**, 1059–1063.

Lisec, J., Meyer, R.C., Steinfath, M., Redestig, H., Becher, M., Witucka-Wall, H., Fiehn, O., Torjek, O., Selbig, J., Altmann, T. and Willmitzer, L. (2008) Identification of metabolic and biomass QTL in *Arabidopsis thaliana* in a parallel analysis of RIL and IL populations. *Plant J.*, **53**, 960–972.

Lizardi, P.M. (2008) Next-generation sequencing-by-hybridization. *Nat. Biotechnol.*, **26**, 649–650.

Matsuda, F., Yonekura-Sakakibara, K., Niida, R., Kuromori, T., Shinozaki, K. and Saito, K. (2009) MS/MS spectral tag-based annotation of non-targeted profile of plant secondary metabolites. *Plant J.*, **57**, 555–577.

Mesnard, F. and Ratcliffe, R.G. (2005) NMR analysis of plant nitrogen metabolism. *Photosynth. Res.*, **83**, 163–180.

Meyer, R.C., Steinfath, M., Lisec, J., Becher, M., Witucka-Wall, H., Torjek, O., Fiehn, O., Eckardt, A., Willmitzer, L., Selbig, J. and Altmann, T. (2007) The metabolic signature related to high plant growth rate in *Arabidopsis thaliana*. *Proc. Natl. Acad. Sci. USA*, **104**, 4759–4764.

Minami, Y., Kasukawa, T., Kakazu, Y., Iigo, M., Sugimoto, M., Ikeda, S., Yasui, A., Van Der Horst, G.T., Soga, T. and Ueda, H.R. (2009) Measurement of internal body time by blood metabolomics. *Proc. Natl. Acad. Sci. USA*, **106**, 9890–9895.

Moco, S., Bino, R.J., Vorst, O., Verhoeven, H.A., de Groot, J., van Beek, T.A., Vervoort, J. and de Vos, C.H. (2006) A liquid chromatography-mass spectrometry-based metabolome database for tomato. *Plant Physiol.*, **141**, 1205–1218.

Moco, S., Capanoglu, E., Tikunov, Y., Bino, R.J., Boyacioglu, D., Hall, R.D., Vervoort, J. and De Vos, R.C. (2007) Tissue specialization at the metabolite level is perceived during the development of tomato fruit. *J. Exp. Bot.*, **58**, 4131–4146.

Monton, M.R. and Soga, T. (2007) Metabolome analysis by capillary electrophoresis-mass spectrometry. *J. Chromatogr. A*, **1168**, 237–246; discussion 236.

Mueller, L.A., Zhang, P. and Rhee, S.Y. (2003) AraCyc: a biochemical pathway database for Arabidopsis. *Plant Physiol.*, **132**, 453–460.

Mungur, R., Glass, A.D., Goodenow, D.B. and Lightfoot, D.A. (2005) Metabolite fingerprinting in transgenic *Nicotiana tabacum* altered by the *Escherichia coli* glutamate dehydrogenase gene. *J. Biomed. Biotechnol.*, **2005**, 198–214.

Naoumkina, M., Farag, M.A., Sumner, L.W., Tang, Y., Liu, C.J. and Dixon, R.A. (2007) Different mechanisms for phytoalexin induction by pathogen and wound signals in *Medicago truncatula*. *Proc. Natl. Acad. Sci. USA*, **104**, 17909–17915.

Oikawa, A., Nakamura, Y., Ogura, T., Kimura, A., Suzuki, H., Sakurai, N., Shinbo, Y., Shibata, D., Kanaya, S. and Ohta, D. (2006) Clarification of pathway-specific inhibition by Fourier transform ion cyclotron resonance/mass spectrometry-based metabolic phenotyping studies. *Plant Physiol.*, **142**, 398–413.

Oikawa, A., Matsuda, F., Kusano, M., Okazaki, Y. and Saito, K. (2008) Rice metabolomics. *Rice*, **1**, 63–71.

Okazaki, Y., Shimojima, M., Sawada, Y., Toyooka, K., Narisawa, T., Mochida, K., Tanaka, H., Matsuda, F., Hirai, A., Hirai, M.Y., Ohta, H. and Saito, K. (2009) A chloroplastic UDP-glucose pyrophosphorylase from *Arabidopsis* is the committed enzyme for the first step of sulfolipid biosynthesis. *Plant Cell*, **21**, 892–909.

Oksman-Caldentey, K.M. and Saito, K. (2005) Integrating genomics and metabolomics for engineering plant metabolic pathways. *Curr. Opin. Biotechnol.*, **16**, 174–179.

Parveen, I., Moorby, J.M., Fraser, M.D., Allison, G.G. and Kopka, J. (2007) Application of gas chromatography-mass spectrometry metabolite profiling techniques to the analysis of heathland plant diets of sheep. *J. Agric. Food Chem.*, **55**, 1129–1138.

Persson, S., Wei, H., Milne, J., Page, G.P. and Somerville, C.R. (2005) Identification of genes required for cellulose synthesis by regression analysis of public microarray data sets. *Proc. Natl. Acad. Sci. USA*, **102**, 8633–8638.

Ralston-Hooper, K., Hopf, A., Oh, C., Zhang, X., Adamec, J. and Sepulveda, M.S. (2008) Development of GCxGC/TOF-MS metabolomics for use in ecotoxicological studies with invertebrates. *Aquat. Toxicol.*, **88**, 48–52.

Ramautar, R., Somsen, G.W. and de Jong, G.J. (2009) CE-MS in metabolomics. *Electrophoresis*, **30**, 276–291.

Ratcliffe, R.G. and Shachar-Hill, Y. (2006) Measuring multiple fluxes through plant metabolic networks. *Plant J.*, **45**, 490–511.

Roessner, U., Luedemann, A., Brust, D., Fiehn, O., Linke, T., Willmitzer, L. and Fernie, A. (2001) Metabolic profiling allows comprehensive phenotyping of genetically or environmentally modified plant systems. *Plant Cell*, **13**, 11–29.

Rowe, H.C., Hansen, B.G., Halkier, B.A. and Kliebenstein, D.J. (2008) Biochemical networks and epistasis shape the *Arabidopsis thaliana* metabolome. *Plant Cell*, **20**, 1199–1216.

Saito, K., Hirai, M.Y. and Yonekura-Sakakibara, K. (2008) Decoding genes with coexpression networks and metabolomics – 'majority report by precogs'. *Trends Plant Sci.*, **13**, 36–43.

Saito, K. and Matsuda, F. (2010) Metabolomics for functional genomics, systems biology, and biotechnology. *Ann. Rev. Plant Biol.*, **61**, 463–489.

Sawada, Y., Toyooka, K., Kuwahara, A., Sakata, A., Nagano, M., Saito, K. and Hirai, M.Y. (2009) Arabidopsis bile acid:sodium symporter family protein 5 is involved in methionine-derived glucosinolate biosynthesis. *Plant Cell Physiol.*, **50**, 1579–1586.

Schauer, N., Semel, Y., Roessner, U., Gur, A., Balbo, I., Carrari, F., Pleban, T., Perez-Melis, A., Bruedigam, C., Kopka, J., Willmitzer, L., Zamir, D. and Fernie, A.R. (2006) Comprehensive metabolic profiling and phenotyping of interspecific introgression lines for tomato improvement. *Nat. Biotechnol.*, **24**, 447–454.

Schauer, N., Semel, Y., Balbo, I., Steinfath, M., Repsilber, D., Selbig, J., Pleban, T., Zamir, D. and Fernie, A.R. (2008) Mode of inheritance of primary metabolic traits in tomato. *Plant Cell*, **20**, 509–523.

Sekiyama, Y. and Kikuchi, J. (2007) Towards dynamic metabolic network measurements by multi-dimensional NMR-based fluxomics. *Phytochemistry*, **68**, 2320–2329.

Sekiyama, Y., Chikayama, E. and Kikuchi, J. (2010) Profiling polar and semipolar plant metabolites throughout extraction processes using a combined solution-state and high-resolution magic angle spinning NMR approach. *Anal. Chem.*, **82**, 1643–1652.

Shellie, R.A., Welthagen, W., Zrostlikova, J., Spranger, J., Ristow, M., Fiehn, O. and Zimmermann, R. (2005) Statistical methods for comparing comprehensive two-dimensional gas chromatography-time-of-flight mass spectrometry results: metabolomic analysis of mouse tissue extracts. *J. Chromatogr. A*, **1086**, 83–90.

Shinbo, Y., Nakamura, Y., Altaf-Ul-Amin, M., Asahi, H., Kurokawa, K., Arita, M., Saito, K., Ohta, D., Shibata, D. and Kanaya, S. (2006) KNApSAcK: a comprehensive species-metabolite relationship database, in *Biotechnology in Agriculture and Forestry* (eds K. Saito, R. Dixon and L. Willmitzer), Springer, Heidelberg, pp. 165–184.

Smith, C.A., O'Maille, G., Want, E.J., Qin, C., Trauger, S.A., Brandon, T.R., Custodio, D.E., Abagyan, R. and Siuzdak, G. (2005) METLIN: a metabolite mass spectral database. *Ther. Drug Monit.*, **27**, 747–751.

Smith, C.A., Want, E.J., O'Maille, G., Abagyan, R. and Siuzdak, G. (2006) XCMS: processing mass spectrometry data for metabolite profiling using nonlinear peak alignment, matching, and identification. *Anal. Chem.*, **78**, 779–787.

Soga, T., Ueno, Y., Naraoka, H., Matsuda, K., Tomita, M. and Nishioka, T. (2002a) Pressure-assisted capillary electrophoresis electrospray ionization mass spectrometry for analysis of multivalent anions. *Anal. Chem.*, **74**, 6224–6229.

Soga, T., Ueno, Y., Naraoka, H., Ohashi, Y., Tomita, M. and Nishioka, T. (2002b) Simultaneous determination of anionic intermediates for *Bacillus subtilis* metabolic pathways by capillary electrophoresis electrospray ionization mass spectrometry. *Anal. Chem.*, **74**, 2233–2239.

Soga, T., Baran, R., Suematsu, M., Ueno, Y., Ikeda, S., Sakurakawa, T., Kakazu, Y., Ishikawa, T., Robert, M., Nishioka, T. and Tomita, M. (2006) Differential metabolomics reveals ophthalmic acid as an oxidative stress biomarker indicating hepatic glutathione consumption. *J. Biol. Chem.*, **281**, 16768–16776.

Sreekumar, A., Poisson, L.M., Rajendiran, T.M., Khan, A.P., Cao, Q., Yu, J., Laxman, B., Mehra, R., Lonigro, R.J., Li, Y., Nyati, M.K., Ahsan, A., Kalyana-Sundaram, S., Han, B., Cao, X., Byun, J., Omenn, G.S., Ghosh, D., Pennathur, S., Alexander, D.C., Berger, A., Shuster, J.R., Wei, J.T., Varambally, S., Beecher, C. and Chinnaiyan, A.M. (2009) Metabolomic profiles delineate potential role for sarcosine in prostate cancer progression. *Nature*, **457**, 910–914.

Suzuki, H., Sasaki, R., Ogata, Y., Nakamura, Y., Sakurai, N., Kitajima, M., Takayama, H., Kanaya, S., Aoki, K., Shibata, D. and Saito, K. (2008) Metabolic profiling of flavonoids in *Lotus japonicus* using liquid chromatography Fourier transform ion cyclotron resonance mass spectrometry. *Phytochemistry*, **69**, 99–111.

Taguchi, R., Nishijima, M. and Shimizu, T. (2007) Basic analytical systems for lipidomics by mass spectrometry in Japan. *Methods Enzymol.*, **432**, 185–211.

Takahashi, H., Hotta, Y., Hayashi, M., Kawai-Yamada, M., Komatsu, S. and Uchiyama, H. (2005) High throughput metabolome and proteome analysis of transgenic rice plants (Oryza sativa L.). *Plant Biotechnol.*, **22**, 47–50.

Tarpley, L., Duran, A.L., Kebrom, T.H. and Sumner, L.W. (2005) Biomarker metabolites capturing the metabolite variance present in a rice plant developmental period. *BMC Plant Biol.*, **5**, 8.

Thimm, O., Blasing, O., Gibon, Y., Nagel, A., Meyer, S., Kruger, P., Selbig, J., Muller, L.A., Rhee, S.Y. and Stitt, M. (2004) MAPMAN: a user-driven tool to display genomics data sets onto diagrams of metabolic pathways and other biological processes. *Plant J.*, **37**, 914–939.

Tian, C., Chikayama, E., Tsuboi, Y., Kuromori, T., Shinozaki, K., Kikuchi, J. and Hirayama, T. (2007) Top-down phenomics of *Arabidopsis thaliana*: metabolic profiling by one- and two-dimensional nuclear magnetic resonance spectroscopy and transcriptome analysis of albino mutants. *J. Biol. Chem.*, **282**, 18532–18541.

Tohge, T., Nishiyama, Y., Hirai, M.Y., Yano, M., Nakajima, J., Awazuhara, M., Inoue, E., Takahashi, H., Goodenowe, D.B., Kitayama, M., Noji, M., Yamazaki, M. and Saito, K. (2005) Functional genomics by integrated analysis of metabolome and transcriptome of Arabidopsis plants over-expressing an MYB transcription factor. *Plant J.*, **42**, 218–235.

Tohge, T. and Fernie, A.R. (2009) Web-based resources for mass-spectrometry-based metabolomics: a user's guide. *Phytochemistry*, **70**, 450–456.

Tokimatsu, T., Sakurai, N., Suzuki, H., Ohta, H., Nishitani, K., Koyama, T., Umezawa, T., Misawa, N., Saito, K. and Shibata, D. (2005) KaPPA-view: a web-based analysis tool for integration of transcript and metabolite data on plant metabolic pathway maps. *Plant Physiol.*, **138**, 1289–1300.

Vorst, O., de Vos, C.H., Lommen, A., Staps, R.V., Visser, R.G., Bino, R.J. and Hall, R.D. (2005) A non-directed approach to the differential analysis of multiple LC–MS-derived metabolic profiles. *Metabolomics*, **1**, 169–180.

Wang, Y., Holmes, E., Nicholson, J.K., Cloarec, O., Chollet, J., Tanner, M., Singer, B.H. and Utzinger, J. (2004) Metabolomic investigations in mice infected with *Schistosoma mansoni*: an approach for biomarker identification. *Proc. Natl. Acad. Sci. USA*, **101**, 12676–12681.

Watanabe, M., Kusano, M., Oikawa, A., Fukushima, A., Noji, M. and Saito, K. (2008) Physiological roles of the beta-substituted alanine synthase gene family in Arabidopsis. *Plant Physiol.*, **146**, 310–320.

Wentzell, A.M., Rowe, H.C., Hansen, B.G., Ticconi, C., Halkier, B.A. and Kliebenstein, D.J. (2007) Linking metabolic QTLs with network and cis-eQTLs controlling biosynthetic pathways. *PLoS Genet.*, **3**, 1687–1701.

Wheeler, D.L., Barrett, T., Benson, D.A., Bryant, S.H., Canese, K., Chetvernin, V., Church, D.M., Dicuccio, M., Edgar, R., Federhen, S., Feolo, M., Geer, L.Y., Helmberg, W., Kapustin, Y., Khovayko, O., Landsman, D., Lipman, D.J., Madden, T.L., Maglott, D.R.,

Miller, V., Ostell, J., Pruitt, K.D., Schuler, G.D., Shumway, M., Sequeira, E., Sherry, S.T., Sirotkin, K., Souvorov, A., Starchenko, G., Tatusov, R.L., Tatusova, T.A., Wagner, L. and Yaschenko, E. (2008) Database resources of the National Center for Biotechnology Information. *Nucleic Acids Res.*, **36**, D13–21.

Xu, J., Zhang, J., Cai, S., Dong, J., Yang, J.Y. and Chen, Z. (2009) Metabonomics studies of intact hepatic and renal cortical tissues from diabetic *db/db* mice using high-resolution magic-angle spinning ^{1}H NMR spectroscopy. *Anal. Bioanal. Chem.*, **393**, 1657–1668.

Yasugi, E. and Watanabe, K. (2002) LIPIDBANK for Web, the newly developed lipid database. *Tanpakushitsu Kakusan Koso*, **47**, 837–841.

Yonekura-Sakakibara, K., Tohge, T., Niida, R. and Saito, K. (2007) Identification of a flavonol 7-*O*-rhamnosyltransferase gene determining flavonoid pattern in *Arabidopsis* by transcriptome coexpression analysis and reverse genetics. *J. Biol. Chem.*, **282**, 14932–14941.

Yonekura-Sakakibara, K., Tohge, T., Matsuda, F., Nakabayashi, R., Takayama, H., Niida, R., Watanabe-Takahashi, A., Inoue, E. and Saito, K. (2008) Comprehensive flavonol profiling and transcriptome coexpression analysis leading to decoding gene-metabolite correlations in *Arabidopsis*. *Plant Cell*, **20**, 2160–2176.

Yonekura-Sakakibara, K. and Saito, K. (2009) Functional genomics for plant natural product biosynthesis. *Nat. Prod. Rep.*, **26**, 1466–1487.

Index

Page numbers in **bold** refer to entries in tables. Page numbers in *italics* refer to entries in figures.

Plant Metabolism and Biotechnology, First Edition. Edited by Hiroshi Ashihara, Alan Crozier, and Atsushi Komamine.

Printed and bound by CPI Group (UK) Ltd, Croydon, CR0 4YY
16/09/2026
14963475-0003